T0173399

White Rose Maths
Key Stage 3
Student Book 2

Ian Davies and Caroline Hamilton

William Collins' dream of knowledge for all began with the publication of his first book in 1819. A self-educated mill worker, he not only enriched millions of lives, but also founded a flourishing publishing house. Today, staying true to this spirit, Collins books are packed with inspiration, innovation and practical expertise.

They place you at the centre of a world of possibility and give you exactly what you need to explore it.

Collins. Freedom to teach.

Published by Collins
An imprint of HarperCollins*Publishers*
The News Building
1 London Bridge Street
London
SE1 9GF

HarperCollins Publishers
Macken House
39/40 Mayor Street Upper
Dublin 1, D01 C9W8
Ireland

> **Browse the complete Collins catalogue at www.collins.co.uk**

10 9 8 7 6

ISBN: 978-0-00-840089-7

British Library Cataloguing-in-Publication Data
A catalogue record for this publication is available from the British Library.

Authors and series editors: Ian Davies and
   Caroline Hamilton
Publisher: Katie Sergeant
Product manager: Jennifer Hall
Product developer: Natasha Paul
Content editor: Tina Pietron
Editors: Karl Warsi, Julie Bond, Phil Gallagher,
   Tim Jackson, Eric Pradel
Proofreader: Catherine Dakin
Answer checkers: Laurice Suess and Steven Matchett
Project manager: Karen Williams
Cover designer: Kneath Associates Ltd
Internal designer and illustrator:
   Ken Vail Graphic Design Ltd
Typesetter: Ken Vail Graphic Design Ltd
Production controller: Katharine Willard
Printed and bound by Grafica Veneta in Italy

Photo acknowledgements
The publishers wish to thank the following for permission to reproduce photographs. Every effort has been made to trace copyright holders and to obtain their permission for the use of copyright materials. The publishers will gladly receive any information enabling them to rectify any error or omission at the first opportunity.
p 41l Becky Stares/Shutterstock, p 41r JaneHYork/Shutterstock, p 150 A-R-T/Shutterstock, p 344 SpicyTruffle/Shutterstock, p 381 Rattanapon Ninlapoom/Shutterstock, p 407t Kaja Ni/Shutterstock, p 407b kulyk/Shutterstock, p 416 dead_end/Shutterstock, p 527 Kinga/Shutterstock, p 584 Ondrej Prosicky/Shutterstock, 585 Haver/Shutterstock.

Maps on pages 73, 76–78, 80 © Collins Bartholomew Ltd

Contents

Contents

Introduction

How to use this book

Welcome to the **Collins White Rose Maths Key Stage 3** course. We hope you enjoy your learning journey. Here is a short guide to how to get the most out of this book.

Ian Davies and Caroline Hamilton, authors and series editors

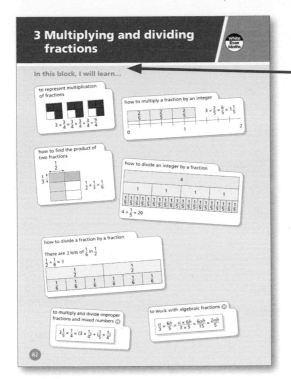

Block overviews Each block of related chapters starts with a visual introduction to the key concepts and learning you will encounter.

Small steps The learning for each chapter is broken down into small steps to ensure progression and understanding. The ⒣ symbol against a chapter or small step indicates more challenging content. The ⓡ symbol against a small step indicates revision content.

Key words Important terms are defined at the start of the chapter, and are highlighted the first time that they appear in the text. Definitions of all key terms are provided in the glossary at the back of the book.

Are you ready? Remind yourself of the maths you already know with these questions, before you move on to the new content of the chapter.

Models and representations Familiarise yourself with the key visual representations that you will use through the chapter.

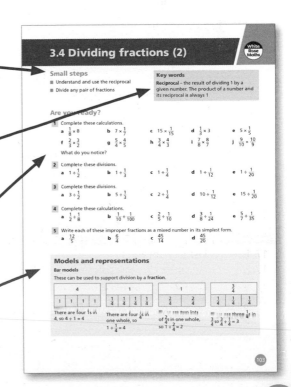

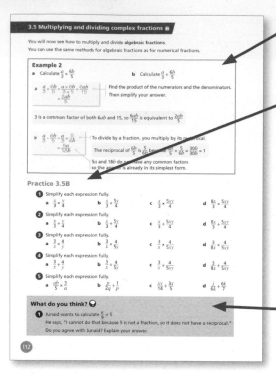

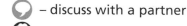

Worked examples Learn how to approach different types of questions with worked examples that clearly walk you through the process of answering, using lots of visual representations.

Practice Put what you've just learned into practice. Icons suggest tools or skills to help you approach a question:

- use manipulatives such as multi-link cubes or Cuisenaire rods
- draw a bar model
- draw a diagram
- discuss with a partner
- think deeply

There is a **What do you think?** section at the end of every practice exercise.

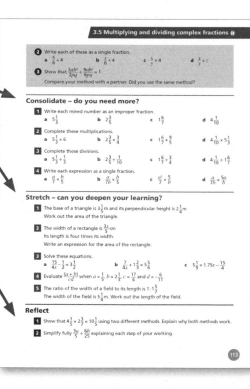

Consolidate Reinforce what you've learned in the chapter with additional practice questions.

Stretch Take the learning further and challenge yourself to apply it in new ways.

Reflect Look back over what you've learned to make sure you understand and remember the key points.

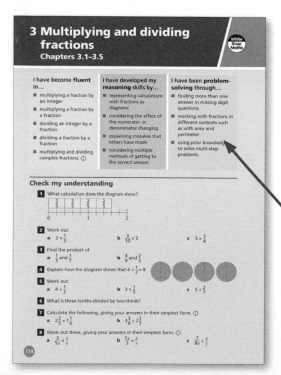

Block summary Bring together the learning at the end of the learning block with fluency, reasoning and problem-solving statements and review questions.

Answers Check your work using the answers provided at the back of the book.

1 Ratio and scale

In this block, I will learn...

how to represent ratios in a variety of forms

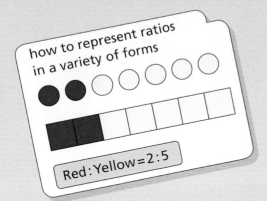

Red : Yellow = 2 : 5

how to solve ratio problems when the total is known

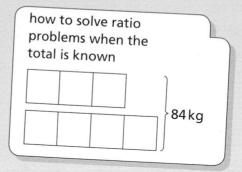

84 kg

how to write ratios in their simplest form

120 : 300   40 : 100

2 : 5   0.4 : 1   1 : 2.5

the links between ratios and fractions

$\frac{2}{3}$ 2 : 5

A

$\frac{3}{2}$

B

2 : 3 $\frac{2}{5}$

how to apply ratios to other areas of mathematics

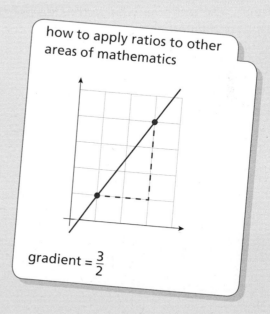

gradient = $\frac{3}{2}$

how to solve ratio problems when one of the shares is known

£40

?

1

1.1 Representing ratios

Small steps

- Understand the meaning and representation of ratio
- Understand and use ratio notation
- Compare ratios and related fractions

Key words

Ratio – a ratio compares the sizes of two or more values

Proportion – a part, share or number considered in relation to a whole

Are you ready?

1 What fraction of each bar model is shaded?

   a   b   c   d

2 Draw a bar model to illustrate each fraction.

   a $\dfrac{1}{4}$ b $\dfrac{3}{5}$ c $\dfrac{2}{7}$

3 In each set, which fraction is the greatest? How do you know?

   a $\dfrac{1}{3}, \dfrac{1}{4}, \dfrac{1}{5}$ b $\dfrac{2}{5}, \dfrac{3}{5}, \dfrac{4}{5}$ c $\dfrac{2}{3}, \dfrac{3}{4}, \dfrac{4}{5}$

Models and representations

You can use real objects or bar models to represent different quantities to help you compare them.

There are three blue counters and four red counters.
You can represent this as a **ratio**.

The ratio of blue counters to red counters is written as 3:4, which you read as "3 to 4".

              3 blue counters

              4 red counters

In the same way, the ratio of red counters to blue counters is 4:3, which you read as "4 to 3".

Here is one possible bar model you could also use to represent this.

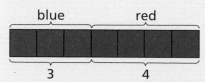

blue   red

3   4

Example 1

Marta has 5 pens and 2 pencils.

a   Show this information as a bar model.

b   Write down the ratio of pens to pencils Marta has.

c   Write down the ratio of pencils to pens Marta has.

a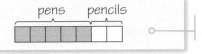

Use 1 section of the bar model to represent each pen and each pencil.

b   5:2   There are 5 pens and 2 pencils, so the ratio is "5 to 2" which you write using a colon as "5:2".

c   2:5   This is the same information, but with the order changed. There are 2 pencils and 5 pens.

It is very important to write the numbers in the correct order when using ratios to represent information.

Example 2

A necklace consists of red beads and white beads.

Are the statements true or false?

a   For every 2 red beads, there is 1 white bead.

b   The ratio of red beads to white beads is 1:2

a   True

In each section of 3 beads, you can see that for every two red beads there is 1 white bead.

b   False   The numbers in the ratio have been written the wrong way round. The ratio of red beads to white beads is 2:1

Practice 1.1A

1

Write down

a the ratio of squares to triangles

b the ratio of triangles to squares.

2

a Write down the ratio of green boxes to yellow boxes shown by the bar model.

b Write down the ratio of yellow boxes to green boxes shown by the bar model.

c Zach says the ratio of green boxes to the total number of boxes in the bar model is 1:5. Is Zach correct?

d Write down the ratio of yellow boxes to the total number of boxes in the bar model.

3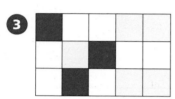

Write down

a the ratio of red squares to white squares

b the ratio of white squares to yellow squares

c the ratio of red squares to yellow squares

d the ratio of white squares to the total number of squares.

4 On a trip, there are 5 boys, 3 girls and 2 teachers.

a Draw a bar model to show this information.

b Which statements are true and which are false?

　A For every 5 boys there are 2 teachers.

　B For every 3 girls there are 5 boys.

　C The ratio of girls to boys is 5:3

　D The ratio of teachers to girls is 2:3

　E For every boy, there is one other person on the trip.

　F The ratio of girls to the total number of people is 3:10

　G The ratio of teachers to girls to boys is 2:5:3

c Rewrite each false statement to make it true.

5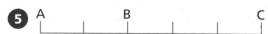

The spaces on the line are equal in length.

Write down the ratio of the lengths

a AB:BC

b AB:AC

c AC:BC

What do you think?

1 Is it possible to have a ratio 1:1? Explain your answer.

2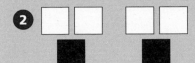

> The ratio of white squares to black squares is 4:2

Benji

> The ratio of white squares to black squares is 2:1

Chloe

Who do you agree with?   You will explore questions like this in more detail in Chapter 1.4

3 $p = 5$ and $q = 3$. Write down the ratios.

   a $q:p$ **b** $p:p+q$ **c** $p+q:p-q$ **d** $3q:2p$

4 The ratio of boys to girls in a class is 3:2

Does this mean there are 5 students in the class?

5 Draw a bar model with 7 equal parts. Shade in the parts using three different colours altogether. How many ratios can you find that compare two of the colours? How many ratios can you find comparing all three of the colours?

Bar models show fractions as well as ratios.

Here are some facts this bar model shows.

■ The ratio of red boxes to blue boxes is 1:3

■ The ratio of red boxes to the total number of boxes is 1:4

■ $\frac{3}{4}$ of the boxes are blue.

■ $\frac{1}{4}$ of the boxes are red.

■ The ratio of blue boxes to red boxes is 3:1

Example 3

a Draw a bar model that is $\frac{5}{8}$ blue, $\frac{1}{8}$ green and the rest is yellow.

b Write down the ratio of

   i the blue area to the green area

   ii the blue area to the yellow area

   iii the green area to the total area

   iv the yellow area to the blue area to the green area.

c What fraction of the bar is yellow?

a  Here is one possible answer. The colours do not have to be in a particular order, and you do not have to have all the blue sections next to each other, but it does make it easier to count and compare parts if you do.

b **i** $5:1$ — There are 5 blue sections and 1 green section.

   ii $5:2$ — There are 5 blue sections and 2 yellow sections.

   iii $1:8$ — 1 of the 8 sections is green.

   iv $2:5:1$ — Ratios often compare two quantities, but you can compare three or more parts as well. You need to take care with the order.

c $\frac{1}{4}$ — 2 of the 8 sections are yellow, so the fraction is $\frac{2}{8}$ which simplifies to $\frac{1}{4}$

Practice 1.1B

1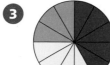

Seb thinks $\frac{2}{3}$ of the bar is white. Explain Seb's mistake.

2 There are 3 apples and 4 bananas in a bowl.

   a What fraction of the total number of fruits are bananas?

   b What is the ratio of the number of bananas to the number of apples?

   c What is the ratio of the number of apples to the total number of fruits?

3

   a What is the ratio of the red area to the yellow area in the pie chart?

   b What fraction of the pie chart is red?

   c What is the ratio of the blue area to the yellow area in the pie chart?

   d What fraction of the pie chart is blue? Give your answer in its simplest form.

4 The ratio of red counters to yellow counters in a bag is 2 : 5

What fraction of the counters are

a yellow **b** red?

5 The fraction of a class that are boys is $\frac{3}{5}$

   a What fraction of the class are girls?

   b Emily says the ratio of boys to girls is 3 : 5. Explain why Emily is wrong, and find the correct ratio of boys to girls in the class.

What do you think?

1 Jakub is making a drink using cordial and water. For every one part cordial, he uses four parts of water.

   a Write the ratio of cordial to water in Jakub's drink.

   b Jakub makes a second drink using cordial to water in ratio 1 : 5. Will this drink be stronger or weaker than his first drink?

   c Which of these ratios of cordial to water will make the strongest drink? Explain your answer.

   1 : 5   2 : 5   2 : 3   3 : 10

2 Which statements are true and which are false?

   A $\frac{2}{3}$ of the shape is blue.

   B $\frac{3}{5}$ of the shape is red.

   C The number of blue squares is $\frac{2}{3}$ of the number of red squares.

   D The number of red squares is $\frac{3}{2}$ of the number of blue squares.

   E The ratio of red squares to the total number of squares is 3 : 5

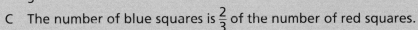

3 Investigate how many different fractions and ratios you can describe using the diagram. Try to include fractions greater than 1 and greater than 2.

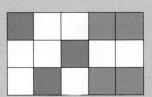

4 Here is some information about a shape.

The shape contains red, white and blue squares.

$\frac{1}{3}$ of the shape is white squares.

The ratio of red to blue squares is 2 : 1

   a Draw a possible shape that fits these conditions.

   b What other ratios and fractions can you see in the shape?

Consolidate – do you need more?

1

   a   What is the ratio of ginger biscuits to chocolate biscuits?

   b   What is the ratio of chocolate biscuits to ginger biscuits?

   c   Represent the relationship between the numbers of biscuits as a bar model.

2

Write down the ratio of

   a   white squares to grey squares

   b   grey squares to white squares

   c   grey squares to the total number of squares.

3

   a   Write down the ratio of the number of 50p coins to the number of 20p coins.

   b   What fraction of the coins are 20p coins?

4   In a board game there are 3 red tokens, 5 blue tokens and 1 gold token.

   a   What fraction of the tokens are

   i blue   **ii** gold   **iii** not gold   **iv** red?

   b   Write down the ratio of

   i the number of gold tokens to the total number of tokens

   ii the number of gold tokens to the number of tokens that are not gold

   iii the number of blue tokens to the number of red tokens to the number of gold
      tokens.

   c   What other fractions and ratios can you find?

5   Draw a bar model where the ratio of white to red sections is 3 : 1

6   Flo spends $\frac{3}{4}$ of her pocket money and saves the rest. What is the ratio of the amount
she saves to the amount she spends?

Stretch – can you deepen your learning?

1 For every pound that Ali has, Flo has 50p. Explain why the ratio of Ali's money to Flo's money is not $1:50$

2 Darius thinks the ratio $2:3$ is the same as the ratio $1:2$, as the numbers have both been increased by 1

    a Explain why Darius is wrong. You can use diagrams and fractions to help you.

    You will explore questions like this in more detail in Chapter 1.4

    b Find a ratio that is the same as $2:3$

3 **a** Given that $x = 3y$, write down

      i the ratio $x:y$

      ii the ratio $y:x$

      iii the ratio of the sum of x and y: the difference between x and y

    b What fraction of x is y?   **c** What fraction of $x+y$ is x?

    d How would your answers to **a**, **b** and **c** change if instead $y = 3x$?

4 Investigate the relationships in question 3 further by exploring other numbers, for example $x = 4y$, $x = 10y$, and so on.

Reflect

1 Explain what is meant by the word "ratio".

2 How many ways can you represent the ratio $2:3$?

1.2 Solving ratio problems

Small steps

- Solve problems involving ratios of the form $1:n$ (or $n:1$)
- Solve proportion problems involving the ratio $m:n$

Are you ready?

1 $\frac{1}{3}$ of a bag of counters are red and the rest are blue.

Write down the ratio of

  a red counters to blue counters

  b blue counters to red counters.

2 The ratio of apples to oranges in a bowl of fruit is $4:3$

What fraction of the fruit are

  a apples

  b oranges?

3 A book costs £8. Find the cost of

  a 3 books

  b 7 books

  c 12 books

  d 500 books.

4 3 metres of cable costs £5.40

  a Find the cost of 1 metre of cable.

  b Find the cost of 30 metres of cable.

  c Compare your methods for finding the answer to part **b** with a partner.

Models and representations

Bead string

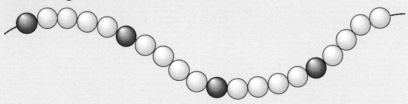

For every red bead on the necklace there are four white beads.

The ratio of red beads to white beads is $1:4$

Double number line

You can represent this information on a **double number line**.

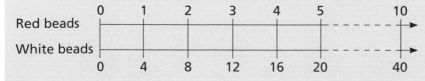

Bar model

You can also use **comparison bar models**, with one bar for each colour. This bar model shows information about the beads on a necklace with 30 beads altogether.

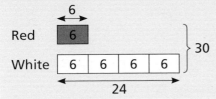

There is 1 box for red beads and there are four boxes for white beads, as the ratio is 1 : 4

Example 1

The ratio of girls to boys in a class is 1 : 3

a   If there are 9 girls in the class, how many boys are there?

b   If there are 9 boys in the class, how many girls are there?

Method A

a

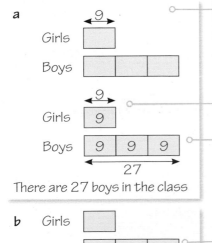

First represent the problem as a comparison bar model, with 1 box for girls and 3 boxes for boys as the ratio is 1 : 3. Label 9 on the box for girls.

As one box represents 9 students, fill in the rest of the boxes representing 9 students each.

There are 3 boxes for boys so there are $3 \times 9 = 27$ boys in the class.

You can also see there there are 36 students altogether, as $9 + 27 = 36$ or $4 \times 9 = 36$

b

You can represent the problem with the same model, this time labelling 9 boys.

As 3 boxes represents 9 boys, each box represents $9 \div 3 = 3$ boys.

Fill in the rest of the boxes representing 3 students each.

There is 1 box for girls so there are 3 girls in the class.

You can also see there there are 12 students altogether in the class, as $3 + 9 = 12$ or $4 \times 3 = 12$

Method B

a 27

You could use a double number number line.

The ratio of girls to boys is 1 to 3

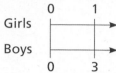

There are 9 girls, so the number of girls has been multiplied by 9

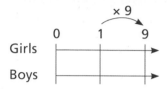

So you also need to multiply the number of boys by 9

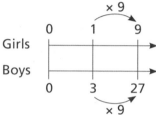

There are 3 × 9 = 27 boys.

b 3

You start with the same model.

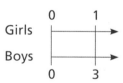

This time there are 9 boys so the number of boys has been multiplied by 3

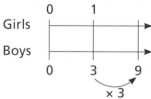

You also need to multiply the number of girls by 3

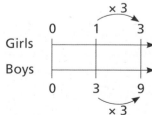

There are 1 × 3 = 3 girls

Practice 1.2A

1 The ratio of adults to children at a nursery is 1 to 4

Adults

Children

   a How many children are there if there are 20 adults?

   b How many adults are there if there are 20 children?

2 Mortar is made by mixing 1 part cement to 7 parts of other materials.

Cement

Other materials

   a What fraction of the mortar is cement?

   b **i** How much other material do you need to mix with 5 kg of cement?

   ii How much cement do you need to mix with 63 kg of other materials?

3 A drink is made by mixing water and cordial in the ratio 5 : 1

   a How much water do you need for 60 ml of cordial?

   b How much cordial do you need for 200 ml of water?

   c What fraction of the drink is water?

4 The double number line shows information about beads on a necklace.

Red: 0 1 2 3 30

Black: 0 4 8 12 ?

   a What is the ratio of red beads to black beads on the necklace?

   b How many black beads will be on the necklace if there are 30 red beads?

   c How many red beads will be on the necklace if there are 40 black beads?

5 The ratio of the width to length of a rectangle is 1 : 4

   a How long is the rectangle if its width is 16 cm?

   b How wide is the rectangle if its length is 20 cm?

   c Find the perimeter of the rectangle if its width is 12 cm.

   d What fraction of the perimeter is one width of the rectangle?

What do you think?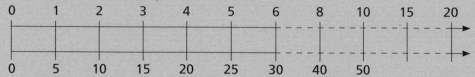

① Explore the relationships between pairs of numbers on the double number line, both vertically and horizontally.

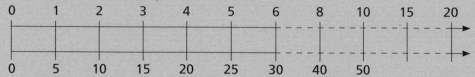

② The points A, B and C lie on a straight line. The line segment AB is 6 cm long.

Find the length of BC if

a AB : BC = 1 : 2   **b** AB : BC = 1 : 3   **c** AB : BC = 1 : 5

d BC : AB = 1 : 2   **e** BC : AB = 1 : 4   **f** BC : AB = 1 : 6

You can also solve problems where the ratios are not of the form $1:n$ or $n:1$

Example 2

Purple paint is made by mixing red paint and blue paint in the ratio 2 : 3

a How much red paint do you need to mix with 12 litres of blue paint?

b How much blue paint do you need to mix with 12 litres of red paint?

a Red

Blue

12 litres

12 ÷ 3 = 4 litres

Red | 4 | 4

Blue | 4 | 4 | 4

12 litres

2 × 4 = 8 litres

First represent the problem as a bar model with 2 boxes for red paint and 3 boxes for blue paint, as the ratio is 2 : 3. Label 12 litres of blue paint.

As 3 boxes represents 12 litres, each box must represent 12 ÷ 3 = 4 litres. Add this information to the diagram.

There are 2 boxes for red, so you need 2 × 4 = 8 litres of red paint.

b 12 litres

Red

Blue

12 ÷ 2 = 6 litres

12 litres

Red | 6 | 6

Blue | 6 | 6 | 6

3 × 6 = 18 litres

This time, there are 12 litres of red paint.

As 2 boxes represents 12 litres, each box must represent 12 ÷ 2 = 6 litres. Add this information to the diagram.

There are 3 boxes for blue, so you need 3 × 6 = 18 litres of blue paint.

Practice 1.2B

1 The ratio of apples to bananas in a bowl is 2:3

    a How many apples are there if there are 12 bananas?

    b How many bananas are there if there are 12 apples?

    c How many pieces of fruit are there altogether if there are 18 apples?

2 The ratio of boys to girls in a class is 3:5

    a If there are 15 boys, how many girls are there?

    b If there are 20 girls, how many boys are there?

    c Explain why there cannot be 10 boys in the class, but there could be 10 girls.

3 $\frac{3}{7}$ of a company's workforce are male.

    a Write down the ratio of female to male in the workforce.

    b 160 females work for the company. How many people does the company employ altogether?

4 Every week, Lydia spends $\frac{3}{5}$ of her pocket money and saves the rest.

    a Write down the ratio of the amount she spends to the amount she saves.

    b One week she saves £8. How much pocket money did she get that week?

    c Another week she spends £9. How much pocket money did she save that week?

5 Huda spends her working week on admin, customer service and marketing in the ratio 2:3:5. She spends 12 hours on customer service a week. How many hours does she spend on each of the other activities?

What do you think? 💭

1 The length and the width of a rectangle are in the ratio 3:2

Draw a sketch of the rectangle if

    a its length is 6 cm

    b its width is 6 cm

    c the length is 6 cm longer than the width.

2 Rectangle X is 5 cm long and 2 cm wide. The ratio of the sides of rectangle X to the sides of rectangle Y is 1:2. Find the ratio of the area of rectangle X to the area of rectangle Y.

Consolidate – do you need more?

1 A pattern consists of squares and triangles in the ratio 1:5

   a How many triangles will be there be if there are 60 squares?

   b How many squares will there be if there are 60 triangles?

   c What fraction of the shapes in the pattern are triangles?

2 **a** In a recipe, for every 3 cups of flour, 1 cup of milk is used.

      i Chloe has 6 cups of milk. How much flour will she need?

      ii Jakub has 12 cups of flour. How much milk will he need?

   b Work out how much flour Chloe will need and how much milk Jakub will need if instead the recipe needs 2 cups of milk for every cup of flour.

3 To make pink paint, white and red paint can be mixed in the ratio 3:4

   a How much red paint is needed to mix with 15 litres of white paint?

   b How much white paint is needed to mix with 10 litres of red paint?

4 The length and width of a parallelogram are in the ratio 8:5. The width of the parallelogram is 15 cm. Work out the perimeter of the parallelogram.

5 For every 2 red counters in a game, there are 3 blue counters and 5 white counters.

   a Write down the ratio of red counters : blue counters : white counters.

   b How many red counters are there if the game contains

      i 30 blue counters   **ii** 30 white counters?

   c How many white counters are there if the game contains

      i 30 red counters   **ii** 30 blue counters?

   d Investigate other possible numbers of counters the game could or could not have.

Stretch – can you deepen your learning?

1 The angles in a triangle are in the ratio 1:2:3

The smallest angle in the triangle is 20°

Beca

The smallest angle in the triangle is 30°

Abdullah

Who is correct? How do you know?

2 ABC is a triangle. The ratio of the size of angle ABC to the size of angle BAC is $4:5$

Find all three angles in the triangle if

    a   Angle ABC = 40°   **b**   Angle BAC = 40°

3 The numbers x and y are in the ratio $5:4$

    a   If $x = 30$, work out $x + 2y$ **b**   If $y = 30$, work out $3y - 2x$

    c   If $x = 50$, work out $x^2 + y^2$ **d**   If $y = 60$, find the mean of x and y

4 A car used 30 litres of fuel to travel 105 miles.

The fuel tank of the car has a capacity of 50 litres.

Can the car complete a 175-mile journey without refuelling?

Reflect

1 What's the same and what's different about solving problems with the ratio $1:3$ and solving problems with the ratio $2:3$?

2 Explain how you can use a bar model or a double number line to solve ratio problems. Which do you prefer?

Small steps

- Divide a value into a given ratio
- Solve problems involving ratios of the form $1:n$ (or $n:1$)
- Solve proportional problems involving the ratio $m:n$

Key words

Divide in a ratio – share a quantity into two or more parts so that the shares are in a given ratio

Are you ready?

1   The ratio of red counters to blue counters in a game is $4:1$. There are 12 blue counters. How many red counters are there?

2   The ratio of fiction to non-fiction books in a library is $1:5$. There are 400 fiction books in the library. How many non-fiction books are there?

3   Ed and Faith share some money in the ratio $3:5$

    a   Ed gets £60. How much does Faith get?

    b   What fraction of the money does Ed get?

4   Work out

    a $\frac{1}{5}$ of 40   **b** $\frac{2}{5}$ of 40   **c** $\frac{2}{5}$ of 90   **d** $\frac{3}{5}$ of 60   **e** $\frac{5}{7}$ of 210

Models and representations

In the last chapter, you worked out one part in a ratio problem when you knew another part. In this chapter, you will also look at problems when you are given the total or the difference between parts. Comparison bar models are very useful for this.

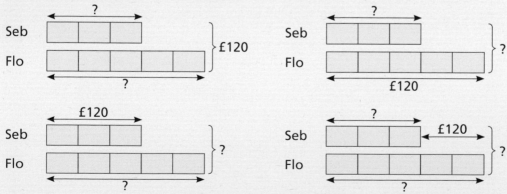

Firstly, let's look at when you know the total. This is often called **"dividing into a given ratio"** or "sharing into a given ratio".

Example 1

Beth and Sven share 40 sweets in the ratio 2:3

How many sweets do they each receive?

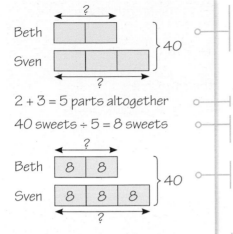

First draw a comparison bar model with two boxes for Beth and three boxes for Sven, labelling the total as 40 sweets. Label what you want to work out with "?".

2 + 3 = 5 parts altogether

You can see the total number of equal parts is 2 + 3 = 5

40 sweets ÷ 5 = 8 sweets

You can work out each equal part by dividing the total by 5

Add this information to each of the boxes representing each equal part.

Beth gets 2 × 8 = 16

Sven gets 3 × 8 = 24

You can now work out how many sweets each person gets.

Example 2

Charlie and Ali share £80 in the ratio 1:4

How much more money does Ali get than Charlie?

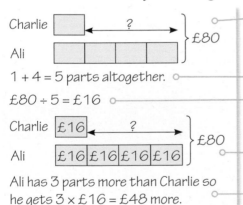

First draw a comparison bar model with one box for Charlie and four boxes for Ali, labelling the total as £80 Label what you want to work out with "?".

1 + 4 = 5 parts altogether.

You can see the total number of equal parts is 1 + 4 = 5

£80 ÷ 5 = £16

You can work out each equal part by dividing the total by 5

Add this information to each of the boxes representing each equal part.

Ali has 3 parts more than Charlie so he gets 3 × £16 = £48 more.

You can now work out the difference between their shares as this is three boxes.

You could also have worked out the difference like this:

Charlie's share = 1 × £16 = £16

Ali's share = 4 × £16 = £64

So Ali gets £64 − £16 = £48 more than Charlie.

Practice 1.3A

1 Share £60 in each of the ratios.

    a 1:2   **b** 1:3   **c** 1:4   **d** 2:3   **e** 1:5   **f** 1:2:3   **g** 1:1

2 A piece of wood, 150 cm long, is cut into two pieces. Find the length of each piece if the ratio of the shorter piece to the longer piece is

    a 1:2   **b** 1:3   **c** 2:3   **d** 3:7   **e** 7:8

3 Pink paint is made by mixing red and white paint in the ratio 1:2

How much of each colour paint do you need to make 15 litres of pink paint?

4 ABC is a straight line 20 cm long.

A   B   C

BC is three times as long as AB. Work out the length of AB.

5 Rob, Marta and Faith do some gardening. Rob works for 2 hours, Marta for 3 hours and Faith for 4 hours. They are paid £72 in total. How much money should they each receive?

6 Abdullah and Faith share £280 in the ratio 5:2

Abdullah gives $\frac{1}{5}$ of his money to Darius, and Faith gives 30% of her money to Darius. How much money does Darius receive altogether?

What do you think?

1 Flo is solving a problem.

"Junaid and Rhys share some money in the ratio 3:5 Rhys gets £40. How much does Junaid get?"

She writes

3 + 5 = 8 parts altogether

£40 ÷ 8 = £5

Junaid gets 3 parts which is 3 × £5 = £15

    a What mistake has Flo made?

    b Find the correct answer to the question.

2 Two children want to share 30 sweets. How many different ratios can you find that could divide the sweets so that each child gets an integer number of sweets?

3 Emily says it is impossible share £10 in the ratio 1:2. Do you agree? Explain your answer.

4 Two people share £28

    a One person gets £8. What ratio do they share the money in? Give your answer in the form $a:b$, where a and b are both less than 10

    b How would your answer change if one person gets £10.50?

Most ratio questions involve multiplication and/or division. You need to read ratio questions carefully to make sure you are multiplying or dividing the right numbers.

Example 3

Ed and Bobbie share some money in the ratio $3:4$. How much money does Ed get if

a there is £350 altogether?   **b** Bobbie gets £350?

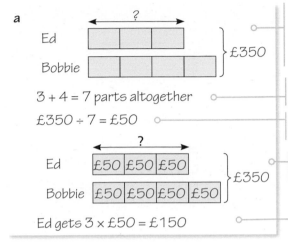

a

$3 + 4 = 7$ parts altogether

$£350 \div 7 = £50$

Ed gets $3 \times £50 = £150$

First draw a comparison bar model with three boxes for Ed and four boxes for Bobbie, labelling the total as £350. Put a "?" for what you want to work out, which is Ed's share.

You can see the total number of equal parts is $3 + 4 = 7$

You can work out each equal part by dividing the total by 7

Add this information to each of the boxes representing each equal part.

You can now work out how much money Ed gets.

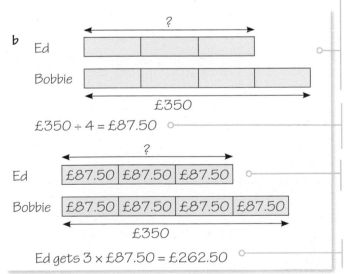

b

$£350 \div 4 = £87.50$

Ed gets $3 \times £87.50 = £262.50$

You start with the same comparison bar model, with three boxes for Ed and four boxes for Bobbie, but this time you label Bobbie's share as £350. Label what you want to work out with "?", which is Ed's share.

This time, 4 boxes represent £350, so you can work out each equal part by dividing £350 by 4

Add this information to each of the boxes representing each equal part.

You can now work out how much money Ed gets.

Practice 1.3B

1 **a** Match the diagram to the situation.

A Seb and Flo share some money. Seb gets £120

B Seb and Flo share £120

C Seb and Flo share some money. Flo gets £120

D Seb and Flo share some money. Flo gets £120 more than Seb.

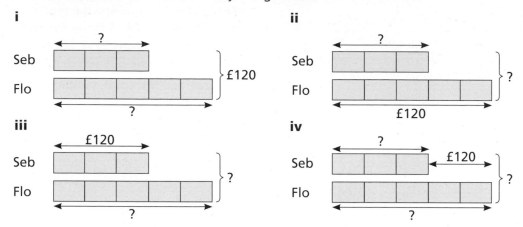

b For each situation, find the amount(s) of money represented by the question marks.

2 The ratio of Rob's height to his father's height is 5:6. Find Rob's height if

a his father is 30 cm taller than Rob

b his father is 186 cm tall.

3 The lengths of the sides of a triangle are in the ratio 3:4:5. Find the length of each side of the triangle if

a the perimeter is 192 cm

b the shortest side is 9 cm

c the longest side is 9 cm

d the difference between the longest and shortest sides is 18 cm.

4 Every year, a grandmother shares £240 between her two grandchildren in the ratio of their ages. Her elder grandchild is 2 years older than her younger grandchild.

a How much money do they each get when

i the younger grandchild is 3

ii the elder grandchild is 3

iii the elder grandchild is 17?

b Why is the problem more difficult to work out when the younger grandchild is 17? How could you overcome this?

c One year, the elder grandchild gets £20 more than the younger grandchild. How much do they each get?

5 Two numbers are in the ratio 7:3. Find the numbers if

a their total is 40

b their difference is 40

6 The points A, B and C lie on a straight line, in the order shown.

```
|——————————|—————————————————————|
A          B                     C
```

a Find AB if AB : BC = 3 : 7 and AC = 20 cm.

b Find AB if AB : BC = 11 : 7 and AC = 27 cm.

c Find AB if AB : BC = 3 : 7 and BC = 28 cm.

d Find AC if AB : BC = 11 : 7 and BC = 28 cm.

e Find AB if AC is 7 times the length of AB and BC = 27 cm.

What do you think? 💭

1 Fertiliser is made by mixing a chemical and water in the ratio 3 : 25

a How much of the chemical is needed for 1 litre of water? Give your answer in ml.

b How much water is needed for 75 ml of chemical? Give your answer in litres.

c How much of the chemical and how much water are needed to make 560 ml of fertiliser?

2 Lydia is making "Sunflower" paint. To get "Sunflower" paint, she needs to mix yellow, orange and white paint in the ratio 5 : 3 : 1

Lydia wants to make 1.8 litres of paint. She has

800 ml of yellow paint

900 ml of orange paint

150 ml of white paint.

a Show that Lydia does not have enough of each colour paint to make 1.8 litres of "Sunflower" paint. Work out how much extra paint she needs to buy and of what colour.

b What is the largest amount of "Sunflower" paint she can make with the paint she already has?

Consolidate – do you need more?

1 Share £90 in each of the ratios.

    a 1 : 2   **b** 1 : 3   **c** 1 : 4   **d** 2 : 3   **e** 1 : 5   **f** 1 : 2 : 3   **g** 1 : 1

2 Three children share 60 sweets in the ratio 3 : 4 : 5

How many sweets do they each get?

3 Emily, Huda and Rhys share £3600. Huda gets twice as much as Emily and Rhys gets three times as much as Huda.

a Explain why the money is **not** shared in the ratio 1 : 2 : 3

b Find the amount of money each of them receives.

4 On a school trip, the ratio of teachers to students is $2:15$. There are 34 teachers on the trip. How many students are there on the trip?

5 Ali and Marta share some sweets in the ratio $3:8$

   a If Marta gets 40 more sweets than Ali, how many sweets does Ali get?

   b If there are 44 sweets altogether, how many sweets does Marta get?

Stretch – can you deepen your learning?

1 **a** The angles in a triangle are in the ratio $1:2:3$. Show that the triangle contains a right angle.

   b The angles in a triangle are in the ratio $2:3:4$. Find the size of each of the angles in the triangle.

   c Now find the angles in a triangle for other ratios of consecutive numbers, for example $3:4:5$, $9:10:11$, and so on. What do you notice?

2

> To share an amount in the ratio $5:3$, you find $\frac{5}{8}$ and $\frac{3}{8}$ of the amount.

Jackson

   a Explain why Jackson is correct.

   b Find the fraction of each share when you divide an amount in the ratios

     i $3:4$ **ii** $3:1$ **iii** $1:2:3$ **iv** $3:4:5$ **v** $a:b$

3

> When you share an amount in the ratio $5:3$, one part is $\frac{3}{5}$ of the other part.

Beca

> No it isn't, one part is $\frac{5}{3}$ of the other part.

Flo

Do you agree with Beca or Flo? Explain your answer.

4 Investigate what fractions you can find when sharing the ratios.

   a $4:5$ **b** $2:3:5$ **c** $x:y$ **d** $x:y:z$

Reflect

What's the same and what's different about solving ratio problems if you know one of the shares or if you know the total?

Small steps

■ Express ratios in their simplest integer form

■ Express ratios in the form 1 : n Ⓗ

Key words

Simplify – rewrite in a simpler form, for example rewrite 8 × *h* as 8*h*

Equivalent – numbers or expressions that are written differently but are always equal in value

Factor – a positive integer that divides exactly into another positive integer

Highest common factor (HCF) – the greatest number that is a factor of every one of a set of numbers

Are you ready?

1 Write the fractions in their simplest form.

  a $\frac{4}{8}$ **b** $\frac{12}{15}$ **c** $\frac{12}{20}$ **d** $\frac{18}{24}$

2 20 out of 300 students wear glasses. Write the fraction of students that wear glasses in its simplest form.

3 **a** How many g are there in a kg?

  b How many cm are there in a m?

  c How many m are there in a km?

4 **a** List the factors of 12

  b List the factors of 20

  c What is the highest common factor of 12 and 20?

Models and representations

Cubes and counters are very useful to represent ratios.

This shows that for every blue cube, there are four green cubes, so the ratio of blue cubes to green cubes is 1 : 4

Here there are 2 blue cubes and 12 green cubes. You could say the ratio of blue cubes to green cubes is 2 : 12

You could rearrange the cubes to show that for every blue cube there are 6 green cubes. So you could also write the ratio of blue cubes to green cubes as 1:6

The ratios 2:12 and 1:6 are **equivalent** as they give the same information.

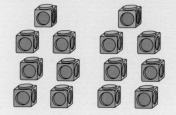

Because the numbers in the ratio 1:6 are smaller than those in the ratio 2:12, they are easier to work with. In this chapter you will be exploring ratios using the smallest numbers possible. This is called "simplifying ratios".

You can find the simplest form of a ratio by using cubes or counters, or by looking for factors of the numbers

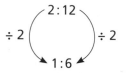

2 is a **factor** of both 2 and 12, so you can divide both numbers by 2

Example 1

Write each ratio in its simplest form.

a 3:12   **b** 4:10   **c** 60:80

a
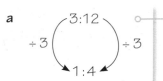

3 is a factor of both 3 and 12, so you can divide both numbers by 3 to **simplify** the ratio.

You could use counters to work out or to check your answer.

b
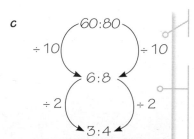

2 is a factor of both 4 and 10, so you can divide both numbers by 2 to simplify the ratio.

You could use counters to work out or to check your answer.

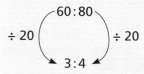

c
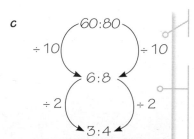

10 is a factor of both 60 and 80, so you can divide both numbers by 10 to simplify the ratio.

The ratio is not yet in its simplest form, as 2 is a factor of both 6 and 8, so you can divide both numbers by 2 to simplify the ratio.

You could do this more quickly by looking for the **highest common factor** of 60 and 80, which is 20

Example 2

Ali has £3 and Faith has 90p. Find the ratio of the amount of money Ali has to the amount of money Faith has.

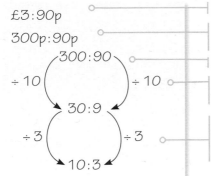

Firstly, write the amounts as a ratio.

To compare them, they need to be in the same units so change £3 to 300p

Now they are in the same form, you do not need the units.

10 is a factor of both 300 and 90, so you can divide both numbers by 10 to simplify the ratio.

The ratio is not yet in its simplest form as 3 is a factor of both 30 and 9, so you can divide both numbers by 3 to simplify the ratio.

As in Example 1, you could do this more quickly by looking for the highest common factor.

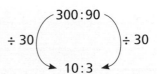

The highest common factor of 300 and 90 is 30, so you can divide both numbers by 30.

Practice 1.4A

1 Use counters or cubes to show that each pair of ratios is equivalent.

a   3:1 and 9:3   b   2:3 and 8:12   c   3:4 and 12:16

2 Match the equivalent ratios.

5:3   20:15   20:25   4:3   25:15   12:20   3:5   4:5

3 Simplify the ratios.

a   16:20   b   15:18   c   300:60   d   10:12

e   10:30:60   f   6:9:27   g   12:18:20   h   35:50:80

4 80% of adults watch sport regularly. Write the ratio of the number of adults who watch sport to the number who do not watch sport, in its simplest form.

5 Marta has these coins

Jackson has these coins

Find the ratio of the amount of money Marta has to the amount of money Jackson has, giving your answer in its simplest form.

6 Simplify the ratios.

   a £3:40p   **b** 2kg:300g   **c** 50cm:2.5m   **d** 750ml:30cl

7

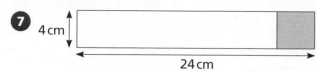

4 cm

24 cm

The shaded part of the rectangle is a square.

Show that the ratio of the white area to the shaded area is 5:1

8 **a** Simplify the ratios.

   i 6:4   **ii** 150:100   **iii** 24:16   **iv** 210:140

  b Work out

   i 6 ÷ 4   **ii** 150 ÷ 100   **iii** 24 ÷ 16   **iv** 210 ÷ 140

  c Look at the answers to **a** and **b**. What do you notice?

What do you think?

1 **a** Use your learning from Chapter 1.3 to share £300 in the ratios

   i 1:2   **ii** 2:4   **iii** 10:20   **iv** 50:100

  b What do you notice about your answers? Explain why this happens.

2 Ms A and Ms B invest £12000 and £18000 into a business venture.

  a Write the ratio of their investments in its simplest form.

  b In the first year the business makes £6600 profit.

   i Explain why Ms A and Ms B should not get £3000 each as their share of the profits.

   ii How much should they each receive?

   iii How much should they receive if their investments had been £20000 and £24000?

3 **a** Simplify the ratio 120:300

  b Simplify the fraction $\frac{120}{300}$

  c What's the same and what's different about your approaches to **a** and **b**?

Although ratios are usually written in their simplest form using integers, sometimes it is useful to write them in the form 1:n or n:1

Example 3

A purple paint is made by mixing 5 tins of red paint with 4 tins of blue paint.

a Write the ratio of red paint to blue paint in the mix in the form $1:n$

b Write the ratio of red paint to blue paint in the mix in the form $n:1$

c How many more times as much red paint is there than blue paint?

a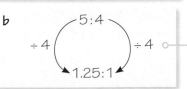

There are 5 tens of red paint and 4 tins of blue paint, so the ratio is 5:4

To change the ratio to the form $1:n$, you need to divide both numbers by 5, as 5 ÷ 5 is 1

b

$$5:4$$
$$\div 4 \quad \div 4$$
$$1.25:1$$

To change the ratio to the form $n:1$, you need to divide both numbers by 4, as 4 ÷ 4 is 1

c 1.25

For every tin of blue paint, you need 1.25 tins of red paint, as worked out in **b**. So there is 1.25 times as much red paint as blue paint.

Practice 1.4B

1 Write each ratio in the form $1:n$. Write n to 2 decimal places if necessary.

    a 5:3   **b** 12:30   **c** 30:40   **d** 17:4   **e** 3:10

2 Write each ratio in the form $n:1$. Write n to 2 decimal places if necessary.

    a 5:3   **b** 12:30   **c** 30:40   **d** 17:4   **e** 3:10

3 One week, it rained for 5 days.

Write the ratio of rainy days to dry days in the form

    a $1:n$ **b** $n:1$

4 Jakub is making a cake. The ratio of sugar to flour should be 1:3. Jakub uses 350 g of sugar and 1 kg of flour. Show that Jakub has not used the correct ratio of sugar to flour, and work out how he can correct his mistake.

5 In a nursery for two-year-olds, the ratio of adults to children should be 1:4. How many adults should the nursery employ if there are

    a 40 children   **b** 60 children   **c** 70 children?

6 $\frac{2}{5}$ of the counters in a bag are red and the rest are yellow.

    a Write the ratio of the number of red counters to the number of yellow counters in the form $1:n$

    b How many yellow counters are in the bag if there are

        i 40 red counters   **ii** 40 counters in total?

What do you think? 💬

1. Write the ratio
   a $1:1.6$ in the form $n:1$
   b $1:\frac{3}{4}$ in the form $n:1$
   c $\frac{b}{a}:1$ in the form $1:n$

2. There are 180 delegates at a conference, all of whom work for either Firm A, Firm B or Firm C.

   One third of the delegates work for Firm A, and there are four times as many delegates from Firm B as there are from Firm C. Find the ratio of the number of delegates from Firm A to the number of delegates from Firm C, giving your answer in the form $n:1$

3. When might it be useful to express a ratio in the form $1:n$ or $n:1$ instead of simplest integer form?

Consolidate – do you need more?

1. Simplify the ratios.
   a $10:2$ b $2:20$ c $5:20$
   d $20:25$ e $12:15:18$

2. Simplify the ratios.
   a $1.5\,m:2.5\,m$ b $3\,km:9\,km$ c $3\,km:500\,m$
   d   £1.50 : £1.80   e $60p:£1$ f $60p:£3$

3. Find the missing numbers so that the ratios are equivalent.
   a $3:5$ and $15:\boxed{}$ b $\boxed{}:12$ and $5:2$ c $8:\boxed{}$ and $40:25$

4. A drink is made by mixing 40 ml of cordial with 100 ml of water. Write down the ratio of water to cordial
   a   in simplest integer form
   b   in the form $1:n$
   c   in the form $n:1$

5. Write these ratios in the form
   i $1:n$ ii $n:1$
   Give n to 2 decimal places if necessary.
   a $3:2$ b $2:5$ c $8:5$ d $5:8$

Stretch – can you deepen your learning?

1 Here are the ingredients to make eight Yorkshire puddings.

Ingredients
120 g flour
3 large eggs
180 ml milk
oil for cooking

a How much flour and how much milk are needed for each egg?

b How much of each ingredient is needed for each pudding?

c Why would it be difficult to make one Yorkshire pudding using the recipe?

d How do you know how much oil to use as you change the amounts of puddings made?

2 Draw a right-angled triangle with sides 3 cm, 4 cm and 5 cm and draw squares on each side of the triangle as shown.

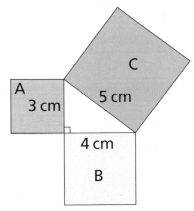

We will explore right-angled triangles in Block 11

Write these ratios in the form $1:n$. Give n to 2 decimal places if necessary.

a The area of square A to the area of square B

b The area of square B to the area of square C

c The area of square C to the total area of squares A and B

3 a Find the ratios if $x = 3y$

   i $x:y$ **ii** $x:2y$ **iii** $x + y:y$

b How would your answers to **a** change if instead $x = \frac{1}{3}y$?

Reflect

1 What's the same and what's different about simplifying a ratio and simplifying a fraction?

2 How do you simplify a ratio with more than two parts?

1.5 Ratios in circles and graphs

Small steps

■ Understand π as the ratio between diameter and circumference

■ Understand gradient of a line as a ratio Ⓗ

Are you ready?

1 Find the perimeter of each shape.

a
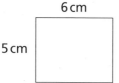
6 cm
5 cm

b
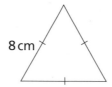
8 cm

c

7 cm

2 Round these numbers to the nearest integer.

a 31.2

b 31.27

c 31.47

d 31.41

e 31.95

f 31.59

3 a Write these ratios in their simplest form.

   i 20:50

   ii 6:15

   iii 140:350

b Work out

   i 20 ÷ 50

   ii 6 ÷ 15

   iii 140 ÷ 350

c What is the connection between your answers to **b** and **c**?

Models and representations

You can use dynamic geometry software to investigate the relationship between the **diameter** and the **circumference** of a circle.

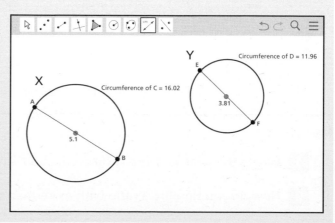

Use the dynamic geometry software to measure the diameter and the circumference of each circle that you draw. For each circle, work out circumference ÷ diameter. What do you notice?

Alternatively, draw some circles on paper, and measure the circumference as accurately as you can, using a piece of string. Again, work out circumference ÷ diameter for each one and see what you notice.

You should get an answer of about 3.14 every time.

For any circle, the diameter will fit around the outside of the circle just over 3 times.

In fact, it will fit 3.14159… times.

3.14159… is a non-recurring decimal that is denoted by the Greek letter π, which is said "pi".

You can think of this relationship in several ways:

- the ratio of the circumference of a circle to its diameter is $\pi : 1$, or
- the circumference of a circle divided by its diameter is π, or
- the circumference of a circle is π times the size of its diameter.

Using C to represent the circumference of a circle and d to represent its diameter, you can write these last two facts as

$$\frac{C}{\pi} = d \qquad \text{and} \qquad C = \pi d$$

$C = \pi d$ is a very useful formula to work out the circumference of a circle when you know its diameter.

Remember, the diameter goes from a point on the circle to another point on the circle and goes through the centre of the circle, but the **radius** goes from the centre to a point on the circle. The diameter is twice the length of the radius.

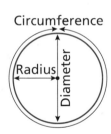

Example 1

Use the formula $C = \pi d$ to work out the circumferences of the circles. Give your answers to the nearest integer.

a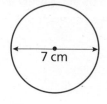
7 cm

b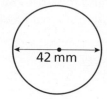
42 mm

c
50 mm

a $\quad C = \pi d$

When $d = 7$

$C = \pi \times 7$

$C = 21.99... \text{cm} \approx 22 \text{cm}$

Substitute d into the formula, remembering that πd means $C = \pi \times d$

Round the answer to the nearest integer and remember to include the correct units in your answer.

As you know that the circumference is just over 3 times its diameter, you also know the answer should be more than $3 \times 7 = 21$ cm. The answer here is 22 cm, which makes sense. You can use $\pi \approx 3$ to estimate answers in this section.

b $\quad C = \pi d$

When $d = 42$

$C = \pi \times 42$

$C = 131.88... \text{mm} \approx 132 \text{mm}$

Substitute d into the formula

Round the answer to the nearest integer.

c $\quad C = \pi d$

When $r = 50, d = 100$

$C = \pi \times 100$

$C = 314.15... \text{mm} \approx 314 \text{mm}$

The diameter is twice the size of the radius.

Substitute d into the formula – not r

Round the answer to the nearest integer.

You will look at different ways of rounding answers in Chapter 12.2

Practice 1.5A

1 Use the diagrams to find approximate values for π

a 15.7 cm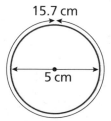
5 cm

b 31.42 cm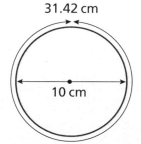
10 cm

c 43.96 cm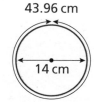
14 cm

2 Use the formula $C = \pi d$ to work out the circumferences of the circles. Use the π key on your calculator or use 3.14 for π. Give your answers to the nearest integer.

a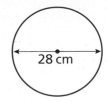
28 cm

b
113 mm

c
49 mm

3
35 mm

a What is the diameter of the circle?

b Work out the circumference of the circle, giving your answer to the nearest integer.

4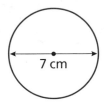
7 cm   7 cm

$C = \pi d$
When $d = 7$
$C = \pi \times 7$
$C = 21.99\ldots \text{cm} \approx 22\,\text{cm}$

a

The circumference of the circle is 22 cm, so the perimeter of the semicircle is 22 cm ÷ 2 = 11 cm

Explain why Ed is wrong.

b Find the correct perimeter of the semicircle.

5 Find the perimeters of these shapes. Give your answers to the nearest integer.

a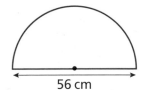
56 cm

b
28 cm

c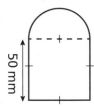
50 mm

6 What's the difference between the circumference of a circle and the perimeter of a circle?

What do you think?

1 **a** When you double the diameter, what happens to the circumference? Why?

b Circle X has circumference 60 cm.

i The ratio of the diameter of circle X to the diameter of circle Y is 1:3. Work out the circumference of circle Y

ii The ratio of the diameter of circle X to the diameter of circle Z is 4:1. Work out the circumference of circle Z

2 **a** What is the ratio of the radius of a circle to its diameter?

b What is the ratio of the radius of a circle to its circumference?

c

Another formula for working out the circumference of a circle is $C = 2\pi r$

Do you agree with Faith? Explain your answer.

The **gradient** of a line measures how steep it is.

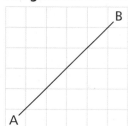 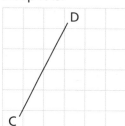

You can see that line segment CD is steeper than line segment AB. You can explore the gradient of lines by drawing right-angled triangles underneath the lines.

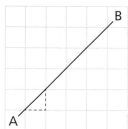 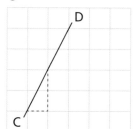

In the line segment AB, for every 1 square moved across to the right, the line has gone up 1 square.

The ratio of distance moved across:distance moved up is 1:1

In the line segment CD, for every 1 square moved across to the right, the line has gone up 2 squares.

The ratio of distance moved across:distance moved up is 1:2

You will look at the gradient of straight lines in more detail in Block 4

The gradient of a line is calculated by using the formula Gradient = $\dfrac{\text{Change in vertical distance}}{\text{Change in horizontal distance}}$

Example 2

Find the gradient of the line segment AB

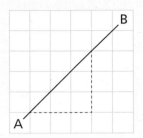

It does not matter where you draw a right-angled triangle under a straight line, the gradient will be the same value.

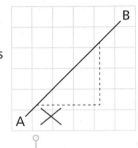

$$\text{Gradient} = \frac{\text{Change in vertical distance}}{\text{Change in horizontal distance}}$$
$$= \frac{3}{3} = 1$$

Substitute the numbers into the formula for the gradient.

Always choose points on the line that lie at intersections of squares on the grid to make the counting easier.

Example 3

a Find the ratio of distance moved across : distance moved up for the line segment XY. Give your answer in the form $1:n$

b Find the gradient of the line segment XY

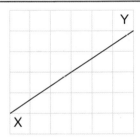

a

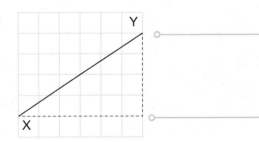

Copy the diagram and draw a right-angled triangle underneath the line segment.

You can find the ratio by counting squares.

distance moved across : distance moved up

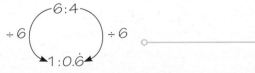

To find the ratio in the form $1:n$ you need to divide both the numbers in the ratio by 6

b Gradient = $\dfrac{\text{Change in vertical distance}}{\text{Change in horizontal distance}}$

$= \dfrac{4}{6} = \dfrac{2}{3}$ ○————————

Substitute the numbers into the formula for the gradient.

Gradient is usually given as a fraction rather than a decimal.

Practice 1.5B

1 For each right-angled triangle, find the ratio of the base : height

a   **b**   **c**   **d**

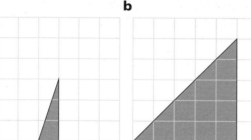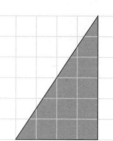

e   **f**   **g**

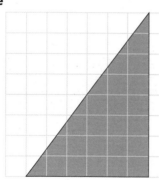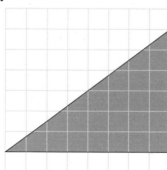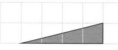

2

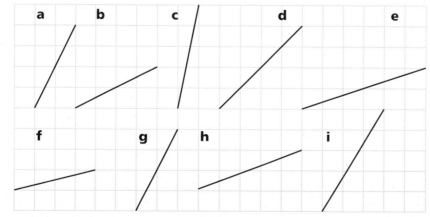

For each line segment, find

    i   the ratio of distance moved across : distance moved up in the form $1:n$

    ii   the gradient of the line segment.

3 On squared paper, draw line segments with the gradients

   a 2   **b** 4   **c** $\frac{3}{4}$

   d $\frac{3}{2}$ **e** $\frac{5}{6}$ **f** $\frac{6}{5}$

Compare your answers with a partner's. What's the same and what's different?

4 On squared paper, draw three different line segments with a gradient of 3

What do you notice about your lines?

What do you think?

1 Darius thinks it's impossible to find the gradient of these lines. What do you think?

2

I find the gradient of lines using the formula Gradient $= \dfrac{\text{up}}{\text{across}}$

Chloe

I find the gradient of lines using the formula Gradient $= \dfrac{\text{rise}}{\text{run}}$

Abdullah

Discuss Chloe's and Abdullah's formulae.

3

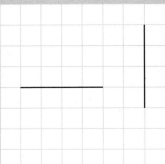

The gradient of PQ is 2

What's the same and what's different about PQ and MN? What do you think the gradient of MN might be?

Consolidate – do you need more?

1 Use the formula $C = \pi d$ to work out the circumferences of the circles. Use the π key on your calculator or use 3.14 for π. Give your answers to the nearest integer.

a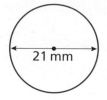
21 mm

b
226 mm

c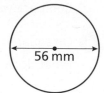
56 mm

2 **a** What is the diameter of the circle?

b Work out the circumference of the circle.

20 mm

3 Find the perimeter of a semicircle with radius 50 cm.

4 A right-angled triangle has base 10 cm and height 40 cm. Find the ratio of the base of the right-angled triangle to the height of the right-angled triangle.

5 Find the gradients of the line segments.

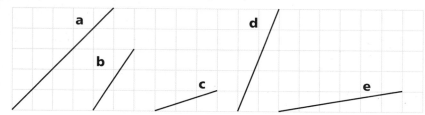

Stretch – can you deepen your learning?

1 **a** Before calculators were widely available, $\frac{22}{7}$ and $\frac{355}{113}$ were used as approximations for π. Investigate how good these approximations are.

b Use $\pi \approx \frac{22}{7}$ to find an approximate value for the circumference of a circle with

i diameter 84 cm

ii radius 0.7 m

2 **a** Flo uses a tape measure to find the circumference of a tree trunk.
She finds the circumference is 150 cm.

i Explain why Flo has divided the circumference by 3 to find her estimate.

ii Is the actual diameter of the tree more or less than 50 cm?

> The diameter of the tree trunk is roughly 150 ÷ 3 = 50 cm

b Use $\pi = 3.14$ to find the diameters of these circles.

i   34.54 cm

ii   282.6 mm

iii   9.42 m

c   What are the radii of the circles?   Radii is the plural of radius

3   Faith goes on a 12 km bike ride. The wheels of her bike each have diameter 55 cm. How many revolutions do her wheels make in the journey?

4   What do you think these road signs mean? How do the percentages relate to ratio and gradient?

a

b

Reflect

1   What is the ratio of the circumference of a circle to its diameter? Explain how you can use this to work out the circumference of a circle.

2   What is the connection between ratios and the gradient of a line?

1 Ratio and scale
Chapters 1.1–1.5

I have become fluent in...

- using ratios to compare quantities
- writing ratios in their simplest form
- writing ratios in the form $1:n$ and $n:1$
- dividing quantities in a given ratio
- expressing the gradient of a line as a ratio. Ⓗ

I have developed my reasoning skills by...

- representing ratios in a variety of forms
- linking ratios to finding the circumference of a circle
- selecting the right lengths when calculating circumference
- linking ratios and fractions.

I have been problem-solving through...

- interpreting ratio problems
- representing ratio problems as bar models
- knowing when to multiply and when to divide in solving ratio problems
- solving problems involving parts of circles.

Check my understanding

1 For every 3 counters in a game, there are 4 cards.

   a The game has 60 counters. How many cards are there?

   b A special edition of the game has 120 cards. How many counters are there?

2 The ratio of Amina's height to her mother's height is $4:5$

   a Amina is 30 cm shorter than her mother. How tall is Amina?

   b Amina's brother is 165 cm tall. Write the ratio of his height to their mother's height in its simplest form.

3 The ratio of chocolate to other ingredients in a recipe for a cake is $2:7$.
What fraction of the cake is chocolate?

4 Filipo, Samira and Kate are collecting game cards.

For every card Filipo has, Samira has 3 cards. The ratio of the number of cards Samira has to the number of cards Kate has is $2:5$

   a If Filipo has 10 cards, how many cards do all three students have altogether?

   b Explain why Samira cannot have 10 cards.

5 Emily and Zach share some money in the ratio $4:5$

   a How much money do they each get if they are sharing £72?

   b How much money does Zach get if Emily gets £20?

6 Explain how to find the circumference of a circle if you know

   a its diameter   **b** its radius.

7 On squared paper, draw a line segment with a gradient of $\frac{2}{3}$. Ⓗ

2 Multiplicative change

In this block, I will learn...

how to solve problems using direct proportion

Here is a recipe to make soup for 4 people.

- 4 tablespoons of butter
- 1 onion
- 600 ml vegetable stock
- 500 g tomatoes

How much of each ingredient do you need to make soup for 6 people?

how to use conversion graphs

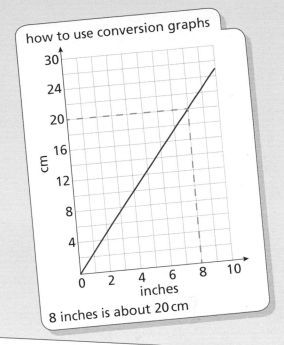

8 inches is about 20 cm

how to convert between different units of currency

£1 = $1.35

I have got £200

I have got $250

Who has more money?

the mathematical meaning of the word "similar"

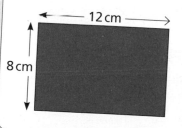

how to draw and interpret scale diagrams

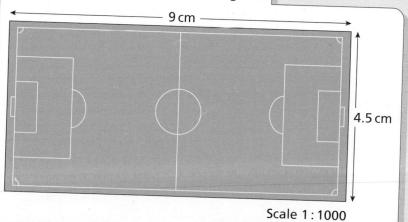

9 cm

4.5 cm

Scale 1 : 1000

how to find distances using maps

SCOTLAND

NEWCASTLE

N.I

IRELAND

? km

ENGLAND

EXETER

Small steps

- Solve problems involving direct proportion
- Explore direct proportion graphs **H**

Key words

Direct proportion – two quantities are in direct proportion when as one increases or decreases, the other increases or decreases at the same rate

Constant – not changing

Multiplier – a number you multiply by

Are you ready?

1 Simplify the ratios.

    a   10:30   **b**   15:24   **c**   64:80

2 Find the missing numbers in these pairs of equivalent ratios.

    a   3:8 = 30:☐   **b**   3:8 = ☐:56   **c**   3:☐ = 15:55

3 What do you multiply 2 by to get these answers?

    a   10   **b**   40   **c**   60   **d**   14   **e**   15

4 A book costs £12.50. Find the cost of

    a   3 books   **b**   30 books   **c**   60 books.

Models and representations

This double number line shows how the volume of water in a tank changes as the height of water in the tank changes.

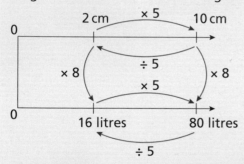

You can see the **multiplier** from the height to the volume is **constant** – it does not change.

You can also see that when you multiply or divide one quantity by a number, you multiply or divide the other quantity by the same number.

This means the volume of the water in the tank and the height of water in the tank are in **direct proportion**.

Another way of saying this is that the volume of water is directly proportional to the height of water.

You can also use ratio tables to work with proportional relationships.

Example 1

The cost of grass seed is proportional to the amount of seed bought. 5 kg of grass seed costs £18

a   How much does 20 kg of grass seed cost?   **b**   How much does 2.5 kg of grass seed cost?

c   How much grass seed does £180 buy?   **d**   How much does 11 kg of grass seed cost?

a $20 \div 5 = 4$

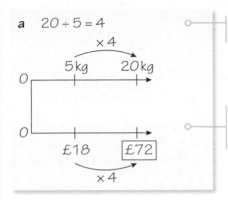

First you work out how many times more grass seed there is by dividing 20 by 5. You might just "spot" this.

As the cost is in direct proportion to the weight, you also multiply the cost by 4. You can show this on a double number line.

b $5 \div 2.5 = 2$

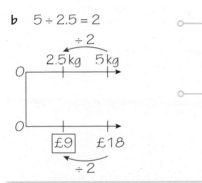

First you work out how many times less grass seed there is by dividing 5 by 2.5. You might just "spot" this.

As the cost is in direct proportion to the weight, you also divide the cost by 2.

c $50 \div 10 = 5$

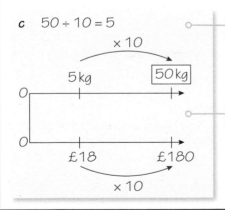

First you work out how many times more grass seed there is by dividing 50 by 5. You might just "spot" this.

As the cost is in direct proportion to the weight, you also multiply the cost by 10.

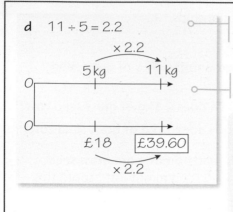

d $11 \div 5 = 2.2$

First you work out how many times more grass seed there is by dividing 11 by 5. It is not as easy to just "spot" this.

As the cost is in direct proportion to the weight, you also multiply the cost by 2.2.

Another way of solving this problem is by finding the cost of 1kg of grass seed first, and then multiplying your answer by 11

This is a useful method because you can find the cost of any number of kg by multiplying the number by £3.60

Example 2

5 kg of grass seed covers an area of 160 m²

a   How much grass seed is needed to cover an area of 480 m²?

b   What area is covered by 45 kg of grass seed?

c   What area is covered by 12 kg of grass seed?

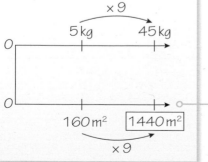

a $15 \div 5 = 3$

Find out what the area has been multiplied by first.

Use the same multiplier for the mass.

b $45 \div 5 = 9$

Find out what the mass has been multiplied by.

5 kg   45 kg

160 m² $\boxed{1440\,m^2}$

×9

Use the same multiplier for the area.

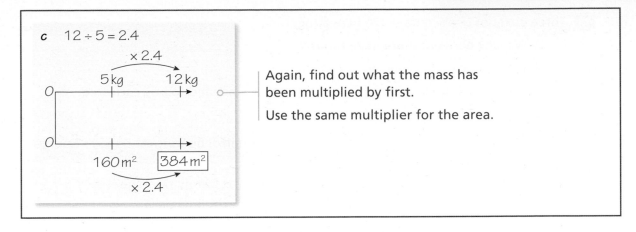

c   12 ÷ 5 = 2.4

Again, find out what the mass has been multiplied by first.

Use the same multiplier for the area.

Practice 2.1A

1 The double number line shows the cost of different lengths of cloth. Work out the missing costs.

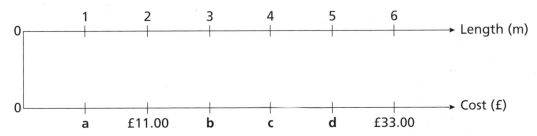

2 The double number line shows the masses of different lengths of rope. Work out the missing quantities.

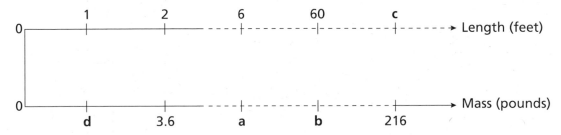

3 Here is a recipe to make soup for 4 people.

> 4 tablespoons butter
> 1 onion
> 600 ml vegetable stock
> 500 g tomatoes

a How much of each ingredient is needed to make soup for 12 people?

b In how many different ways can you work out the amount of each ingredient needed for 6 people?

c Amina has 3 litres of vegetable stock.

   i How many people can she make soup for?

   ii How much of each other ingredient will she need?

4 Three boxes of pens contain 120 pens altogether.

a How many pens are there in 20 boxes?

b How many boxes are needed to buy 600 pens?

c A school has 1000 students. How many boxes are needed to buy each student a pen?

5 The cost of petrol is proportional to the amount of petrol bought. At Danny's Garage, 30 litres costs £34.80

a How much would it cost to buy 10 litres of petrol at Danny's Garage?

b How much would it cost to buy 60 litres of petrol at Danny's Garage?

c How much would it cost to buy 25 litres of petrol at Danny's Garage?

d At Milly's Garage, 20 litres of petrol costs £22.80. Which garage is cheaper? Show workings to explain how you know.

6 20 party poppers cost 90p.

a How much will 80 party poppers cost?

b How many party poppers can you buy for £9?

c Explain why you might not be able to buy an odd number of party poppers.

What do you think?

1 Each of these tanks contains water. For each one, state whether the volume of water in the tank would be directly proportional to the height of the tank.

a **b** **c** **d**

2 A 25 g packet of crisps costs 60p. A 40 g packet of crisps costs 90p.

a Is the cost of crisps proportional to the weight of the packet?

b Suggest some cases where the cost of an item is not directly proportional to the amount bought. Think of reasons why this could happen.

3 **a** For each table, decide whether p and q are in direct proportion.

p	q
5	12
30	72
92	220.8

p	q
3	7
10	21
23	70

p	q
8	20
11	27.5
36	90

b For each table where p and q are in direct proportion, find the values of

i q when $p = 100$ **ii** p when $q = 100$

You can represent proportional relationships on graphs.

The graph of a directly proportional relationship will always be a straight line that goes through the origin (0, 0)

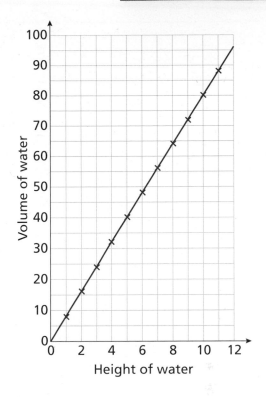

Volume of water / Height of water

Example 3

The cost of electricity (£*C*) to run a heater for *t* hours is shown on the graph.

Use the graph to find

a the cost of running the heater for 5 hours

b how long you can run the heater for £7.50

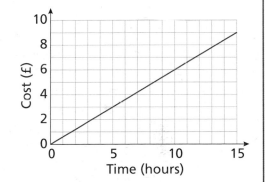

Cost (£) / Time (hours)

a £3

Cost (£) / Time (hours)

Draw a vertical line from 5 hours to the graph.

Draw a horizontal line from the graph to the cost axis and read off the cost.

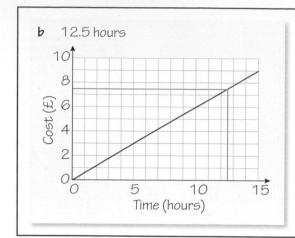

b   12.5 hours

Draw a horizontal line from £7.50 hours to the graph.

Draw a vertical line from the graph to the time axis and read off the time.

Practice 2.1B

1 The cost of ribbon is proportional to its length. Five metres of ribbon costs £1.40

    a Copy and complete the table of lengths and costs.

Length of ribbon (*l* m)	0	5	10	15	20	25
Cost (£*C*)	0	1.40				7.00

    b Suggest a good scale to draw the graph of the values in the table.

    c Draw the graph of *C* against *l*

    d Use your graph to find the cost of 12 m of ribbon.

    e Use your graph to find the length of ribbon you can buy for £3.50

    f Can you use your graph to find the length of ribbon you can buy for £5?

2 The table shows the total cost (£*C*) of buying *n* copies of a textbook.

Number of copies (*n*)	0	2	4	6	8	10
Cost (£*C*)	0	25	50	75	100	125

    a How can you tell that the cost of the textbooks is directly proportional to the number of textbooks bought?

    b What would be a good scale to use for the graph if you wanted *n* to range from 0 to 20?

    c Draw the graph of *C* against *n*, with *n* from 0 to 20

    d Use your graph to find the cost of

        i 3 books   **ii** 12 books.

    e A teacher has a budget of £140 for textbooks. How many books can she buy?

3 The table shows the lengths of the shadows of three trees.

Height of tree (h m)	5	15	20
Length of shadow (s m)	8	24	32

a Calculate $\dfrac{\text{Length of shadow}}{\text{Height of tree}}$ for each tree.

b Explain how you know the length of the shadow is proportional to the height of the tree.

c Draw a graph of s against h, with h from 0 to 30

d Use your graph to find

  i the height of a tree whose shadow is 20 m

  ii the length of the shadow of a tree if its height is 12 m

4 In a science experiment, different masses are attached to the end of a spring and the length of the spring is measured. The results are shown in the table.

Mass (m g)	100	150	200	250	300
Length of spring (l mm)	60	80	100	120	140

a Draw a graph of l against m

b Is the length of the spring directly proportional to the mass of the weight? Explain how you know

  i by looking at the graph

  ii by doing calculations using the numbers in the table.

What do you think?

1 You only need to know one pair of values in order to draw a graph showing a direct proportion relationship.

Investigate Emily's claim.

2 Which of the following show direct proportion relationships?

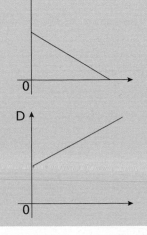
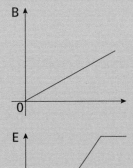
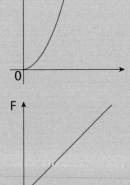

Consolidate – do you need more?

1 The double number line shows the connection between gallons and cups, which are measures of capacity in the US. Work out the missing numbers.

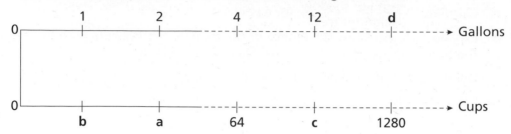

2 The double number line shows the costs of different lengths of rope. Work out the missing quantities.

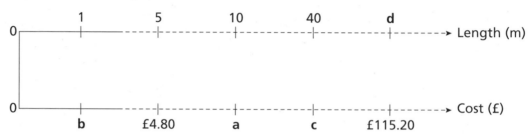

3 5 miles is about the same as 8 km

    a About how many km is the same as 40 miles?

    b About how many miles is the same as 40 km?

4 The total cost of tickets for a theme park is proportional to the number of tickets bought.

    a Copy and complete the table.

Number of tickets (n)	0	2	4	8	10
Cost (£C)	0		28		70

    b Draw a graph of C against n, with n from 0 to 15

    c Use your graph to find the cost of 12 tickets.

Stretch – can you deepen your learning?

1 Which of these pairs of variables are directly proportional to each other?

    A The angles in a pie chart and the frequency that each sector represents

    B The heights of bars in a bar chart and the frequency that each bar represents

    C The number n and the nth term of a linear sequence

    D The number n and the nth term of a geometric sequence

    E The side length of a square and the perimeter of the square

    F The side length of a square and the area of the square

    G The diameter of a circle and the circumference of the circle

2 A 750 ml bottle of water costs 80p. A 2 litre bottle of water costs £2. Is the cost of water directly proportional to the amount of water in the bottle?

3 State which of the following are in direct proportion.

   a The distance (D) travelled in T hours by a car travelling at a constant speed (S)

   b The time taken (T) to cover a distance D miles travelling at a constant speed (S)

   c The speed (S) needed to cover a constant distance D in T hours

4 Find the ratio of the length to width for sheets of A3, A4 and A5 paper. Is the length proportional to the width? Investigate for other shapes of paper (for example, your textbooks and exercise books).

Reflect

1 Explain what is meant by direct proportion.

2 How can you find whether two quantities are in direct proportion?

3 Why do direct proportion graphs go through the origin?

Small steps

- Explore conversion graphs
- Convert between currencies

Key words

Conversion graph – a graph used to change from one unit to another

Axis – a reference line on a graph

Exchange rate – the value of a currency compared to another

Are you ready?

1 Write down the coordinates of the points A, B, C and D.

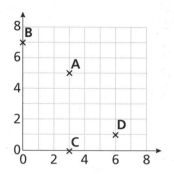

2 Draw a pair of axes from 0 to 6 in both directions. Plot the points (2, 4), (5, 0), (0, 2) and (3, 3).

3 Round these numbers to the nearest integer.

   a 18.76   **b** 32.12   **c** 32.127   **d** 403.5

4 Round these amounts to the nearest penny. Some are done for you.

Amount	To the nearest penny
32.4p	32p
88.7p	
91.2p	
£1.27316	£1.27
£3.464	
£18.125	

Models and representations

A **conversion graph** shows the relationship between two quantities. It is very useful for converting between units of measure.

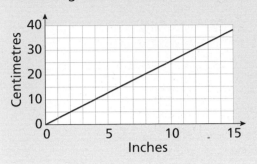

This graph can be used to convert inches to centimetres.

Example 1

Use the graph above to convert

a 6 inches to cm   **b** 20 cm to inches   **c** 200 cm to inches.

a 15 cm

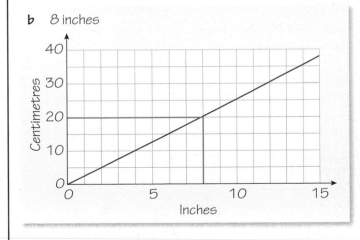

Draw a vertical line from 6 inches to the graph.

Draw a horizontal line from the graph to the centimetres **axis** and read off the number of centimetres.

b 8 inches

Draw a horizontal line from 20 cm to the graph.

Draw a vertical line from the graph to the inches axis and read off the number of inches.

> c $200 \div 20 = 10$ ○————| There are 10 lots of 20 cm in 200 cm
>
> $8 \times 10 = 80$ inches ○——| As the number of inches and number of centimetres are in direct proportion, you can multiply the answer to part **b** by 10
>
> If you do not have copies of graphs, you can line up "by eye" carefully with a ruler, but if you do have copies it is more reliable for drawing accurate lines.

Practice 2.2A

1 You can use this graph to convert between gallons and litres.

   a   Convert 10 gallons to litres.

   b   Convert 30 litres to gallons.

   c   How accurate are your answers?

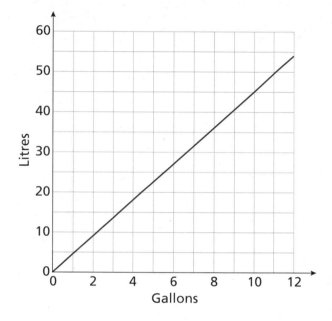

2 **a** Will a conversion graph for stones and kilograms go through (0, 0)? How do you know?

   b   Use your answer to **a**, and the fact that 10 stones is about 63.5 kg, to draw a conversion graph for stones and kilograms. Go from 0 to 15 stones on the horizontal axis and 0 to 100 kg on the vertical axis.

   c   Use your graph to convert

      i 8 stones to kg   **ii** 80 kg to stones.

   d   Ed weighs 41 kg and Jackson weighs 6 stones. Who is heavier, Ed or Jackson?

3 A teacher is drawing a graph to convert marks out of 80 (on the horizontal axis) to percentages (on the vertical axis).

   a   Will the graph pass through the point (0, 0)?

   b   Which of the points (80, 100) and (100, 80) will the graph go through?

   c   Draw the conversion graph and use it to convert these marks out of 80 to percentages.

      i 60   **ii** 32   **iii** 52

4 This graph can be used to convert between temperatures in degrees Celsius and degrees Fahrenheit.

    a Use the graph to convert these temperatures to degrees Celsius.

      i 70°F   **ii** 10°F   **iii** 85°F

    b Use the graph to convert these temperatures to degrees Fahrenheit.

      i 30°C   **ii** 0°C   **iii** −10°C

    c Zach used the graph to convert 10°C to 50°F

> 10°C = 50°F so
> 20°C = 2 × 50°F = 100°F

    Explain why Zach is wrong.

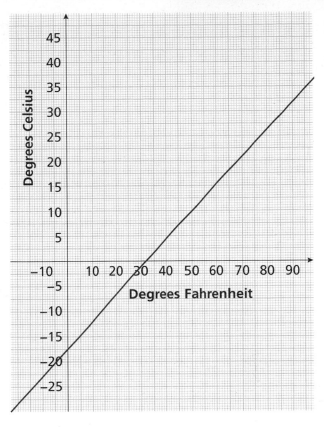

What do you think?

1 Marta and Jakub are drawing graphs to convert miles to kilometres using the fact that 5 miles ≈ 8 km. The miles axis goes from 0 to 60 miles.

> I'm going to use the points (0, 0) and (5, 8) to draw my line.

Marta

> I'm going to use the points (0, 0) and (50, 80) to draw my line.

Jakub

Whose graph is likely to be more accurate to read from, and why?

2 Seb thinks all conversion graphs show directly proportional relationships. Explain why Seb is wrong.

3 At the time of writing this book, £30 = $38 (US dollars).

    a Draw a conversion graph for pounds and dollars, from 0 to £40 on the horizontal axis.

    b Use your graph to convert

      i £20 to US dollars   **ii** $30 to pounds.

4 A conversion graph between pounds and Croatian kuna goes as far as £45

How could you use the graph to convert

a £400 to kuna **b** £100 to kuna?

You can also convert currencies by doing calculations using **exchange rates**.

Example 2

Samira is going on holiday to Brazil. The unit of currency in Brazil is the real.
£1 = 7 Brazilian real

a Samira takes £250 spending money and converts this all to Brazilian real. How many Brazilian real does she receive?

b She spends 294 Brazilian real at a restaurant. How much is this in pounds?

a $250 \times 7 = 1750$ Brazilian real ○— Each pound is worth 7 Brazilian real, so £250 is worth 250 × 7 Brazilian real.

b $294 \div 7 = £42$ ○— Each pound is worth 7 Brazilian real, so to find how many pounds, you need to find how many 7s there are in 294

This is 294 ÷ 7

Practice 2.2B

1 The currency in Australia is the Australian dollar ($). £1 = $1.80

    a How many Australian dollars can you buy with

      i £50   **ii** £200   **iii** £500?

    b Benji converts £3 to Australian dollars on his calculator.

    He writes

    £3 = $5.4 ✗

    Why has Benji's teacher marked his answer wrong?

    c How many pounds can you buy with

      i $180   **ii** $720   **iii** $45.90?

2 The currency in the Czech Republic is the koruna (CZK). £1 = 30 CZK

    a How many Czech koruna can you buy for

      i £40   **ii** £90   **iii** £250?

    b How many pounds can you buy with

      i 9000 CZK   **ii** 15 CZK   **iii** 17 400 CZK?

3 The double number line shows money in British pounds and Thai bhat.
Find the missing numbers.

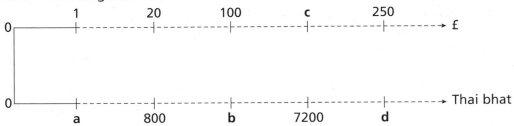

4 The currency in India is the rupee. £1 = 93 rupees

Ed uses £1 = 90 rupees to estimate how many rupees he can buy for £700.

   a What is Ed's estimate for the number of rupees he can buy?

   b Explain how you know his estimate is an underestimate.

   c What is the actual number of rupees he can buy?

   d Explain why Ed cannot get all his money in 200 rupee notes.

   e What is the largest number of 200 rupee notes Ed can buy with his money?

5 Jackson bought a jacket in Belgium. He paid €319. The exchange rate is £1 = €1.10

   a Work out the cost of the jacket in pounds (£).

   b Jackson also bought a belt for €19.80. The same belt costs £17.50 in the UK.
   Is the belt cheaper in Belgium or the UK?

6 A phone costs 6120 krone in Denmark. The same phone costs £799 in the UK.
The exchange rate is £1 = 8.16 Danish krone. Work out the difference in cost of
the phone between Denmark and the UK, giving your answer in pounds (£).

What do you think? 💭

1 Lydia converts £600 to euros (€) at a rate of £1 = €1.08. She spends four-fifths of her
euros and converts the rest back at a rate of £1 = €1.44. How many pounds does she
have now?

2 A company is considering buying office space in Italy and the USA.

The offices in Italy cost €352 000 and the offices in the USA cost $400 000

The exchange rates are £1 = €1.10 and £1 = $1.25

   a Work out which offices are cheaper by converting both prices to pounds.

   b Check your answer by converting both prices to

   i euros   **ii** US dollars.

3 The exchange rate in Paris is €1 = £0.92. The exchange rate in London is £1 = €1.10

In which city would you get more euros for your pounds? Show working to justify
your answer.

Consolidate – do you need more?

1 The conversion graph can be used to change between lengths measured in metres and feet.

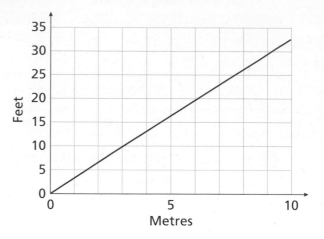

  a Use the graph to find the approximate conversion of

    i 6 metres to feet

    ii 30 feet to metres.

  b How could you use the graph to help convert 50 metres to feet?

2 1 acre is about 4000 square metres.

  a Use the points (0, 0) and (10, 40 000) to draw a conversion graph between acres and square metres.

  b Use your graph to convert

    i 8 acres to square metres   **ii** 10 000 square metres to acres.

3 Flo goes on a trip to Japan. At the time of her trip, £1 = 150 Japanese yen.

  a How many Japanese yen is the same as £20?

  b Draw a conversion graph between pounds and yen, from 0 to £25 on the horizontal axis and 0 to 4000 yen on the vertical axis.

  c Use your graph to convert

    i £15 to Japanese yen   **ii** 2700 Japanese yen to pounds.

4 The currency in Ethiopia is the birr. £1 = 48 birr

  a How many Ethiopian birr can you buy for

    i £50   **ii** £80   **iii** £2000?

  b How many pounds can you buy with

    i 9600 birr   **ii** 12 000 birr   **iii** 240 birr?

5 A family of four is going on holiday to the USA. The flights are £867 each.

They need to book two hotel rooms, which each cost $135 a night, for 14 nights.

They take a total of $1000 spending money, and spend it all.

Using £1 = $1.25, work out the total cost of the holiday in pounds.

Stretch – can you deepen your learning?

1 If you have money left over from a holiday, would you prefer the exchange rate to have gone up or down during your time away? Explain your reasoning.

2 Investigate the costs of the same items (for example, mobile phones) in the UK, the USA and Europe. Where are items cheapest? Write a report on your findings.

Reflect

1 Describe how you would construct and use a conversion graph.

2 How do you use exchange rates to convert from one currency to another? How do you know whether to multiply or divide?

Small steps

■ Explore relationships between similar shapes

■ Understand scale factors as multiplicative representations

Key words

Similar – two shapes are similar if their corresponding sides are in the same ratio

Enlargement – making a shape bigger, or smaller

Scale factor – how much a shape has been enlarged by

Are you ready?

1 Work out

   a $20 \div 2$ **b** $20 \times \frac{1}{2}$ **c** $20 \div 4$ **d** $20 \times \frac{1}{4}$

2 Work out

   a $\frac{3}{5}$ of 60   **b** $\frac{5}{3}$ of 60

3 Simplify the ratios.

   a $12:20$ **b** $15:40$ **c** $10:15$

4 Simplify the fractions.

   a $\frac{18}{30}$ **b** $\frac{25}{40}$ **c** $\frac{48}{80}$

Models and representations

You can use squared paper or dotty paper to explore **similar** shapes.

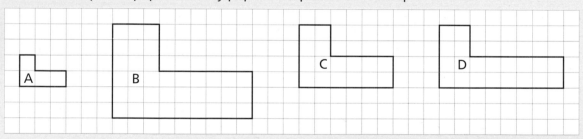

Shapes A, B and C are all similar as their corresponding sides are in the same ratio.

Shape D is not similar to the other three – for example the ratio of D's height to A's height is 4:2 (or 2:1) but the ratio of D's length to A's length is 8:3

The ratio of any pair of corresponding sides in shape A and shape B is 1:3. You can say "Shape B is an enlargement of shape A by scale factor 3". In the same way, "Shape A is an enlargement of shape B by scale factor $\frac{1}{3}$". This is still called an enlargement, even though you are making the shape smaller.

You can also see shape C is an enlargement of shape A by **scale factor** 2 and shape C is an enlargement of shape A by scale factor 1.5, or as a fraction $\frac{3}{2}$.

Example 1

These three triangles are similar.

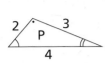 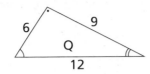 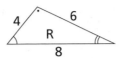

Find the scale factor of the enlargements from

a P to Q   **b** Q to P   **c** R to Q   **d** R to P.

a $12 \div 4 = 3$ ○—| You can use the marked angles to see which sides are corresponding.

You could also have worked out $9 \div 3$ or $6 \div 2$

As the shapes are similar, you will get the same answer.

b $4 \div 12 = \frac{1}{3}$ ○—| It is usual to give scale factors less than 1 as fractions rather than decimals.

c $12 \div 8 = \frac{3}{2}$ ○—| You can leave this as an improper fraction, or give the answer as 1.5

d $4 \div 8 = \frac{1}{2}$ ○—| Be careful to divide in the correct order. The enlargement from P to R would be scale factor 2, but the enlargement from R to P is scale factor $\frac{1}{2}$

Example 2

ABC and PQR are similar.

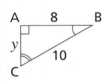 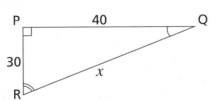

Work out the length of the sides labelled x and y.

Scale factor from ABC to PQR is $40 \div 8 = 5$ ○—| Sides AB and PQ are corresponding.

So $x = 10 \times 5 = 50$ ○————| x is the length of RQ which corresponds to BC

$y = 30 \div 5 = 6$ ○————| y is the length of AC which corresponds to PR. Triangle ABC is smaller than triangle PQR so AC = PR ÷ 5

Alternatively, for the last part, you could use the fact that the scale factor of enlargement from PQR to ABC is $\frac{1}{5}$, so $y = 30 \times \frac{1}{5} = 6$

Practice 2.3A

1 Rectangle A is 4 cm long and 3 cm wide.

Draw rectangle A and the result of enlarging rectangle A by scale factors 2, 2.5 and 3

2 Here are rectangles P and Q

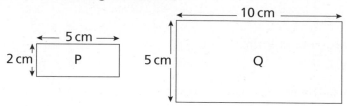

a Find the ratio of the length of rectangle P to the length of rectangle Q.

b Find the ratio of the width of rectangle P to the width of rectangle Q.

3 Here is a hexagon.

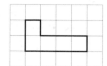

On squared paper, draw this hexagon and three other hexagons that are similar to this one. Also, draw another hexagon that is not similar. Compare your answers with a partner's. Can you spot the "odd one out"?

4 Find the missing side lengths in the pairs of similar shapes.

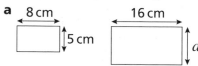

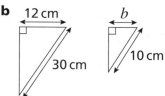

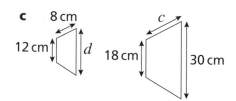

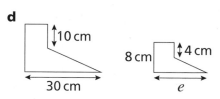

5 An equilateral triangle X has sides 10 cm.

a Construct an equilateral triangle that is an enlargement of X by scale factor $\frac{1}{2}$

b Triangle Y is an enlargement of X by scale factor 4. Find the perimeter of triangle Y.

c The ratio of the side lengths of triangle Z to the side lengths of triangle X is 3 : 5. How long is each side of triangle Z?

d What can you say about the angles in each of the equilateral triangles?

6 The two parallelograms are similar.

a Find the missing side length.

b Find the perimeter of the larger parallelogram.

c Find the perimeter of the smaller parallelogram.

d State the ratio of the perimeter of the smaller parallelogram to the perimeter of the larger parallelogram.

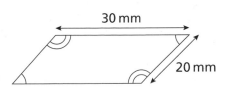
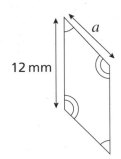

30 mm

20 mm

12 mm

a

What do you think? 💭

1 Why is it important when enlarging a photograph that the enlargement is similar to the original photograph?

2 Faith says these two triangles cannot be similar because they are facing opposite directions.

Do you agree with Faith? Explain your answer.

3 Emily says these two triangles are similar because they are both isosceles.

Do you agree with Emily? Explain your answer.

4 What do you notice about the corresponding angles in similar shapes?

Example 3

Here are two similar trapezia.

a Write the ratio of the side lengths of shape P to the side lengths of shape Q in its simplest form.

b State the scale factor of the enlargement of

i shape P to shape Q   **ii** shape Q to shape P

c Find the lengths labelled x and y

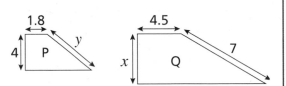

1.8   4.5

4   P y x Q   7

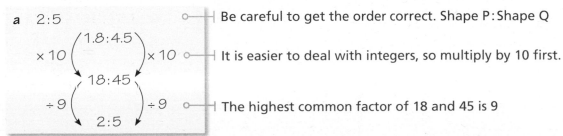

a 2:5   Be careful to get the order correct. Shape P : Shape Q

× 10 $\left(\begin{array}{c}1.8:4.5\end{array}\right)$ × 10   It is easier to deal with integers, so multiply by 10 first.

18:45

÷ 9 $\left(\begin{array}{c}\end{array}\right)$ ÷ 9   The highest common factor of 18 and 45 is 9

2:5

b

   i $\dfrac{5}{2}$ ○———

You can see this from the ratio, or start again with $\dfrac{4.5}{1.8}$

It is common to give a scale factor as a fraction, even if improper, but you could say 2.5

   ii $\dfrac{2}{5}$ ○——— Or you could say 0.4

c $x = 4 \times \dfrac{5}{2} = 10$ ○——— You multiply the corresponding side by the scale factor, either as a fraction or a decimal.

You can do this using the fraction key on your calculator if you like. You will learn how to multiply and divide fractions without a calculator in the next block.

$y = 7 \times \dfrac{2}{5} = 2.8$ ○——— This time you need to use the scale factor of enlargement from Q to P

Practice 2.3B

1 In each pair of similar shapes, which side corresponds to AB?

a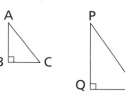

b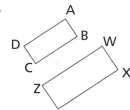

c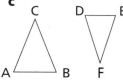

d

e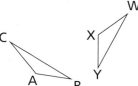

f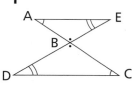

Which ones are easier to spot and which ones are harder? Why?

2 How do you work out $12 \times \dfrac{3}{5}$ on your calculator? Compare your calculator with a partner's.

3 Which of these scale factors will increase the size of a shape and which will decrease the size of a shape?

   A 3   **B** $\dfrac{4}{5}$ **C** $\dfrac{7}{5}$ **D** $\dfrac{5}{7}$ **E** $\dfrac{1}{3}$ **F** $\dfrac{2}{5}$ **G** 2.5   **H** $\dfrac{3}{4}$ **I** $\dfrac{4}{3}$

4 Work out the length of the side labelled x in each of these pairs of similar shapes.

a

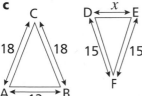
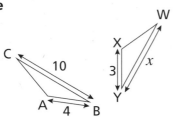

b

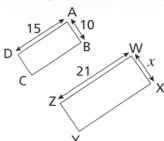
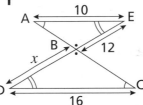

c

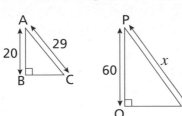

d

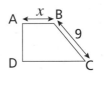

e

f

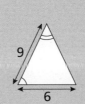

5 A photograph is 8 inches long and 5 inches wide. Which of these could be enlargements of the photograph?

A 16 inches by 10 inches

B 15 inches by 24 inches

C 24 inches by 12.5 inches

D 8 cm by 5 cm

E 5 feet by 8 feet

F 2.5 inches by 4 inches

What do you think? 💭

1 A pair of corresponding sides on two similar shapes have lengths 9 cm and 15 cm.

The scale factor of enlargement is $\frac{5}{3}$

Faith

The scale factor of enlargement is 1.67

Jakub

Who do you agree with?

2 Which of these calculations will find the value of x?

$9 \times \dfrac{14}{6}$ $9 \times \dfrac{7}{3}$ $\dfrac{9 \times 7}{3}$

$9 \div 3 \times 7$ $9 \times 3 \div 7$ $9 \times 7 \div 3$

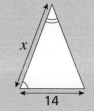

3 Find the multipliers from P to Q in each of these cases.

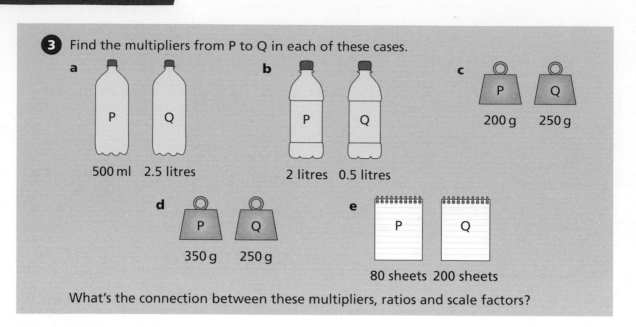

a

P   Q

500 ml   2.5 litres

b

P   Q

2 litres   0.5 litres

c

P   Q

200 g   250 g

d

P   Q

350 g   250 g

e

P   Q

80 sheets   200 sheets

What's the connection between these multipliers, ratios and scale factors?

Consolidate – do you need more?

1 Draw four rectangles that are similar to this one.

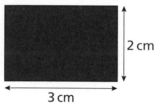

2 cm

3 cm

2 Here are five pairs of similar shapes. Which shape is similar to which other shape?

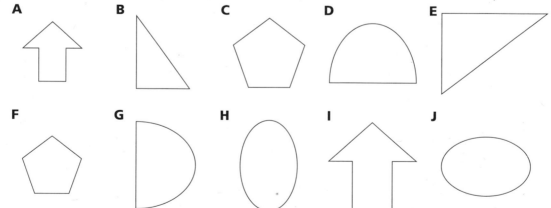

A   B   C   D   E

F   G   H   I   J

3 Each pair of shapes are similar. Find the lengths of the sides labelled with letters.

a
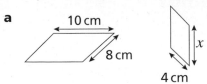
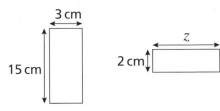

b
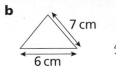
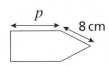

c
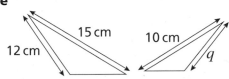

d
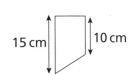

e
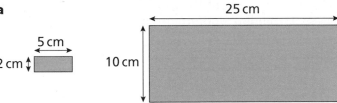

f
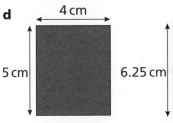

4 Are these pairs of rectangles similar? Explain your answers.

a
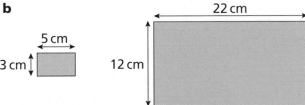

b
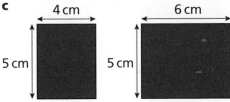

c

d

Stretch – can you deepen your learning?

1 Which of these statements are true?

A | All squares are similar

B | All rectangles are similar

C | All rhombuses are similar

D | All circles are similar

E | All equilateral triangles are similar

F | All isosceles triangles are similar

G | No scalene triangles are similar

Explain your reasoning and give examples/non-examples.

2 Beca thinks that when you enlarge a shape by scale factor 2, the new shape will have double the perimeter and double the area of the original shape. Do you agree? Investigate with rectangles and triangles, and other scale factors.

3 Huda says the two triangles are similar. Do you agree? Why or why not?

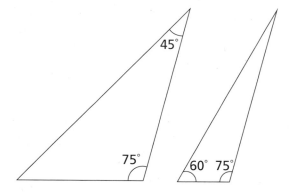

4 a Use your knowledge of angles rules to show that these two triangles have the same angles.

   b Deduce the length of a

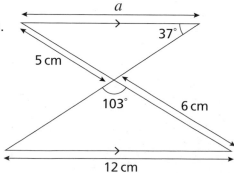

5 The two quadrilaterals are similar.

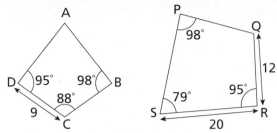

a Which side corresponds to AB?

b Which side corresponds to PQ?

c Work out all the angles in the quadrilaterals.

d Which sides can you work out?

e How much more information do you need to be able to work out the remaining unknown sides?

Reflect

1 What's the same and what's different when you enlarge a shape by

a a scale factor greater than 1

b a scale factor less than 1

c a scale factor of 1?

2 What's the difference between the mathematical meaning of the word "similar" and its everyday meaning?

What do you think? 💭

1. Sometimes diagrams are labelled "Not to scale". What does this mean? Why do you think this label is given to some diagrams?

2. **a** Here are two ways of writing scales, 1:200 and 1cm to 2m. Are they the same or different? How do you know?

   b How else could you write the scales
   - **i** 1:400　　　**ii** 1:150　　　**iii** 2cm to 5m?

3. An old textbook has a diagram labelled "1 inch to 1 foot". Darius thinks he cannot use the diagram as his ruler is only graduated in cm. What do you think?

4. What would be a sensible scale for drawing a diagram of
   - **a** your classroom　　　**b** the school hall　　　**c** the school grounds?

Example 3

The scale of a map is 1:50000

a Two hotels are 4cm apart on the map. What is the actual distance between the hotels? Give your answer in kilometres.

b The distance between two castles is 6.5km. How far apart would they be on the map? Give your answer in centimetres.

a $4\,cm \times 50\,000 = 200\,000\,cm$
$200\,000\,cm \div 100 = 2000\,m$
$2000\,m \div 1000 = 2\,km$

> Each cm is 50000cm, so 4cm is 4 × 50000cm
> To change to metres, divide the number of centimetres by 100
> To change to km, divide the number of metres by 1000

b $6.5\,km = 6500\,m = 650\,000\,cm$
$650\,000 \div 50\,000 = 13\,cm$

> 6.5 ÷ 50000 would give a very small decimal, so it is easier to convert to km to cm first, by multiplying by 100 and then 1000
> Then you can use the scale to see how many times 50000 divides into 650000

Practice 2.4B

1. Here is part of a map. 1cm on the map represents 50km.

   a Measure the distance, in centimetres, between Bedford and Cambridge on the map.

   b Work out the distance, in kilometres, between Bedford and Cambridge.

   💬 **c** Tom is going to drive from Bedford to Cambridge. Will the distance he drives be more or less than your answer to **b**? Why?

d Investigate the distances between other towns and cities labelled on the map. Which are closest together? Which are furthest apart?

e Show that the scale of the map can be written 1:5 000 000

2 The scale of this map is 1:1 000 000

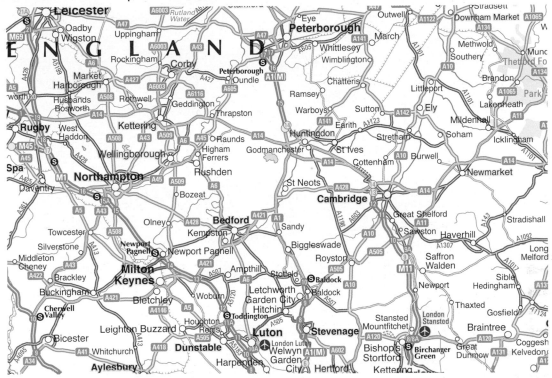

a What distance, in km, does 1 cm on the map represent?

b Measure the distance, in centimetres, between Bedford and Cambridge on the map.

c Work out the distance, in kilometres, between Bedford and Cambridge.

d Are your answers more or less accurate than your answer to question **1**? Why?

e Investigate the distances between other towns and cities labelled on the map. Can you find towns that are exactly 50 km apart? What other questions can you ask?

3 The scale of this map is 1:500 000

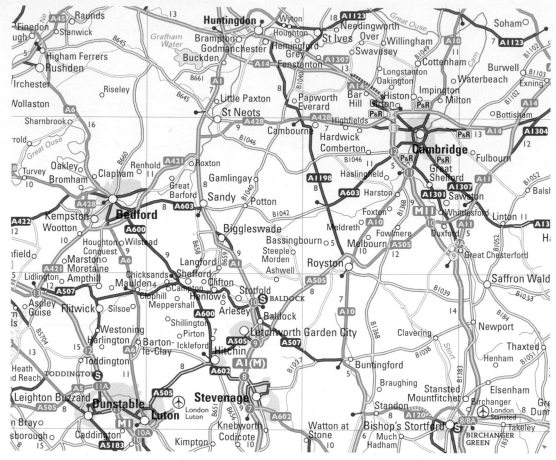

a What distance, in km, does 1 cm on the map represent?

b Measure the distance, in centimetres, between Bedford and Cambridge on the map.

c Work out the distance in kilometres between Bedford and Cambridge.

d Where in Cambridge and where in Bedford did you choose to measure from?

e Investigate the distances between other towns and cities labelled on the map. How accurate are you answers now?

4 The scale of a map is 1:400 000

a On the map, the distance between two towns is 7.2 cm. Find the actual distance between the two towns.

b Find the distance apart on the map of two towns that are actually 100 km apart.

5 Express each scale in the form 1:n

a 1 cm represents 10 m

b 1 cm represents 2 km

c 1 cm represents 40 km

d 2 cm represents 10 m

e 5 cm represents 2 km

f 4 cm represents 10 km

What do you think?

1 Emily and Jackson are going on a hike.

My map is 1:50000 so it'll be more detailed than yours.

Emily

No, my map is more detailed because the scale is 1:25000

Jackson

Who do you agree with? Why?

2 On a map of a town, a street is 12 cm long. The street is actually 240 m long. Work out the scale of the map in the form

   a 1 cm to ☐ m   **b** 1:n

3 The scale of a map is 1:40000. A square field on the map is 3 cm long.

   a Work out the area of the square field as drawn on the map.

   b What is the actual length of the field?

   c What is the actual area of the field?

   d How many times bigger than the area on the map is the actual area of the field?

Consolidate – do you need more?

1 Here is a sketch of a rectangle

   a Make an accurate scale drawing of the rectangle using a scale of

     i 1:10   **ii** 1:20

   b Find the dimensions of a scale drawing of the rectangle if using a scale of

     i 1:4   **ii** 1:100

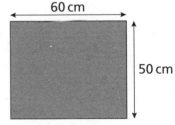

60 cm

50 cm

   c Which is the most appropriate scale to use for a diagram of the rectangle?

2 A warehouse is 40 m long and 15 m wide.

   Work out the length and width of a scale drawing of the warehouse using a scale of

   a 2 cm to 1 m   **b** 1 cm to 1 m   **c** 1 cm to 2 m

3 A map of a town is drawn to a scale of 1:10000

   a What actual distance does 1 cm on the map represent? Give your answer in cm, m and km.

   b What actual distance does 6 cm on the map represent? Give your answer in cm, m and km.

    c   What actual distance does 1 mm on the map represent?
Give your answer in cm, m and km.

    d   What actual distance does 1 inch on the map represent?
Give your answer in inches.

4 A room is 8 m long. On a scale drawing, the length of the room is 4 cm.
Find the scale of the drawing in the form $1:n$

Stretch – can you deepen your learning?

1 Benji, Marta and Ali draw a scale diagram of
their classroom. They all use different scales.

Do you agree with Benji? Justify your answer.

> All our diagrams are mathematically similar.

2 Write each scale in the form "1 cm to ☐ km"

    a 1:200 000   **b** 1:500 000   **c** 1:50 000   **d** 1:25 000   **e** 1:4000

3 On a 1:40 000 map, a section of road is 12 cm long. Find the length of the section of
road on a 1:50 000 map. How many different ways can you find to solve this problem?

4 Here is a map of a large coastal town.

    What might the scale be? What features can you use to decide?

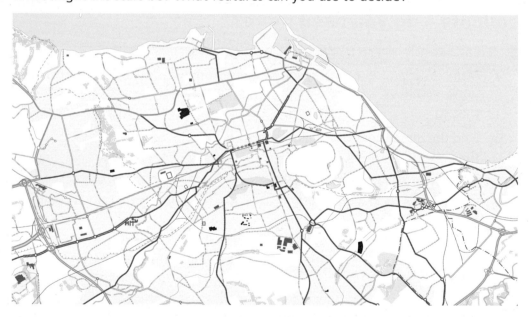

Reflect

1 What's the same and what's different about a scale and a ratio?

2 What's the same and what's different about using a map to find a real distance and
working out what lengths would be on a map?

I have become fluent in...

- using multipliers to work out unknown values
- converting between different currencies
- drawing and interpreting conversion graphs
- recognising similar shapes
- working out unknown sides in similar shapes
- drawing and interpreting scale diagrams.

I have developed my reasoning skills by...

- recognising whether two quantities are directly proportional to each other
- using conversion graphs to find values beyond the range of the graph
- knowing whether to multiply or to divide when converting between currencies
- explaining when pairs of shapes are similar and when they are not
- linking scale factors and multipliers to ratio.

I have been problem-solving through...

- using direct proportion to find out values
- breaking down complex problems into smaller steps
- selecting appropriate methods to answer questions
- working out missing side lengths in similar shapes
- representing problems in different forms including tables and graphs.

Check my understanding

1 A pack of 50 envelopes costs £3.20

   a How much will 250 envelopes cost?

   b How many envelopes can I buy for £64?

2 This graph can be used to convert inches to centimetres.

Use the graph to convert

   a 10 cm to inches   **b** 10 inches to cm

   c 50 inches to cm

3 The currency of Jordan is the dinar. £1 = 0.91 Jordanian dinar

   a Convert 182 dinar to pounds.   **b** Convert £2000 to dinar.

   c Which is worth more, £250 or 230 dinar?

4 Are these triangles similar? Explain how you know.

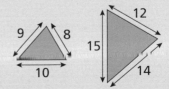

5 A map is drawn to a scale of 1 : 400 000

   a How far, in kilometres, does 7 cm on the map represent?

   b Two towns are 60 km apart. How far apart are they on the map?

3 Multiplying and dividing fractions

In this block, I will learn...

to represent multiplication of fractions

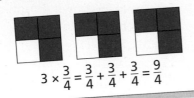

$$3 \times \frac{3}{4} = \frac{3}{4} + \frac{3}{4} + \frac{3}{4} = \frac{9}{4}$$

how to multiply a fraction by an integer

| $\frac{2}{5}$ | $\frac{2}{5}$ | $\frac{2}{5}$ |

$$3 \times \frac{2}{5} = \frac{6}{5} = 1\frac{1}{5}$$

0   1   2

how to find the product of two fractions

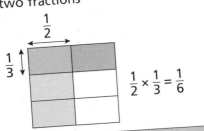

$$\frac{1}{2} \times \frac{1}{3} = \frac{1}{6}$$

how to divide an integer by a fraction

4			
1	1	1	1
$\frac{1}{5}$ $\frac{1}{5}$ $\frac{1}{5}$ $\frac{1}{5}$ $\frac{1}{5}$	$\frac{1}{5}$ $\frac{1}{5}$ $\frac{1}{5}$ $\frac{1}{5}$ $\frac{1}{5}$	$\frac{1}{5}$ $\frac{1}{5}$ $\frac{1}{5}$ $\frac{1}{5}$ $\frac{1}{5}$	$\frac{1}{5}$ $\frac{1}{5}$ $\frac{1}{5}$ $\frac{1}{5}$ $\frac{1}{5}$

$$4 \div \frac{1}{5} = 20$$

how to divide a fraction by a fraction

There are 3 lots of $\frac{1}{6}$ in $\frac{1}{2}$

$$\frac{1}{2} \div \frac{1}{6} = 3$$

$\frac{1}{2}$			$\frac{1}{2}$		
$\frac{1}{6}$	$\frac{1}{6}$	$\frac{1}{6}$	$\frac{1}{6}$	$\frac{1}{6}$	$\frac{1}{6}$

to multiply and divide improper fractions and mixed numbers **H**

$$3\frac{1}{5} \times \frac{1}{4} = (3 \times \frac{1}{4}) + (\frac{1}{5} \times \frac{1}{4})$$

to work with algebraic fractions **H**

$$\frac{a}{3} \times \frac{6b}{5} = \frac{a \times 6b}{3 \times 5} = \frac{6ab}{15} = \frac{2ab}{5}$$

3.1 Multiplying fractions by integers

White Rose Maths

Small steps

■ Represent multiplication of fractions

■ Multiply a fraction by an integer

Key words

Fraction – a number that compares equal parts of a whole

Integer – a whole number

Are you ready?

1 Complete these multiplications.

   a 12×6 **b** 7×9 **c** 3×15 **d** 11×8 **e** 5×26

2 What fraction is represented in each diagram?

   a   **b**   **c**   **d**   **e**   **f**

3 Complete these calculations.

   a $\frac{1}{4} + \frac{1}{4} + \frac{1}{4}$ **b** $\frac{1}{7} + \frac{1}{7} + \frac{1}{7}$ **c** $\frac{2}{11} + \frac{2}{11} + \frac{2}{11}$ **d** $\frac{1}{3} + \frac{1}{3} + \frac{1}{3}$

4 Convert these improper fractions to mixed numbers.

   a $\frac{12}{5}$ **b** $\frac{17}{4}$ **c** $\frac{28}{3}$ **d** $\frac{87}{11}$

5 Convert these mixed numbers to improper fractions.

   a $3\frac{2}{5}$ **b** $6\frac{1}{4}$ **c** $8\frac{1}{3}$ **d** $7\frac{6}{11}$

Models and representations

Bar models

These can be used to represent **fractions** and multiplication of fractions.

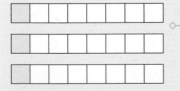

Each bar model represents $\frac{1}{8}$. There are 3 sections shaded altogether, and each section represents one eighth so $3 \times \frac{1}{8} = \frac{3}{8}$

Number lines

These are useful for showing the relative sizes of fractions and for representing multiplication of fractions.

Using a bar model above a number line can help when converting between improper fractions and mixed numbers.

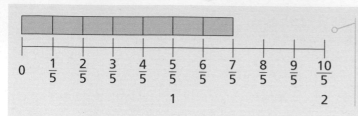

The bar model above the number line shows 7 squares, each representing $\frac{1}{5}$ and represents the calculation $7 \times \frac{1}{5}$. You can see from the number line that $7 \times \frac{1}{5} = \frac{7}{5}$ or $1\frac{2}{5}$

In this chapter, you will look at multiplying a fraction by an **integer**.

You can think of multiplication as repeated addition.

You know that $3 \times 5 = 5 + 5 + 5$ so, similarly, $3 \times \frac{1}{5} = \frac{1}{5} + \frac{1}{5} + \frac{1}{5}$

Example 1

a Draw a diagram to represent $\frac{2}{9} \times 4$ **b** Use your diagram to calculate $\frac{2}{9} \times 4$

a
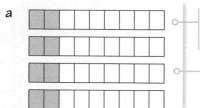

Each bar model is split into nine equal parts and two of these parts are shaded. So $\frac{2}{9}$ of each bar model is shaded.

There are 4 bar models, so there are 4 lots of $\frac{2}{9}$

The diagram represents $4 \times \frac{2}{9}$ or $\frac{2}{9} \times 4$

b $\frac{8}{9}$

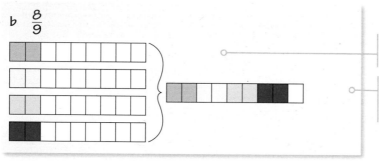

Each of the $\frac{2}{9}$ are shown here in different colours.

You can put them all on the same bar model to show that there are $\frac{8}{9}$ shaded in total.

Do not be tempted to think that there are now 36 equal parts altogether and that the answer is $\frac{8}{36}$. This is not correct because each *whole* is split into ninths, so you are working with ninths.

Practice 3.1A

1 Explain how this diagram shows that $3 \times \frac{1}{7} = \frac{1}{7} + \frac{1}{7} + \frac{1}{7} = \frac{3}{7}$

2 What multiplication is represented in each diagram?

You do not need to work out the answers.

a **b** **c** 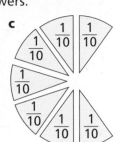 **d**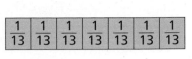

3 Write down the multiplication represented in each diagram and work out the answer.

a **b** **c** **d**

4 Work out

a 5×3 **b** 5×3 ones **c** 5×3 tens

d 5×3 thousands **e** 5×3 cm **f** 5×3 trees

g 5×3 eighteenths **h** $5 \times \frac{3}{18}$ **i** $5 \times 3x$

What do you notice?

5 Complete these calculations.

a $\frac{1}{5} \times 2$ **b** $2 \times \frac{1}{7}$ **c** $\frac{1}{16} \times 3$ **d** $7 \times \frac{1}{19}$

e $\frac{1}{100} \times 9$ **f** $\frac{2}{5} \times 2$ **g** $\frac{3}{7} \times 2$ **h** $3 \times \frac{5}{16}$

i $\frac{2}{19} \times 7$ **j** $9 \times \frac{11}{100}$

Which calculations are easy to represent pictorially? Which are not?

6 Explain how the diagram shows that both calculations are correct.

$3 \times \frac{2}{5} = \frac{6}{5}$ $3 \times \frac{2}{5} = 1\frac{1}{5}$

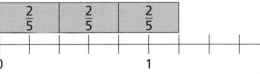

7 Complete these calculations. Give each answer as an improper fraction and as a mixed number in its simplest form.

a $\frac{1}{2} \times 3$ **b** $2 \times \frac{2}{3}$ **c** $7 \times \frac{1}{4}$ **d** $\frac{2}{5} \times 4$

e $5 \times \frac{3}{10}$ **f** $\frac{9}{10} \times 8$ **g** $7 \times \frac{10}{11}$ **h** $\frac{5}{6} \times 7$

i $\frac{21}{25} \times 5$ **j** $8 \times \frac{99}{100}$

😃 **8** Work out the missing numbers.

a $\frac{1}{7} \times \square = \frac{5}{7}$ **b** $\square \times \frac{1}{7} = 1$ **c** $6 \times \frac{\square}{7} = \frac{18}{7}$ **d** $\frac{4}{7} \times \square = 5\frac{1}{7}$

What do you think? 😃

1 Ed has drawn a diagram to work out $\frac{3}{5} \times 4$

Here is his working and answer.

$$\frac{3}{5} \times 4 = \frac{12}{20}$$

a Simplify $\frac{12}{20}$

What do you notice? How does this show that Ed has made a mistake?

b Explain Ed's mistake.   **c** Work out $\frac{3}{5} \times 4$

2 a Show that the answer to $6 \times \frac{2}{3}$ is an integer.

b Write another calculation involving an integer multiplied by a fraction that gives an integer answer.

3 Beca, Chloe, Abdullah and Seb have each written down the calculation they think is represented by the diagram.

Beca

$5 \times \frac{3}{8}$

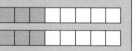

Abdullah

$\frac{3}{8} + \frac{3}{8} + \frac{3}{8} + \frac{3}{8} + \frac{3}{8}$

Chloe

$5 \times \frac{5}{8}$

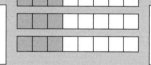

Seb

$\frac{5}{8} + \frac{5}{8} + \frac{5}{8} + \frac{5}{8} + \frac{5}{8}$

a Explain why each person could be correct.

b Work out the answer to each person's calculation. What do you notice?

4 Work out the missing number.

$10 \times \frac{?}{4} = 7\frac{1}{2}$

Consolidate – do you need more?

1 What calculation is represented in each diagram? Work out each answer.

a   **b**   **c**   **d**

 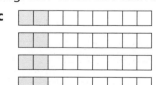 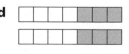

2 Draw a diagram to show that

a $3 \times \frac{1}{10} = \frac{3}{10}$ **b** $3 \times \frac{7}{10} = \frac{21}{10}$

c $5 \times \frac{7}{10} = 3\frac{1}{2}$ **d** $5 \times \frac{8}{10} = 4$

3 Complete these calculations.

a $\frac{1}{9} \times 8$ **b** $5 \times \frac{1}{7}$ **c** $\frac{2}{9} \times 4$ **d** $\frac{2}{7} \times 2$

e $8 \times \frac{2}{9}$ **f** $4 \times \frac{2}{7}$ **g** $\frac{2}{9} \times 9$ **h** $7 \times \frac{1}{7}$

Stretch – can you deepen your learning?

1 There are 9 identical pieces of ribbon, each $\frac{7}{20}$ m long.

The 9 pieces of ribbon are placed end to end.

What is the total length of the ribbon? Give your answer as a mixed number.

2 Tom spends $\frac{3}{4}$ of an hour doing homework 5 times a week.

How long does he spend doing homework in 6 weeks?

3 Evaluate $24y$ when $y = \frac{5}{8}$

Give your answer in its simplest form.

4 The area of the rectangle is $10\,\text{m}^2$

What fraction of a metre is the width of
the rectangle?

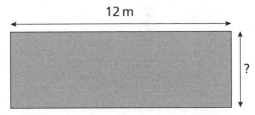

12 m

?

5 Solve these equations.

a $\frac{x}{9} = \frac{2}{5}$ **b** $\frac{x}{9} - 1 = \frac{2}{5}$ **c** $\frac{x}{9} - 1 = 1\frac{2}{5}$

6 Simplify $a \times \frac{b}{c}$

7 These trapezia are similar.

a What is the scale factor of the enlargement from A to B?

b Find the value of x, giving your answer as a mixed number.

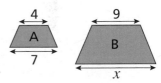

4

A

7

9

B

x

Reflect

1 The diagram represents $2 \times \frac{3}{7}$

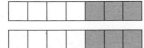

Explain where each digit in the calculation is represented in the diagram.

2 In your own words, explain how to multiply a fraction by an integer.

Small steps

- Represent multiplication of fractions
- Find the product of a pair of unit fractions
- Find the product of any pair of fractions

Key words

Numerator – the top number in a fraction that shows the number of parts

Denominator – the bottom number in a fraction; it shows how many equal parts one whole has been divided into

Unit fraction – a fraction with a numerator of 1

Non-unit fraction – a fraction with a numerator that is not 1

Are you ready?

1 Complete these multiplications.

a $5 \times \frac{1}{7}$

b $\frac{1}{9} \times 4$

c $11 \times \frac{1}{15}$

d $99 \times \frac{1}{100}$

e $5 \times \frac{2}{17}$

f $\frac{5}{17} \times 2$

g $\frac{9}{49} \times 5$

h $\frac{13}{100} \times 3$

2 Sort the fractions into unit and non-unit fractions.

$\frac{5}{17}$ $\frac{17}{5}$ $\frac{1}{17}$ $5\frac{1}{17}$ $\frac{1}{5}$ $\frac{5}{1}$ $\frac{1}{517}$ $17\frac{1}{5}$

3 Simplify these fractions.

a $\frac{15}{20}$

b $\frac{12}{15}$

c $\frac{24}{40}$

d $\frac{36}{90}$

4 Work out

a 4^2

b 7^2

c 2^3

5 Calculate

a 3×-5

b -2×-6

c -8×3

Models and representations

Area model

An area model can be used to support understanding of multiplying fractions.

 This is a unit square.

The area is 1 square unit because $1 \times 1 = 1$

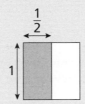 This is still the unit square, but this time it has been split into two equal parts.

The shaded area is $\frac{1}{2}$ square unit because $1 \times \frac{1}{2} = \frac{1}{2}$

 This time the unit square has been split into three equal parts.

The shaded area is $\frac{1}{3}$ square unit because $\frac{1}{3} \times 1 = \frac{1}{3}$

 The diagram shows $\frac{1}{3} \times \frac{1}{2}$.

The overlapping section represents the product of the two fractions.

 $\frac{1}{3} \times \frac{1}{2} = \frac{1}{6}$

Example 1

a What multiplication does this represent? **b** Calculate $\frac{1}{3} \times \frac{1}{5}$

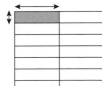

a

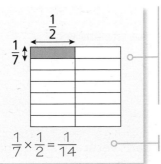

Horizontally, the square has been split into 7 equal parts, so each row represents $\frac{1}{7}$

Vertically, the square has been split into 2 equal parts so each column represents $\frac{1}{2}$

There are 14 equal parts and 1 of these is shaded.

$\frac{1}{7} \times \frac{1}{2} = \frac{1}{14}$

The multiplication represented is $\frac{1}{7} \times \frac{1}{2} = \frac{1}{14}$

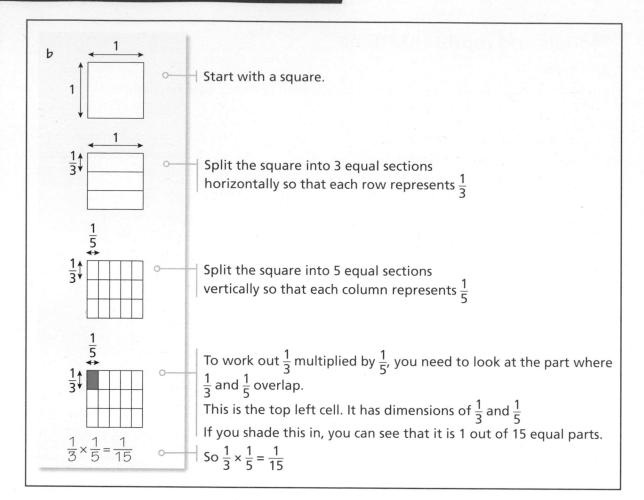

b

Start with a square.

Split the square into 3 equal sections horizontally so that each row represents $\frac{1}{3}$

Split the square into 5 equal sections vertically so that each column represents $\frac{1}{5}$

To work out $\frac{1}{3}$ multiplied by $\frac{1}{5}$, you need to look at the part where $\frac{1}{3}$ and $\frac{1}{5}$ overlap.

This is the top left cell. It has dimensions of $\frac{1}{3}$ and $\frac{1}{5}$

If you shade this in, you can see that it is 1 out of 15 equal parts.

$\frac{1}{3} \times \frac{1}{5} = \frac{1}{15}$

So $\frac{1}{3} \times \frac{1}{5} = \frac{1}{15}$

Practice 3.2A

1 Write the calculation and the answer represented by each diagram.

a   b   c   d

2 Draw a diagram to represent each calculation. Then work out the answer.

a $\frac{1}{2} \times \frac{1}{2}$ b $\frac{1}{4} \times \frac{1}{3}$ c $\frac{1}{2} \times \frac{1}{3}$ d $\frac{1}{10} \times \frac{1}{5}$

What do you notice about the questions and answers?

3 Beca is working out $\frac{1}{15} \times \frac{1}{20}$

She says, "I am going to draw a square and split it into 15 parts vertically and 20 parts horizontally."

a Explain why Beca's method is not very efficient.

b Work out $\frac{1}{15} \times \frac{1}{20}$ and explain your method.

4 Complete these multiplications.

a $\frac{1}{9} \times \frac{1}{5}$

b $\frac{1}{8} \times \frac{1}{3}$

c $\frac{1}{2} \times \frac{1}{40}$

d $\frac{1}{6} \times \frac{1}{7}$

e $\frac{1}{12} \times \frac{1}{4}$

f $\frac{1}{27} \times \frac{1}{20}$

g $\frac{1}{11} \times \frac{1}{80}$

h $\frac{1}{19} \times \frac{1}{50}$

5 Work out the missing number in each calculation.

a $\frac{1}{8} \times \frac{1}{\square} = \frac{1}{80}$

b $\frac{1}{63} = \frac{1}{\square} \times \frac{1}{9}$

c $\frac{1}{75} = \frac{1}{25} \times \frac{\square}{3}$

d $\frac{1}{200} \times \frac{1}{8} = \frac{1}{\square}$

6 Complete these calculations. Give your answer to each part as a fraction.

a $(\frac{1}{5})^2$

b $(\frac{1}{6})^2$

c $\frac{1}{2} \times \frac{1}{3} \times \frac{1}{4}$

d $\frac{1}{5} \times (\frac{1}{8})^2$

e $(\frac{1}{4})^3$

f $(\frac{1}{5} \times \frac{1}{4})^2$

g 0.1^2

h 0.01^2

7 Work out the answer to each calculation.

a $-\frac{1}{4} \times \frac{1}{3}$

b $-\frac{1}{5} \times -\frac{1}{9}$

c $(-\frac{1}{14})^2$

What do you think?

1 Do you agree with Faith? Explain your reasoning.

The product of any pair of unit fractions is always a unit fraction.

2 Amina exercises for $\frac{1}{2}$ an hour.

She spends $\frac{1}{5}$ of this time walking and the rest of it running.

What fraction of an hour does Amina spend walking?

You will now look at multiplying two or more **non-unit fractions**. In the previous exercise, you focused on **unit fractions** only. You will now develop this knowledge further to see what happens when the **numerator** of one or both fractions is greater than 1

Example 2

a Write the multiplication represented by this diagram.

b Calculate $\frac{2}{3} \times \frac{3}{5}$

a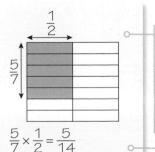

Horizontally, the square has been split into 7 equal parts, so each row represents $\frac{1}{7}$. The arrow spans 5 out of the 7 parts, so has length $\frac{5}{7}$

Vertically, the square has been split into 2 equal parts so each column represents $\frac{1}{2}$

There are 5 out of 14 equal parts shaded.

The multiplication represented is $\frac{5}{7} \times \frac{1}{2} = \frac{5}{14}$

$\frac{5}{7} \times \frac{1}{2} = \frac{5}{14}$

b

Start with a square.

Split the square into 3 equal sections horizontally so that each row represents $\frac{1}{3}$. Draw an arrow of length $\frac{2}{3}$

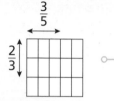

Split the square into 5 equal sections vertically so that each column represents $\frac{1}{5}$. Draw an arrow of length $\frac{3}{5}$

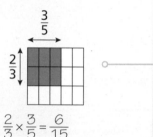

$\frac{2}{3} \times \frac{3}{5} = \frac{6}{15}$

To work out $\frac{2}{3}$ multiplied by $\frac{3}{5}$, you need to look at the part of the representation where $\frac{2}{3}$ and $\frac{3}{5}$ overlap.

This is the shaded part of the diagram.

6 out of 15 equal parts are shaded.

So $\frac{2}{3} \times \frac{3}{5} = \frac{6}{15}$

Using prior knowledge of fractions, you can then simplify your answer.

3 is a common factor of 6 and 15, so you can divide both the numerator and the **denominator** by 3

In this diagram it is clearer to see that $\frac{6}{15}$ and $\frac{2}{5}$ are equivalent

Practice 3.2B

1 Write the calculation and the answer represented by each diagram.

a

b

c

d

2 Draw a diagram to represent each calculation. Then work out the answer.

a $\frac{1}{2} \times \frac{3}{4}$

b $\frac{1}{4} \times \frac{2}{3}$

c $\frac{1}{2} \times \frac{2}{3}$

d $\frac{3}{10} \times \frac{4}{5}$

3 **a** Discuss with a partner how to multiply two fractions without drawing a diagram.

b Without drawing a diagram, calculate $\frac{7}{15} \times \frac{11}{20}$

c Explain why drawing a diagram is not efficient in this case.

4 Complete these multiplications.

a $\frac{2}{9} \times \frac{4}{5}$ **b** $\frac{7}{8} \times \frac{1}{3}$ **c** $\frac{1}{2} \times \frac{13}{40}$ **d** $\frac{5}{6} \times \frac{5}{7}$

e $\frac{7}{12} \times \frac{1}{4}$ **f** $\frac{13}{27} \times \frac{19}{20}$ **g** $\frac{7}{11} \times \frac{5}{80}$ **h** $\frac{13}{19} \times \frac{11}{50}$

5 Work out the missing numbers in each calculation.

a $\frac{\square}{8} \times \frac{1}{\square} = \frac{7}{80}$ **b** $\frac{5}{63} = \frac{1}{\square} \times \frac{\square}{9}$

c $\frac{11}{75} = \frac{11}{\square} \times \frac{\square}{3}$ **d** $\frac{\square}{20} \times \frac{7}{8} = \frac{49}{\square}$

6 Complete these calculations. Give each answer in its simplest form.

a $\frac{2}{9} \times \frac{3}{5}$ **b** $\frac{3}{8} \times \frac{2}{3}$ **c** $\frac{1}{2} \times \frac{21}{40}$ **d** $\frac{1}{6} \times \frac{3}{7}$

e $\frac{11}{12} \times \frac{3}{4}$ **f** $\frac{4}{27} \times \frac{17}{20}$ **g** $\frac{8}{11} \times \frac{33}{80}$ **h** $\frac{15}{19} \times \frac{19}{50}$

7 Complete these calculations. Give each answer as a fraction in its simplest form.

a $\left(\frac{2}{5}\right)^2$ **b** $\left(\frac{5}{6}\right)^2$ **c** $\frac{1}{2} \times \frac{2}{3} \times \frac{3}{4}$ **d** $\frac{4}{5} \times \left(\frac{5}{8}\right)^2$

e $\left(\frac{3}{4}\right)^3$ **f** $\left(\frac{2}{5} \times \frac{3}{4}\right)^2$ **g** 0.3^2 **h** 0.07^2

8 Work out the answers to these calculations. Give each answer in its simplest form.

a $\frac{7}{9} \times -\frac{2}{3}$ **b** $-\frac{4}{5} \times -\frac{5}{8}$ **c** $\left(-\frac{1}{15}\right)^2$

What do you think?

1 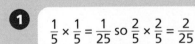 $\frac{1}{5} \times \frac{1}{5} = \frac{1}{25}$ so $\frac{2}{5} \times \frac{2}{5} = \frac{2}{25}$

a Explain why Ed thinks this.

b Explain why Ed is incorrect.

c Calculate $\frac{2}{5} \times \frac{2}{5}$ and explain your method.

2 Calculate $\frac{2}{2} \times \frac{3}{4}$ and give your answer in its simplest form.
What do you notice? Why does this happen?

3 $a = \dfrac{1}{3}$ $b = \dfrac{5}{7}$ $c = \dfrac{2}{9}$

Work out the value of

a ab **b** ac **c** bc **d** abc

4 $\dfrac{7}{12} \times \dfrac{?}{?} = \dfrac{7}{18}$

Seb wants to work out the missing fraction.

He says, "It is impossible because 12 doesn't go into 18."

Show that Seb is incorrect.

Consolidate – do you need more?

1 Write the calculation and the answer represented by each diagram.

a
b
c
d

e
f
g
h

2 Complete these calculations. You may wish to copy and shade the diagrams to help you.

a $\dfrac{1}{3} \times \dfrac{1}{7}$ **b** $\dfrac{1}{5} \times \dfrac{1}{2}$ **c** $\dfrac{1}{4} \times \dfrac{1}{3}$ **d** $\dfrac{1}{6} \times \dfrac{1}{2}$

e $\dfrac{2}{3} \times \dfrac{5}{7}$ **f** $\dfrac{3}{5} \times \dfrac{1}{2}$ **g** $\dfrac{3}{4} \times \dfrac{2}{3}$ **h** $\dfrac{5}{6} \times \dfrac{1}{2}$

Can any of your answers be simplified?

3 Complete these calculations. Simplify your answers where possible.

a $\dfrac{1}{10} \times \dfrac{1}{15}$ **b** $\dfrac{1}{12} \times \dfrac{1}{7}$ **c** $\dfrac{1}{20} \times \dfrac{1}{6}$ **d** $\dfrac{1}{100} \times \dfrac{1}{4}$

e $\dfrac{7}{10} \times \dfrac{2}{15}$ **f** $\dfrac{7}{12} \times \dfrac{2}{7}$ **g** $\dfrac{13}{20} \times \dfrac{5}{6}$ **h** $\dfrac{99}{100} \times \dfrac{3}{4}$

Stretch – can you deepen your learning?

1 Work out $\frac{a}{b} \times \frac{c}{d}$ giving your answer as a single fraction.

2 A piece of ribbon is $\frac{5}{8}$ m long.

Chloe uses one ninth of the ribbon to wrap around a present.

a What fraction of a metre of ribbon does she wrap around the present?

b What fraction of a metre of ribbon does she have remaining?

Compare your method with a partner.

3

a Show that both calculations give the same answer.

Why does this happen? Which method do you prefer?

b **i** Give an example of a calculation where it is easier to work with decimals.

ii Give an example of a calculation where it is easier to work with fractions.

4 A rectangular field has dimensions $\frac{4}{5}$ km by $\frac{9}{10}$ km

a What is the area of the field? Give your answer as a fraction of a square kilometre.

b A tub of fertiliser covers $100\,\text{m}^2$ and costs £2.79. How much will it cost to fertilise the whole field?

5 Tommy is calculating $\frac{2}{5} \times \frac{1}{4}$

He says, "I am going to find a common denominator first."

a Explain why Tommy might want to do this.

b Explain why this will give the correct answer.

c List the pros and cons of finding a common denominator first.

Reflect

In your own words, explain how to multiply any pair of fractions.

Give examples and draw diagrams to support your reasoning.

3.3 Dividing fractions (1)

White Rose Maths

Small steps

- Divide an integer by a fraction
- Divide a fraction by a unit fraction

Key words

Equivalent – numbers or expressions that are written differently but are always equal in value

Divide – to split into equal groups or parts

Are you ready?

1 Complete these multiplications.

a $\frac{1}{5} \times 2$ b $7 \times \frac{1}{50}$ c $11 \times \frac{1}{100}$ d $\frac{1}{9} \times 4$

e $\frac{2}{5} \times 2$ f $7 \times \frac{3}{50}$ g $11 \times \frac{9}{100}$ h $\frac{2}{9} \times 4$

2 Complete these multiplications.

a $\frac{1}{3} \times \frac{1}{5}$ b $\frac{1}{2} \times \frac{1}{9}$ c $\frac{1}{10} \times \frac{1}{5}$ d $\frac{1}{7} \times \frac{1}{4}$

3 Which two division calculations can be seen from each bar model? The first one has been done for you.

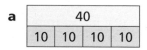

a	40			
	10	10	10	10

b	40				
	8	8	8	8	8

c	54					
	9	9	9	9	9	9

d	8							
	1	1	1	1	1	1	1	1

$40 \div 4 = 10$
$40 \div 10 = 4$

4 Complete each fraction so that it is equivalent to one whole.

a $\frac{\square}{5}$ b $\frac{7}{\square}$ c $\frac{\square}{10}$ d $\frac{\square}{19}$ e $\frac{135}{\square}$ f $\frac{\square}{999}$

5 Complete each equivalent fraction.

a $\frac{7}{8} = \frac{\square}{16}$ b $\frac{5}{9} = \frac{\square}{27}$ c $\frac{\square}{6} = \frac{5}{30}$ d $\frac{\square}{21} = \frac{3}{7}$ e $\frac{3}{4} = \frac{\square}{20}$ f $\frac{\square}{10} = \frac{35}{50}$

Models and representations

Bar models

These can be used to support division by a fraction.

4			
1	1	1	1

There are four 1s in 4, so $4 \div 1 = 4$

1			
$\frac{1}{4}$	$\frac{1}{4}$	$\frac{1}{4}$	$\frac{1}{4}$

There are four $\frac{1}{4}$s in one whole, so
$1 \div \frac{1}{4} = 4$

1	
$\frac{2}{4}$	$\frac{2}{4}$

There are two lots of $\frac{2}{4}$s in one whole, so $1 \div \frac{2}{4} = 2$

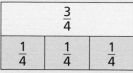

$\frac{3}{4}$		
$\frac{1}{4}$	$\frac{1}{4}$	$\frac{1}{4}$

There are three $\frac{1}{4}$s in $\frac{3}{4}$ so $\frac{3}{4} \div \frac{1}{4} = 3$

In this chapter you will learn how to **divide** an integer by a fraction.

Remember: when the numerator and denominator of a fraction are equal, then the fraction is **equivalent** to one whole.

Example 1

Complete these calculations.

a $6 \div \frac{1}{4}$ **b** $6 \div \frac{3}{4}$

a

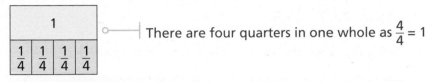

	1		
$\frac{1}{4}$	$\frac{1}{4}$	$\frac{1}{4}$	$\frac{1}{4}$

There are four quarters in one whole as $\frac{4}{4} = 1$

$$6 \div \frac{1}{4} = 6 \times 4 = 24$$

In 6 wholes, there are 6 lots of 1 whole. This means that there are 6 lots of 4 quarters.

$6 \times 4 = 24$ so $6 \div \frac{1}{4} = 24$

b

$$6 \div \frac{3}{4} = (6 \times 4) \div 3 = 8$$

You want to work out how many lots of three-quarters there are in 6 wholes.

You already know that there are 24 lots of one quarter in 6 wholes.

You can group these quarters into threes to make three-quarters.

$24 \div 3 = 8$ so $6 \div \frac{3}{4} = 8$

Practice 3.3A

1 **a** How many fifths are there in

     i 1 whole   **ii** 2 wholes

     iii 4 wholes   **iv** 10 wholes?

1				
$\frac{1}{5}$	$\frac{1}{5}$	$\frac{1}{5}$	$\frac{1}{5}$	$\frac{1}{5}$

   b Complete these calculations.

     i $1 \div \frac{1}{5}$ **ii** $2 \div \frac{1}{5}$

     iii $4 \div \frac{1}{5}$ **iv** $10 \div \frac{1}{5}$

2 Draw a diagram to show that $2 \div \frac{1}{3} = 6$

3 Write a multiplication calculation that is equivalent to each division. The first one has been done for you.

   a $6 \div \frac{1}{5} = 6 \times 5$ **b** $9 \div \frac{1}{4}$ **c** $4 \div \frac{1}{9}$ **d** $10 \div \frac{1}{8}$

   e $24 \div \frac{1}{7}$ **f** $15 \div \frac{1}{6}$ **g** $100 \div \frac{1}{26}$ **h** $14 \div \frac{1}{15}$

Explain why each pair of calculations is equivalent.

4 Complete these calculations.

   a $6 \div \frac{1}{5}$ **b** $9 \div \frac{1}{4}$ **c** $4 \div \frac{1}{9}$ **d** $10 \div \frac{1}{8}$

   e $24 \div \frac{1}{7}$ **f** $15 \div \frac{1}{6}$ **g** $100 \div \frac{1}{26}$ **h** $14 \div \frac{1}{15}$

5

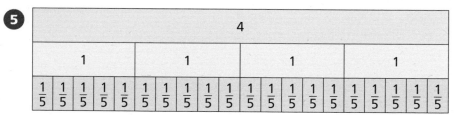

   a How many fifths are there in four wholes?

   b How many lots of two-fifths are there in four wholes?

6 **a** Explain why there are 36 quarters in 9 wholes.

   b How many lots of three-quarters are there in 36 quarters?

   c Calculate $9 \div \frac{3}{4}$

7 Draw a diagram to show that $2 \div \frac{2}{3} = 2 \times 3 \div 2 = 3$

8 Complete these calculations.

   a $6 \div \frac{2}{5}$ **b** $9 \div \frac{3}{4}$ **c** $4 \div \frac{4}{9}$ **d** $10 \div \frac{5}{8}$

   e $24 \div \frac{6}{7}$ **f** $15 \div \frac{5}{6}$ **g** $100 \div \frac{5}{26}$ **h** $14 \div \frac{7}{15}$

What do you think?

1 Complete these calculations.

$5 \div \frac{5}{9}$ $5 \times \frac{9}{5}$

What do you notice? Why does this happen?

2 Benji is calculating $14 \div \frac{7}{10}$

Here is his method.

$$14 = \frac{140}{10} \qquad \frac{140}{10} \div \frac{7}{10} = \frac{140 \div 7}{10 \div 10} = \frac{20}{1} = 20$$

Explain Benji's method.

Will it always work?

3 A piece of rope is 20 m long.

It is cut into strips so that each strip is $\frac{4}{5}$ m long.

How many strips is the rope cut into?

Example 2

Complete these calculations.

a $\frac{7}{20} \div \frac{1}{20}$ **b** $\frac{7}{10} \div \frac{1}{20}$

Method A

a

		$\frac{7}{20}$				
$\frac{1}{20}$	$\frac{1}{20}$	$\frac{1}{20}$	$\frac{1}{20}$	$\frac{1}{20}$	$\frac{1}{20}$	$\frac{1}{20}$

$\frac{7}{20} \div \frac{1}{20} = 7$

○——| There are seven lots of $\frac{1}{20}$ in $\frac{7}{20}$

○——| This means that $\frac{7}{20} \div \frac{1}{20} = 7$

b

			$\frac{7}{10}$		
$\frac{1}{10}$	$\frac{1}{10}$	$\frac{1}{10}$	$\frac{1}{10}$	$\frac{1}{10}$	$\frac{1}{10}$

$\frac{1}{20}$	$\frac{1}{20}$	$\frac{1}{20}$	$\frac{1}{20}$	$\frac{1}{20}$	$\frac{1}{20}$	$\frac{1}{20}$	$\frac{1}{20}$	$\frac{1}{20}$	$\frac{1}{20}$	$\frac{1}{20}$	$\frac{1}{20}$

$\frac{7}{10} \div \frac{1}{20} = \frac{14}{20} \div \frac{1}{20} = 14$

There are seven lots of $\frac{1}{10}$ in $\frac{7}{10}$

You need to know how many twentieths there are in $\frac{7}{10}$

Remember: $\frac{1}{10}$ is equivalent to $\frac{2}{20}$

$\frac{7}{10}$ is equivalent to $\frac{14}{20}$

You can rewrite the calculation as $\frac{14}{20} \div \frac{1}{20}$

There are fourteen lots of $\frac{1}{20}$ in $\frac{14}{20}$

so $\frac{14}{20} \div \frac{1}{20} = 14$

Method B

a $\dfrac{7}{20} \div \dfrac{1}{20} = 7 \div 1 = 7$ ○─── As both fractions have a denominator of 20 you can just consider the numerators.

As both fractions are working with twentieths, you can think of this as how many lots of 1 twentieth are there in 7 twentieths, which is 7 divided by 1

b $\dfrac{7}{10} = \dfrac{14}{20}$ ○─── The fractions $\dfrac{7}{10}$ and $\dfrac{1}{20}$ do not have the same denominator.

$\dfrac{7}{10} \div \dfrac{1}{20} = \dfrac{14}{20} \div \dfrac{1}{20} = 14$ ○─── $\dfrac{7}{10}$ is equivalent to $\dfrac{14}{20}$ so you can rewrite the calculation as $\dfrac{14}{20} \div \dfrac{1}{20}$ which is 14

You need to find a common denominator. The lowest common multiple of 10 and 20 is 20, so the common denominator is 20

Practice 3.3B

1 **a** How many lots of one-tenth are there in three-tenths?

b Calculate $\dfrac{3}{10} \div \dfrac{1}{10}$

$\dfrac{3}{10}$		
$\dfrac{1}{10}$	$\dfrac{1}{10}$	$\dfrac{1}{10}$

2 Draw a diagram to show that $\dfrac{5}{16} \div \dfrac{1}{16} = 5$

3 Here is a fraction wall.

a Copy and complete the sentences. Use the fraction wall to help you.

   i There are ☐ lots of $\dfrac{1}{6}$ in $\dfrac{1}{2}$

   ii There are ☐ lots of $\dfrac{1}{12}$ in $\dfrac{1}{6}$

   iii There are ☐ lots of $\dfrac{1}{12}$ in $\dfrac{1}{2}$

   iv There are ☐ lots of $\dfrac{1}{12}$ in $\dfrac{5}{6}$

1											
$\frac{1}{2}$						$\frac{1}{2}$					
$\frac{1}{6}$		$\frac{1}{6}$		$\frac{1}{6}$		$\frac{1}{6}$		$\frac{1}{6}$		$\frac{1}{6}$	
$\frac{1}{12}$	$\frac{1}{12}$	$\frac{1}{12}$	$\frac{1}{12}$	$\frac{1}{12}$	$\frac{1}{12}$	$\frac{1}{12}$	$\frac{1}{12}$	$\frac{1}{12}$	$\frac{1}{12}$	$\frac{1}{12}$	$\frac{1}{12}$

b Complete these calculations.

   i $\dfrac{1}{2} \div \dfrac{1}{6}$ **ii** $\dfrac{1}{6} \div \dfrac{1}{12}$ **iii** $\dfrac{1}{2} \div \dfrac{1}{12}$ **iv** $\dfrac{5}{6} \div \dfrac{1}{12}$

How could you have completed the calculations without a fraction wall?

4 **a** Complete the equivalent fraction $\dfrac{3}{5} = \dfrac{?}{10}$

b Calculate $\dfrac{3}{5} \div \dfrac{1}{10}$

Compare your method with a partner.

5 Complete these calculations.

a $\dfrac{7}{8} \div \dfrac{1}{16}$ **b** $\dfrac{5}{9} \div \dfrac{1}{27}$ **c** $\dfrac{1}{6} \div \dfrac{1}{30}$ **d** $\dfrac{3}{7} \div \dfrac{1}{21}$

e $\dfrac{3}{4} \div \dfrac{1}{20}$ **f** $\dfrac{7}{10} \div \dfrac{1}{50}$ **g** $\dfrac{7}{100} \div \dfrac{1}{500}$ **h** $\dfrac{19}{20} \div \dfrac{1}{100}$

6 Work out the length of the side marked x

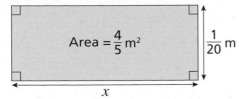

Area $= \dfrac{4}{5}$ m²

$\dfrac{1}{20}$ m

x

What do you think? 💭

1 Complete these calculations.

a $\dfrac{3}{5} \div 0.1$ **b** $0.9 \div \dfrac{1}{20}$ **c** $0.96 \div \dfrac{1}{50}$ **d** $0.75 \div \dfrac{1}{48}$

2 Here are some calculation cards.

$$\dfrac{15}{20} \div \dfrac{1}{40} \qquad \dfrac{7}{15} \div \dfrac{1}{60} \qquad \dfrac{2}{3} \div \dfrac{1}{9} \qquad \dfrac{3}{10} \div \dfrac{1}{50}$$

Write the calculations in ascending order of value.

Consolidate – do you need more?

1 8 eighths = one whole. Use this fact to complete these calculations.

a $1 \div \dfrac{1}{8}$ **b** $2 \div \dfrac{1}{8}$ **c** $3 \div \dfrac{1}{8}$ **d** $7 \div \dfrac{1}{8}$ **e** $10 \div \dfrac{1}{8}$ **f** $15 \div \dfrac{1}{8}$

2 Complete these calculations.

a $1 \div \dfrac{1}{7}$ **b** $2 \div \dfrac{1}{5}$ **c** $3 \div \dfrac{1}{4}$ **d** $7 \div \dfrac{1}{9}$ **e** $10 \div \dfrac{1}{3}$ **f** $15 \div \dfrac{1}{2}$

3 Here is a fraction wall.

a Copy and complete these sentences. Use the fraction wall to help you.

  i There are ☐ lots of $\dfrac{1}{8}$ in $\dfrac{1}{4}$

  ii There are ☐ lots of $\dfrac{1}{8}$ in $\dfrac{3}{4}$

  iii There are ☐ lots of $\dfrac{1}{9}$ in $\dfrac{1}{3}$

  iv There are ☐ lots of $\dfrac{1}{9}$ in $\dfrac{2}{3}$

  v There are ☐ lots of $\dfrac{1}{10}$ in $\dfrac{1}{5}$

  vi There are ☐ lots of $\dfrac{1}{10}$ in $\dfrac{4}{5}$

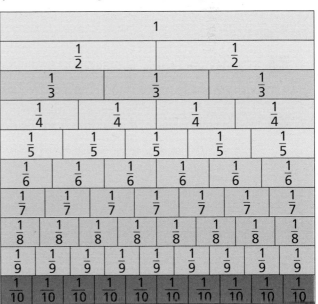

b Complete these calculations.

i $\frac{1}{4} \div \frac{1}{8}$

ii $\frac{3}{4} \div \frac{1}{8}$

iii $\frac{1}{3} \div \frac{1}{9}$

iv $\frac{2}{3} \div \frac{1}{9}$

v $\frac{1}{5} \div \frac{1}{10}$

vi $\frac{4}{5} \div \frac{1}{10}$

What other divisions can you see from the fraction wall?

Stretch – can you deepen your learning?

1 Solve these equations.

a $\frac{1}{5}a = 12$ **b** $60 = \frac{2}{3}t$ **c** $\frac{4}{5} = \frac{1}{10}p$ **d** $\frac{1}{27}q = \frac{2}{3}$ **e** $5.8 = \frac{29}{300}y$

2 A, B and C are similar trapezia.

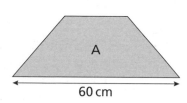
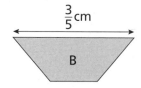

Work out the scale factor of enlargement

a from C to A

b from B to A

c from C to B

3 $a = 27x$ and $y = \frac{b}{27}$

Given $x = 80$ and $y = \frac{5}{9}$, work out the value of ab

4 Simplify these expressions.

a $a \div \frac{1}{b}$

b $a \div \frac{c}{b}$

c $\frac{x}{y} \div \frac{1}{b}$

Reflect

1 In your own words, explain how to divide an integer by a fraction.
Give examples and draw diagrams to support your reasoning.

2 In your own words, explain how to divide a fraction by a unit fraction.
Give examples and draw diagrams to support your reasoning.

3.4 Dividing fractions (2)

Small steps

- Understand and use the reciprocal
- Divide any pair of fractions

Key words

Reciprocal – the result of dividing 1 by a given number. The product of a number and its reciprocal is always 1

Are you ready?

1 Complete these calculations.

   a $\frac{1}{8} \times 8$ **b** $7 \times \frac{1}{7}$ **c** $15 \times \frac{1}{15}$ **d** $\frac{1}{3} \times 3$ **e** $5 \times \frac{1}{5}$

   f $\frac{2}{3} \times \frac{3}{2}$ **g** $\frac{5}{4} \times \frac{4}{5}$ **h** $\frac{3}{4} \times \frac{4}{3}$ **i** $\frac{7}{8} \times \frac{8}{7}$ **j** $\frac{9}{10} \times \frac{10}{9}$

   What do you notice?

2 Complete these divisions.

   a $1 \div \frac{1}{2}$ **b** $1 \div \frac{1}{3}$ **c** $1 \div \frac{1}{4}$ **d** $1 \div \frac{1}{12}$ **e** $1 \div \frac{1}{20}$

3 Complete these divisions.

   a $3 \div \frac{1}{2}$ **b** $5 \div \frac{1}{3}$ **c** $2 \div \frac{1}{4}$ **d** $10 \div \frac{1}{12}$ **e** $15 \div \frac{1}{20}$

4 Complete these calculations.

   a $\frac{1}{2} \div \frac{1}{8}$ **b** $\frac{1}{10} \div \frac{1}{100}$ **c** $\frac{2}{5} \div \frac{1}{10}$ **d** $\frac{3}{8} \div \frac{1}{24}$ **e** $\frac{5}{7} \div \frac{1}{35}$

5 Write each of these improper fractions as a mixed number in its simplest form.

   a $\frac{12}{5}$ **b** $\frac{6}{4}$ **c** $\frac{45}{14}$ **d** $\frac{45}{20}$

Models and representations

Bar models

These can be used to support division by a **fraction**.

4			
1	1	1	1

There are four 1s in 4, so $4 \div 1 = 4$

1			
$\frac{1}{4}$	$\frac{1}{4}$	$\frac{1}{4}$	$\frac{1}{4}$

There are four $\frac{1}{4}$s in one whole, so $1 \div \frac{1}{4} = 4$

1	
$\frac{2}{4}$	$\frac{2}{4}$

There are two lots of $\frac{2}{4}$s in one whole, so $1 \div \frac{2}{4} = 2$

$\frac{3}{4}$		
$\frac{1}{4}$	$\frac{1}{4}$	$\frac{1}{4}$

There are three $\frac{1}{4}$s in $\frac{3}{4}$ so $\frac{3}{4} \div \frac{1}{4} = 3$

Dividing by a fraction is the same as multiplying by its **reciprocal**. You will use this idea throughout this chapter.

Example 1

1 **a** Work out the missing number. $\frac{1}{6} \times \boxed{} = 1$

  b Write down the reciprocal of $\frac{1}{6}$

2 Calculate $\frac{3}{4} \div \frac{2}{7}$

---

1 **a** $\frac{1}{6} \times \boxed{6} = 1$ ○——— There are 6 sixths in one whole.

b The reciprocal of $\frac{1}{6}$ is 6

> The reciprocal of a number is its multiplicative inverse. This means that the product of a number and its reciprocal is always 1

2 $\frac{3}{4} \div \frac{2}{7} = \frac{3}{4} \times \frac{7}{2}$ To divide by a fraction, you multiply by its reciprocal.

$\frac{3 \times 7}{4 \times 2} = \frac{21}{8}$ ○——— You can rewrite the calculation as a multiplication.

$\frac{21}{8} = 2\frac{5}{8}$ ○——— Write $\frac{21}{8}$ as a mixed number.

$\frac{3}{4} \div \frac{2}{7} = 2\frac{5}{8}$ You learned how to convert improper fractions into mixed numbers in Book 1

Practice 3.4A

1 **a** Work out the missing numbers.

  i $\frac{1}{8} \times \boxed{} = 1$ **ii** $\boxed{} \times \frac{1}{7} = 1$ **iii** $15 \times \boxed{} = 1$

  iv $\boxed{} \times 3 = 1$ **v** $\frac{1}{5} \times \boxed{} = 1$ **vi** $5 \times \boxed{} = 1$

  b Write down the reciprocal of each number.

  i $\frac{1}{8}$ **ii** $\frac{1}{7}$ **iii** 15

  iv 3   **v** $\frac{1}{5}$ **vi** 5

2 **a** Work out the missing fractions.

$\mathbf{i}\ \dfrac{2}{3} \times \boxed{} = \dfrac{6}{6}$ $\mathbf{ii}\ \boxed{} \times \dfrac{4}{5} = \dfrac{20}{20}$ $\mathbf{iii}\ \dfrac{4}{3} \times \boxed{} = \dfrac{12}{12}$

$\mathbf{iv}\ \boxed{} \times \dfrac{7}{8} = \dfrac{56}{56}$ $\mathbf{v}\ \dfrac{9}{10} \times \boxed{} = \dfrac{90}{90}$ $\mathbf{vi}\ \dfrac{10}{9} \times \boxed{} = \dfrac{90}{90}$

b Work out the missing fractions.

$\mathbf{i}\ \dfrac{2}{3} \times \boxed{} = 1$ $\mathbf{ii}\ \boxed{} \times \dfrac{4}{5} = 1$ $\mathbf{iii}\ \dfrac{4}{3} \times \boxed{} = 1$

$\mathbf{iv}\ \boxed{} \times \dfrac{7}{8} = 1$ $\mathbf{v}\ \dfrac{9}{10} \times \boxed{} = 1$ $\mathbf{vi}\ \dfrac{10}{9} \times \boxed{} = 1$

c Write down the reciprocal of each number.

$\mathbf{i}\ \dfrac{2}{3}$ $\mathbf{ii}\ \dfrac{4}{5}$ $\mathbf{iii}\ \dfrac{4}{3}$

$\mathbf{iv}\ \dfrac{7}{8}$ $\mathbf{v}\ \dfrac{9}{10}$ $\mathbf{vi}\ \dfrac{10}{9}$

3 **a** Complete each pair of calculations

$\mathbf{i}\ 1 \div \dfrac{1}{2}$ 1×2 $\mathbf{ii}\ 2 \div \dfrac{1}{4}$ 2×4

$\mathbf{iii}\ \dfrac{1}{2} \div \dfrac{1}{6}$ $\dfrac{1}{2} \times 6$ $\mathbf{iv}\ 6 \div \dfrac{3}{4}$ $6 \times \dfrac{4}{3}$

What do you notice?

b Copy and complete the sentence.

To divide by a fraction, you _____ by its _____.

4 Rewrite each division as a multiplication that gives the same answer. The first one has been done for you.

a $5 \div \dfrac{3}{4} = 5 \times \dfrac{4}{3}$ **b** $7 \div \dfrac{2}{3}$ **c** $10 \div \dfrac{1}{9}$

d $21 \div \dfrac{2}{5}$ **e** $\dfrac{4}{5} \div \dfrac{1}{3}$ **f** $\dfrac{3}{4} \div \dfrac{1}{8}$

g $\dfrac{5}{7} \div \dfrac{1}{10}$ **h** $\dfrac{5}{6} \div \dfrac{1}{5}$ **i** $\dfrac{4}{5} \div \dfrac{2}{3}$

j $\dfrac{3}{4} \div \dfrac{5}{8}$ **k** $\dfrac{5}{7} \div \dfrac{3}{10}$ **l** $\dfrac{5}{6} \div \dfrac{4}{5}$

5 Complete these calculations. Give each answer as an improper fraction in its simplest form.

a $5 \div \frac{3}{4}$ b $7 \div \frac{2}{3}$ c $10 \div \frac{1}{9}$ d $21 \div \frac{2}{5}$ e $\frac{4}{5} \div \frac{1}{3}$ f $\frac{3}{4} \div \frac{1}{8}$

g $\frac{5}{7} \div \frac{1}{10}$ h $\frac{5}{6} \div \frac{1}{5}$ i $\frac{4}{5} \div \frac{2}{3}$ j $\frac{3}{4} \div \frac{5}{8}$ k $\frac{5}{7} \div \frac{3}{10}$ l $\frac{5}{6} \div \frac{4}{5}$

6 Complete these divisions. Give each answer as a mixed number in its simplest form.

a $\frac{4}{5} \div \frac{1}{3}$ b $\frac{3}{4} \div \frac{1}{2}$ c $\frac{5}{7} \div \frac{2}{9}$ d $\frac{9}{10} \div \frac{2}{5}$

e $\frac{7}{8} \div \frac{3}{4}$ f $\frac{2}{5} \div \frac{1}{9}$ g $\frac{5}{6} \div \frac{4}{7}$ h $\frac{14}{15} \div \frac{1}{7}$

7 Solve these equations.

a $\frac{2}{3}x = \frac{5}{7}$ b $\frac{4}{5}y = \frac{2}{9}$ c $\frac{7}{8}p = 5$ d $\frac{3}{4} = \frac{2}{11}m$

What do you think? 💭

1 a Write 0.4 as a fraction in its simplest form.

b Find the reciprocal of 0.4

c Work out $\frac{3}{7} \div 0.4$

2 If a number is negative, then so is its reciprocal.

Do you agree with Abdullah? Explain your answer.

3 If $a = \frac{1}{5}$, $b = \frac{2}{3}$ and $c = -\frac{3}{4}$ work out the value of

a $\frac{a}{b}$ b $\frac{b}{a}$ c $\frac{a}{c}$ d $\frac{c}{a}$ e $\frac{ab}{c}$

4 Here are three different methods for calculating $\frac{5}{6} \div \frac{2}{9}$

Method A	Method B	Method C
$\frac{5}{6} \div \frac{2}{9} = \frac{15}{18} \div \frac{4}{18}$	$\frac{5}{6} \div \frac{2}{9} = \frac{5}{6} \times \frac{9}{2}$	$\frac{5}{6} \div \frac{2}{9} = p$
$= \frac{15}{4}$	$= \frac{45}{12}$	$\frac{5}{6} = \frac{2}{9}p$
$= 3\frac{3}{4}$	$= \frac{15}{4}$	$\frac{45}{6} = 2p$
	$= 3\frac{3}{4}$	$p = 3\frac{3}{4}$

Explain each method.

Consolidate – do you need more?

1 **a** Work out the missing numbers.

i $\frac{1}{5} \times \boxed{} = 1$ **ii** $\boxed{} \times \frac{1}{9} = 1$ **iii** $\frac{1}{11} \times \boxed{} = 1$

iv $\frac{1}{10} \times \boxed{} = 1$ **v** $\boxed{} \times \frac{1}{3} = 1$ **vi** $1 = \boxed{} \times \frac{1}{2}$

b Write down the reciprocal of each of these fractions.

i $\frac{1}{5}$ **ii** $\frac{1}{9}$ **iii** $\frac{1}{11}$ **iv** $\frac{1}{10}$ **v** $\frac{1}{3}$ **vi** $\frac{1}{2}$

2 Complete these divisions.

a $\frac{2}{31} \div \frac{1}{5}$ **b** $\frac{3}{29} \div \frac{1}{9}$ **c** $\frac{4}{51} \div \frac{1}{11}$ **d** $\frac{1}{6} \div \frac{1}{3}$ **e** $\frac{2}{5} \div \frac{1}{10}$ **f** $\frac{5}{6} \div \frac{1}{2}$

3 **a** Work out the missing fractions.

i $\frac{2}{5} \times \boxed{} = \frac{10}{10}$ **ii** $\boxed{} \times \frac{4}{9} = \frac{36}{36}$ **iii** $\frac{7}{11} \times \boxed{} = \frac{77}{77}$

iv $\frac{3}{10} \times \boxed{} = \frac{30}{30}$ **v** $\boxed{} \times \frac{2}{3} = \frac{6}{6}$ **vi** $\frac{3}{8} \times \boxed{} = \frac{24}{24}$

b Write down the reciprocal of each of these fractions.

i $\frac{2}{5}$ **ii** $\frac{4}{9}$ **iii** $\frac{7}{11}$ **iv** $\frac{3}{10}$ **v** $\frac{2}{3}$ **vi** $\frac{3}{8}$

4 Complete these divisions. Give each answer as a mixed number in its simplest form.

a $\frac{3}{4} \div \frac{2}{5}$ **b** $\frac{2}{3} \div \frac{4}{9}$ **c** $\frac{5}{6} \div \frac{7}{11}$ **d** $\frac{4}{5} \div \frac{3}{10}$ **e** $\frac{7}{8} \div \frac{2}{3}$ **f** $\frac{1}{2} \div \frac{3}{8}$

Stretch – can you deepen your learning?

1 Write down the reciprocal of each term.

a $-\frac{1}{w}$ **b** p **c** pq **d** $\frac{j}{k}$ **e** $\frac{xy}{z}$ **f** $-\frac{z}{xy}$

2 Work out the reciprocal of each number.

a 2.5   **b** 1.25   **c** 0.125   **d** 0.83   **e** 5.75   **f** 6.8

3 Simplify each expression.

a $a \div \frac{1}{b}$ **b** $a \div \frac{b}{c}$ **c** $\frac{a}{b} \div \frac{1}{c}$ **d** $\frac{a}{b} \div \frac{c}{d}$

4 Write each of these ratios in the form $1:n$

a $\frac{2}{3}:5$ **b** $\frac{3}{4}:\frac{5}{7}$ **c** $\frac{8}{9}:\frac{2}{3}$ **d** $\frac{9}{10}:1.5$

5 **a** Find three pairs of values such that $10 \div \frac{p}{q}$ is an integer.

b Generalise your findings to describe the values of p and q that make $10 \div \frac{p}{q}$ an integer.

6 **a** Find values of a, b, c and d such that $\frac{a}{b} \div \frac{c}{d}$ is an improper fraction.

  b Find values of a, b, c and d such that $\frac{a}{b} \div \frac{c}{d}$ is a proper fraction.

  c Under what conditions is $\frac{a}{b} \div \frac{c}{d}$ an improper fraction?

Reflect

1 In your own words, explain what is meant by "the reciprocal".

2 In your own words, explain how to divide by a fraction.

🅗 3.5 Multiplying and dividing complex fractions

Small steps

- Multiply and divide improper and mixed fractions 🅗
- Multiply and divide algebraic fractions 🅗

Key words

Improper fraction – a fraction in which the numerator is greater than the denominator

Mixed number – a number presented as an integer and a proper fraction

Algebraic fraction – a fraction whose numerator and/or denominator are algebraic expressions

Are you ready?

1. Write each of these improper fractions as a mixed number.

   a $\frac{17}{5}$ b $\frac{5}{3}$ c $\frac{15}{9}$ d $\frac{15}{7}$

2. Write each of these mixed numbers as an improper fraction.

   a $2\frac{1}{2}$ b $5\frac{1}{4}$ c $6\frac{1}{3}$ d $4\frac{1}{2}$

3. Work out these without using a calculator. Give each answer in its simplest form.

   a $\frac{2}{3} \times \frac{1}{5}$ b $\frac{4}{7} \times \frac{8}{9}$ c $\frac{1}{10} \times \frac{3}{4}$ d $\frac{12}{15} \times \frac{7}{8}$

   e $\frac{2}{3} \div \frac{1}{5}$ f $\frac{4}{7} \div \frac{8}{9}$ g $\frac{1}{10} \div \frac{3}{4}$ h $\frac{12}{15} \div \frac{7}{8}$

4. Simplify these expressions.

   a $a \times b^2 \times c$

   b $3a + 4a$

   c $2a \times 3a$

   d $12ab - 3ab + 4a^2b$

Models and representations

Grid method

	5	$\frac{1}{4}$
3	15	$\frac{3}{4}$
$\frac{2}{5}$	$\frac{10}{5}$	$\frac{2}{20}$

This grid shows that $3\frac{2}{5} \times 5\frac{1}{4}$ can be calculated by partitioning each **mixed number** so that the calculation becomes

$(3 \times 5) + (3 \times \frac{1}{4}) + (\frac{2}{5} \times 5) + (\frac{2}{5} \times \frac{1}{4})$

Bar model

A bar model is useful when converting between **improper fractions** and mixed numbers.

This bar model shows that $2\frac{5}{8} = \frac{21}{8}$

In this chapter, you will multiply and divide complex fractions.

When calculating with improper fractions and mixed numbers, it is essential that you can convert fluently between the two.

You will build on your knowledge of multiplication and division of fractions from Chapters 3.3 and 3.4.

Example 1

Without using a calculator, work out

a $3\frac{2}{5} \times \frac{21}{4}$, giving your answer as a mixed number **b** $3\frac{2}{5} \div \frac{21}{4}$

Method A

a $3\frac{2}{5} \times \frac{21}{4} = \frac{17}{5} \times \frac{21}{4}$ —○— Convert $3\frac{2}{5}$ into an improper fraction and rewrite the calculation.

$\frac{17}{5} \times \frac{21}{4} = \frac{357}{20}$ —○— Find the product of the fractions.

$\frac{357}{20} = 17\frac{17}{20}$ —○— Convert your answer into a mixed number.

b $3\frac{2}{5} \div \frac{21}{4} = \frac{17}{5} \div \frac{21}{4}$ —○— Convert $3\frac{2}{5}$ into an improper fraction and rewrite the calculation.

$\frac{17}{5} \div \frac{21}{4} = \frac{17}{5} \times \frac{4}{21}$ —○— To divide by a fraction, you multiply by its reciprocal.

$\frac{17}{5} \times \frac{4}{21} = \frac{68}{105}$

so $3\frac{2}{5} \div \frac{21}{4} = \frac{68}{105}$

Method B

a $3\frac{2}{5} \times \frac{21}{4} = 3\frac{2}{5} \times 5\frac{1}{4}$ —○— Convert $\frac{21}{4}$ into a mixed number and rewrite the calculation.

	5	$\frac{1}{4}$
3	15	$\frac{3}{4}$
$\frac{2}{5}$	$\frac{10}{5}$	$\frac{2}{20}$

$3\frac{2}{5} \times \frac{21}{4} = 17\frac{17}{20}$

Partition each mixed number and use the grid method of multiplication to support your working.

This shows that $3\frac{2}{5} \times 5\frac{1}{4} = 15 + \frac{3}{4} + \frac{10}{5} + \frac{2}{20}$

So $3\frac{2}{5} \times 5\frac{1}{4} = 15 + \frac{3}{4} + 2 + \frac{2}{20}$

$= 17 + \frac{3}{4} + \frac{2}{20} = 17 + \frac{15}{20} + \frac{2}{20} = 17\frac{17}{20}$

b $3\frac{2}{5} \div \frac{21}{4} = 3.4 \div 5.25$ —○— You can rewrite $3\frac{2}{5}$ and $\frac{21}{4}$ as decimals.

$3.4 \div 5.25 = \frac{3.4}{5.25} = \frac{340}{525} = \frac{68}{105}$ —○— This calculation can then be written as a single fraction, and simplified to work out the answer.

Multiply the numerator and denominator by 100 so that the denominator is an integer.

Practice 3.5A

1 Complete these calculations.

a $3 \times 2\frac{1}{2}$ **b** $5\frac{1}{4} \times 7$ **c** $6\frac{1}{3} \times 2$ **d** $5 \times 4\frac{1}{2}$

e $\frac{5}{2} \times 6$ **f** $4 \times \frac{21}{4}$ **g** $5 \times \frac{19}{3}$ **h** $7 \times \frac{9}{2}$

2 a Explain why $3\frac{1}{5} \times \frac{1}{4} = (3 \times \frac{1}{4}) + (\frac{1}{5} \times \frac{1}{4})$ **b** Calculate $3\frac{1}{5} \times \frac{1}{4}$

3 Complete these calculations.

a $\frac{1}{3} \times 2\frac{1}{2}$ **b** $5\frac{1}{4} \times \frac{1}{5}$ **c** $6\frac{1}{3} \times \frac{1}{2}$ **d** $4\frac{1}{2} \times \frac{1}{8}$

4 Copy and complete both methods to work out $2\frac{1}{3} \times 3\frac{1}{5}$

Method A

$2\frac{1}{3} \times 3\frac{1}{5}$

$\times$	3	$\frac{1}{5}$
2	6	
$\frac{1}{3}$		$\frac{1}{15}$

Method B

$2\frac{1}{3} = \frac{\Box}{3}$

$3\frac{1}{5} = \frac{\Box}{5}$

$2\frac{1}{3} \times 3\frac{1}{5} = \frac{\Box}{3} \times \frac{\Box}{5}$

Which method do you prefer and why?

5 Complete these calculations.

a $4\frac{1}{3} \times 2\frac{3}{4}$ **b** $\frac{17}{5} \times \frac{5}{3}$ **c** $3\frac{1}{4} \times \frac{15}{9}$ **d** $\frac{15}{7} \times 8\frac{3}{10}$

6 Explain how the diagram shows that $5\frac{1}{3} \div \frac{1}{3} = 16$

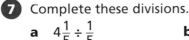

7 Complete these divisions.

a $4\frac{1}{5} \div \frac{1}{5}$ **b** $6\frac{1}{2} \div \frac{1}{2}$ **c** $5\frac{2}{3} \div \frac{1}{3}$ **d** $2\frac{7}{8} \div \frac{1}{8}$

e $4\frac{1}{5} \div \frac{1}{10}$ **f** $6\frac{1}{2} \div \frac{1}{8}$ **g** $5\frac{2}{3} \div \frac{1}{6}$ **h** $2\frac{7}{8} \div \frac{1}{40}$

8 a Write $3\frac{1}{2}$ as an improper fraction.   **b** Write $2\frac{4}{5}$ as an improper fraction.

c Hence or otherwise, complete these calculations.

i $3\frac{1}{2} \div 2\frac{4}{5}$ **ii** $2\frac{4}{5} \div 3\frac{1}{2}$

9 Complete these calculations.

a $4\frac{1}{3} \div 2\frac{3}{4}$ **b** $\frac{17}{5} \div \frac{5}{3}$ **c** $3\frac{1}{4} \div \frac{15}{9}$ **d** $\frac{15}{7} \div 8\frac{3}{10}$

What do you think?

1 What's the same and what's different about these calculations?

$5\frac{6}{7} \times \frac{41}{5}$ $5 \times \frac{6}{7} \times \frac{41}{5}$

2 Complete these calculations. Give your answers as fractions in their simplest form.

a 5.75×2.4 **b** 3.4×1.5 **c** $4.7 \div 0.9$ **d** $3.125 \div 1.25$

You will now see how to multiply and divide **algebraic fractions**.

You can use the same methods for algebraic fractions as for numerical fractions.

Example 2

a Calculate $\dfrac{a}{3} \times \dfrac{6b}{5}$

b Calculate $\dfrac{a}{3} \div \dfrac{6b}{5}$

a $\dfrac{a}{3} \times \dfrac{6b}{5} = \dfrac{a \times 6b}{3 \times 5} = \dfrac{6ab}{15}$
$= \dfrac{2ab}{5}$

Find the product of the numerators and the denominators.
Then simplify your answer.

3 is a common factor of both $6ab$ and 15, so $\dfrac{6ab}{15}$ is equivalent to $\dfrac{2ab}{5}$

b $\dfrac{a}{3} \div \dfrac{6b}{5} = \dfrac{a}{3} \times \dfrac{5}{6b}$
$= \dfrac{5a}{18b}$

To divide by a fraction, you multiply by its reciprocal.

The reciprocal of $\dfrac{6b}{5}$ is $\dfrac{5}{6b}$ because $\dfrac{6b}{5} \times \dfrac{5}{6b} = \dfrac{30b}{30b} = 1$

$5a$ and $18b$ do not have any common factors so the answer is already in its simplest form.

Practice 3.5B

1 Simplify each expression fully.

a $\dfrac{x}{3} \times \dfrac{y}{4}$ **b** $\dfrac{x}{3} \times \dfrac{5y}{4}$ **c** $\dfrac{x}{3} \times \dfrac{5xy}{4}$ **d** $\dfrac{8x}{3} \times \dfrac{5xy}{4}$

2 Simplify each expression fully.

a $\dfrac{x}{3} \div \dfrac{y}{4}$ **b** $\dfrac{x}{3} \div \dfrac{5y}{4}$ **c** $\dfrac{x}{3} \div \dfrac{5xy}{4}$ **d** $\dfrac{8x}{3} \div \dfrac{5xy}{4}$

3 Simplify each expression fully.

a $\dfrac{3}{x} \times \dfrac{4}{y}$ **b** $\dfrac{3}{x} \times \dfrac{4}{5y}$ **c** $\dfrac{3}{x} \times \dfrac{4}{5xy}$ **d** $\dfrac{3}{8x} \times \dfrac{4}{5xy}$

4 Simplify each expression fully.

a $\dfrac{3}{x} \div \dfrac{4}{y}$ **b** $\dfrac{3}{x} \div \dfrac{4}{5y}$ **c** $\dfrac{3}{x} \div \dfrac{4}{5xy}$ **d** $\dfrac{3}{8x} \div \dfrac{4}{5xy}$

5 Simplify each expression fully.

a $\dfrac{ab}{5} \times \dfrac{3}{a}$ **b** $\dfrac{p}{2q} \div \dfrac{1}{p}$ **c** $\dfrac{xy}{14} \div \dfrac{3x}{5}$ **d** $\dfrac{j}{6k} \div \dfrac{6k}{7}$

What do you think? 💭

1 Junaid wants to calculate $\dfrac{w}{9} \div 5$

He says, "I cannot do that because 5 is not a fraction, so it does not have a reciprocal."

Do you agree with Junaid? Explain your answer.

2 Write each of these as a single fraction.

a $\dfrac{5}{6} \div 4$　　　　b $\dfrac{x}{6} \div 4$　　　　c $\dfrac{x}{y} \div 4$　　　　d $\dfrac{x}{y} \div z$

3 Show that $\dfrac{3ab^2}{2pq} \div \dfrac{9ab^2}{6pq} = 1$

Compare your method with a partner. Did you use the same method?

Consolidate – do you need more?

1 Write each mixed number as an improper fraction.

a $5\dfrac{1}{3}$　　　　b $2\dfrac{3}{5}$　　　　c $1\dfrac{6}{7}$　　　　d $4\dfrac{1}{10}$

2 Complete these multiplications.

a $5\dfrac{1}{3} \times 6$　　　b $2\dfrac{3}{5} \times \dfrac{3}{4}$　　　c $1\dfrac{6}{7} \times \dfrac{9}{5}$　　　d $4\dfrac{1}{10} \times 5\dfrac{1}{3}$

3 Complete these divisions.

a $5\dfrac{1}{3} \div \dfrac{1}{3}$　　　b $2\dfrac{3}{5} \div \dfrac{1}{10}$　　　c $1\dfrac{6}{7} \div \dfrac{3}{4}$　　　d $4\dfrac{1}{10} \div 1\dfrac{6}{7}$

4 Write each expression as a single fraction.

a $\dfrac{a}{7} \times \dfrac{b}{5}$　　　b $\dfrac{a}{7b} \times \dfrac{b}{5}$　　　c $\dfrac{a^2}{7} \times \dfrac{5}{b}$　　　d $\dfrac{a}{7b} \times \dfrac{5a}{b}$

Stretch – can you deepen your learning?

1 The base of a triangle is $3\dfrac{1}{5}$ m and its perpendicular height is $2\dfrac{1}{4}$ m

Work out the area of the triangle.

2 The width of a rectangle is $\dfrac{3x}{5}$ cm

Its length is four times its width.

Write an expression for the area of the rectangle.

3 Solve these equations.

a $\dfrac{15}{4x} - \dfrac{1}{2} = 3\dfrac{1}{2}$　　　　b $\dfrac{7}{4x} + 1\dfrac{2}{3} = 5\dfrac{5}{6}$　　　　c $5\dfrac{1}{9} = 1.75x - \dfrac{15}{4}$

4 Evaluate $\dfrac{(a+b)}{cd}$ when $a = \dfrac{1}{5}$, $b = 2\dfrac{1}{3}$, $c = \dfrac{17}{6}$ and $d = -\dfrac{6}{11}$

5 The ratio of the width of a field to its length is $1 : 1\dfrac{5}{7}$

The width of the field is $5\dfrac{1}{4}$ m. Work out the length of the field.

Reflect

1 Show that $4\dfrac{1}{5} \times 2\dfrac{3}{7} = 10\dfrac{1}{5}$ using two different methods. Explain why both methods work.

2 Simplify fully $\dfrac{3a}{5} \div \dfrac{6b}{25}$ explaining each step of your working.

3 Multiplying and dividing fractions

Chapters 3.1–3.5

I have become fluent in...

- multiplying a fraction by an integer
- multiplying a fraction by a fraction
- dividing an integer by a fraction
- dividing a fraction by a fraction
- multiplying and dividing complex fractions. Ⓗ

I have developed my reasoning skills by...

- representing calculations with fractions as diagrams
- considering the effect of the numerator or denominator changing
- explaining mistakes that others have made
- considering multiple methods of getting to the correct answer.

I have been problem-solving through...

- finding more than one answer in missing digit questions
- working with fractions in different contexts such as with area and perimeter
- using prior knowledge to solve multi-step problems.

Check my understanding

1 What calculation does the diagram show?

2 Work out

    a $2 \times \frac{1}{3}$ **b** $\frac{3}{10} \times 3$ **c** $5 \times \frac{3}{4}$

3 Find the product of

    a $\frac{1}{5}$ and $\frac{1}{7}$ **b** $\frac{4}{9}$ and $\frac{2}{3}$

4 Explain how the diagram shows that $4 \div \frac{1}{2} = 8$

5 Work out

    a $4 \div \frac{1}{3}$ **b** $3 \div \frac{1}{5}$ **c** $5 \div \frac{2}{7}$

6 What is three tenths-divided by two-thirds?

7 Calculate the following, giving your answers in their simplest form. Ⓗ

    a $2\frac{2}{5} \times 1\frac{1}{6}$ **b** $3\frac{3}{4} \div 2\frac{3}{7}$

8 Work out these, giving your answers in their simplest form. Ⓗ

    a $\frac{3}{5x} \times \frac{x}{y}$ **b** $\frac{5x}{3} \times \frac{x}{y}$ **c** $\frac{7}{8x} \div \frac{x}{y}$

4 Working in the Cartesian plane

In this block, I will learn...

how to work with lines in the form $y = kx$

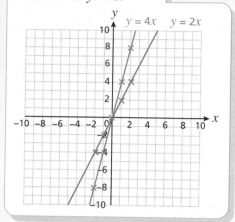

about lines parallel to the axes

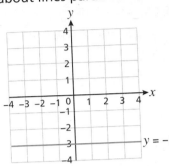

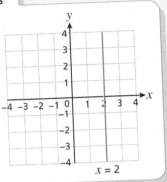

$y = -3$

$x = 2$

to link graphs to direct proportion problems

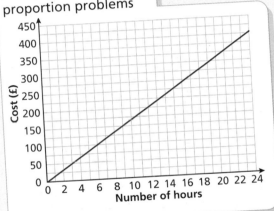

how to work with graphs that have equations of the form $y = x + a$

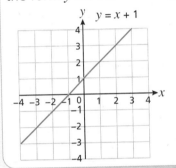

$y = x + 1$

about graphs with a negative gradient

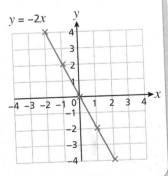

$y = -2x$

about non-linear graphs Ⓗ

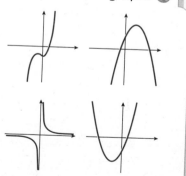

how to find the midpoint of a line segment Ⓗ

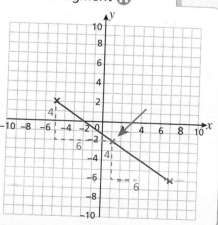

Small steps

- Work with coordinates in all four quadrants
- Identify and draw lines that are parallel to the axes

Key words

Quadrant – one of the four sections made by dividing an area with an x-axis and a y-axis

Origin – the point where the x-axis and y-axis meet

Coordinate – an ordered pair used to describe the position of a point

Parallel – always the same distance apart and never meeting

Are you ready?

1 Write down the coordinates of each point.

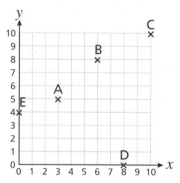

2 Copy these axes. Then plot and label the coordinates of each point, A to E

    A (3, 7)   B (2, 9)   C (9, 0)
    D (0, 0)   E (0, 10)

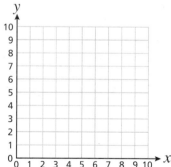

3 Which of these coordinates have an x value equal to 2?

    (3, 2)   (2, 0)   (7.45, –2)   (2.3, 9)   (–2, –5)

4 Which of these coordinates have a y value equal to –5?

    (0, –5)   (5, 11)   (–5, 14)   (12, 5)   (–5, 5)

Models and representations

Coordinate grid

You should be familiar with the position of each of the four **quadrants** on the **coordinate** grid.

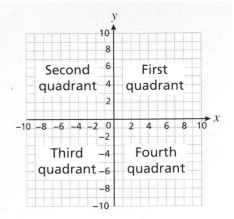

In this chapter, you will look at coordinates in all four quadrants.

Coordinates are used to describe the position of a point in a plane. In a pair of coordinates, the first number is the x value and tells you the horizontal position of a point; the second number is the y value and tells you the vertical position of the point in relation to a starting point, called the **origin**.

Example 1

a Write the coordinates of point A

b In which quadrant does the point (–7, 3) lie?

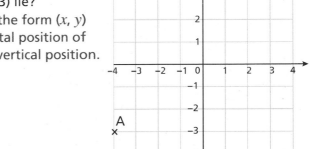

a (–4, –3) ○── All coordinates are of the form (x, y) where x is the horizontal position of the point and y is the vertical position.

b (–7, 3) is in the second quadrant

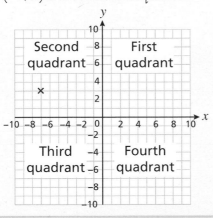

Practice 4.1A

1 Write down the coordinates of each point.

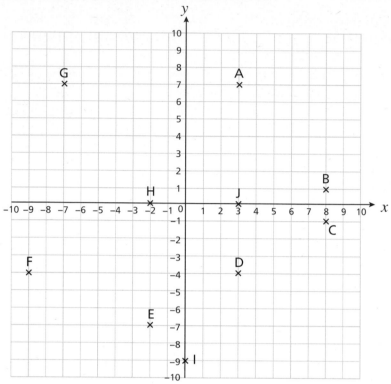

2 Copy the coordinate grid. Then plot and label each of these points.

A (7, –4)   B (0, 7)

C (7, 0)   D (–7, 4)

E (0, –7)   F (–7, 0)

G (–7, –4)   H (0, 4)

I (4, 0)   J (7, 4)

K (0, –4)   L (–4, 0)

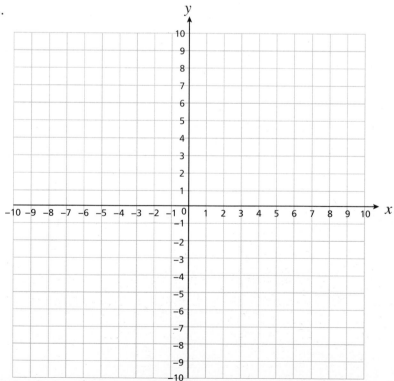

3 Chloe, Beca and Ed have each written down the coordinates of point P

Chloe (4, −3)

Beca (−3, 4)

Ed (3, 4)

a Who is correct?

b Explain the mistakes the other two have made.

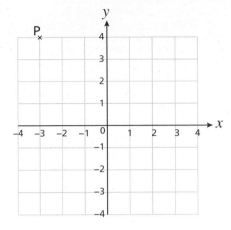

4 Ajmal says that the coordinates of point Q are (−2, −3)

Do you agree? Explain your answer.

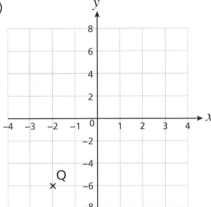

5 Here are the coordinates of the vertices of shape A.

(3, 7)   (−4, 7)   (−4, −2)   (3, −2)

a What can you say about shape A without drawing it?

b By drawing or otherwise, what is the mathematical name of shape A?

What do you think?

1 Without drawing, state the quadrant in which each point lies.

a (2, 5)   b (3, −2)   c (4, −6)   d (−9, −1)   e (7, 3)

f (−5, 2.2)   g (−1, −1)   h (5, −7.3)   i (2.78, 0.001)   j (−0.09, 0.05)

Compare your method and answers with a partner.

2 ABCD is a square.

Points A and B have coordinates (−1, 4) and (6, 4), respectively.

What could be the coordinates of points C and D?

3 **a** A and B lie on the same horizontal line.

    i What could be the coordinates of point B?

    ii What cannot be the coordinates of point B?

    iii What can you say for sure about the coordinates of point B?

  b A and C lie on the same vertical line.

    i What could be the coordinates of point C?

    ii What cannot be the coordinates of point C?

    iii What can you say for sure about the coordinates of point C?

Example 2

a **i** Plot these points on a grid: $(2, -3)$, $(2, 0)$, $(2, 1)$ and $(2, 4)$

   ii Draw a straight line through these points.

b Explain why the equation of this line is $x = 2$

a

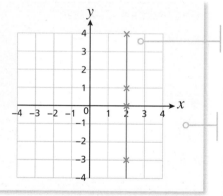

At all points on this line, the x value is equal to 2

Any point on the line has coordinates of the form $(2, a)$ where a can take any value.

The coordinates of all the points in between also have x value equal to 2

This would continue forever in both directions, for example, $(2, 100\,000)$ and $(2, -9.946\,27)$ would lie on the same vertical line.

b All the x values are equal to 2 therefore the line is called $x = 2$

Practice 4.1B

1 Here are the coordinates of some points.

| (4, −3) | (2, 0) | (9, −1) | (4, 0.5) | (7, −1) |

| (−1, 4) | (−4, 9) | (12, 1) | (4, 175) | (5.2, −1) |

a Which of the points have the x value equal to 4?

b Write down the coordinates of three more points that have their x value equal to 4

c Which of the points have a y value equal to −1?

d Write down the coordinates of three more points that have their y value equal to −1

2 a i Draw a coordinate grid from −10 to 10 then plot the points (−6, 9), (−6, 3), (−6, 0) and (−6, −5)

ii What do you notice?

iii Draw a straight line through the points. This is the line $x = -6$

Copy and complete this sentence to describe the points on the line $x = -6$

At every point on the line $x = -6$ the ☐-coordinate is equal to ☐

b i On the same coordinate grid, plot the points (−7, −2), (−5, −2) (0, −2) and (9, −2)

ii What do you notice?

iii Draw a straight line through the points. This is the line $y = -2$

Copy and complete the sentence to describe the line $y = -2$

At every point on the line $y = -2$ the ☐-coordinate is equal to ☐

3 Here are the equations of five lines, A to E

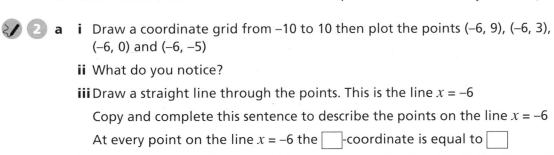

A $y = -4$ B $y = -1$ C $y = 3$ D $y = 7$ E $y = -9$

a Write down the coordinates of three points that lie on each line.

b On a coordinate grid, plot your points from part **a** and draw and label each line.

c What do you notice?

d Copy and complete the sentences to describe lines of the form $y = a$

Any line of the form $y = a$ is a _____ line that goes through ☐ on the ☐-axis.

Any line of the form $y = a$ is parallel to the ☐-axis.

At any point on the line $y = a$, the ☐-coordinate is equal to ☐.

4 Here are the equations of five lines, A to E

A $x = -8$ B $x = -4$ C $x = 2$ D $x = 5$ E $x = -3$

a Write down the coordinates of three points that lie on each line.

b Copy the coordinate grid from question 2. Plot your points from part **a**. Join each set of points to draw lines A to E. Label each line.

c What do you notice?

d Copy and complete the sentences to describe lines of the form $x = b$.

Any line of the form $x = b$ is a _____ line that goes through ☐ on the ☐-axis.

Any line of the form $x = b$ is parallel to the ☐-axis.

At any point on the line $x = b$, the ☐-coordinate is equal to ☐.

5 **a** Find the equation of each line.

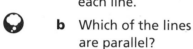

b Which of the lines are parallel?

What do you notice? Why does this happen?

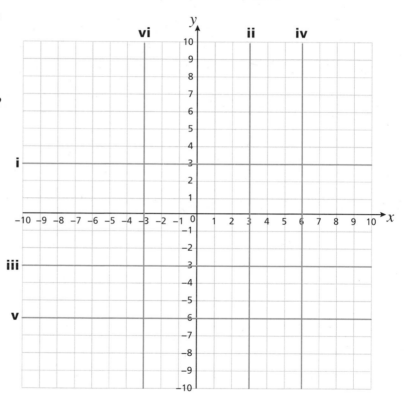

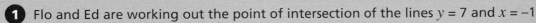

What do you think?

1 Flo and Ed are working out the point of intersection of the lines $y = 7$ and $x = -1$

   a Use Flo's method to find the coordinates of the point of intersection of $y = 7$ and $x = -1$

> I'll draw both lines to find the coordinates of the point of intersection.

   b

> I didn't need to draw either of the lines, I just know the coordinates.

    Explain how Ed knows this.

   c Without drawing, find the coordinates of the point of intersection of the lines $x = -8$ and $y = 17$

2 Beca and Jakub are finding the equation of this line.

   a

> The equation of the line is $x = 2$

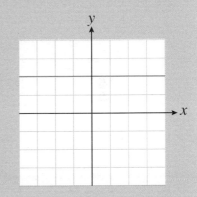

    Explain why Beca cannot be correct.

   b

> The equation of the line is $y = 12$

    Explain why Jakub could be correct.

   c Explain why you cannot tell the exact equation of the line.

3 **a** Plot and label the line $y = 0$

      What do you notice?

    b Plot and label the line $x = 0$

       What do you notice?

Consolidate – do you need more?

1 Write down the coordinates of each of the points on the grid.

2 On a blank copy of the coordinate grid for question 1, plot and label each of these points.

A (5, –3) B (–2, 5)

C (10, 9) D (–8, –1)

E (7, 0) F (0, –9)

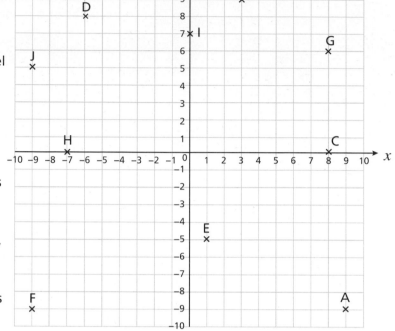

3 **a** Write down five pairs of coordinates with x value equal to 5

b On a coordinate grid, draw and label the graph $x = 5$

c Write down five pairs of coordinates with y value equal to –2

d On the same coordinate grid as for part **b**, draw and label the graph of $y = -2$

4 Find the equation of each of these lines.

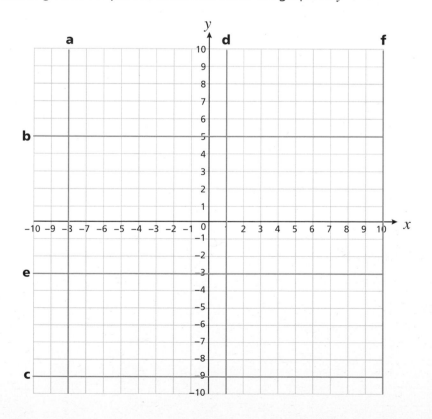

Stretch – can you deepen your learning?

1 A square has two vertices at (–5, –4) and (–1, 0), diagonally opposite each other.

Find the coordinates of the other two vertices of the square.

2 A triangle has two vertices at (–1, 4) and (7, 4)

Given that the area of the triangle is 20 square units, what could the coordinates of the third vertex be?

3 The diagram shows four identical right-angled triangles joined to form a square.

Work out the coordinates of points A–F

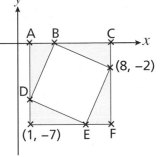

4 **a** An isosceles triangle has vertices at (–5, 8), (–5, 0) and (0, 4)

Find the equation of the line of symmetry of the triangle.

b Another isosceles triangle has $x = -1$ as a line of symmetry.

Suggest the coordinates of the vertices of this triangle.

You will look at symmetry in more detail in Block 15

c A rectangle has $x = -4$ and $y = -3$ as lines of symmetry.

Suggest the coordinates of the vertices of the rectangle.

Reflect

1 In your own words, explain how to plot and read coordinates.

2 What do all points on the line $y = 9$ have in common?

3 Describe the key feature of the line $x = -5$

4.2 Lines of the form $y = kx$

Small steps

- Recognise and use the line $y = x$
- Recognise and use lines of the form $y = kx$
- Link $y = kx$ to direct proportion problems
- Explore the gradient of the line $y = kx$ **H**

Key words

Linear – forming a straight line

Direct proportion – two quantities are in direct proportion when as one increases or decreases, the other increases or decreases at the same rate

Gradient – the steepness of a line

Are you ready?

1 Write down the coordinates of five points that lie on each of these lines.

   a $y = 5$ **b** $x = -3$ **c** $y = -3$ **d** $x = 5$

2 Find the equation of each line.

   a 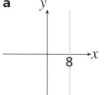   **b** 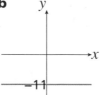   **c**   **d**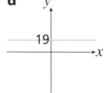

3 Sketch each graph and label any points of intersection with the axes.

   a $x = 7$ **b** $y = -9$ **c** $x = -20$ **d** $y = 100$

4 Evaluate each expression when

     i $x = 2$ **ii** $x = -2$

   a $6x$ **b** $-x$ **c** $-3x$ **d** $20x$ **e** $-100x$

Models and representations

Coordinate grids

It is important to look carefully at the scale on each axis. Although all these graphs look the same, they are actually different because of the scales on the axes.

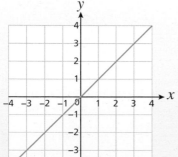 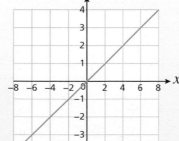 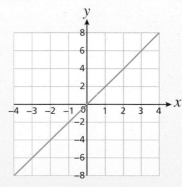

Table of values

You can use a table of values for working out points that lie on a given line.

The top rows tells you the values of x that you need to substitute into the equation of the line.

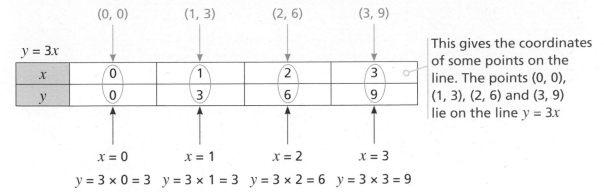

(0, 0)   (1, 3)   (2, 6)   (3, 9)

$y = 3x$

x	0	1	2	3
y	0	3	6	9

This gives the coordinates of some points on the line. The points (0, 0), (1, 3), (2, 6) and (3, 9) lie on the line $y = 3x$

$x = 0$ $x = 1$ $x = 2$ $x = 3$

$y = 3 \times 0 = 3$ $y = 3 \times 1 = 3$ $y = 3 \times 2 = 6$ $y = 3 \times 3 = 9$

The bottom row tells you the corresponding value of y for each x value.

In Chapter 4.1 you looked at horizontal and vertical lines that were parallel to one of the axes.

In this chapter, you will learn to plot and recognise diagonal lines of the form $y = kx$, where k is a number (constant).

Remember: the equation of a line describes the relationship between the x and y values of the coordinates at any point on that line.

Example 1

a Complete the table of values for $y = 4x$

x	−2	−1	0	1	2
y					

b Draw the graph of $y = 4x$

This shows the coordinate (−2, −8)

a $y = 4x$

$\times 4$

x	−2	−1	0	1	2
y	−8	−4	0	4	8

$\times 4$

$x = -2$ $x = -1$ $x = 0$ $x = 1$ $x = 2$

$y = 4 \times -2$ $y = 4 \times -1$ $y = 4 \times 0$ $y = 4 \times 1$ $y = 4 \times 2$

$= -8$ $= -4$ $= 0$ $= 4$ $= 8$

Substitute each x value into the equation of the line to find the corresponding y value.

b

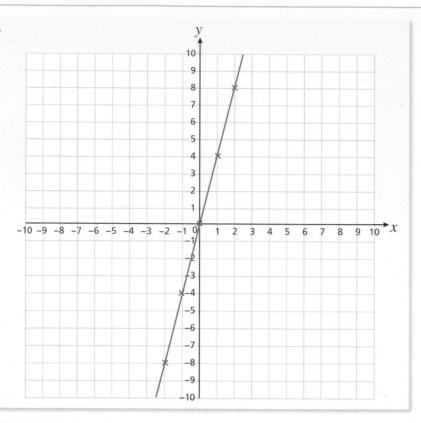

> $y = 4x$ is an example of a **linear** graph, so the points should lie in a straight line.

> Although your points are only for values of x between –2 and 2, the line would extend infinitely in both directions.

You can draw a graph of the form $y = kx$ using only two pairs of coordinates, but you would not be able to tell if you had made a mistake, so it is always better to work with more than two points.

Practice 4.2A

1 **a** In which of these coordinate pairs is the y value equal to the x value?

(2, –2)   (0, 0)   (4, 4)   (–9, –9)   (–5, 5)

b Write the coordinates of three more points where the y value is equal to the x value.

c Write the coordinates of five points that lie on the line $y = x$

2 **a** Copy and complete the table of values for $y = x$

x	–2	–1	0	1	2
y					

b On a copy of these axes, draw the graph of $y = x$

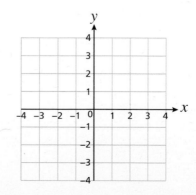

3 **a** In which of these coordinate pairs is the y value twice the x value?

(2, 4)   (4, 2)   (6, −3)   (3, −6)   (−3, −6)

(0, 0)   (19, 38)   (−12, −6)   (50, 25)   (−20, −40)

b Write the coordinates of three more points where the y value is twice the x value.

c Write the coordinates of five points that lie on the line $y = 2x$

4 **a** Copy and complete the table of values for $y = 2x$

x	−2	−1	0	1	2
y					

b On a copy of these axes, draw the graph of $y = 2x$

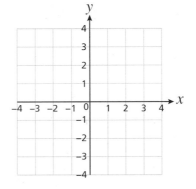

5 **a** Copy and complete a table of values for each of these equations.

  i $y = 3x$ **ii** $y = 5x$ **iii** $y = 0.5x$ **iv** $y = 6x$

x	−2	−1	0	1	2
y					

b On a copy of these axes, draw and label the graph of each equation in part **a**.

c What do you notice?

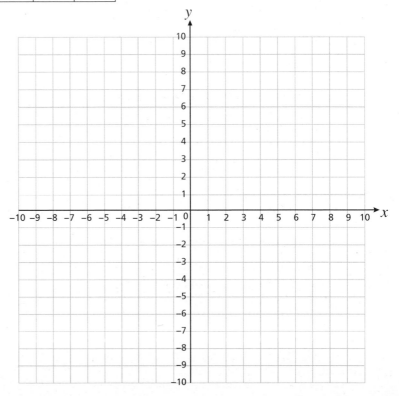

What do you think?

1 Rob has drawn the graph of $y = 2x$

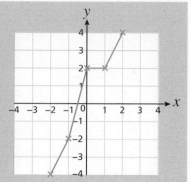

    a How do you know, by looking at the graph, that Rob has made a mistake?

    b Which of his points is wrong?

2 Draw the graph of $y = x$ on a copy of each of these axes.

a 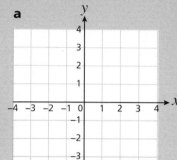   **b** 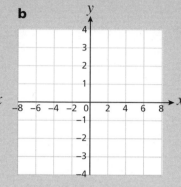   **c**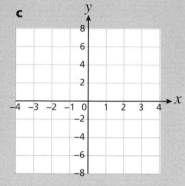

3 All of these points lie on the same line.

    (1, 10)   (3, 30)   (−10, −100)   (−5, −50)

    a Copy and complete this sentence to describe the relationship between the x and y values for each coordinate pair.

    In each pair of coordinates, the y value is _____ the x value.

    b Write the equation of the straight line that passes through all four points.

4 Any line of the form $y = kx$ will go through the origin.

    Do you agree with Emily? Explain your answer.

In the next section, you will discover the link between lines of the form $y = kx$ and **direct proportion**.

You have already considered direct proportion as multiplicative change. You will now formally explore the connection between direct proportion and graphs of the form $y = kx$

Example 2

A machine makes 50 greetings cards an hour.

a Complete the table of values to show how many greetings cards could be made in an hour for different numbers of machines.

Number of machines	0	1	2	3	4	5	10
Number of greetings cards							

b Draw a graph to represent the relationship between the number of machines and the number of greetings cards.

c   Find the equation of the line drawn in part **b**.

For every 1 machine, there are 50 greetings cards.

The number of greetings cards is 50 times the number of machines.

a

$\times 50$

Number of machines	0	1	2	3	4	5	10
Number of greetings cards	0	50	100	150	200	250	500

$\times 50$

b

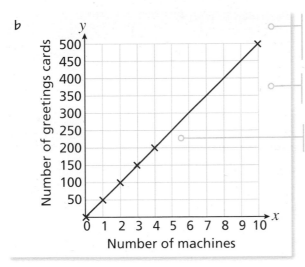

Using your table from part **a** you can generate coordinates of points that will lie on the line.

(0, 0), (1, 50), (2, 100), (3, 150), (4, 200), (5, 250) and (10, 500) all lie on the line.

Plot these points and join them with a straight line.

This shows that the variables are in direct proportion.

c $y = 50x$

$\times 50$	$\times 50$	$\times 50$	$\times 50$	$\times 50$	$\times 50$	$\times 50$
$x \quad y$	$x \quad y$	$x \quad y$	$x \quad y$	$x \quad y$	$x \quad y$	$x \quad y$
(0, 0)	(1, 50)	(2, 100)	(3, 150)	(4, 200)	(5, 250)	(10, 500)

At each point on the graph, the y value is 50 times the x value.

For every 1 machine, there are 50 greetings cards. So to get the number of greetings cards, you multiply the number of machines by 50

Practice 4.2B

1   One bag of marbles costs £2

a   Copy and complete the table of values to show the cost of buying bags of marbles.

Number of bags	0	1	2	4	5	10	20
Cost (£)		2					

b   Draw the straight line graph that represents the values in the table.

c   How does your graph indicate that the number of bags of marbles and the cost are in direct proportion?

d   Write the equation of the straight line drawn in part **b**.

2 One tennis ball costs 50p

    a Copy and complete the table of values to show the cost of buying tennis balls.

Number of tennis balls	0	1	2	4	5	10	20
Cost (£)							

    b Draw the straight line graph that represents the values.

    c How does your graph indicate that the number of tennis balls and the cost are in direct proportion?

    d Write the equation of the straight line drawn in part **b**.

3 The cost of two pens is 48p

    a Copy and complete the table of values to show the cost of buying pens.

Number of pens	0	1	2	4	5	10	20
Cost (pence)							

    b Draw the straight line graph that represents the values.

    c How does your graph indicate that the number of pens and the cost are in direct proportion?

    d Write the equation of the straight line drawn in part **b**.

4 The graph shows the cost of hiring a bouncy castle for a given number of hours.

    a Explain how the graph shows that the cost is directly proportional to the number of hours.

    b How much does it cost to hire the bouncy castle for 10 hours?

    c Use your answer to part **a** to work out the cost of hiring the bouncy castle for 1 hour.

    d Write the equation of the line shown in the graph.

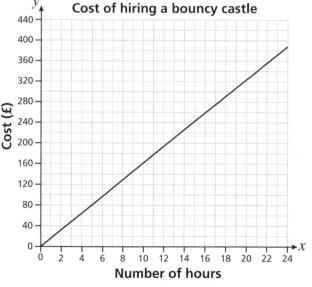

Cost of hiring a bouncy castle

What do you think? 💭

1 In a workshop, the number of decorations made in one day is directly proportional to the number of people making decorations. When there are 4 people working, the number of decorations made is 100

When represented graphically, the line goes through the point (4, 100)

    a How many decorations can one person make in one day? How do you know?

    b Write the equation of the line that represents the relationship between the number of decorations made and the number of people.

You will now formally explore the **gradient** of lines of the form $y = kx$

You will look at the connection between the equation of a line and its gradient.

The gradient of a line is a measure of how steep it is.

Example 3

a Complete the table of values for $y = 2x$ and $y = 4x$

x	−2	−1	0	1	2
y					

b On the same grid, draw the graphs of $y = 2x$ and $y = 4x$

c What's the same and what's different about $y = 2x$ and $y = 4x$?

a $y = 2x$

$\times 2$
x	−2	−1	0	1	2
y	−4	−2	0	2	4
$\times 2$

$y = 4x$

$\times 4$
x	−2	−1	0	1	2
y	−8	−4	0	4	8
$\times 4$

b

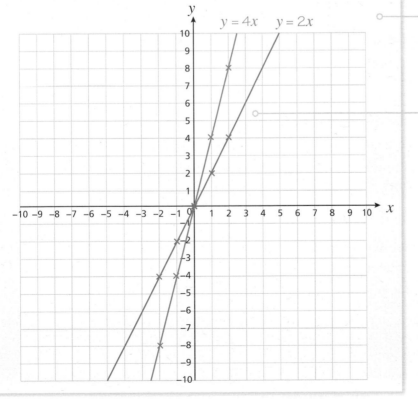

Use your tables from part **a** to generate the coordinates for the points on each line.

Join the points to give two straight lines.

It is important to label each line so that you know which one is which.

You could find the gradient of the lines using triangles like you did in Chapter 1.5

> c Both graphs go through the origin.
> The graph of $y = 4x$ is steeper than the graph of $y = 2x$

The gradient of the line is the value of k when the equation is written in the form $y = kx$

Practice 4.2C

1 **a** Draw each of these graphs for values of x from –3 to 3

   i $y = x$ **ii** $y = 5x$ **iii** $y = 3x$ **iv** $y = 1.5x$

   b Explain the key features of each graph. What's the same? What's different?

   c Put the equations in order, from least steep to most steep.

   Could you have done this without drawing?

2 **a** **i** Copy and complete the table of values for $y = 2x$

x	–2	–1	0	1	2
y					

   ii Look at the sequence formed by the y values. What do you notice?

   How is this linked to the gradient of the line?

   b **i** Copy and complete the table of values for $y = 10x$

x	–2	–1	0	1	2
y					

   ii Look at the sequence formed by the y values. What do you notice?

   How is this linked to the gradient of the line?

   c Compare your answers to parts **a** and **b**.

   How does this explain the fact that $y = 10x$ is steeper than $y = 2x$?

3 Here is a sketch of $y = 12x$

   a On a copy of the graph, sketch

   i the graph of $y = 4x$ **ii** the graph of $y = 15x$

   b Compare your answers with a partner. What's the same? What's different?

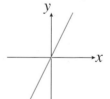

4 Copy and complete these sentences.

On a graph of the form $y = kx$, the number k is the _____ of the line.

The greater the value of k, the _____ the line.

The lower the value of k, the _____ the line.

What do you think?

1 **a** Explain why Chloe thinks this.

None of these lines are $y = x$

b Explain why Chloe could be wrong.

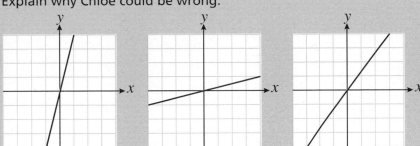

Consolidate – do you need more?

1 **a** Copy and complete this sentence to describe the relationship between the x and y values at each point on the line $y = 3x$

At each point on the line $y = 3x$, the ☐-coordinate is three times the ☐-coordinate.

b Copy and complete the table of values for $y = 3x$

x	–2	–1	0	1	2
y					

c Draw the graph of $y = 3x$

2 **a** Copy and complete this sentence to describe the relationship between the x and y values at each point on the line $y = 8x$

At each point on the line $y = 8x$, the ☐-coordinate is _____ times the ☐-coordinate.

b Copy and complete the table of values for $y = 8x$

x	–2	–1	0	1	2
y					

c Draw the graph of $y = 8x$

3 One bag of sand costs £5

a Copy and complete the table of values to show the cost of buying bags of sand.

Number of bags	0	1	2	3	4	5	10
Cost (£)		5					

b Draw the straight line graph that represents the values.

c How does your graph indicate that the number of bags of sand and cost are in direct proportion?

d Write the equation of the straight line drawn in part **b**.

Stretch – can you deepen your learning?

1 A straight line goes through the points (0, 0) and (15, 52.5).

What is the equation of this line?

2 Find the equation of each of these lines.

a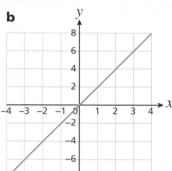

b

c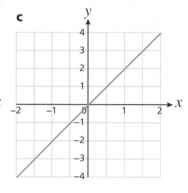

3 Here is the graph of $y = 12x$

Find the equation of the horizontal line that goes through point B

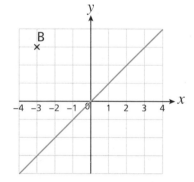

4 The point (a, b) lies on the line $y = 7x$

a Write an expression for b in terms of a

b Write an expression for a in terms of b

5 The graphs of $x = 7$, $y = 59.5$ and $y = px$ intersect at a single point.

Work out the value of p

Reflect

1 Explain the relationship between the x and y values at any point on the line $y = 20x$

2 Explain the key features of a graph of the form $y = kx$

3 Explain why any graph showing direct proportion is of the form $y = kx$

4.3 Lines of the form $y = x + a$

Small steps

■ Recognise and use lines of the form $y = x + a$

Key words

y-intercept – the point at which a graph crosses or intersects the y-axis

Gradient – the steepness of a line

Are you ready?

1 Describe the meaning of each expression in words.

    a $5a$ **b** $-5a$ **c** $a + 5$ **d** $a - 5$ **e** $5 + a$ **f** $5 - a$

2 Evaluate each expression when $a = 3$

    a $5a$ **b** $-5a$ **c** $a + 5$ **d** $a - 5$ **e** $5 + a$ **f** $5 - a$

3 Evaluate each expression when $a = -3$

    a $5a$ **b** $-5a$ **c** $a + 5$ **d** $a - 5$ **e** $5 + a$ **f** $5 - a$

4 Write the coordinates of three points that lie on each line.

    a $y = 5x$ **b** $y = -5x$

Models and representations

Coordinate grid

It is important to look carefully at the scale on each axis. Although all these graphs look the same, they are actually different because of the scales on the axes.

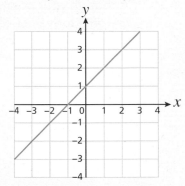 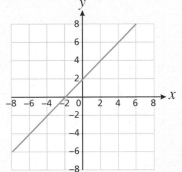 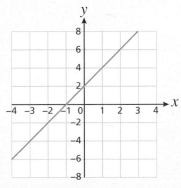

Table of values

You can use a table of values for working out points that lie on a given line.

$y = x + 4$

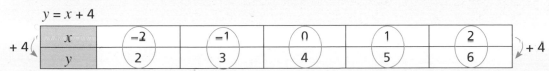

In Chapter 4.2 you looked at the effect of changing the **gradient** of a line. Is this chapter, you will focus on what happens to a graph when the **y-intercept** changes.

In this chapter, you will look at lines of the form $y = x + a$

Example 1

a Complete the table of values for $y = x - 3$ **b** Draw the graph of $y = x - 3$

x	−2	−1	0	1	2
y					

a $y = x - 3$

The equation of the line is $y = x - 3$

This means "y is equal to x subtract 3"

x	−2	−1	0	1	2
y	−5	−4	−3	−2	−1

-3 -3

At any point on the line, the y value will be 3 less than the x value.

b

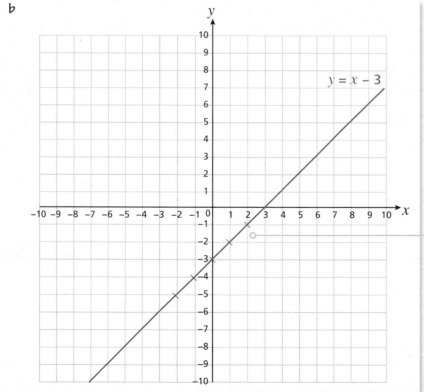

$y = x - 3$

Although your points are only for values of x between −2 and 2, this line would extend infinitely in both directions.

Practice 4.3A

1 a Copy and complete these sentences to describe the relationship between the x and y values of each point on the line.

    i $y = x + 1$ The y value is _____ than the x value.

    ii $y = x - 1$ The y value is _____ than the x value.

    iii $y = x - 4$ The y value is _____ than the x value.

    iv $y = x + 4$ The y value is _____ than the x value.

b Find the coordinates of three points that lie on each line.

    i $y = x + 1$ **ii** $y = x - 1$ **iii** $y = x - 4$ **iv** $y = x + 4$

2 Here are some pairs of coordinates.

(4, 19)	(19, 4)	(0, −15)	(15, 0)	(−15, 0)

(23, 38)	(3.8, 2.3)	(100, 85)	(−17, −2)	(12, −3)

a Which of the points lie on the line $y = x + 15$? How do you know?

b Which of the points lie on the line $y = x - 15$? How do you know?

3 a Copy and complete the table of values for each of these lines.

x	−2	−1	0	1	2
y					

    i $y = x + 1$ **ii** $y = x + 2$ **iii** $y = x + 3$ **iv** $y = x + 4$

b Draw the graph of each of the lines in part **a**.

c What do you notice?

d Without using a table of values, draw the graph of $y = x + 9$ on the same axes.

4 a Copy and complete the table of values for each of these lines.

x	−2	−1	0	1	2
y					

    i $y = x - 1$ **ii** $y = x - 2$ **iii** $y = x - 3$ **iv** $y = x - 4$

b Draw the graph of each of the lines in part **a**.

c What do you notice?

d Without using a table of values, draw the graph of $y = x - 7$ on the same axes.

5 Here are three straight line graphs. Each graph has an equation of the form $y = x + a$

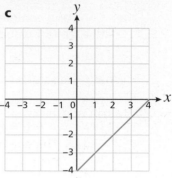

For each line

    i find the coordinates of three points that lie on the line.

    ii copy and complete the sentence

       At each point on the line, the y value is _____ than the x value.

    iii find the equation of the line.

6 Find the equation of each of these lines.

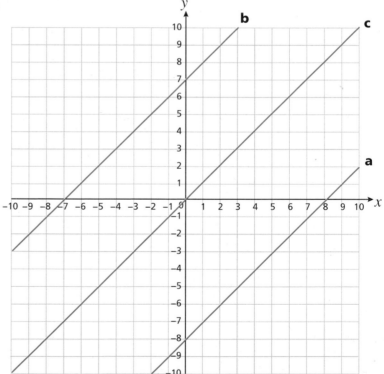

What do you think?

1. Draw each of these graphs on the same axes.

   a $y = x$ **b** $y = x + 2$ **c** $y = x - 2$

   What's the same and what's different?

2. Marta and Jackson are working out the equation of this line.

 The equation is $y = x + 7$

The equation is $y = 7 + x$

Marta   Jackson

Who do you agree with? Explain your answer.

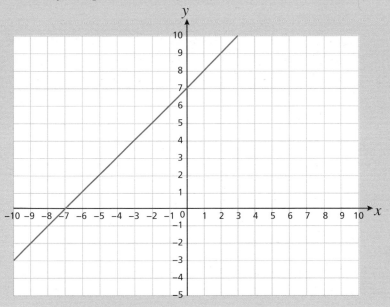

Consolidate – do you need more?

1. Copy and complete these function machines.

a Input → Output
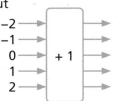
−2, −1, 0, 1, 2 → + 1

b Input → Output
−2, −1, 0, 1, 2 → − 1

c Input → Output
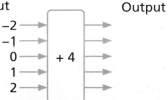
−2, −1, 0, 1, 2 → + 4

d Input → Output
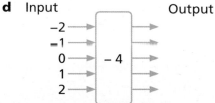
−2, −1, 0, 1, 2 → − 4

2 **a** Copy and complete the table of values for each of these lines.

x	−2	−1	0	1	2
y					

i $y = x + 1$ **ii** $y = x - 1$ **iii** $y = x + 4$ **iv** $y = x - 4$

b Draw the graph of each of the lines in part **a**.

Stretch – can you deepen your learning?

1 Here is the graph of $y = x + a$

Is the value of a positive or negative?
How do you know?

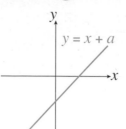

2 The line drawn has equation $y = x$

On a copy of this graph, draw the lines
with these equations.

a $y = x + \dfrac{3}{2}$

b $y = x - 1\dfrac{1}{2}$

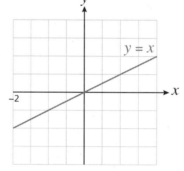

3 The graphs of $y = x + 5$, $y = 2$ and $x = 3$ border a triangle.

Work out the area of the triangle.

4 A straight line with equation $y = x + b$ intersects the x-axis at $(18, c)$

Work out the values of b and c

Reflect

In your own words, explain how a line of the form $y = x + a$ changes for different values of a

4.4 Lines with a negative gradient

Small steps

■ Explore graphs with a negative gradient
($y = -kx$, $y = a - x$, $x + y = a$)

Key words

Gradient – the steepness of a line

Negative – less than zero

Are you ready?

1 Evaluate each expression when $w = 8$

   a $2w$ **b** $5w$ **c** $2 + w$ **d** $-5 + w$

   e $-2w$ **f** $-5w$ **g** $2 - w$ **h** $-5 - w$

2 Solve these equations.

   a $j + 15 = 23$ **b** $11 + k = 34$ **c** $10 + l = 0$ **d** $m + 7 = 5$

3 Copy and complete the table of values for $y = 15x$

x	−2	−1	0	1	2
y					

Models and representations

Double sided counters

    This shows 4 groups of −2

Number lines

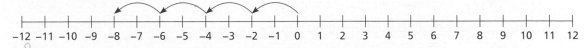

A number line is useful for showing repeated addition or subtraction.

Double sided counters and number lines can support you when working with negative numbers.

In Chapters 4.2 and 4.3 you saw linear graphs of different forms, all with a positive gradient.

In this chapter, you will look at graphs that have a **negative gradient**.

If a line has a negative gradient, as the x value increases, the y value decreases.

On a graph, a line with a negative gradient slopes "downwards".

Example 1

a Complete the table of values for $y = -2x$ **b** Draw the graph of $y = -2x$

x	−2	−1	0	1	2
y					

a $y = -2x$

$\times -2$

x	−2	−1	0	1	2
y	4	2	0	−2	−4

$\times -2$

The equation of the line is $y = -2x$

This means "y is equal to −2 lots of x"

b

Practice 4.4A

1 **a** Copy and complete the table of values for $y = x$

x	−2	−1	0	1	2
y					

b On a copy of these axes, draw the graph of $y = x$

c Copy and complete the table of values for $y = -x$

x	−2	−1	0	1	2
y					

d On the same axes, draw the graph of $y = -x$

e What's the same and what's different about the lines $y = x$ and $y = -x$?

f Which of the graphs has a negative gradient? How can you tell?

2 **a** **i** Copy and complete the table of values for $y = -4x$

x	-2	-1	0	1	2
y					

ii On a set of axes, draw and label the graph of $y = -4x$

b **i** Copy and complete the table of values for $y = 5 - x$

x	-2	-1	0	1	2
y					

ii On a set of axes, draw and label the graph of $y = 5 - x$

c **i** Copy and complete the table of values for $x + y = 6$

x	-2	-1	0	1	2
y					

ii On a set of axes, draw and label the graph of $x + y = 6$

d What's the same and what's different about your graphs?

3 **a** Copy and complete the table of values for each of these graphs.

x	-2	-1	0	1	2
y					

i $y = 7 - x$ **ii** $y = -5x$ **iii** $x + y = 8$ **iv** $10 = x + y$ **v** $y = 1 - x$

b On the same set of axes, draw and label each of your graphs from part **a**.

c Each of the lines has a negative gradient.

How can you tell this by looking at the graph? How can you tell from the equation?

4 Here are eight linear graphs.

A   B   C   D

E   F   G   H

Which of the graphs have a negative gradient? How do you know?

5 Here are the equations of eight linear graphs.

$y = 3x - 2$

$y = 2 - 3x$

$y + 3x = 2$

$2y = -3x$

$y = x + \dfrac{2}{3}$

$y = -3 + 2x$

$y + x = \dfrac{3}{2}$

$y = \dfrac{2}{3}x$

Which of the equations would have a graph with a negative gradient?
How do you know?

What do you think?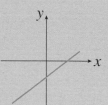

1 Ed says that this graph has a negative gradient.

Explain why this cannot be true.

2 Here are the equation, graph and table of values of a straight line.

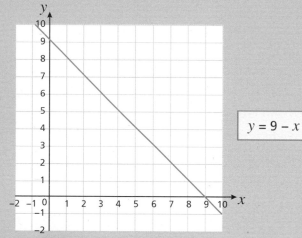

$y = 9 - x$

x	0	1	2	3	4
y	9	8	7	6	5

a I can tell that the gradient of this line is negative from the graph.   How can Abdullah tell this?

b I can tell that the gradient of this line is negative from the equation.   How can Faith tell this?

c I can tell that the gradient of this line is negative from the table of values.   How can Benji tell this?

Consolidate – do you need more?

1 Copy and complete these function machines.

a Input 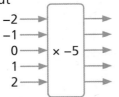 Output

b Input 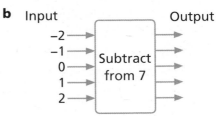 Output

c Input 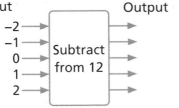 Output

d Input 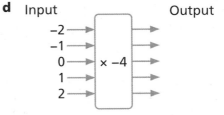 Output

2 Copy and complete tables of values for each of these lines.

x	−2	−1	0	1	2
y					

a $y = -5x$ **b** $y = 7 - x$ **c** $y = 12 - x$ **d** $y = -4x$

3 On a copy of the axes, draw each of these straight lines.

a $y = -5x$

b $y = 6 - x$

c $y = 12 - x$

d $y = -4x$

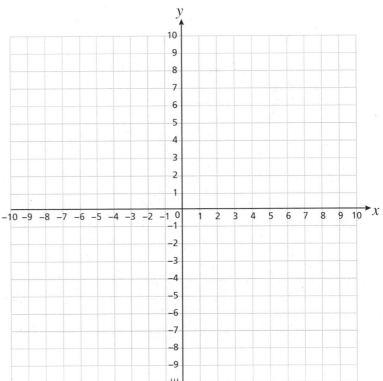

Stretch – can you deepen your learning?

1 Here is the graph of $y = ax$

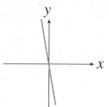

  a Is the gradient positive or negative? How do you know?

  b On a copy of the axes, sketch the graph of $y = -ax$

2 Is this statement always true, sometimes true or never true? Give examples to support your answer.

A graph with a negative gradient goes through the first quadrant.

3 Sven says that this table of values is for a graph with a negative gradient because the y values are decreasing.

x	2	1	0	−1	−2
y	4	3	2	1	0

Do you agree with Sven? Explain your answer.

4 Match each line on the graph with its correct equation.

$y = x$	$y = -x$
$y = 4$	$x = -5$
$y = 4 - x$	$y = -x - 3$
$x + y = 7$	$y = 4 + x$

Talk with a partner about how you decided.

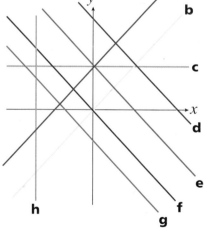

Reflect

In your own words, explain how you can tell whether a line has a positive or negative gradient.

4.5 Lines of the form $y = mx + c$

White Rose Maths

Small steps

- Plot graphs of the form $y = mx + c$
- Link graphs to linear sequences

Key words

Gradient – the steepness of a line

y-intercept – the point at which a graph crosses or intersects the y-axis

Linear – forming a straight line

Are you ready?

1 Find the output of each function machine.

a Input → Output

$3 \longrightarrow \boxed{\times 2} \longrightarrow \boxed{+ 5} \longrightarrow ?$

b Input → Output

$-3 \longrightarrow \boxed{\times 2} \longrightarrow \boxed{+ 5} \longrightarrow ?$

c Input → Output

$0 \longrightarrow \boxed{\times 4} \longrightarrow \boxed{- 2} \longrightarrow ?$

d Input → Output

$-1 \longrightarrow \boxed{\times 5} \longrightarrow \boxed{- 4} \longrightarrow ?$

e Input → Output

$1 \longrightarrow \boxed{\times 3} \longrightarrow \boxed{+ 1} \longrightarrow ?$

f Input → Output

$-2 \longrightarrow \boxed{\times 3} \longrightarrow \boxed{- 3} \longrightarrow ?$

2 Evaluate each of these expressions when $x = 2$

    a $5x + 3$ **b** $2x + 1$ **c** $4x - 1$ **d** $-3x - 2$ **e** $\frac{1}{2}x + 4$

3 Decide whether each sequence is linear or non-linear. Explain your answer.

    a 100, 50, 25, 12.5 …   **b** 60, 65, 70, 75 …

    c 4, 8, 16, 32 …   **d** 12, 9, 6, 3 …

4 Find the term-to-term rule for each sequence.

    a 100, 50, 25, 12.5 …   **b** 60, 65, 70, 75 …

    c 4, 8, 16, 32 …   **d** 12, 9, 6, 3 …

Models and representations

Sequences can be represented in lots of different ways.

You can use a list.   2   5   8   11   …

You can use objects.

You can use a table.

Position	1	2	3	4
Term	2	5	8	11

You can use a **graph**.

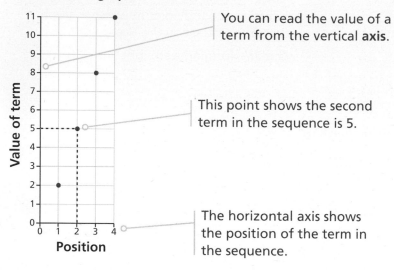

You can read the value of a term from the vertical **axis**.

This point shows the second term in the sequence is 5.

The horizontal axis shows the position of the term in the sequence.

You can use graphing software to plot the points of sequences.

So far, you have looked at graphs of the form $y = kx$ (or $y = mx$) which have different gradients, and $y = x + a$ (or $y = x + c$) which have different **y-intercepts**; you will now bring this together.

In this chapter, you will plot graphs of the form $y = mx + c$ where m is the **gradient** and c is the y-intercept. You will discover more about this in Book 3

Example 1

a Complete the table of values for $y = 2x + 5$

x	−2	−1	0	1	2
y					

b Draw the graph of $y = 2x + 5$

a Input   Output

$-3 \longrightarrow \boxed{\times 2} \longrightarrow \boxed{+5} \longrightarrow ?$

The equation $y = 2x + 5$ means that at each point on the line, the y value is 5 more than twice the x value.

You can draw a function machine to represent the relationship between the x and y values at any point on the line.

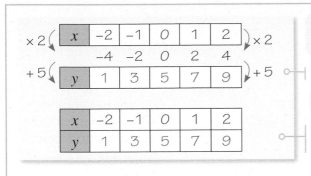

Rather than performing just one operation, you now need to perform two.

Multiply x by 2 and then add 5

It can be helpful to write down each step.

Notice that there is a sequence formed by the y values.

The y values increase by 2 each time the x value increases by 1.

b

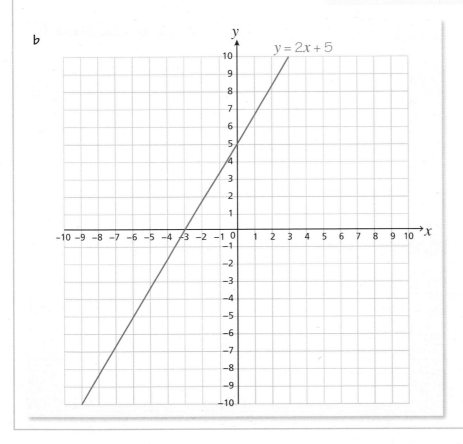

Practice 4.5A

1. Junaid and Bobbie are talking about straight line graphs.

   a. Junaid says, "At each point on the line $y = 5x$, the y value is 5 times the x value."

      Do you agree with Junaid? Explain your answer.

   b. Bobbie says, "At each point on the line $y = x + 3$, the y value is 3 more than the x value."

      Do you agree with Bobbie? Explain your answer.

c What is the relationship between the x and y values at each point on the line $y = 5x + 3$?

d Describe the relationship between the x and y values at each point on the lines with these equations.

   i $y = 2x + 1$ **ii** $y = 4x - 1$ **iii** $y = -3x - 2$ **iv** $y = \frac{1}{2}x + 4$

2 **a** Copy and complete a table of values for each of these lines.

x	−2	−1	0	1	2
y					

   i $y = 2x + 1$ **ii** $y = 4x - 1$

   iii $y = -3x - 2$ **iv** $y = \frac{1}{2}x + 4$

b Draw each line on a copy of this coordinate grid.

   i $y = 2x + 1$ **ii** $y = 4x - 1$

   iii $y = -3x - 2$ **iv** $y = \frac{1}{2}x + 4$

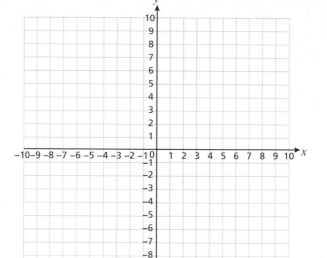

3 **a** Copy and complete a table of values for each of these lines.

x	−2	−1	0	1	2
y					

   i $y = 3x + 1$ **ii** $y = 4 + 3x$

   iii $y = 3x - 1$ **iv** $y = 3x - 7$

b Draw each line on a copy of this coordinate grid.

   i $y = 3x + 1$ **ii** $y = 4 + 3x$

   iii $y = 3x - 1$ **iv** $y = 3x - 7$

c What do you notice?

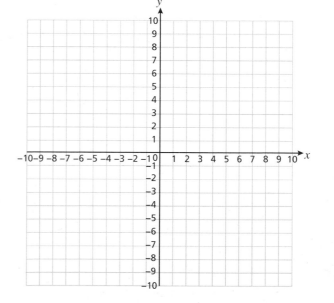

4 **a** Copy and complete a table of values for each of these lines.

x	−2	−1	0	1	2
y					

    i $y = 5x + 4$ **ii** $y = -2x + 4$

    iii $y = 3x + 4$ **iv** $y = 4 - 5x$

  b Draw each line on a copy of this coordinate grid.

    i $y = 5x + 4$ **ii** $y = -2x + 4$

    iii $y = 3x + 4$ **iv** $y = 4 - 5x$

  c What do you notice?

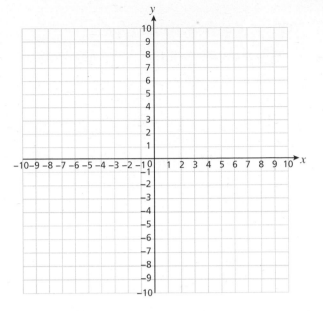

5 Here are the equations of four straight lines.

  A $y = 3x - 7$ B $y = -2x + 3$ C $y = 1 - x$ D $y = \frac{1}{2}x - 7$

  a Copy and complete a table of values for each line.

x	−2	−1	0	1	2
y					

  b On the same coordinate grid, draw the graphs of A, B, C and D

  c Identify the coordinates of two points that lie on exactly two of the lines. The coordinates should be whole numbers. On which lines do these points lie?

  d Identify the coordinates of one point that lies on exactly three of the lines. On which lines does this point lie?

What do you think? 🫧

1 Samira has drawn the graph of $y = 2x + 4$

  a Explain how you know that Samira must have made a mistake.

  b On a copy of the grid, draw the graph of $y = 2x + 4$

  c Which points has Samira plotted correctly?

  d Which points has Samira plotted incorrectly?

  e What mistake do you think she has made?

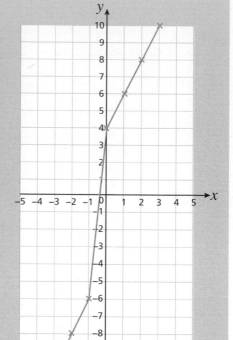

2 Here are three straight lines.

$y = 3 - 2x$ $\qquad\qquad$ $y = x - 2$ $\qquad\qquad$ $y = \dfrac{1}{2}x + 4$

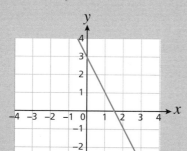 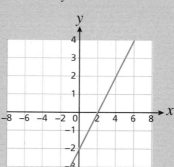 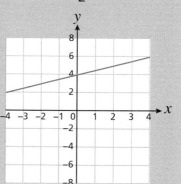

Copy and complete the table of values for each of the lines.

x	−2	−1	0	1	2
y					

You have looked at sequences in detail earlier in the curriculum and should be able to identify linear and non-linear sequences.

You will now link straight line graphs to **linear** sequences.

It is important that you understand "linear" in both of these contexts and can see connections between the two.

The coordinates of each point on a straight line relate to the position and term of a linear sequence.

Example 2

Here are the first four terms of an increasing linear pattern sequence.

Linear means that the numbers in the sequence increase or decrease by a constant amount.

a Complete the table.

Position	1	2	3	4
Term				

b Represent the sequence on a graph.

a

Position	1	2	3	4
Term	3	5	7	9

The first term has 3 counters, the second term has 5, the third has 7 and the fourth has 9

The sequence increases by 2 counters (or 2 circles) each time

b

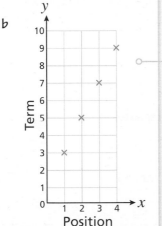

When representing a sequence graphically, you plot the position on the x-axis and the term on the y-axis.

Plot the points generated in the table in part **a**.

The points lie on a straight line. This is because the sequence is linear.

Do not draw a line drawn through these points – although there is a straight line that goes through them, the line would not represent the sequence because it only exists where the position is an integer.

Practice 4.5B

1 Here are the first four terms of an increasing linear sequence.

1st term   2nd term   3rd term   4th term

a Draw the next two terms in the sequence.

b Describe how the sequence is increasing.

c Copy and complete the table.

Position	1	2	3	4	5	6
Term						

d Plot the points from the table on a copy of these axes.

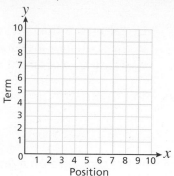

2 **a** Complete the table of values for $y = x + 1$

x	1	2	3	4	5	6
y						

b On a copy of the axes in question 1, draw the graph of $y = x + 1$

3 Compare your answers to questions 1 and 2

What's the same? What's different?

4 Here are the first four terms of a pattern sequence.

1st term   2nd term   3rd term   4th term

a Explain why, when plotted on a graph, the points of the sequence will lie on a straight line.

b The equation of the line that goes through points generated by this sequence is $y = 2x - 2$

How is the equation linked to the sequence?

5 Here are the first four terms of a pattern sequence.

1st term   2nd term   3rd term   4th term

a Explain why, when plotted on a graph, the points of the sequence will lie on a straight line.

b The equation of the line that goes through points generated by this sequence is $y = 2x + 1$

How is the equation linked to the sequence?

6 Here are the first four terms of a pattern sequence.

1st term   2nd term   3rd term   4th term

a Explain why, when plotted on a graph, the points of the sequence will lie on a straight line.

b The equation of the line that goes through points generated by this sequence is $y = 7 - x$

How is the equation linked to the sequence?

What do you think?

1 Here are the first four terms of a linear pattern sequence.

1st term   2nd term   3rd term   4th term

What is the equation of the straight line that would go through the points generated by the sequence?

2 The straight line $y = 3x + 4$ goes through the points generated by a sequence.

a Explain why the sequence must be linear.

b Is the sequence increasing or decreasing? How do you know?

c Draw what the sequence could look like.

3 Here is the graph of a sequence.

I can use the graph to find the 2.5th term of the sequence.

Explain why Flo is incorrect.

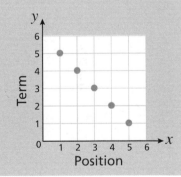

157

Consolidate – do you need more?

1 Copy and complete these function machines.

a Input → ×4 → −1 → Output
Input values: −2, −1, 0, 1, 2

b Input → ×−2 → +3 → Output
Input values: −2, −1, 0, 1, 2

c Input → ×2 → +7 → Output
Input values: −2, −1, 0, 1, 2

d Input → ×$\frac{1}{2}$ → +3 → Output
Input values: −2, −1, 0, 1, 2

2 Copy and complete a table of values for each of these lines.

x	−2	−1	0	1	2
y					

a $y = 4x - 1$ **b** $y = -2x + 3$ **c** $y = 2x + 7$ **d** $y = \frac{1}{2}x + 3$

3 Draw each line on a copy of the axes.

a $y = 4x - 1$ **b** $y = -2x + 3$ **c** $y = 2x + 7$ **d** $y = \frac{1}{2}x + 3$

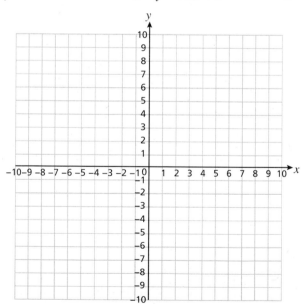

Stretch – can you deepen your learning?

1 Match each equation with the correct line.

$y = \frac{1}{2}x$ $y = 5x + 2$

$y = 7 - 3x$ $y = 5x + 7$

$y = -2x - 1$ $y = 5x - 4$

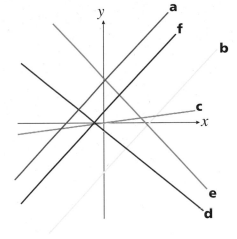

2 **a** Draw the graph of $y = 8x - 3$

   b Without calculation, explain why the point $(12, -15)$ does not lie on the graph.

3 The point $(0, p)$ lies on the line with equation $y = 7x - 1$

   Write down the value of p. Explain how you worked out your answer.

4 The points $(\frac{3}{5}, a)$ and $(b, 2.5)$ lie on the line $y = 12x + 7$

   a Work out the value of a **b** Work out the value of b

5 The point $(\frac{3}{8}, -2)$ lies on the line $y = 16x + w$

   Find the value of w

Reflect

1 In your own words, explain how to draw a graph of the form $y = mx + c$

2 Explain why you should always find at least three pairs of coordinates when plotting a line.

Small steps

■ Explore non-linear graphs Ⓗ

Key words

Non-linear – not forming a straight line

Curve – a line on a graph showing how one quantity varies with respect to another

Are you ready?

1 Decide whether each sequence is linear or non-linear.

Explain how you know.

    a 15, 19, 23, 27 …   **b** 5, 10, 20, 40 …   **c** 1, 4, 9, 16 …   **d** 10, 7.5, 5, 2.5 …

2 Find the next three terms in each sequence.

    a 15, 19, 23, 27 …   **b** 5, 10, 20, 40 …   **c** 1, 4, 9, 16 …   **d** 10, 7.5, 5, 2.5 …

3 Calculate

    a 5^2 **b** 3^3 **c** $(-2)^2$ **d** $(-1)^3$ **e** 15^2

4 Evaluate each expression when $t = 6$

    a t^2 **b** $\frac{t}{3}$ **c** $3t + 5$ **d** $\frac{600}{t}$ **e** $3t^2$

Models and representations

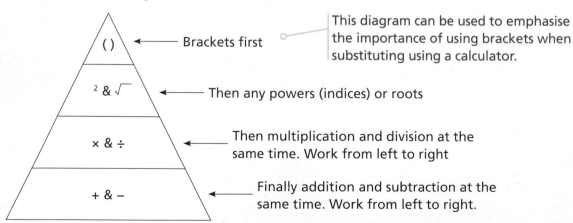

This diagram can be used to emphasise the importance of using brackets when substituting using a calculator.

( ) ← Brackets first

² & √ ← Then any powers (indices) or roots

× & ÷ ← Then multiplication and division at the same time. Work from left to right

+ & − ← Finally addition and subtraction at the same time. Work from left to right.

So far, all the graphs that you have looked at have formed straight lines (linear).

In this chapter you will look at **non-linear** graphs.

Non-linear graphs do not increase or decrease by the same amount for equal increases in x

They form smooth **curves** rather than straight lines.

This is similar to non-linear sequences, which do not increase/decrease by the same amount each time.

Example 1

a Complete the table of values for $y = 3x^2$ **b** Plot the graph of $y = 3x^2$

x	−2	−1	0	1	2
y					

The equation $y = 3x^2$ means that, at each point on the line, the y value is 3 times the x value squared.

a

square ↘
× 3 ↘

x	−2	−1	0	1	2
	4	1	0	1	4
y	12	3	0	3	12

square
× 3

Square each value of x and then multiply by 3

b

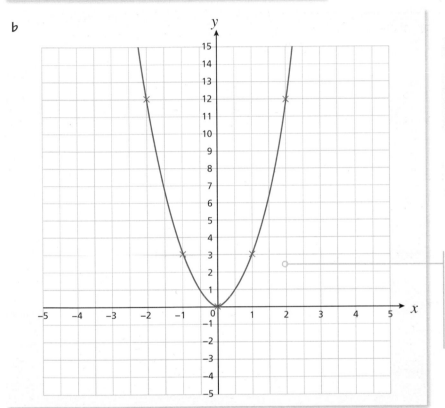

Unlike the graphs you drew in the previous chapters, the points do not lie in a straight line.

$y = 3x^2$ is a non-linear graph.

Instead of using a ruler to draw a straight line, you now need to draw a smooth curve through the points.

Practice 4.6A

1 **a** Copy and complete the table of values for $y = x$

x	−2	−1	0	1	2
y					

b Draw the graph of $y = x$

c How can you tell from the graph that $y = x$ is linear?

2 **a** Copy and complete the table of values for $y = x^2$

x	−2	−1	0	1	2
y					

b Draw the graph of $y = x^2$

c How can you tell from the graph that $y = x^2$ is non-linear?

3 **a** Copy and complete the table of values for $y = x^3$

x	−2	−1	0	1	2
y					

b Draw the graph of $y = x^3$

c Is $y = x^3$ linear or non-linear? How do you know?

4 **a** Copy and complete a table of values for each equation.

x	−2	−1	0	1	2
y					

  i $y = 2x^2$ **ii** $y = 5x^2$ **iii** $y = -x^2$ **iv** $y = -3x^2$

b Draw each of these graphs.

  i $y = 2x^2$ **ii** $y = 5x^2$ **iii** $y = -x^2$ **iv** $y = -3x^2$

c What do you notice?

5 **a** Copy and complete a table of values for equation.

x	−2	−1	0	1	2
y					

  i $y = x^2 + 3$ **ii** $y = 5 + x^2$

  iii $y = -x^2 + 4$ **iv** $y = -2 - x^2$

b Draw each of these graphs.

  i $y = x^2 + 3$ **ii** $y = 5 + x^2$

  iii $y = -x^2 + 4$ **iv** $y = -2 - x^2$

c What do you notice?

6 **a** Copy and complete a table of values for each equation.

x	−2	−1	0	1	2
y					

  i $y = 2x^3 + 3$ **ii** $y = x^3 + 4$

  iii $y = 0.5x^3$ **iv** $y = -6 + x^3$

b Draw each of these graphs.

  i $y = 2x^3 + 3$ **ii** $y = x^3 + 4$

  iii $y = 0.5x^3$ **iv** $y = -6 + x^3$

c What do you notice?

What do you think? 💭

1 Rhys has drawn the graph of $y = x^2$

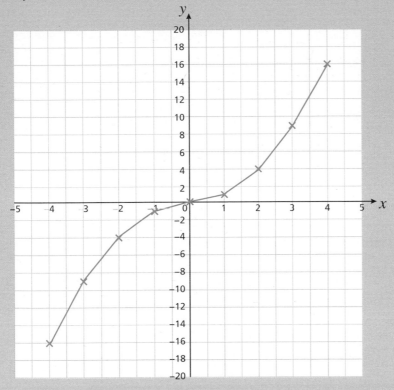

Identify two mistakes that Rhys has made.

2 Which of these will produce non-linear graphs? How do you know?

$$y = \frac{5x}{2} \qquad y = \frac{x^2}{2} \qquad y = 3x^8 - 10 \qquad y = \frac{15}{4x} \qquad y = 3x^2 + 4x - 2x^2 - x^2$$

3 Which of these graphs are non-linear? How do you know?

A   B   C   D   E

Consolidate – do you need more?

1 Complete these function machines.

a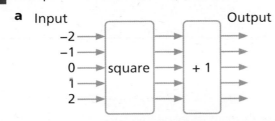

b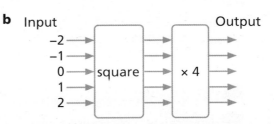

c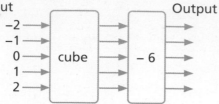

d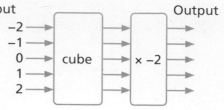

2 Copy and complete a table of values for each equation.

x	–2	–1	0	1	2
y					

    a $y = x^2 + 1$ **b** $y = 4x^2$ **c** $y = x^3 - 6$ **d** $y = -2x^3$

3 Plot each graph.

    a $y = x^2 + 1$ **b** $y = 4x^2$ **c** $y = x^3 - 6$ **d** $y = -2x^3$

Stretch – can you deepen your learning?

1 Match each graph with its correct equation.

A   B   C   D

    $y = x^2$ $y = 5x^2$ $y = 5 - x^2$ $y = -x^2$

How did you decide?

2 Match each graph with its correct equation.

A   B   C   D

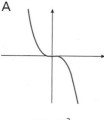   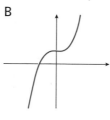   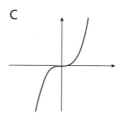   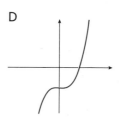

    $y = x^3$ $y = 5 + x^3$ $y = x^3 - 7$ $y = -x^3$

How did you decide?

3 The graph of $y = ax^2$ intersects the line $y = 75$ when $x = 5$

    Work out the value of a

Reflect

1 Explain what is meant by a non-linear graph.

2 How can you recognise a non-linear graph from its equation?

ⓗ 4.7 Midpoint of a line segment

Small steps

- Find the midpoint of a line segment ⓗ

Key words

Midpoint – the point halfway between two others

Line segment – part of a line that connects two points

Are you ready?

1 What number is each arrow pointing to?

a

16   18

b

16   19

c

16   20

d

16   40

e

11   25

f

23   38

g

99   135

h

−10   −6

2 Write the number that is exactly halfway between each pair.

a 3 and 5   **b** 7 and 10   **c** 10 and 20   **d** 35 and 41   **e** 42 and 45

f 54 and 60   **g** −2 and 0   **h** −4 and 4   **i** 30 and 0   **j** −15 and −8

3 Show using three different methods that 24.5 is halfway between 19 and 30

Models and representations

Number line

A number line can help to find the **midpoint** of two numbers, which could be the coordinates of the end points of a **line segment**.

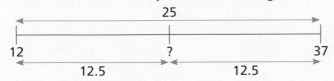

In this chapter, you will find the coordinates of the midpoint of a line segment.

Example 1

Find the midpoint of the line segment joining (–5, 2) and (7, –6)

Method A

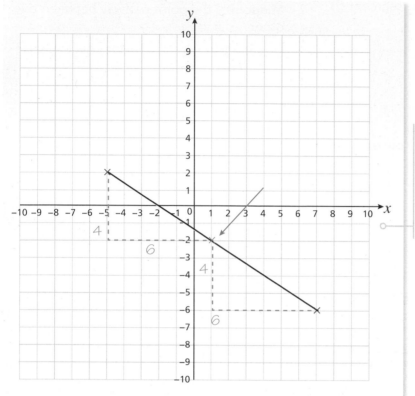

To get from each end point to the midpoint you must move half of the total distance between the points, so 6 units horizontally and 4 units vertically.

Moving 6 units horizontally and 4 units vertically from each end point will bring you to the same point on the line segment. This is the midpoint. It has coordinates (1, –2)

The coordinates of the midpoint of (–5, 2) and (7, –6) are (1, –2)

Method B

You can find the midpoint of the line segment by finding the midpoint of the x values and the midpoint of the y values.

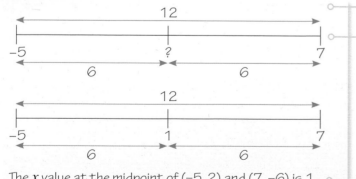

Start with the x values.

You can position –5 and 7 on a number line.

The total difference is 12, therefore half the difference must be 6

The x value at the midpoint of (–5, 2) and (7, –6) is 1.

The coordinates of the midpoint are of the form (1, __)

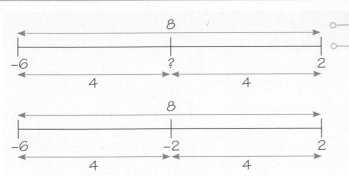

Now look at the y values.

Position –6 and 2 on a number line.

The total difference is 8, therefore half the difference must be 4

The y value of the midpoint of (–5, 2) and (7, –6) is –2

The coordinates of the midpoint are (1, –2)

Method C

$$\frac{-5+7}{2} = \frac{2}{2} = 1$$

The mean of –5 and 7 is 1

$$\frac{2+-6}{2} = \frac{-4}{2} = -2$$

The mean of 2 and –6 is –2

The coordinates of the midpoint of (–5, 2) and (7, –6) are (1, –2)

So the x-cordinate at the midpoint is 1

Practice 4.7A

1 Eight line segments are drawn on the grid.

For each line segment, write down the coordinates of

   i the end points

   ii the midpoint.

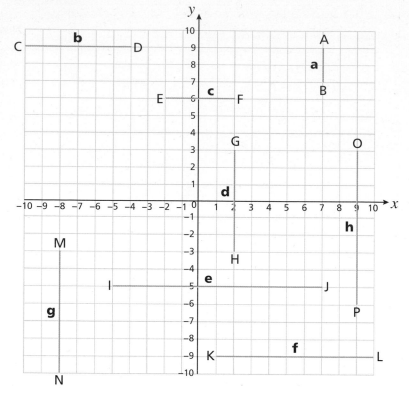

2 Find the coordinates of the midpoints of the line segments with these end points.

   a (7, 3) and (7, 9)

   b (–5, 2) and (–1, 2)

   c (10, 2) and (2, 2)

   d (–9, 7) and (–9, –1)

   e (3, –5) and (3, 5)

   f (2, 0) and (8, 0)

   g (6, –1) and (11, –1)

   h (–2, 5) and (–2, –4)

What do you notice?

3 Four line segments are drawn on the grid.

For each line segment, write down the coordinates of

   i the end points

   ii the midpoint.

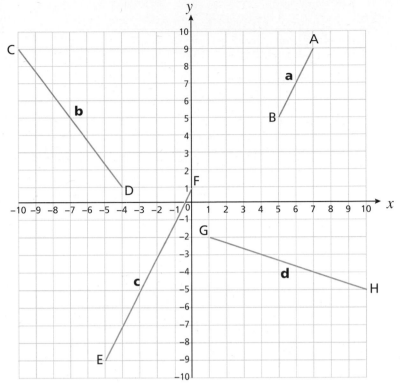

4 Find the midpoints of the line segments with these end points.

  a (2, 7) and (8, 9)   **b** (3, 5) and (−1, 1)   **c** (10, 9) and (4, 1)

  d (0, 5) and (8, 1)   **e** (−2, −6) and (1, −2)   **f** (0, 0) and (9, 3)

  g (−7, 9) and (−2, 4)   **h** (−9, −3) and (9, 6)

5 Find the midpoints of the line segments with these end points.

  a (7, 3) and (5, 5)   **b** (10, −4) and (7, −4)

  c (10, 42) and (20, 42)   **d** (15, 16) and (15, 18)

  e (16, 22) and (19, 22)   **f** (20, 16) and (16, 40)

  g (11, 25) and (23, 38)   **h** (−2, −15) and (0, −8)

  i (60, 20) and (54, 10)   **j** (0, 35) and (30, 41)

  k (99, −10) and (135, −6)   **l** (42, 60) and (45, 54)

6 Points A, B, C and D have coordinates (2, −9), (−5, −1), (4, 2) and (−7, −12), respectively.

Show that line segments AB and CD share a midpoint.

What do you think? 💭

1 Point A has coordinates (–6, 3)

Find the coordinates of point B if the midpoint of line segment AB is

   a (0, 3)   **b** (–6, 12)   **c** (–1, 5)   **d** (–20, 20)   **e** (–100, –100)

2 The midpoint of line segment XY has coordinates (17, –21)

Find three possible pairs of coordinates for points X and Y

Consolidate – do you need more?

1 By plotting them on a coordinate grid, find the midpoint of each pair of coordinates.

   a (0, 4) and (6, 4)   **b** (2, 5) and (2, 10)   **c** (6, –3) and (0, –3)

   d (–9, –4) and (–9, –8)   **e** (–7, –5) and (–7, –4)   **f** (5, 8) and (10, 8)

2 By plotting them on a coordinate grid, find the midpoint of each pair of coordinates.

   a (0, 4) and (6, 6)   **b** (2, 5) and (8, 10)   **c** (6, –3) and (0, –5)

   d (–9, –4) and (1, –8)   **e** (–7, –5) and (–4, –4)   **f** (5, 8) and (10, 0)

Stretch – can you deepen your learning?

1 Find the midpoint of each pair of coordinates.

   a (2.5, 3) and (4, –7)   **b** (6.25, –1) and (8, 2.4)

   c (–10, 0.2) and (3.8, 9.5)   **d** $(\frac{1}{2}, 0)$ and (5, 3)

   e $(-4, \frac{1}{4})$ and $(\frac{3}{10}, \frac{1}{2})$ **f** $(9.75, -\frac{3}{5})$ and $(\frac{27}{4}, -12.9)$

2 Find an expression for the midpoint of each pair of coordinates.

   a $(c + 8, d - 1)$ and $(c + 12, d - 5)$ **b** $(2c, 5d)$ and $(4c, 8d)$

   c $(2c + 8, 5d - 1)$ and $(4c + 12, 8d - 5)$ **d** $(9c + 7, 2 - d)$ and $(15 - 4c, 8d + 9)$

3 The coordinates of the end points of a line segment are (p, q) and (w, z)

Find an expression for the coordinates of the midpoint.

4 $a : b = 3 : 5$ $a + b = 24$

Find the midpoint of $(2a, 7b)$ and $(3b, -4a)$

Compare your method with a partner.

5 A square has vertices at (–7, 1), (–1, 7), (5, 1) and (–1, –5)

The midpoints of the edges of the square are joined to form another square.

   a Find the area of the new square.

   b What percentage of the original square does the new square cover?

Reflect

1 Explain why the midpoint of a horizontal line will have the same y value as the end points.

2 In your own words, explain how to find the midpoint of any pair of coordinates.

4 Working in the Cartesian plane

Chapters 4.1–4.7

I have become **fluent in…**	I have developed my **reasoning** skills by…	I have been **problem-solving** through…
■ working with coordinates in all four quadrants ■ finding coordinates of points that lie on a line ■ drawing straight line graphs ■ finding the coordinates of the midpoint of a line segment. Ⓗ	■ explaining the relationship between the x and y values at any point on a line ■ linking straight line graphs to arithmetic sequences ■ linking straight line graphs to direct proportion.	■ finding coordinates from given information ■ forming and solving equations to complete given coordinates ■ comparing gradients and y-intercepts to match the sketch of a line to its equation.

Check my understanding

1 Draw a coordinate grid going from –5 to 5 on both the x- and y-axes.

Plot each of these points.

A (3, 2)   B (2, –4)   C (–5, 1)   D (0, –3)   E (4, 0)

2 Draw a coordinate grid going from –5 to 5 on both the x- and y-axes.

Draw and label each of these lines on the grid.

a $x = 3$ **b** $y = –4$ **c** $x = –1$ **d** $y = x$ **e** $y = –x$

3 a Copy and complete a table of values for each of these lines.

x	–2	–1	0	1	2
y					

     i $y = 3x$ **ii** $y = –4x$ **iii** $y = x + 2$

   b Draw the graph of each line for values of x from –2 to 2

     i $y = 3x$ **ii** $y = –4x$ **iii** $y = x + 2$

4 a Copy and complete a table of values for each of these lines.

x	–2	–1	0	1	2
y					

     i $y = 3x + 1$ **ii** $y = 2 – 4x$ **iii** $y = 2x – 3$

   b Draw the graph of each line for values of x from –2 to 2.

     i $y = 3x + 1$ **ii** $y = 2 – 4x$ **iii** $y = 2x – 3$

5 Draw the graph of Ⓗ

   a $y = x^2$ **b** $y = x^3$

5 Representing data

In this block, I will learn...

how to draw and interpret scatter graphs

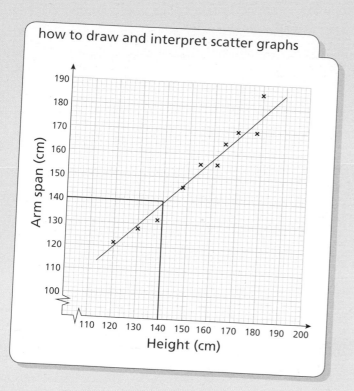

how to read and interpret frequency tables

Number of siblings	Frequency
0	5
1	3
2	4
3	7
4	1

Mass, m (kg)	Frequency
$30 < m \leqslant 40$	15
$40 < m \leqslant 50$	12
$50 < m \leqslant 60$	9
$60 < m \leqslant 70$	17

how to represent data in two-way tables

	Year 7	Year 8	Year 9	Total
hot lunch	12	24	22	58
packed lunch	18	8	6	32
Total	30	32	28	90

about different types of data

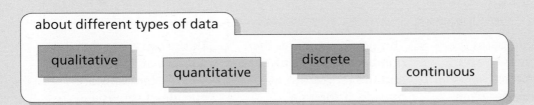

qualitative

quantitative

discrete

continuous

5.1 Scatter graphs

Small steps

- Draw and interpret scatter graphs
- Understand and describe linear correlation
- Draw and use lines of best fit
- Identify non-linear relationships

Key words

Linear – forming a straight line

Correlation – a connection between two or more things

Origin – the point where the x-axis and y-axis meet

Outlier – a value that differs significantly from the others in a data set

Are you ready?

1 Write the coordinates of each point marked on the grid.

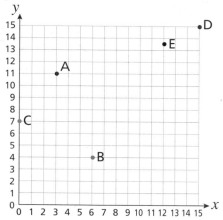

2 Draw a grid with x and y values from 0 to 10

Plot the following points on your grid.

A (3, 5) B (0, 8) C (4, 0) D (6.5, 9.5)

Models and representations

Scatter graph

Scatter graphs are a good way of displaying two sets of data to see if there is a correlation or connection.

This graph compares the height of a person and their IQ. It shows there is **no correlation** between the height of a person and their IQ.

This graph compares the number of ice creams sold and the temperature outside. It shows a **positive correlation**. The warmer it is, the more ice creams are sold.

This graph compares the number of scarves sold and the temperature outside. It shows a **negative correlation**. The warmer it is, the fewer scarves are sold.

Sometimes scatter graphs don't start at the **origin** so the graph is drawn using a **broken scale**.

A **broken scale** is used when values close to 0 are not required. In this case, you can start the x-axis at 8 and the y-axis at 15

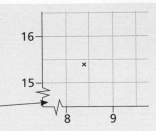

Example 1

Here is a scatter graph comparing temperature and the number of ice creams sold in a park.

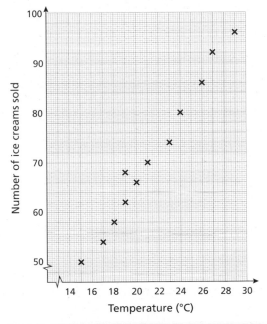

a Show this information on the graph.

Temperature (°C)	22	25
Ice creams sold	78	77

b Describe the relationship between the temperature and the number of ice creams sold.

a

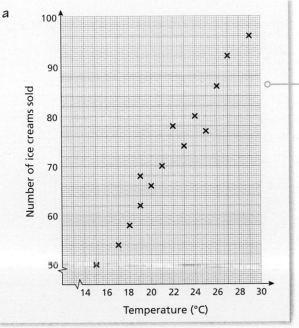

The temperature is on the horizontal axis and the number of ice creams is on the vertical axis so you can write the information in the table as the coordinates (22, 78) and (25, 77)

Always ensure that you check the scale carefully.

> **b** The higher the temperature the greater the number of ice creams sold.
>
> Or, as the temperature increases, the number of ice creams sold increases.

Often the relationship shown on a scatter graph is quite obvious. In this case, it makes sense that the warmer the weather, the more people buy ice creams.

There is usually more than one way of describing a relationship.

Practice 5.1A

1 The table shows the scores obtained by 8 people in two quizzes.

Quiz 1	6	18	16	12	14	8	10	8
Quiz 2	10	20	16	15	18	14	14	12

a On a copy of the grid, plot the information from the table.

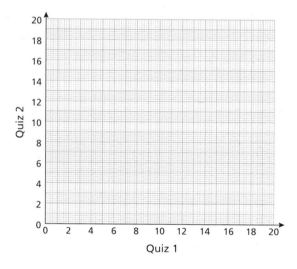

b Describe the relationship between the scores in quiz 1 and the scores in quiz 2

c What else can you say about quiz 1 and quiz 2?

2 The scatter graph shows the age of a car and its value.

a What is the value of the car when it is brand new?

b What is the value of the car when it is 4 years old?

c How old is the car when its value is £3000?

d Describe the relationship between the age of the car and its value.

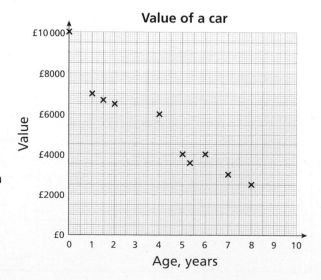

3 **a** Draw a scatter graph for each set of data.

i

Test 1	12	14	18	20	19	13	10	17	20	11
Test 2	16	15	15	20	18	11	12	18	18	11

ii

Temperature (°C)	28	25	12	17	3	30	6	10	22	15
Number of hot chocolates sold	1	5	25	20	50	0	28	35	18	30

iii

Height (cm)	140	145	152	163	160	165	169	172	175	180
Shoe size	2	3	4	5	6	7	8	9	10	11

b Describe the relationship shown by each scatter graph.

4 This scatter graph shows the house number and distance from the city centre of 12 houses.

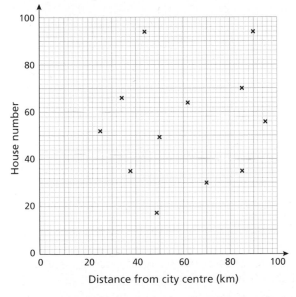

a Describe the relationship between the house number and the distance from the city centre.

b Think of two other things that would show this type of relationship.

What do you think?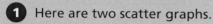

1 Here are two scatter graphs.

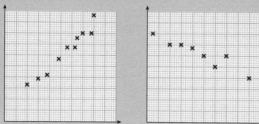

    a   Suggest what each scatter graph could represent.

         Talk about this with a partner.

    b   For each of your suggestions in part **a**, what would be the scale on the axes?

2 This scatter graph shows that students who performed well on a calculator mathematics paper also performed well on a non-calculator paper.

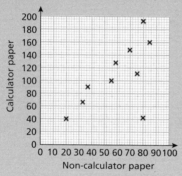

Which result does not follow the trend? Suggest a reason for this.

You will now learn how correlation can help you to describe the relationship between two sets of data. Two sets of data may have a positive correlation, a negative correlation or no correlation. If there is a correlation, then you can draw a line of best fit to highlight the correlation and help you to make predictions.

Example 2

The scatter graph shows the height and arm span of 10 people.

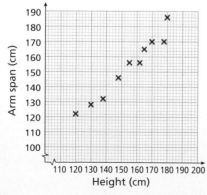

    a   Describe the correlation.

    b   Draw a line of best fit.

    c   Estimate the arm span of a person who is 145 cm tall.

a   Positive correlation

As the height increases, the arm span increases. Therefore this shows a positive correlation.

b

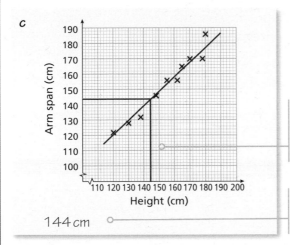

The line of best fit sits between the points. There should be approximately the same number of points on each side of the line.

The line of best fit does not need to meet the axes but instead "floats" on the graph in the appropriate place.

c

To estimate the arm span from your scatter graph, draw a line from 145 cm up to the line of best fit and then draw another line across to the arm span axis.

144 cm

This is only an estimate as the line of best fit could have been drawn in a slightly different place.

Practice 5.1B

1   Here are three graphs.

A   B   C

Which graph shows

a   a positive correlation

b   a negative correlation

c   no correlation?

2 What type of correlation do you think the following sets of data would show?

a Temperature and number of scarves sold.

b Maths test result and the amount of time spent revising.

c Height and shoe size.

d House number and shoe size.

3 Huda, Filipo and Lydia have all drawn a line of best fit.

Huda   Filipo   Lydia

Who has drawn their line correctly? Explain your answer.

4 13 people take part in a talent competition.

There are two judges, who each give the contestants a score out of 10

The scatter graph shows the results.

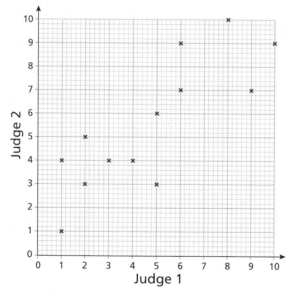

a What type of correlation does the scatter graph show?

b On a copy of the graph, draw a line of best fit.

c Another contestant gets a score of 8 from judge 1. Estimate the score that judge 2 will give them.

5 The scatter graph shows the cost of rent per month for office space and the distance from the city centre.

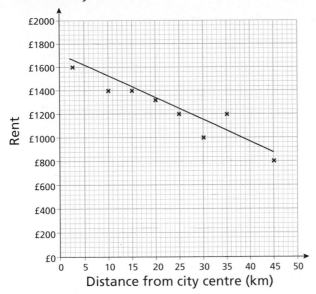

Rent

Distance from city centre (km)

a What type of correlation does the scatter graph show?

b Estimate the distance from the city centre for an office space that costs £1000 per month.

 c If the graph was extended to include a distance of up to 100 km from the city centre, do you think the trend shown by the line of best fit would continue? Explain your answer.

What do you think?

1 The scatter graph shows the mass and height of nine students.

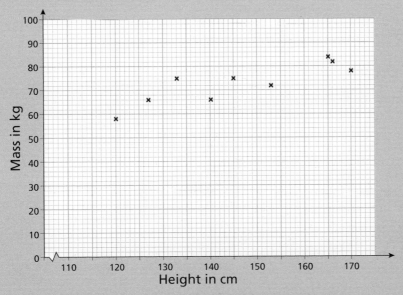

Mass in kg

Height in cm

a Estimate the mass of a student who is 158 cm tall.

b A lift has a maximum capacity of 650 kg. Can all nine students go in the lift at the same time?

c Find the mean, median and range of the heights of the nine students.

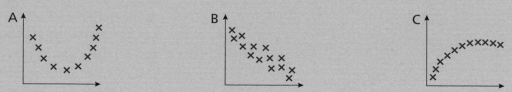

2 **a** Which of these graphs show a non-linear relationship? How do you know?

b Draw another example of a graph that shows a non-linear relationship.

Consolidate – do you need more?

1 **a** On a copy of the grid, show the information in the table.

Length of a carrot (cm)	8	12	14.5	11	13	13.5	10	10.5	14	15
Mass of a carrot (g)	15.5	19	21	17.5	19	20	17	16.5	20.5	19

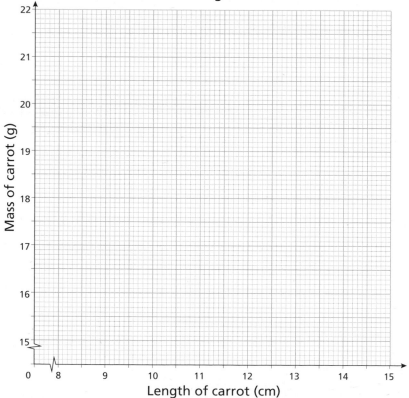

Mass and length of carrots

b Describe the relationship between the length of a carrot and the mass of a carrot.

c What type of correlation does the scatter graph show?

d Why might a longer carrot have a lower mass than a shorter carrot?

2 Here are some scatter graphs.

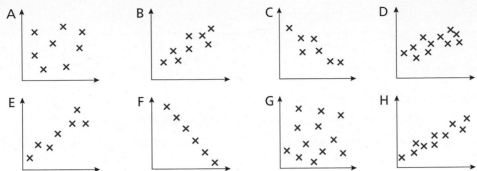

a Write the letters of all the graphs that show

   i a positive correlation

   ii a negative correlation

   iii no correlation.

b Which graph shows the strongest negative correlation?

c Which graph shows the weakest positive correlation?

d Work with a partner to suggest a context for each graph.

Stretch – can you deepen your learning?

1 The scatter graph shows the cost of a horse-riding magazine over the last few years.

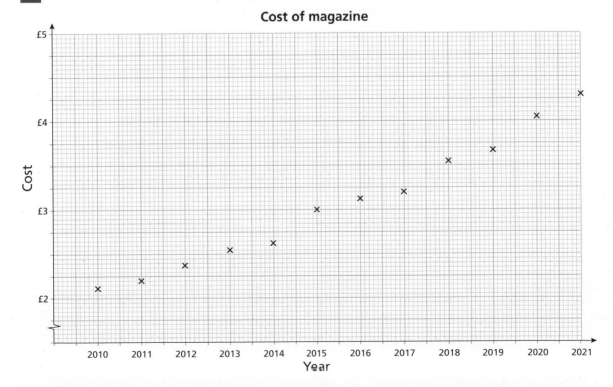

Assuming that the trend continues, estimate the cost of the horse-riding magazine in 2025

2 15 students sat an end-of-year exam.

Their results, along with their attendance, are shown in the table.

Attendance (%)	100	98	98	97	96	96	94	93	92	92	91	90	88	87	85
Exam score (%)	75	82	81	77	72	65	15	78	77	64	62	67	55	62	60

a Plot these points on a scatter graph.

b Which result do you think is an outlier? Explain your reason.

c Another student attended 95% of lessons. What is their percentage score likely to be? How confident are you about your answer?

d Another student only attended 40% of lessons. Explain why the scatter diagram should not be used to estimate their result.

e The school wants to use the results as evidence to show that "Students who attend school do better in their end-of-year exams." Do the results in the scatter graph support this?

Reflect

1 Explain why a scatter graph can be a useful way to display data.

2 Explain why a scatter graph sometimes isn't a useful way to display data.

Small steps

■ Identify different types of data

■ Read and interpret ungrouped frequency tables

Key words

Discrete – data that can only take certain values

Continuous – data that is measured

Qualitative – data that describes characteristics

Quantitative – numerical data

Frequency – the number of times something happens

Are you ready?

1 Copy and complete the tally chart showing people's favourite animals.

Animal	Tally	Frequency
dog	ⵜⵜ ⵜⵜ ////	
cat	ⵜⵜ ////	
horse		11
other		

2 Here are the ages of 20 teenagers at a youth club.

14	12	14	15	12	13	12	14	15	14
13	13	14	12	14	12	15	14	15	13

Arrange the data in a tally chart.

Models and representations

Charts and graphs

Line graph

Pie chart

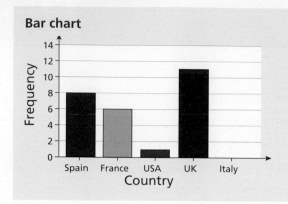

Bar chart

Frequency table

Shoe size	Frequency
4	12
5	14
6	7

In this chapter, you will learn about different types of data and how to interpret **frequency** tables. Data is regularly used in everyday life. You will find that information is frequently organised in tables or using a graph to make it easier to read and interpret.

Example 1

a Write down an example of

   i discrete data   **ii** continuous data

   iii qualitative data   **iv** quantitative data

b Here is a list of some people's eye colour.

blue, blue, blue, green, brown, brown, blue, brown, blue, brown, brown, green, brown, brown

   i Arrange the data into a frequency table.

   ii How many people were asked altogether?

a i The number of bedrooms in a house.

> **Discrete** data can be counted. To find out how many bedrooms there are in a house you count them.

ii The mass of a dog.

> **Continuous** data can take any value; therefore it is a measure. The mass of a dog could be 16 kg, for example. The dog may actually weigh 15.834 662 kg but often the figure is rounded to make it more readable.

iii Your favourite colour.

> **Qualitative** data describes characteristics so the answer will be a word.

iv The number of students in a school.

> **Quantitative** data can be discrete or continuous. The results will be numerical.

b i

Eye colour	Frequency
blue	5
brown	7
green	2

> The categories in the table need to represent the results of the data. In this case the categories are "blue", "brown" and "green". Count the number of times, or frequency, of each category and record it in the frequency column.

ii 14 people

> Add up the total of the frequency column to find out how many people are represented.

Example 2

The table shows the numbers of goals scored by a football team in 25 games.

Number of goals	Frequency
0	4
1	11
2	7
3	1
4	2

How many goals were scored in total?

36 goals

A common mistake here is to just add up the frequency column. You want to know the number of goals scored, not the number of games.

Number of goals	Frequency		Total number of goals
0	4	In 4 games there were no goals	0
1	11	In 11 games there was 1 goal	11
2	7	In 7 games there were 2 goals	14
3	1	In 1 game there were 3 goals	3
4	2	In 2 games there were 4 goals	8

$0 + 11 + 14 + 3 + 8 = 36$

Practice 5.2A

1 For each of the following, state whether it is discrete or continuous data.

  a The number of people on a bus   **b** The number of pages in a book

  c The time taken to watch a film   **d** The distance between two towns

  e The size of an angle   **f** The size of a shoe

  g The length of a foot   **h** The mass of a parcel

2 Write down two more examples of discrete data.

3 Write down two more examples of continuous data.

4 Faith and Chloe are talking about the difference between qualitative data and quantitative data.

Qualitative data describes characteristics. It reminds me of describing the qualities of a person.

Faith

Quantitative data reminds me of the word quantity so it helps me to remember that it is about numerical values.

Chloe

For each of the following, state whether the data is qualitative or quantitative.

a Age of a pet

b Favourite ice cream flavour

c Eye colour

d Volume of liquid in a glass

e Type of car

f The length of a plank of wood

5 State the type of data that each chart represents.

a

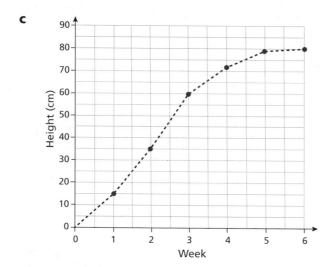

b

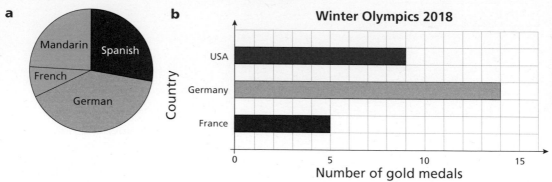

c

6 Darius records the colours of the cars that pass his house in 30 minutes.

Car colour	Frequency
red	9
silver	15
white	28
black	31
other	7

a How many cars passed his house in total?

b What colour car is the most popular?

c Why do you think Darius has included an "other" row?

d Darius says, "I have collected quantitative data because I have used numbers."

Do you agree with Darius?

e How many fewer silver cars passed Darius's house than black cars?

f What type of chart could Darius use to display his data?

7 Charlie collects data on the shoe size of his classmates.

5	5	4	5	3	3	2	5	6	6
4	6	6	6	6	3	3	8	6	6
4	4	5	3	5	5	5	5	4	5

a Create a frequency table to display the data.

b Compare your table with a partner's. Is it the same or different?

8 The table shows the number of passengers in each car entering a car park.

Number of passengers	Frequency
1	7
2	21
3	11
4	5

Ed says, "There are 44 passengers in total."

a Explain why Ed is wrong.

b Work out the total number of passengers in the cars.

Explain your method to a partner.

9 A class records how many pens each student has.

Number of pens	Frequency
0	1
1	4
2	7
3	12
4	0
5	3

a How many pens do the class have in total?

b How many students do not have a pen?

What do you think? 🗨

1 Here are some examples of data.

Age

Surname

Number of text messages received yesterday

Height

Method of travel to school

Number of days absent

a Sort the data into two groups.

b Describe your two groups.

c Compare your groups with a partner.

2 Mario makes a list of words. He counts the number of letters in each word. The table shows the results.

Number of letters in word	Number of words
3	23
4	25
5	34
6	40
7	28
8	16
9	4

Which of the following pieces of information can you work out from the data?

The length of the longest word.	The longest word.	The number of words that Mario counts.

The total number of letter Es in the words.	The most likely length of a word.	The total number of letters counted.

The number of words that had more than six letters.

Consolidate – do you need more?

1 Copy and complete the following sentences.

    a The number of students in a class is discrete data because…

    b The height of students in a class is continuous data because…

    c The eye colour of students in a class is qualitative data because…

    d The number of pens students in a class have is quantitative data because…

2 The table shows data about some people's favourite fruit.

Type of fruit	Frequency
apple	9
banana	12
strawberries	18
grapes	15
other	4

    a How many people chose banana?

    b What is the most popular type of fruit?

    c How many more people chose grapes than chose apple?

    d How many people were asked in total?

3 Beth and Sven each collect some data about how many pets are owned by people in their classes.

Here are their results.

Beth's results

Number of pets	Frequency
0	3
1	19
2	1
3	7

Sven's results

Number of pets	Frequency
0	10
1	13
2	5
3	2

    a How many students are in each class?

    b Show that Sven's class has 29 pets in total.

    c Whose class has more pets? Show working to support your answer.

Stretch – can you deepen your learning?

1 Zach says, "Age is an example of discrete data."

Ed says, "Age is continuous data."

Suggest why each person thinks what they do.

2 A football team plays 20 games. The team scores 31 goals in total.

Copy and complete the table.

Number of goals scored	Number of games
0	
1	7
2	
3	4

3 Faith is doing a charity swim after school each day for one week.

She wants to swim 5 km altogether. Each length is 25 metres long.

The table shows the number of lengths she swims on Monday to Thursday.

Monday	Tuesday	Wednesday	Thursday	Friday
32 lengths	45 lengths	58 lengths	41 lengths	

a How many lengths does Faith need to swim on Friday to meet her target?

b Find three ways of working out the answer.

Reflect

Explain why the method for finding the total number of goals would be different for each table.

Number of goals scored	Frequency
1	4
2	6
3	5

Player	Number of goals scored
Ryan	4
Jenny	6
Saif	5

5.3 Grouping data

Small steps

- Read and interpret grouped frequency tables
- Represent continuous data grouped into equal classes

Key words

Grouped data – data that has been ordered and sorted into groups called classes

Class interval – the range of data in each group

Are you ready?

1 Give an example of continuous data.

2 Give an example of discrete data.

3 Work out the range of this data.

12 cm, 29 cm, 13 cm, 22 cm, 25 cm

4 Write the meaning of each inequality in words.

    a $t > 4$ **b** $h < 163$ **c** $15.1 \geqslant w$ **d** $a \leqslant 0.2$

5 Use h for height and suitable inequality symbols to show that someone's height is less than 158 cm but greater than 151 cm.

Models and representations

Grouped frequency tables

Grouped discrete data

House number	Number of houses
0 to 9	3
10 to 19	8
20 to 29	8
30 to 39	9

Grouped continuous data

Time taken, t (hours)	Frequency
$2 < t \leqslant 4$	15
$4 < t \leqslant 6$	26
$6 < t \leqslant 8$	38
$8 < t \leqslant 10$	40

In this chapter, you will explore grouped frequency tables using both discrete data and continuous data. It is important to understand when data needs to be grouped and when it can be left ungrouped.

Before you start, make sure that you know the meaning of the inequality symbols as this will help you with some of the questions in the chapter. You will learn more about inequalities in Block 7

Example 1

Here is some data about the ages of people who live on a street.

11, 54, 38, 28, 33, 17, 15, 41, 37, 35, 50, 21, 49, 44, 52

a Why would an ungrouped frequency table be unsuitable for this data?

b Put the data into a grouped frequency table.

a All of the values are different. You would end up with 15 rows with a frequency of 1 in each row.

b

Age	Frequency
10–25	4
26–40	5
41–55	6

Before you group data, it helps to put it in order.

11, 15, 17, 21, 28, 33, 35, 37, 38, 41, 44, 49, 50, 52, 54

You can find the range of the data by working out

54 − 11 = 43

range = largest data value − smallest data value

Knowing the range can help you to group the data.

Groups could be made in a variety of ways. Here it makes sense to have three equal groups with a range of 15 in each group.

Practice 5.3A

1 Which of these are grouped frequency tables?

Table A

Number of people in house	Number of houses
1	4
2	12
3	9
4	3

Table B

Number of cats	Frequency
0–2	182
3–5	43
6–10	70
10+	29

Table C

Number of cakes sold	0 to 5	6 to 10	11 to 15	16 to 20	21 to 25
Number of days	4	7	8	7	5

Table D

Number of siblings	0	1	2	3	4
Frequency	5	3	4	7	1

2 For each set of data, decide whether the values should be put into a frequency table or a grouped frequency table.

    a   1, 1, 1, 1, 1, 1, 1, 2, 2, 2, 2, 3, 3, 3, 3, 3, 3, 3

    b   red, green, yellow, yellow, green, red, green, yellow

    c   1, 2, 3, 4, 5, 6, 7, 8, 9, 10, 11, 12, 13, 14, 15, 16, 17, 18

    d   18, 52, 43, 55, 24, 38, 32, 41, 16, 51, 29, 35, 30, 20, 45, 54, 19, 28, 44, 40

Explain your reasons.

3 Here are the heights in mm of 20 plants.

| 880 | 875 | 909 | 773 | 892 | 900 | 903 | 850 | 723 | 921 |
| 750 | 736 | 985 | 960 | 805 | 732 | 849 | 810 | 790 | 775 |

Copy and complete the grouped frequency table for the heights of the plants.

Height	Frequency
700 to 799	7
800 to 899	

4 The table shows the numbers of laps completed by people who attended an ice-skating rink last Sunday.

Number of laps	0–19	20–39	40–59	60–79	80–100
Number of people	17	48	21	36	23

    a   How many people completed 60–79 laps?

    b   How many people attended the ice-skating rink in total?

    c   How many people completed fewer than 40 laps?

    d   Why is it impossible to tell how many people completed more than 90 laps?

What do you think? 💭

1 Here are the ages of some visitors to a coffee shop.

23	32	29	18	16	45	36	42	43	59
29	38	72	19	26	28	32	32	28	36
55	45	30	26	17	22	22	34	56	65
27	39	18	40	20	33	71	65	25	20

a Design your own table to summarise the data.

b Compare your group intervals with those of others in your class.

What's the same and what's different? Why did you choose your intervals?

c Write down three facts about the data.

Was it easier to get this information from the table or the original list?

So far in this chapter, you have been working with discrete data. You will now begin to group continuous data. Continuous data is data that can take any value. Examples are mass, height, distance and time. You will use inequality symbols to group the data into **class interval**s.

Example 2

Here are the masses, in kilograms, of 24 horses.

423	525	496	500	620	415
561	584	510	508	470	611
401	498	553	569	519	600
522	570	587	576	635	602

a Put the data into a grouped frequency table.

b In which class interval is the horse with a mass of 500 kg?

c How many horses have a mass greater than 550 kg?

a

Mass, m (kg)	Frequency
$400 < m \leqslant 450$	3
$450 < m \leqslant 500$	4
$500 < m \leqslant 550$	5
$550 < m \leqslant 600$	8
$600 < m \leqslant 650$	4

The smallest mass is 401 kg and the greatest mass is 635 kg: therefore, a sensible range is from 400 kg to 650 kg.

The first class interval, $400 < m \leqslant 450$, means that the mass is greater than 400 kg but less than or equal to 450 kg.

b $450 < m \leqslant 500$

There are two class intervals that use the number 500. This interval contains masses greater than 450 kg but less than or **equal to** 500 kg, so a horse with a mass of 500 kg is included in this class interval.

c   12

There are 8 horses with a mass greater than 550 kg but less than or equal to 600 kg and 4 horses with a mass greater than 600 kg but less than or equal to 650 kg. Therefore, in total, 12 horses have a mass greater than 550 kg.

Practice 5.3B

1 Which statement matches each inequality?

a $10 < t \leqslant 60$

A t is greater than 10 and less than 60

B t is less than 10 and greater than or equal to 60

C t is greater than 10 and less than or equal to 60

b $7.5 \leqslant l < 10$

A l is greater than or equal to 7.5 and less than 10

B l is greater than 7.5 and less than or equal to 10

C l is less than or equal to 7.5 and greater than 10

2 Junaid has made a table to record the heights of people in his class.

Height, h (cm)	Frequency
140 to 145	4
145 to 150	7
150 to 155	15
155 to 160	2

a Explain why Junaid's table would be difficult to use.

b Suggest a more suitable table that Junaid could use.

c Why is it not possible for you to complete your table with Junaid's frequencies?

3 The table shows the times taken, in minutes, to complete a puzzle.

Time taken, t (minutes)	Frequency
$2 < t \leqslant 4$	16
$4 < t \leqslant 6$	26
$6 < t \leqslant 8$	38
$8 < t \leqslant 10$	40

a Which class interval has the lowest frequency?

b How many people took longer than 8 minutes to complete the puzzle?

c How many people in total completed the puzzle?

d What fraction of the people took longer than 6 minutes to complete the puzzle?

4 Here are the heights of some plants. The heights are in centimetres to 1 decimal place.

15.6	25.4	13.9	12.8	31.1
8.6	21.6	20.8	30.0	22.0
47.1	43.5	27.6	9.5	28.0
20.0	47.9	32.5	4.3	17.0

a Copy and complete the grouped frequency table.

Height, h (cm)	Frequency
$0 < h \leq 10$	
$10 < h \leq 20$	

b Which class interval has the highest frequency?

c What percentage of the plants are less than or equal to 10 cm tall?

What do you think?

1 The table shows the times it takes 200 children to run a 100-metre race.

Time taken (seconds)	Frequency
$10 < t \leq 15$	37
$15 < t \leq 20$	85
$20 < t \leq 25$	46
$25 < t \leq 30$	32

a Why can't you work out exactly how many children took longer than 18 seconds?

b Estimate how many children took longer than 18 seconds to complete the race.

2 Flo wants to measure the heights of some plants.

She makes the following table to record the data.

Height (cm)	Tally	Frequency
0–10		
11–20		
21–30		
31–40		

Comment on the table that Flo is planning to use. Would you improve it in any way?

Consolidate – do you need more?

1 The table shows the numbers of achievement points earned by students in Year 8.

Number of achievement points	0–30	31–60	61–90	91–120	121–150
Number of students	10	45	65	110	35

    a How many students have between 61 and 90 achievement points?

    b How many students are in Year 8?

    c Another student joins Year 8 and quickly earns 30 achievement points. In which group interval should the number of students change?

2 Match each statement to the correct inequality.

    a

      t is greater than or equal to 10 and less than 20 $10 < t \leqslant 20$

    b

      t is greater than 10 and less than or equal to 20 $10 \leqslant t < 20$

    c

      t is greater than 10 and less than 20 $10 < t < 20$

3 Write down an inequality to represent the statement: "h is greater than 5 and less than or equal to 15".

4 Here are the masses, in kilograms, of 20 parcels.

    55   61   35   70   50   42   39   64   53   47

    40   31   52   67   49   38   61   60   55   34

    Copy and complete the table to show the data.

Mass, m (kg)	Frequency
$30 < m \leqslant 40$	
$40 < m \leqslant 50$	
$50 < m \leqslant 60$	
$60 < m \leqslant 70$	

Stretch – can you deepen your learning?

1 80 people are asked how many emails they received yesterday.

- 30 people said they received 10 or fewer emails.
- 45 people received 20 or fewer emails
- 26 people received more than 30 emails
- 2 people received more than 40 emails

Use the data to complete this table.

Number of emails received	Frequency
$0 < x \leqslant 10$	
$10 < x \leqslant 20$	
$20 < x \leqslant 30$	
$30 < x \leqslant 40$	
$40 < x \leqslant 50$	

2 What's the same and what's different about these two frequency tables?

Mass, m (kg)	Frequency
0–10	
11–20	
21–30	
31–40	

Mass, m (kg)	Frequency
$0 < m \leqslant 10$	
$10 < m \leqslant 20$	
$20 < m \leqslant 30$	
$30 < m \leqslant 40$	

Which table should be used to record mass? Why?

Reflect

What's the same and what's different about organising data in grouped and ungrouped tables?

5.4 Two-way tables

Small steps

- Represent data in two-way tables

Key words

Two-way table – this displays two sets of data in rows and columns

Are you ready?

1 Work out

  a 17 + 15 + 13   **b** 234 + 156   **c** 59 + 103

2 Work out

  a 257 – 134   **b** 192 – 147   **c** 203 – 89

3 Some students were asked how they travel to school.

Mode of travel	Frequency
walk	87
car	123
bus	250
other	5

  a How many students travel by bus?

  b How many students either walk or travel by car?

  c How many students were asked altogether?

Models and representations

Two-way table

A two-way table has two categories.

In this case, they are whether you own a pet and year group.

	Owns a pet	Does not own a pet	Total
Year 7			
Year 8			
Total			

In this chapter, you will extend your knowledge of representing data in tables. As well as interpreting completed tables you will use your addition and subtraction skills to find missing information

Example 1

The two-way table shows the number of students in Years 7, 8 and 9 who have hot lunches or packed lunches.

Complete the two-way table.

	Year 7	Year 8	Year 9	Total
Hot lunch	12			58
Packed lunch		8		
Total	30	32		90

i

	Year 7	Year 8	Year 9	Total
Hot lunch	12	24		58
Packed lunch		8		
Total	30	32		90

There are usually several places to start with questions like this.

To work out the number of Year 8 students who have a hot lunch, subtract the number who have a packed lunch from the total number of Year 8 students:

$32 - 8 = 24$

ii

	Year 7	Year 8	Year 9	Total
Hot lunch	12	24		58
Packed lunch	18	8		
Total	30	32		90

To work out the number of Year 7 students who have a packed lunch calculate $30 - 12 = 18$

iii

	Year 7	Year 8	Year 9	Total
Hot lunch	12	24		58
Packed lunch	18	8		
Total	30	32	28	90

You can work out the Year 9 total by adding together the Year 7 and Year 8 totals and subtracting the answer from the total number of students in all three year groups.

$30 + 32 = 62$

$90 - 62 = 28$

iv

	Year 7	Year 8	Year 9	Total
Hot lunch	12	24	22	58
Packed lunch	18	8	6	32
Total	30	32	28	90

Once you have completed the table, check that all the rows and columns add up.

Practice 5.4A

1 The two-way table shows information about the colours of two types of shapes.

	Trapezium	Parallelogram	Total
Blue	3	7	10
Purple	8	2	10
Total	11	9	20

  a How many trapezia are there in total?

  b How many shapes are there in total?

  c How many blue parallelograms are there?

  d How many purple shapes are there in total?

2 A class has been rewarded with a trip and are deciding where to go.

  7 girls and 10 boys want to watch a film.

  9 girls and 4 boys want to go bowling.

  a Copy the two-way table and use this information to complete it.

	Girls	Boys	Total
Film			
Bowling			
Total			

  b How many girls are in the class?

  c How many students in total are in the class?

3 Copy and complete each two-way table.

a

	Year 7	Year 8	Total
Left-handed		13	19
Right-handed	42	39	
Total	48		100

b

	Football	Hockey	Netball	Total
Boys	7		6	24
Girls		3		
Total	12			40

c

	Maths	English	Total
Year 7	81	19	
Year 8			58
Year 9	67		
Total	175	125	

4 The table shows information about 50 people at a restaurant.

	Wear glasses	Do not wear glasses	Total
Males	15	8	
Females	9		
Total			

a Copy and complete the table.

b What fraction of the males wear glasses?

c What percentage of the people at the restaurant wear glasses?

5 60 people visited an ice rink one day.

13 out of the 19 people who wore a scarf were adults. There were 15 children.

a Draw a two-way table to show this information.

b How many adults did not wear a scarf?

c What fraction of the children wore a scarf?

What do you think?

1 The table shows information about the numbers of adults and children on a beach on a Saturday morning and a Sunday morning.

	Adults	Children
Saturday morning	73	112
Sunday morning	55	78

Which of the following pieces of information can you work out from the table?

The number of adults on the beach on Saturday morning

The total number of children on the beach on both days

The number of adults that were on the beach on both Saturday morning and Sunday morning

How many more children were on the beach on Saturday morning compared with Sunday morning

The number of adults on the beach at 9 a.m.

The number of children on the beach on Saturday afternoon

Explain your answers.

2 A new game is reviewed by some boys and girls.

Here is a summary of the ratings.

	Boys	Girls
5 stars	121	59
4 stars	218	27
3 stars	192	10
2 stars	87	3
1 star	26	1

It's clear that boys like this game more than girls do.

Zach

I'm not sure that you're right. I think girls like it more than boys do.

Beca

Why do you think each of them made their claim?

Who do you agree with?

3 Can the information in a two-way table always, sometimes or never be represented as a frequency tree instead? Discuss with a partner.

Consolidate – do you need more?

1 Here are some red and blue circles and squares.

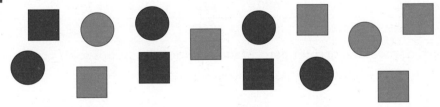

a Copy and complete the two-way table to show the information about the circles and squares.

	Circle	Square	Total
Red			
Blue		5	
Total	6		

b How many shapes are there in total?

c What fraction of the shapes are circles?

d What fraction of the squares are red?

2 The two-way table shows some information about the favourite sports of some Year 9 students. The table is incomplete.

	Rugby	Football	Netball	Badminton	Other	Total
Girls			19		49	190
Boys		41	11	35		
Total		91		76	74	

a Explain why the number of girls who chose football can be calculated using 91 – 41

b Explain why the total number of students who chose netball can be calculated using 19 + 11

c There are 350 students in Year 9

Copy and complete the two-way table.

Stretch – can you deepen your learning?

1 120 children are asked how they travel to school.

- ■ $\frac{2}{3}$ of the those asked are boys.
- ■ 26 girls walk to school.
- ■ Twice as many boys don't walk to school as girls.

Use a two-way table to work out how many boys walk to school.

2 The table shows the number of scoops of different flavours of ice cream sold on Monday and Tuesday.

	Salted caramel	Mint choc chip	Chocolate	Vanilla	Strawberry	Total
Monday	135	218	415	395	128	1291
Tuesday	306	158	480	420	183	1547
Total	441	376	895	815	311	2838

a On which day was the greater percentage of chocolate scoops sold?

The shops sells the ice creams in three different sizes:

- ■ single scoop
- ■ double scoops
- ■ triple scoops

The table shows many ice creams of each size were sold.

	Single scoop	Double scoop	Triple scoop	Total
Monday	313		118	
Tuesday			128	905
Total		698		

b Complete the table.

3 In a golf club, two-thirds of the players are over 60

■ There are 120 females under 60

■ $\frac{2}{5}$ of the over 60s are female.

■ There are 300 females in the club.

How many people are members of the golf club?

Reflect

1 Explain why you can't always start in the same place when you are completing a two-way table.

2 What strategy can you use to check that you have completed a two-way table correctly?

5 Representing data
Chapters 5.1–5.4

White Rose Maths

I have become **fluent** in…	I have developed my **reasoning** skills by…	I have been **problem-solving** through…
■ drawing and interpreting scatter graphs ■ identifying different types of data ■ working with grouped and ungrouped data ■ completing two-way tables.	■ describing relationships and trends in data ■ using graphs to make estimations and predictions ■ making decisions about how to group data ■ considering reasons for outliers.	■ finding an appropriate way to display data given in words ■ working backwards to extract information from a graph into a table ■ using prior knowledge to solve multi-step problems.

Check my understanding

1 What type of correlation is shown by each graph?

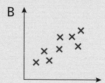

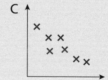

2 Write down an example of

   a qualitative data   **b** quantitative data

   c discrete data   **d** continuous data

3 The table shows the numbers of televisions in 12 homes.

Number of televisions	0	1	2	3
Frequency	1	5	4	2

Find the total number of televisions in these 12 homes.

4 Explain why you would use a grouped frequency table for continuous data.

5 There are 600 students in Years 7, 8 and 9 at a school. Complete the two-way table.

	Year 7	Year 8	Year 9	Total
Boys		95		
Girls			92	307
Total	210		195	

6 Tables and probability

In this block, I will learn...

how to construct a sample space diagram for one or more events

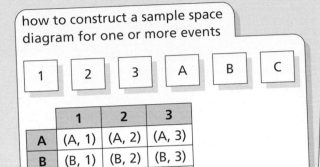

| 1 | 2 | 3 | A | B | C |

	1	2	3
A	(A, 1)	(A, 2)	(A, 3)
B	(B, 1)	(B, 2)	(B, 3)
C	(C, 1)	(C, 2)	(C, 3)

how to find probabilities from two-way tables and Venn diagrams

	Wears glasses	Does not wear glasses	Total
Boys	27	23	50
Girls	31	19	50
Total	58	42	100

The probability that a girl wears glasses is $\frac{31}{50}$

how to use the product rule to find the total number of possible outcomes **H**

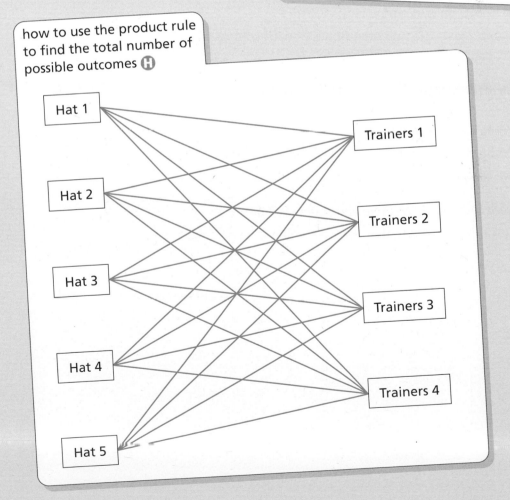

6.1 Sample spaces and probability

Small steps

- Construct sample spaces for 1 or more events
- Find probabilities from a sample space

Key words

Sample space – the set of all possible outcomes or results of an experiment

Probability – how likely an event is to occur

Random – happening without method or conscious decision; each outcome is equally likely to occur

Are you ready?

1 A fair 6-sided dice is rolled.

   What is the probability that the score is

   a 5

   b an even number

   c greater than 3

   d 0

   e a multiple of 2 or a multiple of 3?

2 There are 20 balls in a box.

   The table shows the number of balls of each colour.

Colour	Green	Red	Orange	Blue
Number of balls	5	8	6	

   A ball is selected at random.

   a What is the probability that the ball is red?

   b What is the probability that the ball is orange or green?

   c What is the probability that the ball is blue?

3 The probability of a train being on time is 0.8

   What is the probability that the train will not be on time?

4 Which of these cannot be used to write a probability?

   Fraction   Decimal   Percentage   Ratio

Models and representations

Sample space diagram

A **sample space** diagram helps you to record all the possible outcomes from two events.

The diagram on the right represents the outcomes from spinning these two spinners.

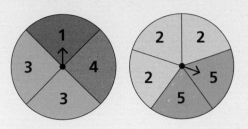

		Spinner 2				
		2	2	2	5	5
Spinner 1	1					
	2					
	3					
	4					

Probability notation will be used throughout this chapter. For example, when talking about the probability a coin will land on heads, it will be written as P(heads).

Example 1

Here are two sets of three cards.

Amir chooses one card from each set.

a Draw a sample space diagram for the possible outcomes.

b How many possible outcomes are there?

c How many of the outcomes contain an even number?

a

	1	2	3
A	(A, 1)	(A, 2)	(A, 3)
B	(B, 1)	(B, 2)	(B, 3)
C	(C, 1)	(C, 2)	(C, 3)

The possible outcomes are written in the sample space diagram. We have written them here using brackets but it is also acceptable to write them without brackets.

b 9

You can see from the sample space diagram that there are 9 possible outcomes in total.

c 3

Of the 9 possible outcomes, 3 contain an even number: (A, 2), (B, 2) and (C, 2)

Practice 6.1A

1 Benji has these 4 coins.

Marta has these 3 coins.

Benji chooses a coin at random.

Marta chooses a coin at random.

a Copy and complete the sample space diagram to show all the possible combinations of coins.

	10p	5p	2p	1p
50p	(50p, 10p)	(50p, 5p)		
20p				
10p				

b Copy and complete the table to show the possible coin totals.

	10p	5p	2p	1p
50p	60p	55p		
20p	30p			
10p				

c Use the most appropriate table to work out each of these probabilities.

i P(Benji chooses a 10p coin and Marta chooses a 20p coin)

ii P(they both choose silver coins)

iii P(they choose a coin with the same value)

iv P(the total is greater than 40p)

v P(the total is greater than £1)

d Which table did you use to help you answer each question? What other questions could you answer?

2 Here are two fair spinners.

A player can choose any spinner.

They win a prize if the pointer lands on the word "Win".

> I will choose Spinner B as it has 2 sections labelled "Win" and the other only has 1

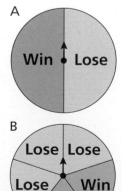

a Explain why Beca is more likely to win if she chooses Spinner A. What mistake has Beca made?

To win a large prize, a player has to spin both spinners and they both have to land on win.

b Copy and complete the sample space diagram to show all the possible outcomes.

	Win	Win	Lose	Lose	Lose
Win	(W, W)				
Lose	(L, W)				

c Use the sample space diagram to work out the probability that a player wins a large prize.

3 Two fair dice are rolled.

The numbers of the dice are added together.

a Copy and complete the table of total scores.

b What is the probability of rolling a score of 8?

c What is the probability of rolling a score greater than or equal to 10?

d Which is more likely, rolling a score of 7 or rolling a score less than 4? Explain your answer.

+	1	2	3	4	5	6
1	2	3	4	5	6	7
2						
3						
4						
5						
6						

4 Here are two fair spinners. Each spinner is spun once.

a Draw a sample space diagram to show all the possible outcomes.

b Use your sample space diagram to work out

 i P(Spinning X and an even number)

 ii P(Spinning X or Y and a number greater than 4)

 iii P(Spinning Z or Spinning the number 3)

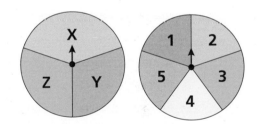

5 Here are 5 number cards.

Rhys chooses a card and keeps hold of it.

Amina then chooses one of the remaining cards.

They draw a sample space diagram to record the results.

| 1 | 2 | 3 | 4 | 5 |

	1	2	3	4	5
1	(1, 1)	(1, 2)	(1, 3)	(1, 4)	(1, 5)
2	(2, 1)	(2, 2)	(2, 3)	(2, 4)	(2, 5)
3	(3, 1)	(3, 2)	(3, 3)	(3, 4)	(3, 5)
4	(4, 1)	(4, 2)	(4, 3)	(4, 4)	(4, 5)
5	(5, 1)	(5, 2)	(5, 3)	(5, 4)	(5, 5)

a Explain the mistake that Rhys and Amina have made with their sample space diagram.

b How many different possible outcomes are there?

c What is the probability that both Rhys and Amina choose an even number?

What do you think?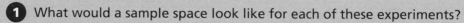

1 What would a sample space look like for each of these experiments?

a Flipping a single coin   **b** Flipping 2 coins

c Rolling a 6-sided dice and then flipping a coin

What's the same and what's different about your sample spaces?
When would you use a table?

2 Here is a sample space diagram.

a What do you think it could represent?

b What can you work out from the diagram?

		Mario		
		Rock	Paper	Scissors
Kate	Rock	(R, R)	(R, P)	(R, S)
	Paper	(P, R)	(P, P)	(P, S)
	Scissors	(S, R)	(S, P)	(S, S)

3 Here are some number cards.

0	2	4	5	8

Here are the instructions for a game using the number cards.

■ Put the cards into a bag and mix them up.

■ Choose a card at random, write down its value then replace it.

■ Choose a second card at random, write down its value then replace it.

■ Find the product of your two values.

■ If the product is zero, you win!

a Do you agree?

I don't think it's important that you choose the card at random.

b Do you agree?

You've got a 20% chance of winning as one of the five cards is a 0

c Do you agree?

It doesn't matter if you don't put the card back in the bag – the probability of winning will be the same.

Consolidate – do you need more?

1 Two fair coins are thrown.

The sample space diagram shows all the possible outcomes.

		Second coin	
		H	T
First coin	H	(H, H)	(H, T)
	T	(T, H)	(T, T)

a What is the probability that both coins show heads?

b What is the probability that both coins land the same way up?

c What is the probability that at least one coin shows heads?

2 A fair 4-sided dice labelled 5, 6, 7, 8 is rolled twice and the scores are added.

a Draw a sample space diagram to show all the possible total scores.

b What is the probability of getting a total score that is greater than 12?

Stretch – can you deepen your learning?

1 180 students in a school were asked how many cats and dogs they had at home.

A student is chosen at random.

Work out the probability that

		Dogs			
		0	1	2	3
Cats	0	46	32	12	2
	1	33	12	10	1
	2	12	17	1	0
	3	0	2	0	0

a they have no cats or dogs

b they have the same number of cats and dogs

c they have more cats than dogs

d they have three pets in total.

2 Here are two unlabelled spinners.

Each spinner is spun once and their values are added.

A number is written on each section.

The probability of getting a total score greater than 10 is $\frac{1}{6}$

The probability of getting an odd score is $\frac{1}{3}$

What could be the numbers on the spinners?

3 Three fair coins are flipped.

What is the probability that you get 3 heads?

Explain how you approached the question.

Reflect

1 How can you use a sample space diagram to help you to work out probabilities?

6.2 Using diagrams to find probabilities

Small steps

- Find probabilities from two-way tables
- Find probabilities from Venn diagrams
- Use the product rule for finding the total number of possible outcomes (H)

Key words

Two-way table – this displays two sets of data in rows and columns

Venn diagram – a diagram used for sorting data

Probability – how likely an event is to occur

Outcome – the possible result of an experiment

Are you ready?

1 **a** Copy and complete the two-way table about some shapes.

	Red	Blue	Total
Square	7		20
Triangle			12
Total		17	32

   b How many shapes are there in total?

   c How many squares are there?

   d How many red shapes are there?

   e How many blue triangles are there?

2 The Venn diagram shows information about 8 people (A to H) who took two tests.

   a Which person passed both tests?

   b Which people passed just test 2?

   c Which people passed just one of the tests?

   d Which people passed neither of the tests?

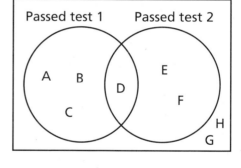

3 Mr and Mrs Patel go to the local Italian restaurant.

   They want to choose a starter and then a main meal.

   Here are the possible choices.

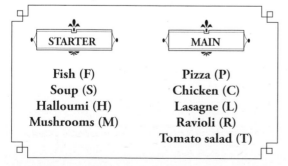

STARTER	MAIN
Fish (F)	Pizza (P)
Soup (S)	Chicken (C)
Halloumi (H)	Lasagne (L)
Mushrooms (M)	Ravioli (R)
	Tomato salad (T)

   Make a list of all the possible combinations of starter and main meal they can choose.

Models and representations

Two-way table

A two-way table has two categories.

In this case, they are whether you own a pet and year group.

	Owns a pet	Does not own a pet	Total
Year 7			
Year 8			
Total			

Venn diagram

Venn diagrams are a useful way of representing data.

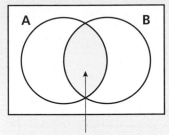

The intersection shows
what is in both sets.

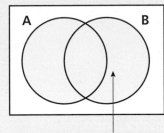

The union shows what
is in at least one set.

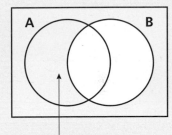

The complement of set B
shows what is not in set B.

In this chapter you will use diagrams to find probabilities. You have already looked at completing Venn diagrams and two-way tables, and you will now build on this knowledge to calculate probabilities. It is important to pay close attention to the wording in the question to make sure that you write each **probability** out of the correct total.

Example 1

The table shows some information about students in a school.

	Wears glasses	Does not wear glasses	Total
Boys	27	23	50
Girls	31	19	50
Total	58	42	100

a A student is chosen at random. What is the probability that this student wears glasses?

b A student is chosen at random. What is the probability that this student is a girl who wears glasses?

c A girl is chosen at random. What is the probability that she wears glasses?

a

	Wears glasses	Does not wear glasses	Total
Boys	27	23	50
Girls	31	19	50
Total	58	42	100

	Wears glasses	Does not wear glasses	Total
Boys	27	23	50
Girls	31	19	50
Total	58	42	100

The probability that this student wears glasses is $\frac{58}{100}$

Pay close attention to the wording used in the question.

"A **student** is chosen at random."

This means that you are considering **all** of the students, so the probability is out of 100

"What is the probability that this student wears glasses?"

58 out of the 100 students wear glasses.

b

	Wears glasses	Does not wear glasses	Total
Boys	27	23	50
Girls	31	19	50
Total	58	42	100

	Wears glasses	Does not wear glasses	Total
Boys	27	23	50
Girls	31	19	50
Total	58	42	100

Out of the 100 students, 31 of them are girls who wear glasses.

The probability that this student is a girl who wears glasses is $\frac{31}{100}$

c

	Wears glasses	Does not wear glasses	Total
Boys	27	23	50
Girls	31	19	50
Total	58	42	100

	Wears glasses	Does not wear glasses	Total
Boys	27	23	50
Girls	31	19	50
Total	58	42	100

This time "a **girl** is chosen at random".

This means that you are considering **only** the girls, not all of the students, so the probability is out of 50

31 out of the 50 girls wear glasses.

The probability that this girl wears glasses is $\frac{31}{50}$

Practice 6.2A

1 The table shows the numbers of boys and girls in Classes A and B.

	Class A	Class B
Boys	18	10
Girls	12	15

  a A student from Class A is chosen at random. What is the probability that the student chosen is a boy?

  b A student from Class B is chosen at random. What is the probability that the student is a girl?

  c A boy is chosen at random. What is the probability that the boy is from Class B?

  d A student is chosen at random. What is the probability that

   i the student is a boy

   ii the student is in Class B

   iii the student is a girl from Class A?

2 The table shows the numbers of Year 10 and Year 11 students who study French and Spanish.

	Year 10	Year 11
French	56	20
Spanish	40	75

  No student studies both French and Spanish.

  One of the students is chosen at random to be the school's language ambassador.

  a What is the probability that the student chosen is from Year 10?

  b What is the probability that the student chosen studies French?

  c What is the probability that the student chosen studies Spanish and is from Year 11?

  d There are 200 students in Year 11. One of these students is chosen at random.

   i What is the probability that the student studies Spanish?

   ii What is the probability that the student chosen does not study either French or Spanish?

3 Two pencil cases contain some coloured pencils.

  They each contain the same number of pencils.

  The table shows how many pencils of each colour there are.

  a Copy and complete the table.

  b A pencil is chosen at random from all the pencils.

	Yellow	Green	Red	Total
Pencil case A	17			
Pencil case B			19	
Total		55	48	150

   i Explain why the probability that the pencil is from pencil case A is 0.5.

   ii What is the probability that the pencil is red?

   iii What is the probability that the pencil is red or green?

   iv What is the probability that the pencil is not yellow?

  Mario chooses a yellow pencil

  c What is the probability that the pencil Mario chooses is from pencil case A? Explain your reasons.

Mario chooses another pencil.

d What is the probability that the pencil is red or from pencil case A?

What mistakes do you think people might make when answering this question?

4 Some children are asked whether they own a cat or a dog.

The Venn diagram shows the results.

a Write down the name of the child who owns a cat and a dog.

b Does Jamil own a cat? Does he own a dog?

c A child is chosen at random.

  i What is the probability that the child chosen is Charlotte?

  ii What is the probability that the child owns both a dog and a cat?

  iii What is the probability that the child just owns a dog?

  iv What is the probability that the child does not own a dog or a cat?

Owns a dog   Owns a cat

Yasmin
Karl
Tez   George   Abdullah
Peter   Charlotte
David   Jamil
                     Sam

5 Two blockbuster films were released this week.

200 people were asked whether they have seen the films.

The Venn diagram shows the results.

One of the people is chosen at random. Find

a P(the person has seen film A)

b P(the person has seen both film A and film B)

c P (the person has seen film A or film B)

d P (the person has not seen either film)

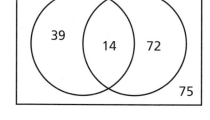

Film A   Film B

39   14   72

             75

6 50 people are asked whether they own a car or a motorcycle.

The Venn diagram shows their responses.

One of the people surveyed is chosen at random.

Explain why it is impossible that this person owns both a car and motorcycle.

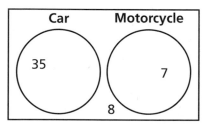

Car   Motorcycle

35   7

        8

7 100 students in Years 7 and 8 are asked how they came to school one day.

Each student walked, came by bus or came by car.

■ 55 of the 100 students are in Year 7

■ 20 of the Year 7 students came by car.

■ 16 Year 8 students walked.

■ 7 of the 23 students who came by bus are from Year 8

A student is chosen at random.

Work out the probability that the chosen student walked to school that day.

What do you think? 💭

1 Do you find working out probabilities from a Venn diagram easier or harder than working out probabilities from a two-way table? Discuss your reasons.

2 The table gives the colours of the balls in two boxes.

a Write down **three** pieces of information about the balls.

b A ball is chosen at random from one of the boxes.

    i What is the probability the ball is orange?

    ii If you know that the ball is from box A, does your answer to part **ii** stay the same or change? Why?

	Green	Orange	Total
Box A	3	15	18
Box B	8	4	12
Total	11	19	30

c Jackson says that the probability that a ball is orange or from box A is $\frac{37}{30}$.

    i How do you know that Jackson is incorrect without checking the table?

    ii What has Jackson done to get his answer?

    iii What should he have done?

You will now look at using the product rule for finding the total number of possible **outcomes**. You will consider different combinations and look at an efficient way of calculating how many different combinations there are.

This is useful when calculating probabilities, as you need to know the total number of possible outcomes.

Example 2

Tommy has 5 hats and 4 pairs of trainers.

How many different combinations of a hat and trainers does he have?

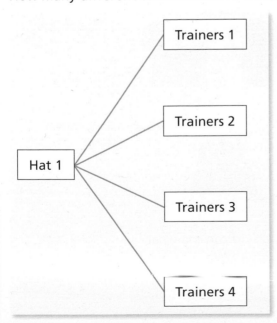

If you number Tommy's hats 1–5 and his pairs of trainers 1–4 you can start to consider the different combinations.

Start by considering hat number 1

He has four choices of pairs of trainers to pair with this hat.

There are four possible combinations for hat 1

You can then repeat this with the other hats.

But, at this point, the diagram becomes quite messy.

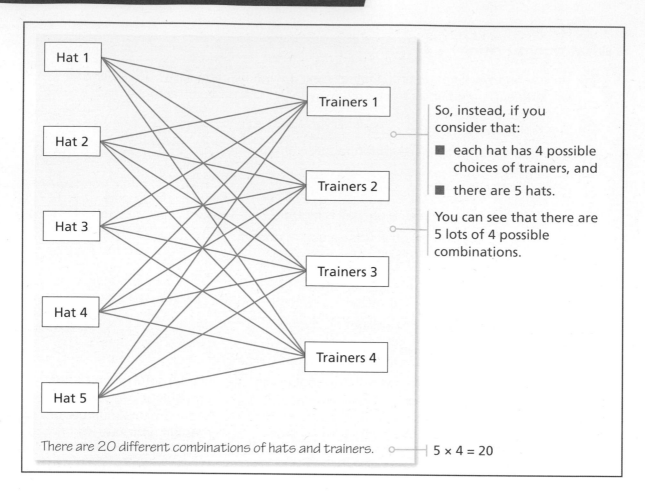

So, instead, if you consider that:

■ each hat has 4 possible choices of trainers, and

■ there are 5 hats.

You can see that there are 5 lots of 4 possible combinations.

There are 20 different combinations of hats and trainers.

$5 \times 4 = 20$

Practice 6.2B

1 Emily has 5 tops.

She also has 3 skirts.

Emily chooses a top and skirt.

How many different combinations of top and skirt can she wear?

2 A student council is made up of a Head Boy and Head Girl.

■ There are 5 choices for the Head Boy

■ There are 7 choices for the Head Girl

Explain why there are 35 possible combinations for the student council.

3 A game uses 12 character cards and 19 spell cards.

A character and spell card are chosen at random.

How many different combinations of character card and spell card are possible?

4 Seb has 3 different erasers.

He also has 4 different rulers.

He also has 5 different pens.

Convince Seb that this is possible.

> I want to choose a different combination of eraser, ruler and pen each day at school for the next 10 weeks.

5 A Year 12 and a Year 13 student are chosen to represent the school.

There are 11 possible Year 12 students and x possible Year 13 students.

There are 132 possible combinations.

How many Year 13 students are there to choose from?

What do you think?

1 Benji has some scarves and gloves.

He chooses a scarf and a pair of gloves to wear.

He has 48 different possible combinations of scarf and gloves.

How many different scarves could Benji have? How many pairs of gloves could he have? Is there more than one answer to each of these questions?

2 Here are 5 cards.

a A card is chosen at random. It is replaced and then another card is chosen.

How many different combinations of card are there?

b A card is chosen at random. It is not replaced. Another card is chosen.

How many different combinations of card are there?

c What's the same and what's different about parts **a** and **b**?

3 Chloe has 8 cards. Each card has a different number on it.

She picks 3 cards at random and correctly works out that there are 336 possible outcomes.

Were the cards replaced between each pick? Explain how you know.

Consolidate – do you need more?

1 A meal deal involves buying a sandwich and a drink.

The two-way table shows what 40 people chose for their meal deals.

A person is chosen at random. What is the probability that they chose

	Cheese	Tuna	Total
Water	5	12	17
Juice	7	16	23
Total	12	28	40

a a cheese sandwich

b juice

c juice and tuna

d juice or tuna?

2 An outdoor game is made up of two tasks.

Each player has to try to complete each task in under a minute.

The Venn diagram shows information about the time it takes a group of people to complete the tasks.

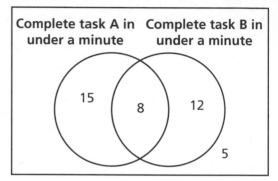

a How many people attempted the tasks?

b How many people completed task A in under a minute?

c How many people didn't complete either task in under a minute?

d A person who completed both tasks is chosen at random. Work out

i P(completed both tasks in under a minute)

ii P(completed just one task in under a minute)

iii P(did not complete either of the tasks in under a minute)

Stretch – can you deepen your learning?

1 Two money boxes contain some 50p, 20p and 10p coins.

There is the same amount of money in each box.

The two-way table shows information about the coins in the boxes.

	50p	20p	10p
Box A	5	4	3
Box B	2	7	

All the coins are emptied into one tub.

A coin is then chosen at random.

What is the probability that the coin is a 10p coin?

2 Ali travels to work each day.

She first catches a bus and then travels on a train.

The Venn diagram shows the numbers of times in the last 100 days that the train and bus were on time.

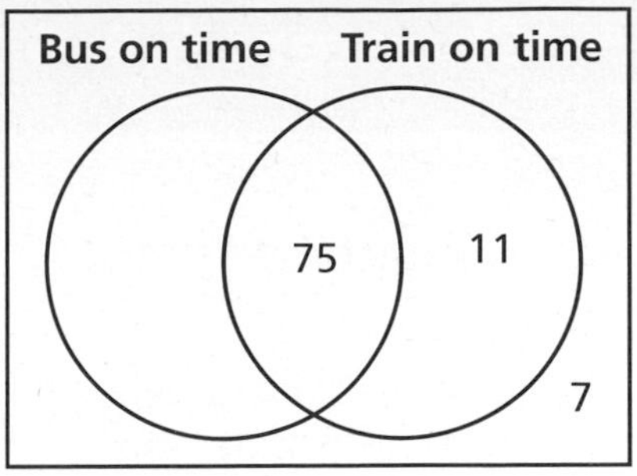

a How many times was the bus on time and the train late?

b Estimate the probability that tomorrow the bus and train will be on time.

c Why is your probability an estimate?

d In order to get to work on time, only one of either the bus or the train can be late. What is the probability that Ali is late for work?

Bus on time   Train on time

75   11

7

3 A school holds a sports day.

10 children take part in a race.

a How many possible combinations of first, second and third place are there?

For the next race, the children are put into pairs.

b How many possible pairs of children are there?

For the final race, some children are paired up so that there is one boy and one girl in each pair.

Reflect

1 In your own words, explain how to calculate probabilities from two-way tables.

2 In your own words, explain how to calculate probabilities from Venn diagrams.

3 In your own words, explain how to find the number of possible combinations of two or more events.

6 Tables and probability
Chapters 6.1–6.2

White Rose Maths

I have become **fluent** in…	I have developed my **reasoning** skills by…	I have been **problem-solving** through…
■ constructing a sample space for one or more events ■ finding probabilities from a sample space ■ finding probabilities from two-way tables ■ finding probabilities from Venn diagrams.	■ thinking about what a diagram could and could not represent ■ correcting mistakes others have made ■ considering the difference between something being fair and something being biased.	■ using the product rule for finding the total number of possible outcomes Ⓗ ■ working backwards to find missing information ■ using prior knowledge to solve multi-step problems.

Check my understanding

1 Copy and complete the sample space diagram for all the different types of sandwich fillings.

	Tomato	Mayo	Lettuce
Cheese			
Egg			
Salmon			

2 A spinner is split into five equal sections labelled A–E

Tommy rolls a fair 6-sided dice and spins the spinner.

a Construct a sample space diagram to show all the possible outcomes.

b Find the probability that Tommy rolls an even number and the spinner lands on E

3 The two-way table shows how many people attend either a dance or a theatre group on Saturday and Sunday.

Find the probability that a randomly chosen person attends a dance on Sunday.

	Dance	Theatre	Total
Saturday	24		60
Sunday		21	
Total	43		100

4 A group of students were asked whether they enjoy watching films and playing computer games. The Venn diagram shows the results of the survey.

A student is chosen at random. What is the probability that they do not like watching films or playing computer games?

Enjoys films   Enjoys games

25   39   18

26

5 Mr McCoy has 5 shirts, 4 ties and 3 jackets. How many different combinations of shirt, tie and jacket can he choose between? Ⓗ

7 Brackets, equations and inequalities

In this block, I will learn...

how to form and use algebraic expressions

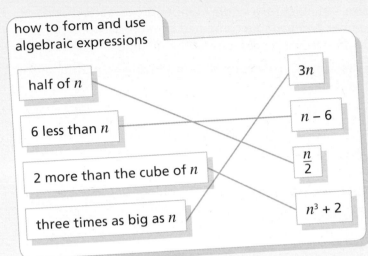

half of n

6 less than n

2 more than the cube of n

three times as big as n

$3n$

$n - 6$

$\dfrac{n}{2}$

$n^3 + 2$

how to expand single brackets

$3(a + 5)$

$5(x + 7) - 3(x - 4)$

$8(2a + 3b - 5)$

how to expand a pair of binomials Ⓗ

$(a + 5)(b + 6)$

$(5 - x)(x - 3)$

how to factorise an expression

$5 + 10x \equiv 5(1 + 2x)$

$3b^2 - 6bc \equiv 3b(b - 2c)$

how to form and solve equations and inequalities

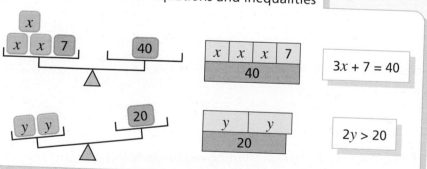

x | x | x | 7
40

$3x + 7 = 40$

y | y
20

$2y > 20$

how to solve an equation involving brackets

I think of a number, add 7 and multiply the result by 5. The answer is 15.

$3(n + 7) = 15$

how to solve equations and inequalities with unknowns on both sides Ⓗ

$5a + 12 = 2a + 30$

$3a + 7 > 5a + 1$

| a | a | a | a | a | 12 |
| a | a | | 30 | | |

the difference between formulae, equations and identities

$F = 1.8C + 32$

$70 = 1.8C + 32$

$2(C + 15) \equiv 2C + 32$

Small steps

- Form algebraic expressions
- Use directed numbers with algebra

Key words

Term – in algebra, a single number or variable, or a number and variable combined by multiplication or division; in sequences, one of the members of a sequence

Expression – a collection of terms involving mathematical operations

Coefficient – a number in front of a variable, for example for $4x$ the coefficient of x is 4

Simplify – rewrite in a simpler form, for example rewrite $8 \times h$ as $8h$

Substitute – to replace letters with numerical values

Are you ready?

1 Write each term without mathematical operations.

   a $3 \times m$ **b** $m \times 3$ **c** $m \div 3$ **d** $3 \div m$ **e** $m \times m$

2 Find the value of each expression when $t = 40$

   a $\dfrac{t}{10}$ **b** $10t$ **c** $\dfrac{400}{t}$ **d** $t - 17$ **e** $80 - t$

   f t^2 **g** $2t - 80$

3 Work out

   a $3 + -5$ **b** $3 - -5$ **c** $-3 + 5$ **d** $-3 - 5$

4 Work out

   a 8×-2 **b** -8×2 **c** -8×-2 **d** $-8 \div 2$ **e** $-8 \div -2$

Models and representations

In Book 1, you met several ways of making and representing **expressions** including:

Function machines

Input ⟶ $\boxed{+\ 7}$ ⟶ Output

Cubes and counters

Algebra tiles

▢ represents 1 ▮ represents x and ⬛ represents x^2

Bar models

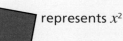

Words

"Double a number and add 3"

Symbols

$2x + 3$

Remember, you can use double-sided counters to represent and solve problems with directed numbers, for example: $2 + -5 = -3$

Example 1

Represent each statement using objects or models and using symbols.

a 2 more than a number

b 3 less than a number

c 1 more than double a number

d double "1 less than a number"

e 5 more than half a number

a Let a cube represent the number.

Let the number be k

$k + 2$

You are adding 2 ones to the number. You could use any object or a bar model.

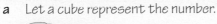

You could also write $2 + k$, but it is more usual to start with the variable.

b

$k - 3$

Use a different representation for -1 than for $+1$

c

$2k + 1$

For double the number, use two cubes.

Again you could use a bar model: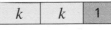

Remember you write $2k$, not $2 \times k$ or $k \times 2$

d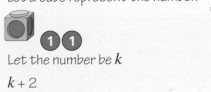

$2k - 2$

One less than a number is  Doubling gives us two cubes and two -1s

Here is one way you could show this using a bar model:

e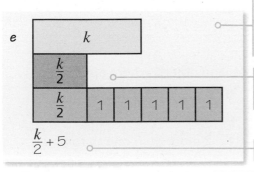

$\dfrac{k}{2} + 5$

It is more difficult to use concrete objects to represent fractions, as you have to break the objects. You could use two cubes to stand for k, but that would affect the answers to parts **a** to **d**.

Remember: when dealing with symbols you write $\dfrac{k}{2}$ to mean $k \div 2$

Five more is $\dfrac{k}{2} + 5$

Example 2

Simplify the expressions.

a $3k + 4p + 2k$ **b** $5a - 3 + 2a + 5$ **c** $x^2 + 3x + x^2 - x$

a $3k + 4p + 2k \equiv 3k + 2k + 4p$
$\equiv 5k + 4p$

You can collect like **terms** ($3k$ and $2k$) together. As the two terms in k are like terms you just add their coefficients.

You cannot simplify further as $5k$ and $4p$ are unlike terms.

You can show this using objects.

b $5a - 3 + 2a + 5 \equiv 5a + 2a - 3 + 5$
$\equiv 7a + 2$

Collect like terms first.

Simplify by adding or subtracting the coefficients or numbers.

You can show this using objects.

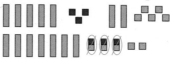

c $x^2 + 3x + x^2 - x \equiv x^2 + x^2 + 3x - x$
$\equiv 2x^2 + 2x$

Collect like terms first.

Simplify by adding or subtracting the **coefficients**.

You cannot simplify further as x^2 and x are unlike.

You can show this using objects.

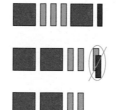

remember $\equiv$ means "is equivalent to". A statement using $\equiv$ is true for all values of the variable or variables.

Practice 7.1A

 1 n is a number. Represent each of these in symbols and objects where appropriate.

    a 5 more than n **b** 2 less than n

    c 3 less than double n **d** double "1 more than n"

    e one third of n **f** 2 more than one third of n

    g one less than half of n

2 **a** Ed is 12 years old. How old will he be

   i in 2 years' time   **ii** in x years' time?

b Faith is y years old. How old will she be

   i in 2 years' time   **ii** in x years' time   **iii** in y years' time?

c How old were Ed and Faith five years ago?

d Find the sum of Ed's age and Faith's age three years from now.

3 Apples cost x pence and bananas cost y pence

a Write an expression for the cost of

   i 5 apples   **ii** 3 bananas   **iii** 5 apples and 3 bananas.

b Write an expression for the difference between the cost of 5 apples and 3 bananas.

4 Sven gets £a pocket money each month. He spends £15 and saves the rest.
Write expressions for

a how much money Sven saves each month

b the total amount of pocket money he receives in a year.

Sven's younger sister, Tiff, gets half as much pocket money as Sven. Write expressions for how much pocket money Tiff gets

c every month

d in a year.

5 **a** Write an expression for the area of a rectangle 5 cm long and t cm wide.

b **i** Which is the correct simplified expression for the area of this rectangle?

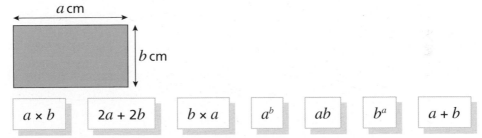

| $a \times b$ | $2a + 2b$ | $b \times a$ | a^b | ab | b^a | $a + b$ |

   ii Write a different simplified expression that could be used for the area of the rectangle.

c Which is the correct simplified expression for the area of this square?

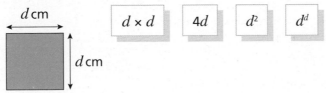

| $d \times d$ | $4d$ | d^2 | d^d |

d Write simplified expressions for the areas of these shapes.

i
2g cm
3 cm

ii
2h cm
k cm

iii
5m cm
4n cm

iv
5d cm
5d cm

v
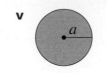
a

vi

4b

6 Write expressions for

  a the sum of p and q

  b the product of p and q

  c the quotient of p and q

  d 5 more than the quotient of q and p

  e the product of 5, p and q

7 Simplify the expressions.

  a $5a + 4b + 3a + 7b$

  b $2a + 7b - a$

  c $5x^2 + 3x + 2x^2 - 4x$

  d $4x^2 + 3x - x^2 - 3x^2$

  e $pq + 2p + 2q + qp$

What do you think?

1 Write expressions for

  a half the difference between a and b

  b 5 more than the product of x and y

  c twice the product of m and n subtracted from 12

  d the product of the square of f and $5g$

2 m questions in an exam are worth p marks each and n questions are worth q marks each. Filipo scored 25% in the exam. Write an expression, in terms of m, n, p and q, for the number of marks Filipo scored.

3 Mario is twice as old as his brother and one third the age of his father. Mario's mother is 5 years older than his father. Write an expression for the total age of Mario, his father, his mother and his brother if

  a Mario is $2t$ years old

  b Mario is p years old

  c Mario's father is $6x$ years old.

4 Explain why these simplifications are incorrect.

  a $6t - t \equiv 6$

  b $\dfrac{p^2}{p} \equiv 2$

  c $\dfrac{mn}{nm} \equiv 0$

You can evaluate algebraic expressions by substituting numbers for the variables.

Example 3

Find the value of $3p - q$ when

a $p = 4$ and $q = 5$ **b** $p = -4$ and $q = 5$ **c** $p = -2$ and $q = -7$

a $3p - q = 3 \times 4 - 5$ ○─── **Substitute** in the values, remembering that $3p$ means $3 \times p$

$\qquad\quad = 12 - 5$ ○─── Work out 3×4 first, then subtract the 5

$\qquad\quad = 7$

b $3p - q = 3 \times -4 - 5$

$\qquad\quad = -12 - 5$ ○─── Because $3 \times 4 = 12$, $3 \times -4 = -12$

$\qquad\quad = -17$ ○─── Starting at –12, subtracting another 5 is –17

c $3p - q = 3 \times -2 - -7$ ○─┤ Be careful not to "lose" a – sign here.
The correct substitution is $3 \times -2 - -7$,

$\qquad\quad = -6 - -7$ not $3 \times -2 - 7$

$\qquad\quad = -6 + 7$ ○─── Remember: subtracting a negative is the same as adding.

$\qquad\quad = 1$ ○─── Starting at –6 and adding 7 takes you to 1

Example 4

Find the value of a^2b when

a $a = 4$ and $b = 5$ **b** $a = -4$ and $b = 5$ **c** $a = -4$ and $b = -5$

a $a^2b = 4^2 \times 5$

$\qquad\quad = 16 \times 5$ ○─┤ You would work out the squared term first even if the
calculation was 5×4^2 because of the order of operations.

$\qquad\quad = 80$

b $a^2b = (-4)^2 \times 5$ ○─┤ It is best to put –4 in brackets to help you to remember that
the calculation is -4×-4 and not -4×4

$\qquad\quad = 16 \times 5$

$\qquad\quad = 80$ Remember: the product of two negative numbers is positive.

c $a^2b = (-4)^2 \times -5$

$\qquad\quad = 16 \times -5$ Remember: the product of a negative

$\qquad\quad = -80$ ○─┤ number and a positive number is negative.

Practice 7.1B

1 **a** Find the value of $3m + 4n$ when

    i $m = 2$ and $n = 5$ **ii** $m = -2$ and $n = 5$

    iii $m = 2$ and $n = -5$ **iv** $m = -2$ and $n = -5$

    b Find the value of $3m - 4n$ when

    i $m = 2$ and $n = 5$ **ii** $m = -2$ and $n = 5$

    iii $m = 2$ and $n = -5$ **iv** $m = -2$ and $n = -5$

2 $p = 8$ and $q = -12$. Work out

    a $p + q$ **b** $q - p$ **c** pq

    d $p - 2q$ **e** $\dfrac{2q}{p}$ **f** $\dfrac{pq}{2}$

    g $p - \dfrac{q}{2}$ **h** $\dfrac{p - q}{2}$ **i** $\dfrac{q - p}{4}$

3 $x = 3$ and $y = 4$. Ali works out xy^2

$$xy^2 = 3 \times 4^2 = 12^2 = 144$$

    a Explain why Ali is wrong.   **b** Find the correct value of xy^2

    c Would the value of xy^2 change if

    i $x = -3$ and $y = 4$ **ii** $x = 3$ and $y = -4$ **iii** $x = -3$ and $y = -4$?

4 Solve the equations.

    a $g + 5 = 20$ **b** $g - 5 = 20$ **c** $g + 5 = -20$ **d** $g - 5 = -20$

    e $5g = 20$ **f** $5g = -20$ **g** $\dfrac{g}{5} = 20$ **h** $\dfrac{g}{5} = -20$

5 Write these as single terms, without mathematical operations.

    a $a \times 2$ **b** $a \times -2$ **c** $a \times a$ **d** $a \times -a$

    e $a \times b \times 2$ **f** $a \times -2 \times b$ **g** $a \times -b \times -2$

6 Simplify the expressions.

    a $3x + 4y - 2x - 2y$ **b** $3x + 4y - 6x - 8y$

    c $3x - 4y + 2x - 5y$ **d** $3x - 4y - 6x + y$

    e $x^2 + 5x - x^2 - x$ **f** $x^2 - 3x - 2x^2 + x$

    g $x^2 - 3x - 2x^2 - x$ **h** $-x^2 - 3x - 2x^2 - x$

What do you think? 💭

1 $x = 6$ and $y = -4$

How many expressions involving both x and y can you find with a value of

a −2   **b** 12?

2 The product of m negative numbers is positive, and the product of n negative numbers is negative.

a Which of these statements are definitely true?

A | m is even   B | n is even   C | m is odd   D | n is odd

b What can you deduce about the values of

i mn **ii** $m - n$ **iii** $n - m$?

3 How many expressions can you find that simplify to

a $3ab$ **b** $a - 3b$?

Consolidate – do you need more?

1 n is a number. Represent each of these in symbols and objects where appropriate.

a 7 less than n **b** 3 more than n

c 2 more than double n **d** double "3 less than n"

e one quarter of n **f** 2 more than half of n

g 1 less than a third of n

2 Draw bar models to show these expressions.

a $2x + 3$ **b** $2x - 3$ **c** $\frac{x}{2} + 3$ **d** $\frac{x}{2} - 3$

3 a How many days are there in p weeks?

b How many eggs are there altogether in m boxes if each box contains

i 6 eggs   **ii** 9 eggs   **iii** n eggs?

4 a Find the value of $5p + 2q$ when

i $p = 2$ and $q = 4$ **ii** $p = 2$ and $q = -4$

iii $p = -2$ and $q = -4$ **iv** $p = -2$ and $q = 4$

b Find the value of $5p - 2q$ when

i $p = 2$ and $q = 4$ **ii** $p = 2$ and $q = -4$

iii $p = -2$ and $q = -4$ **iv** $p = -2$ and $q = 4$

5 $a = -6$ and $b = 8$. Work out

a $a - b$ **b** $b - a$ **c** $4a + 3b$

d $3a - 4b$ **e** $\dfrac{3b}{a}$ **f** $\dfrac{ab}{3}$

g $\dfrac{b}{2} - a$ **h** $b - \dfrac{a}{2}$ **i** $\dfrac{2b + a}{4}$

Stretch – can you deepen your learning?

1 A cube is used to represent $5p$ and a counter is used to represent $3q$

a What do each of the diagrams represent? Give your answers in simplified form.

i

ii

b How do your answers change if the cube is used to represent $5p + 1$ and the counter is used to represent $3q - 2$?

2 a Write as a single fraction.

i $\dfrac{1}{3} + \dfrac{1}{2}$ **ii** $\dfrac{1}{3}p + \dfrac{1}{2}q$ **iii** $\dfrac{1}{p} + \dfrac{1}{q}$ **iv** $\dfrac{1}{p} - \dfrac{1}{q}$

b Investigate values of p and q that will give positive and negative values for the expressions found in **ii**, **iii** and **iv**

3 Write expressions in one variable that are

a always positive   **b** always negative.

What will the graphs of your expressions look like?

Reflect

How do you substitute negative numbers into expressions? What errors could happen?

7.2 Using brackets

Small steps

- Multiply out a single bracket
- Factorise into a single bracket
- Expand multiple single brackets and simplify

Key words

Expand – multiply to remove brackets from an expression

Factorise – find the factors you need to multiply to make an expression

Simplify – rewrite in a simpler form, for example rewrite $8 \times h$ as $8h$

Are you ready?

1 Find the output of each function machine for the given input.

a Input   Output
$4 \longrightarrow \boxed{\times 2} \longrightarrow \boxed{+ 3} \longrightarrow ?$

b Input   Output
$4 \longrightarrow \boxed{+ 3} \longrightarrow \boxed{\times 2} \longrightarrow ?$

c Input   Output
$x \longrightarrow \boxed{\times 2} \longrightarrow \boxed{+ 3} \longrightarrow ?$

d Input   Output
$x \longrightarrow \boxed{+ 3} \longrightarrow \boxed{\times 2} \longrightarrow ?$

What's the same and what's different?

2 Simplify the expressions.

a $a + 4 + a + 4 + a + 4 + a + 4 + a + 4$

b $a - 4 + a - 4 + a - 4 + a - 4 + a - 4$

3 Write as a single term.

a $3 \times p$ **b** $3 \times 2p$ **c** $3 \times -p$ **d** $-3 \times p$ **e** $-3 \times 2p$

f $p \times p$ **g** $p \times -p$ **h** $p \times 2p$ **i** $p \times -2p$ **j** $3p \times 2p$

4 Find the highest common factor of

a 6 and 3   **b** 6 and 9   **c** 6 and 12   **d** 12 and 20   **e** 18 and 30

Models and representations

You can model multiplying brackets using algebra tiles.

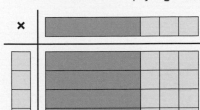

$4(x + 3) \equiv 4x + 12$

Or you can use an area model.

The expression $4(x + 3)$ means "4 lots of $(x + 3)$" or "4 multiplied by $(x + 3)$". This is like finding the area of a rectangle with dimensions 4 and $(x + 3)$.

Both models show that the result is $4x + 12$, which can be found by multiplying both x and 3 by 4

You can write this with symbols:

$4(x + 3) \equiv 4 \times x + 4 \times 3 \equiv 4x + 12$

Each term inside the bracket has been multiplied by 4. This can also be shown using arrows.

$4(x + 3) \equiv 4x + 12$

The shorter arrow shows you multiply 4 by x, and the longer arrow shows you multiply 4 by 3

This process is known as "multiplying out brackets" or "expanding brackets".

Example 1

a Multiply out the brackets in the expression $5(a - 2)$

b By substituting $a = 100$, work out 5×98

a $5(a - 2) \equiv 5a - 10$

Multiply each term in the bracket by 5

$5 \times a = 5a$ and $5 \times -2 = -10$

You can also use algebra tiles to show this.

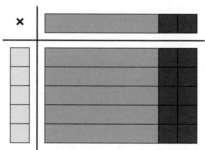

b When $a = 100$

$5(100 - 2) = 500 - 10 = 490$

If $a = 100$, then $a - 2 = 98$

Alternatively, you could substitute $a = 100$ into $5a - 10$

Example 2

Expand the brackets and simplify the answers if possible.

a $6(x + 2y - 3)$ **b** $p(p + 5)$ **c** $3(m + 2n) - 2(4n - 3m)$

a $6(x + 2y - 3) \equiv 6x + 12y - 18$

There are three terms in the bracket, so you multiply them all by 6

$6 \times x = 6x$, $6 \times 2y = 12y$, $6 \times -3 = -18$

The three terms are unlike, so you cannot **simplify** your answer.

b $\quad p(p + 5) \equiv p^2 + 5p$ ○─┤ Remember $p \times p$ is written p^2, not pp

c $\quad 3(m + 2n) - 2(4n - 3m)$ ○── First, **expand** both sets of brackets.
○─ Be careful with the second bracket:
$\quad\quad\quad\quad \equiv 3m + 6n - 8n + 6m$ ─ $-2 \times 4n$ gives $-8n$ but $-2 \times -3m$ gives $+6m$
$\quad\quad\quad\quad \equiv 9m - 2n$ ○── Then you collect the like terms, being careful with any negative terms.

Practice 7.2A

1 Expand the brackets.

a $\quad 3(x + 2)$ **b** $\quad 5(x - 1)$ **c** $\quad 4(2x + 1)$ **d** $\quad 3(2x - 3)$

2

I'm going to work out 8×99 as $8 \times 90 + 8 \times 9$

Jakub

Chloe

I'm going to work out 8×99 as $8 \times 100 - 8 \times 1$

a Compare Jakub's and Chloe's methods. Do they give the same answer? Which method do you prefer?

b Expand the brackets.

i $\quad 8(x + 9)$ **ii** $\quad 8(y - 1)$

c How are the answers to parts **a** and **b** connected?

3 a A monthly magazine costs £4.99. Work out the cost of buying the magazine for a year by expanding the brackets in $12 \times$ (5 pounds – 1 pence).

b Expand the brackets.

i $\quad 12(a - 1)$ **ii** $\quad 12(a + 1)$
iii $12(a + 3)$ **iv** $12(a + b - 2)$
v $\quad 12(a - b + 2)$ **vi** $12(a - b - 2)$
vii $12(2 - a + b)$ **viii** $12(2 - a - b)$

4 Expand the brackets.

a $\quad 3(x + 4)$ **b** $\quad 3(x - 4)$ **c** $\quad 3(2x + 4)$
d $\quad 3(2x - 4)$ **e** $\quad 3(2x - 5y + 4)$

What's the same and what's different?

5 Expand the brackets.

a $\quad a(a + 4)$ **b** $\quad a(b + 4)$ **c** $\quad a(b - 4)$ **d** $\quad a(4 - b)$ **e** $\quad a(4 - a)$
f $\quad a(3a - 4)$ **g** $\quad a(4 - 3a)$ **h** $\quad 3a(a - 4)$ **i** $\quad 3a(2a - 4)$ **j** $\quad 3a(4 - 2a + 5b)$

6 Expand and simplify.

a $3(x + 2) + 2(x + 3)$ **b** $3(x + 2) + 2(x - 3)$

c $3(x + 2) - 2(x + 3)$ **d** $3(x + 2) - 2(x - 3)$

e $p(p + 3) + 5(p + 3)$ **f** $p(p - 3) + p(p + 3)$

g $p(3 - 2p) - 3p(p - 2)$

What do you think?

1 a

 Faith: $3x - (x + 2) \equiv 2x + 2$

Jackson: $3x - (x + 2) \equiv 2x - 2$

When $x = 10$, $3x - (x + 2) = 3 \times 10 - (10 + 2) = 30 - 12 = 18$

By substituting $x = 10$ into Faith's simplification and Jackson's simplification, find out which one of them is correct.

b Expand and simplify.

  i $5x - (2x + 3)$ **ii** $5x - (2x - 3)$ **iii** $3 - (5x + 2)$

  iv $3 - (5x - 2)$ **v** $3 - (2 - 5x)$ **vi** $3 - (2 + 5x)$

2 a Expand the brackets $7(x + a)$

b What values of x and a could you substitute into your expansion to work out

  i 7×103 **ii** 7×99 **iii** 7×9.8 **iv** 7×19.9 **v** 18.9×7?

If an expression is made as a result of expanding brackets, then you can find the original expression with brackets.

For example, $6x + 9$

can be rearranged as a rectangle, like this

The dimensions of the rectangle are 3 and $2x + 3$, and you know $3(2x + 3) \equiv 6x + 9$

This process is called factorising, as you are looking for the factors of $6x + 9$

3 is a factor of $6x$ and 3 is also a factor of 9

$$6x \div 3 = 2x$$

$$6x + 9 \equiv 3(2x + 3)$$

common factor of 6 and 9 $9 \div 3 = 3$

Example 3

Factorise these expressions.

a $3x + 6$
b $10y + 5$
c $6p + 15q + 9$

d $8a + 12$
e $3xy + x^2$
f $5ab + 15bc + 10b^2$

The terms 3x and 6 have a common factor of 3
so $3x + 6 \equiv 3(\square + \square)$

"What do I need to multiply 3 by to get $3x$?"; the answer is x
"What do I need to multiply 3 by to get 6?"; the answer is 2
$3x + 6 \equiv 3(x + 2)$

a $3x + 6 \equiv 3(x + 2)$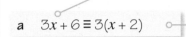

You could also think of this as $3x \div 3$ and $6 \div 3$,
as the inverse of multiplication is division.

b $10y + 5 \equiv 5(2y + 1)$ ○─┤ The common factor is 5

c $6p + 15q + 9 \equiv 3(2p + 5q + 3)$ ○─┤ There are three terms here but the process is the same.
You need a factor that is common to all three terms.

d $8a + 12 \equiv 4(2a + 3)$ ○─┤ Although 2 is a factor of both $8a$ and 12, their highest
common factor is 4, so that is the factor to choose.

e $3xy + x^2 \equiv x(3y + x)$ ○─┤ The common factor can be a variable.

f $5ab + 15bc + 10b^2 \equiv 5b(a + 3c + 2b)$ ○─┤ Both 5 and b are factors of all three terms.
The highest common factor is $5b$

Practice 7.2B

1 Write these as a single term.

 a $6x \div 2$
 b $6x \div 3$
 c $x^2 \div x$
 d $2x^2 \div x$
 e $6x^2 \div 2x$

2 **a** The area of a rectangle is $15\,\text{cm}^2$. The side lengths are integers.
 What might the sides be?

 b The area of a rectangle is $6y\,\text{cm}^2$. The side lengths are single terms.
 What might the sides be?

 c The area of a rectangle is $4x + 12\,\text{cm}^2$. One side length is $4\,\text{cm}$.
 What is the length of the other side?

3 Find the highest common factor of each set.

 a 8 and 12
 b $8a$ and $12b$
 c $8ab$ and $12bc$
 d $8a^2$ and $12a$
 e $8a^2$ and $12ab$

4 Copy and complete the following.

 a $2x + 6 \equiv 2(\square + \square)$
 b $4x + 8 \equiv 4(\square + \square)$
 c $20 + 5x \equiv \square(\square + x)$
 d $6x - 6 \equiv 6(\square - \square)$
 e $8x + 16 \equiv \square(\square + \square)$
 f $x^2 + 3x \equiv x(\square + \square)$
 g $x^2 - 3x \equiv \square(\square - \square)$

5 Factorise these expressions.

   a $6a - 18b + 9c$ **b** $6a + 18b - 12c$ **c** $10a + ab$

   d $10a + ab + a^2$ **e** $5a^2 + 6ab + 7ac$ **f** $5pq - 10qr + 15q^2$

   g $5pq - 10qr + 30rs$

6 Factorise these expressions as fully as you can.

   a $8x - xy$ **b** $8x - 2xy$ **c** $8x - 4xy$

   d $8x - x^2$ **e** $x^2 - 8x$ **f** $3x^2 - 12x$

   g $3x^2 - 12x + xy$ **h** $3x^2 - 12x + 6xy$ **i** $9x^2 + 3x - 12xy$

What do you think?

1 Work these out without using a calculator.

   a $57 \times 64 + 57 \times 36$ **b** $83 \times 54 - 83 \times 44$

   c $62 \times 87 + 13 \times 62$ **d** $48 \times 37 + 24 \times 26$

2 Factorise

   a $a(x + 4) + b(x + 4)$ **b** $x(x - 5) + 3(x - 5)$ **c** $3(p + 5) + 4(2p + 10)$

3 Explain why $xy + xz + yz$ cannot be factorised.

4 Find the missing expressions.

   a $8(2x + 1) - 3(\boxed{}) \equiv x + 11$ **b** $5(\boxed{}) - 3(4x - 3) \equiv 2(9x + 12)$

Consolidate – do you need more?

1 Work out

   a 6×103 **b** 7×99

2 Expand the brackets.

   a $5(a + 3)$ **b** $5(a + 2b)$

   c $5(2a - 3b)$ **d** $7(2a - 3b + 4)$

   e $x(x + 2)$ **f** $y(y - 3)$

   g $p(p + q + r)$ **h** $q(2q - 3r)$

3 Expand and simplify.

   a $3(x + 2) + 2(x + 3)$ **b** $3(x + 4) + 2(x - 1)$ **c** $2(2x + 1) + 3(x - 3)$

4 Factorise

   a $8x - 6$ **b** $6 + 9x$ **c** $4 - 2x$ **d** $x^2 + 6x$ **e** $5x - x^2$

5 Factorise fully

    a $15ab + 20bc$

    b $8xy + 10x^2$

    c $12pq - 16qr$

    d $6de + 12d^2 - 3d$

Stretch – can you deepen your learning?

1 **a** Match the shapes with the expressions that show their perimeters.

    i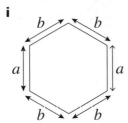

    ii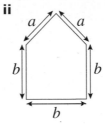

    iii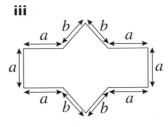

    A $\boxed{2(a + b) + b}$

    B $\boxed{2(3a + 2b)}$

    C $\boxed{2(a + 2b)}$

    b Design polygons with these perimeters.

      i $3(a + b)$

      ii $4(a + 2b)$

      iii $2(a + b + 2c)$

2 Expand and simplify.

    a $x(2x + 3) + 2x(x + 4)$

    b $x(3x + 2) - 2x(x + 4)$

    c $x(3x + 2) - 2x(4 - x)$

    d $x(3x + 2) - 2x(x - 4)$

    e $x(x + y) + y(x + y)$

    f $2x(4x + y) - y(y + 2x)$

    g $3x(2y - x) - 2y(3x - y)$

3 **a** Use factorisation to work out $0.72 \times 47.5 + 4.75 \times 2.8$

    b Use your knowledge of number bonds and powers of 10 to create puzzles similar to that in **a**. Challenge a partner.

Reflect

1 What is the difference between expanding brackets and factorising into brackets?

2 What does it mean if a question asks you to "factorise fully"?

Small steps

- Expand a pair of binomials **H**
- Expand multiple single brackets and simplify

Key words

Binomial – expression with two terms

Are you ready?

1 Expand the brackets.

   a $3(x + 2)$ **b** $x(x + 2)$ **c** $3(x - 2)$ **d** $x(x - 2)$

2 Simplify the expressions.

   a $3a + 4a$ **b** $3a - 4a$ **c** $-3a + 4a$ **d** $-3a - 4a$

3 Write as a single term without mathematical operations.

   a $-3 \times a$ **b** $-3 \times -a$ **c** $-3 \times -2x$ **d** $4 \times -5x$ **e** $-4 \times -5x$

4 Simplify the expressions.

   a $x^2 + 3x + 4x$ **b** $x^2 + 3x + 4x + 5$

   c $x^2 - 3x + 4x + 5$ **d** $x^2 - 3x - 4x - 5$

Models and representations

You can use algebra tiles and also the area model to represent multiplying more than one set of brackets.

In the last chapter you looked at multiplying out a single bracket. In this chapter you will explore multiplying two sets of brackets together, such as $(x + 1)(x + 3)$.

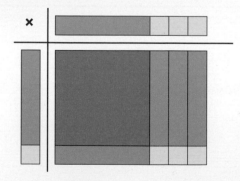

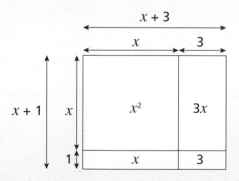

As both of these sets of brackets have two terms in them, they are called **binomials**. (In the same way a single term is a uninomial, an expression like $x + y + z$ is a trinomial, and so on.) Multiplying two sets of brackets, such as $(a + 3)(b + 4)$, is often called "expanding a pair of binomials".

Example 1

Expand $(a + 3)(b + 4)$

Method A

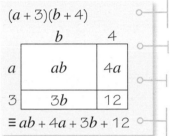

$$(a + 3)(b + 4) \equiv a(b + 4) + 3(b + 4)$$
$$\equiv ab + 4a + 3b + 12$$

$(a + 3)(b + 4)$ means $(a + 3)$ lots of $(b + 4)$.

Split this into a lots of $(b + 4)$ and 3 lots of $(b + 4)$, and then multiply out a single bracket as you did in Chapter 7.2

Notice the four terms are unlike, so the expression cannot be simplified.

Method B

$(a + 3)(b + 4)$

	b	4
a	ab	$4a$
3	$3b$	12

$\equiv ab + 4a + 3b + 12$

Think of $(a + 3) \times (b + 4)$ as finding the area of a rectangle with dimensions $a + 3$ and $b + 4$

Divide the sides into sections of a and 3, and b and 4

Find the area of each section.

Then add to find the total area and so the total value of the expanded expression.

Method C

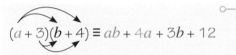

$$(a + 3)(b + 4) \equiv ab + 4a + 3b + 12$$

This is the same as the first method, but shown in a different way.

The arrows above the expression indicate multiplying a by $b + 4$ and the arrows below indicate multiplying 3 by $b + 4$

Example 2

Expand and simplify.

a $(x + 2)(x + 5)$ **b** $(x + 2)(x - 3)$

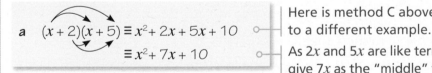

a $(x + 2)(x + 5) \equiv x^2 + 2x + 5x + 10$
$$\equiv x^2 + 7x + 10$$

Here is method C above applied to a different example.

As $2x$ and $5x$ are like terms they can be added to give $7x$ as the "middle" term in the final answer.

You could also use the area model or algebra tiles to show this expansion.

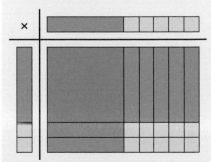

Again there is one x^2 tile, seven x tiles and ten 1s in the expansion.

b $(x + 2)(x - 3) \equiv x^2 - 3x + 2x - 6$

$\equiv x^2 - x - 6$

You need to be careful with positive and negative numbers.

Simplifying $-3x + 2x = -1x$, which you write as $-x$

Using algebra tiles you can see

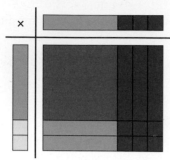

You can simplify by making zero pairs.

You can see the answer is $x^2 - x - 6$

Practice 7.3A

1 Copy and complete the area model to work out 87×46

	80	7
40	3200	280
6	_____	____

$3200 + 280 = 3480$

____ + ____ = ____ + ____

$87 \times 46 = $____

2 **a** Copy and complete an area model to work out $(a + 2)(b + 3)$

	b	3
a	ab	____
2	_____	6

$ab + $____$ + $____$ + 6$

b Use $a = 20$ and $b = 70$ to show that $22 \times 73 = 1606$

c Compare the expansions of $(a + 5)(b + 2)$ and $(a + 2)(b + 5)$
What's the same and what's different?

3 Expand

a $(p + 4)(q + 6)$ **b** $(p - 4)(q + 6)$ **c** $(p - 4)(q - 6)$

4 **a** Expand and simplify.

  i $(x + 2)(x + 6)$ **ii** $(x + 6)(x + 2)$ **iii** $(x + 2)(x - 6)$ **iv** $(x - 2)(x + 6)$

b Verify your answers are correct by substituting some values for x (for example, $x = 10$) into the binomials and into your expansions.

5 **a** Rhys thinks that $(x - 3)(x - 4) \equiv x^2 - 7x - 12$

Explain why Rhys is wrong, and find the correct expansion of $(x - 3)(x - 4)$

b Expand and simplify.

    i $(x - 1)(x - 5)$ **ii** $(x - 6)(x - 2)$ **iii** $(x - 8)(x - 10)$

c Compare the expansions of $(x + 5)(x + 2)$ and $(x - 2)(x - 5)$
What's the same and what's different?

What do you think?

1 **a** Bev is expanding the pairs of binomials $(x + 4)(x + 5)$ and $(2x + 4)(x + 5)$

$$(x + 4)(x + 5) \equiv x^2 + 5x + 4x + 20$$
$$\equiv x^2 + 9x + 20$$

So, $(2x + 4)(x + 5) \equiv 2x^2 + 18x + 40$

Explain why Bev is wrong and find the correct expansion of $(2x + 4)(x + 5)$

b Expand and simplify.

    i $(2x + 1)(x + 3)$ **ii** $(2x + 3)(x + 1)$ **iii** $(2x + 1)(2x + 3)$ **iv** $(2x + 1)(x - 3)$

    v $(2x - 3)(x + 1)$ **vi** $(2x - 1)(2x + 3)$ **vii** $(2x + 1)(2x - 3)$

c **i** Expand and simplify $(2x - 1)(2x - 3)$

    ii Explain why $(2x - 3)(2x - 1)$ will have the same expansion as $(2x - 1)(2x - 3)$

2 **a** Bobbie thinks $(x + 3)^2 \equiv x^2 + 9$. By writing $(x + 3)^2$ as $(x + 3)(x + 3)$, show that Bobbie is wrong.

b Expand and simplify.

    i $(x + 4)^2$ **ii** $(x - 4)^2$ **iii** $(x + a)^2$ **iv** $(2x + 5)^2$

c Find the difference between $(2x + 1)^2$ and $(2x - 1)^2$

Consolidate – do you need more?

1 **a** Work out 82×57 by finding the expansion of $(80 + 2)(50 + 7)$

Check your answer using a different method of multiplication.

b Compare working out 74×89 by finding the expansions of $(70 + 4)(80 + 9)$ and of $(70 + 4)(90 - 1)$. Which do you prefer?

2 **a** Use an area model to expand $(m + 3)(n + 6)$

b Use your preferred method to find

    i $(m + 4)(n + 6)$ **ii** $(m + 2)(n + 7)$

c Explain whether the expansions of $(m + 2)(n + 5)$ and $(m + 5)(n + 2)$ will be the same or different.

3 Expand and simplify.

   a $(a + 4)(a + 8)$ **b** $(x + 4)(x + 8)$ **c** $(y + 4)(y + 8)$

   d $(p + 3)(p + 7)$ **e** $(q + 10)(q + 2)$

4 Expand and simplify.

   a $(a + 4)(a - 8)$ **b** $(x - 4)(x + 8)$ **c** $(y - 4)(y - 8)$

   d $(p - 3)(p + 7)$ **e** $(q - 10)(q + 2)$

Stretch – can you deepen your learning?

1 **a** Expand and simplify.

     i $(x + 3)(x - 3)$ **ii** $(x + 5)(x - 5)$ **iii** $(x + 8)(x - 8)$

     Can you generalise your findings?

   b Use your generalisation to write down the answers to

     i $(x + 10)(x - 10)$ **ii** $(x - 7)(x + 7)$

   c By looking at your answer to part **a** again, work out $101^2 - 99^2$ without using a calculator.

2 All these expressions are equivalent. Find the values of all the letters, if possible.

   $(x + 5)(x + 4)$ $(x + 3)(x + 6) + a$ $(x + 2)(x + 10) + bx$

   $(x + 5)^2 + cx + d$ $(x + 1)(x + e) + fx + g$

3 Expand each pair of binomials in this grid.

$(x + 1)(x + 1)$	$(x + 1)(x + 2)$	$(x + 1)(x + 3)$	$(x + 1)(x + 4)$
$(x + 2)(x + 1)$	$(x + 2)(x + 2)$	$(x + 2)(x + 3)$	$(x + 2)(x + 4)$
$(x + 3)(x + 1)$	$(x + 3)(x + 2)$	$(x + 3)(x + 3)$	$(x + 3)(x + 4)$

   What patterns do you see in the answers? What can you predict about the answers to other binomials if the grid were extended horizontally and/or vertically?

4 How might you find the product of three binomials, for example $(x + 1)(x + 2)(x + 3)$ or $(x + 3)^3$?

Reflect

Explain how you find the product of two binomials.

7.4 Brackets and equations

Small steps

- Solve equations, including with brackets
- Form and solve equations with brackets

Key words

Solve – find a value that makes an equation true

Solution – a value you can substitute in place of the unknown in an equation to make it true

Are you ready?

1 Solve the equations.

   a $2x = 40$ **b** $\frac{x}{2} = 40$ **c** $x + 2 = 40$

   d $x + 40 = 2$ **e** $x - 40 = 2$ **f** $40 - x = 2$

2 Solve the equations.

   a $4y = 68$ **b** $4y + 1 = 68$ **c** $4y - 1 = 78$ **d** $78 = 4y - 1$

3 Solve the equations.

   a $\frac{z}{3} = 15$ **b** $\frac{z}{3} - 4 = 15$ **c** $\frac{z}{3} + 4 = 15$ **d** $\frac{z - 3}{4} = 15$ **e** $\frac{z + 3}{4} = 15$

4 Expand the brackets and simplify where possible.

   a $4(x + 3)$ **b** $5(y - 3)$

   c $7(3 - p)$ **d** $4(a - 3) + 2(3a - 5)$

Models and representations

In Book 1, you met several different ways of representing equations.

Cups and counters

Balance

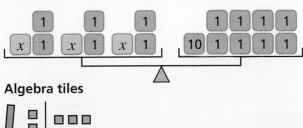

Algebra tiles

Bar model

x	2	x	2	x	2
\multicolumn{6}{c}{18}

These all show the equation $3(x + 2) = 18$

Example 1

Solve the equation $3(x + 2) = 18$

Method A

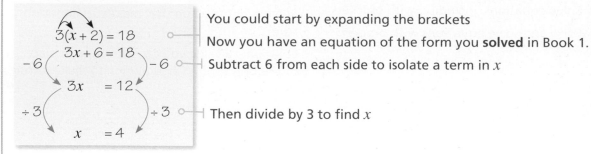

You could start by expanding the brackets

Now you have an equation of the form you **solved** in Book 1.

Subtract 6 from each side to isolate a term in x

Then divide by 3 to find x

You can also use bar models or other representations if you prefer.

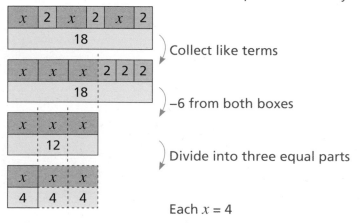

Collect like terms

−6 from both boxes

Divide into three equal parts

Each $x = 4$

Method B

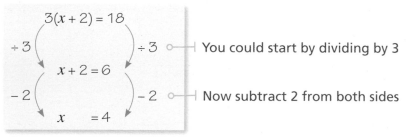

You could start by dividing by 3

Now subtract 2 from both sides

Here is how this method looks using bar models.

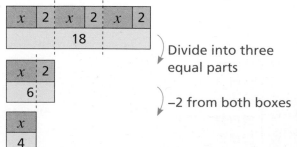

Divide into three equal parts

−2 from both boxes

Remember: you can check the answer to an equation by substituting the answer back into the original equation.

$$3(x + 2) = 18$$

Substitute $x = 4$

$$3(4 + 2) = 18$$

$3 \times 6 = 18$, which is correct.

Example 2

Solve the equations.

a $5(y - 4) = 12$ **b** $30 = 5(7 + t)$

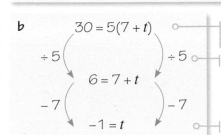

a

$$5(y - 4) = 12$$
$$5y - 20 = 12$$
$+20 \big(\qquad \big) +20$
$$5y = 32$$
$\div 5 \big(\qquad \big) \div 5$
$$y = 6.4$$

It is probably best to start with expanding the brackets here, as 5 doesn't divide exactly into 12

To isolate $5y$ you need to add 20 to both sides.

Then divide both sides by 5

The answer is 6.4. Remember, answers don't have to be integers. You can check by substitution.

b

$$30 = 5(7 + t)$$
$\div 5 \big(\qquad \big) \div 5$
$$6 = 7 + t$$
$-7 \big(\qquad \big) -7$
$$-1 = t$$

It doesn't matter that the unknown is on the right-hand side; you can work in the same way. As 30 is divisible by 5 you could use either method easily.

You can leave the answer as $-1 = t$ or rewrite it as $t = -1$

Practice 7.4A

1 Compare solving $2x + 5 = 18$ with $2(x + 5) = 18$

2 Solve the equations.

  a $4(a + 12) = 60$ **b** $24 = 3(b - 5)$ **c** $6(c + 1.5) = 18$

3 What's the same and what's different about solving $3(x + 4) = 15$ and $3(2x + 4) = 15$?

4 Solve the equations.

  a $2(3x - 1) + 4x = 12$ **b** $2(3x + 1) - 4x = 12$

  c $2(3x - 1) - 4 = 12$ **d** $2(3x - 1) + 4 = 12$

5 Solve the equations.

  a $\dfrac{y}{2} + 7 = 19$ **b** $\dfrac{3y}{2} + 7 = 19$

  c $3(\dfrac{y}{2} + 7) = 19$ **d** $3(\dfrac{y}{2} - 7) = 19$

6 Here are Beca's and Jackson's methods for solving the equation $10 - 4x = 8$

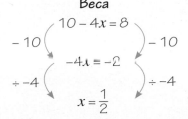

         Beca   *Jackson*

$$10 - 4x = 8$$
$-10 \big(\qquad \big) -10$
$$-4x = -2$$
$\div -4 \big(\qquad \big) \div -4$
$$x = \frac{1}{2}$$

$$10 - 4x = 8$$
$+4x \big(\qquad \big) +4x$
$$10 = 8 + 4x$$
$-8 \big(\qquad \big) -8$
$$2 = 4x$$
$\div 4 \big(\qquad \big) \div 4$
$$\frac{1}{2} = x$$

a Whose method do you prefer?

b Solve the equations.

i $10 - 3x = 7$

ii $10 - 3(x - 1) = 7$

iii $10 - 3(x + 1) = 7$

iv $10 - \dfrac{x + 1}{3} = 7$

What do you think?

1 Seb thinks of a number. He adds 4 and then multiplies by 5. His result is 70

a Calling Seb's number n, which of the equations represents Seb's number puzzle?

$n + 20 = 70$ $5n + 4 = 70$ $5(n + 4) = 70$

b Solve the equation. Check your answer by substituting back into the steps in the puzzle.

2 Here are two number puzzles.

Ed

I think of a number. I multiply by 2 and add 7. My answer is 23

Faith

I think of a number. I add 7 and multiply by 2. My answer is 23

a Do you think Ed's and Faith's numbers are the same or different? Why?

b Find Ed's and Faith's numbers by

i writing them as function machines and working backwards

ii forming and solving equations.

c Find the number Abdullah is thinking of.

I think of a number. I multiply by 4, add 8 and divide by 2. My answer is 1

d Make up your own number puzzles and challenge a partner.

Equations can be formed to help solve puzzles, real-life problems and problems within other areas of mathematics.

Example 3

Twice Huda's age in 3 years' time is her mother's age now. Huda's mother is 36. How old is Huda now?

Let Huda be n years old. — Choose a letter to represent Huda's age.

In 3 years she will be $n + 3$ years old. — Work through the information in the question, forming expressions as you go.

Twice this is $2(n + 3)$

This is equal to her mother's age.

$$2(n + 3) = 36$$ — You can then form an equation.

$\div 2 \quad\quad\quad \div 2$

$$n + 3 = 18$$ — Solve the equation. You could expand the brackets first if you prefer.

$-3 \quad\quad\quad -3$

$$n = 15$$

Example 4

The sum of three consecutive numbers is 42. Find the smallest number.

Let the smallest number be x — Choose a letter to represent the smallest number.

The other numbers will be $x + 1$ and $x + 2$ — Consecutive means following on from each other. One more than x is $x + 1$, and one more than $x + 1$ is $x + 2$

$$x + (x + 1) + (x + 2) = 42$$

$$3x + 3 = 42$$ — You can form an equation as you know the sum of the numbers is 42. Notice the brackets are just to make the three numbers clear, and there is nothing to expand.

$-3 \quad\quad\quad -3$

$$3x = 39$$

$\div 3 \quad\quad\quad \div 3$

$$x = 13$$

You can check the answer by adding 13, 14 and 15
The sum is 42, so 13 is the smallest number.

Practice 7.4B

1 If you treble the age Rhys was four years ago, you get 27. Form and solve an equation to find Rhys' age now.

2 If you double the age Ali will be in six years' time, you will get 40. How old is Ali now?

3 a In a bag of marbles there are three times as many green marbles as red marbles, and twice as many red marbles as blue marbles. There are 135 marbles in the bag.

    i In forming an equation, why is it easier to call the number of blue marbles x instead of calling the number of green marbles x?

    ii Form and solve an equation to find the number of marbles of each colour in the bag.

b In another bag of marbles there are two more white marbles than yellow marbles, and three times as many pink marbles as white marbles. There are 108 marbles in the bag.

Calling the number of yellow marbles x, form and solve an equation to find the number of marbles of each colour in the bag.

4 In a shop, there are three more £10 notes than £5 notes. Altogether there is £210 made up of £5 and £10 notes in the till. Compare forming and solving equations to find the number of each type of note if

a the number of £5 notes is x **b** the number of £10 notes is x

5 The perimeter of a hexagon is given by the formula $P = 3(2x + y)$

a Find y if $x = 5$ and the perimeter is 51 units.

b Find x if $y = 5$ and the perimeter is 51 units.

6 The perimeter of a rectangle is 52 cm. The length is 4 cm greater than the width. Find the area of the rectangle.

What do you think? 💭

1 Amina, Darius and Junaid are playing a game. Junaid scores 4 points more than Darius and Amina scores twice as many points as Junaid. They score 132 points altogether. Find the difference between the number of points scored by Amina and Darius.

2 In how many different ways can you find the values of the letters by forming and solving equations? Which ones involve brackets? What angle facts are you using?

a
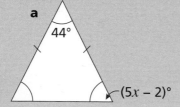

b
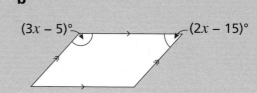

Consolidate – do you need more?

1 Solve the equations.

a $6x + 3 = 45$ **b** $6(x + 3) = 45$

c $6x - 3 = 45$ **d** $6(x - 3) = 45$

2 Solve the equations.

a $3(2y - 1) = 27$ **b** $40 = 5(4p - 2)$ **c** $6(2q + 7) = 18$

3 Solve the equations.

a $3(4a - 2) - 10 = 44$ **b** $3(4b - 2) - 10b = 44$

c $4(2 + 3c) - 4(3 + 2c) = 11$

4 Flo scores 5 marks more than Emily in a test. Kath scores twice as many marks as Flo. Between them, they score 99 marks. Find their marks in the test.

Stretch – can you deepen your learning?

1 Solve the equations.

a $\frac{2}{3}(a - 6) = 2$ **b** $\frac{2(b + 7)}{5} = 4$

c $\frac{1}{4}(7 - c) = 3$ **d** $\frac{4}{3}(5 - 2d) = 1$

2 a What's the same and what's different about solving these equations?

$2n + 3 = 4$ $2n + 3 = 4n$

You will explore solving equations like $2n + 3 = 4n$, where the letter appears on both sides of the equation, in Chapter 7.5

b Compare methods for solving the equations $2(n + 3) = 4$ and $2(n + 3) = 4n$

3 Form and solve equations to solve these number puzzles.

a I think of a number, double it and add 5. The answer is the number I started with.

b I think of a number, add 5 and then double the result. The answer is the number I started with.

Reflect

1 How do you go about solving an equation with brackets?

2 How do you use the information in a question to form an equation?

3 How can you check if your answer to an equation is correct?

7.5 Inequalities

Small steps

- Understand and solve simple inequalities
- Form and solve inequalities

Key words

Inequality – a comparison between two quantities that are not equal to each other

Solution set – a range of values for which a statement is true

Satisfy – make an equation or inequality true

Are you ready?

1 Put each set of numbers in order, starting with the lowest.

    a 7, 3, –5   **b** –4, 2, –6   **c** 8, –3, –8, 3

2 Choose a number to complete each statement correctly.

    a $5 > \square$ **b** $5 < \square$ **c** $-3 > \square$ **d** $\square < -4$

3 Solve the equations.

    a $p - 7 = 18$ **b** $12 = q + 4$ **c** $\frac{r}{5} = 10$ **d** $10t = 5$

4 Solve the equations.

    a $2m - 5 = 21$ **b** $6 + 4n = 38$ **c** $10 + \frac{x}{2} = 7$ **d** $\frac{y - 3}{4} = 5$

Models and representations

In the same way that you can use a balance to show an equation, a balance can also show inequalities.

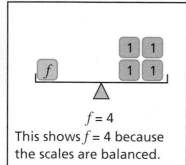

$f = 4$
This shows $f = 4$ because the scales are balanced.

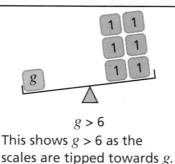

$g > 6$
This shows $g > 6$ as the scales are tipped towards g.

You could write this as $6 < g$.

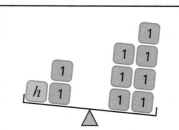

This shows $h + 2 < 7$. As the scales are tipped towards 7, this must be heavier.

You could write this as $7 > h + 2$

You have probably seen the signs > (greater than) and < (less than) before. There are two other signs that are used when working with inequalities.

⩾ means "greater than or equal to". If you are told $a \geqslant 5$, then a could have the value 5 or any value greater than 5, such as 5.1, 6, 150, and so on.

In the same way ⩽ means "less than or equal to".

Example 1

If $x = 7$, which of the statements are true?

| $x > 6$ | $x \geqslant 7$ | $x < 7$ | $x \leqslant 8$ |

$x > 6$ is true ○———┤ 7 is greater than 6, so the statement is true.

$x \geqslant 7$ is true ○———┤ 7 is equal to 7, so it is "greater than or equal to" 7.
Therefore the statement is true.

$x < 7$ is not true ○———┤ 7 is not less than 7

$x \leqslant 8$ is true ○———┤ 7 is less than 8 so it is "less than or equal to 8"

We say that $x = 7$ **satisfies** the inequalities $x > 6$, $x \geqslant 7$ and $x \leqslant 8$

Example 2

Given that $y > 10$, what can you say about

a $y + 1$ **b** $y - 4$ **c** $3y$ **d** $\dfrac{y}{2} + 6$?

a
$$y > 10$$
$+1 \quad \quad +1$
$$y + 1 > 11$$

○———┤ If $y > 10$, then 1 more than y must be greater than 11
Like an equation, if you add the same number to both sides of an inequality it remains true.

You can see this from the bar models.

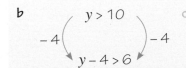

$y > 10$ $\qquad$ $y + 1 > 11$

b
$$y > 10$$
$-4 \quad \quad -4$
$$y - 4 > 6$$

○———┤ The same is true for subtraction.
You could draw a bar model to check.

c
$$y > 10$$
$\times 3 \quad \quad \times 3$
$$3y > 30$$

You can also multiply both sides of an **inequality** by a positive number, as these bar models show.

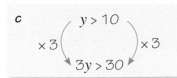

$y > 10$ $\qquad$ $3y > 30$

d
$$y > 10$$
$\div 2 \quad \quad \div 2$
$$\frac{y}{2} > 5$$
$+6 \quad \quad +6$
$$\frac{y}{2} + 6 > 11$$

So $\dfrac{y}{2} > 5$

So $\dfrac{y}{2} + 6 > 11$

○———┤ This example uses the rules discussed in the previous two problems.

Example 3

Find the sets of values for which

a $m + 3 > 8$ **b** $n - 5 \leqslant 8.2$ **c** $11 \geqslant 2p$ **d** $\frac{q}{3} + 4 < 5$

a
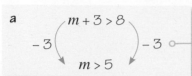
$m + 3 > 8$
$-3 \quad\quad -3$
$m > 5$

By subtracting 3 from both sides of the inequality, you see it is true for any value of m which is greater than 5

b

$n - 5 \leqslant 8.2$
$+5 \quad\quad +5$
$n \leqslant 13.2$

By adding 5 to both sides of the inequality, you see it is true for any value of n which is less than or equal to 13.2

c
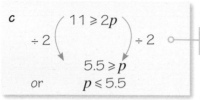
$11 \geqslant 2p$
$\div 2 \quad\quad \div 2$
$5.5 \geqslant p$
or $\quad p \leqslant 5.5$

By dividing both sides of the inequality by 2, you see it is true when $5.5 \geqslant p$. This is the same as saying $p \leqslant 5.5$

d
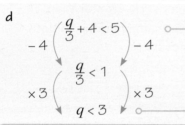
$\frac{q}{3} + 4 < 5$
$-4 \quad\quad -4$
$\frac{q}{3} < 1$
$\times 3 \quad\quad \times 3$
$q < 3$

This is a two-step inequality.

The inequality is true for values of q less than 3

Practice 7.5A

1   **a**   Write these in words.

     i $a > 7$ **ii** $b < 4$ **iii** $c \leqslant 0$ **iv** $d \geqslant -2$

  b   Write these using symbols.

     i x is less than 6   **ii** y is greater than or equal to -3

     iii   10 is greater than z

2   **a**   Write an integer that satisfies each inequality.

     i $w > 7$ **ii** $x < 3$ **iii** $y + 4 < 10$ **iv** $2t \geqslant 11$

  b   Write a number that is **not** an integer that satisfies each inequality.

     i $a < 1$ **ii** $b \geqslant 13$ **iii** $15 < c - 6$ **iv** $8 \leqslant \frac{d}{2}$

3   **a**   Write the inequalities shown by these balance diagrams.

     i

     ii

iii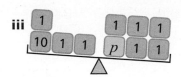

iv

b Draw a balance diagram to represent each inequality.

   i $a > 4$ **ii** $6 < b$ **iii** $2c < 10$ **iv** $15 > 3d + 1$

c Why is a balance not a good model to show $w \geqslant 8$?

4 Which of these are the same as $x > 5$?

| $5 > x$ | $x \geqslant 5$ | $5 < x$ | $5 \leqslant x$ |

5 Solve the inequalities.

a $x + 5 \leqslant 8$ **b** $y + 8 \leqslant 5$ **c** $\dfrac{p}{2} > 30$ **d** $10 \geqslant q - 5$

e $2w - 9 \geqslant 0$ **f** $6 + 4m < 18$ **g** $\dfrac{n + 5}{4} > -3$ **h** $14 > 5m - 7$

6 a Find the smallest integer that satisfies these inequalities.

   i $3p > 20$ **ii** $20 < 4q - 3$ **iii** $2t + 7 > 4$

b Find the greatest integer that satisfies these inequalities.

   i $6a < 42$ **ii** $15 \geqslant 3b$ **iii** $12 + 3p < 9$

What do you think? 💭

1 a Find the smallest integers that satisfy these inequalities.

   i $x > \pi$ **ii** $2x > \pi$ **iii** $\dfrac{x}{2} > \pi$ **iv** $\dfrac{x}{3} - 5 > 2\pi$

b Find the greatest integers that satisfy these inequalities.

   i $x < \pi$ **ii** $x + 4 \leqslant \pi$ **iii** $\dfrac{x}{5} + 3 \leqslant \pi$ **iv** $2x - 1 < \pi$

2 a Benji is trying to solve the inequality $5 - x > 2$

Show by substituting any number greater than 3 into the inequality that Benji's answer is incorrect.

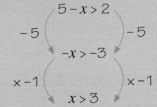

b Copy and complete this working to solve the inequality $5 - x > 2$

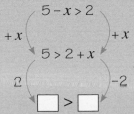

c Solve the inequalities.

   i $10 > 4 - x$ **ii** $7 \leqslant 4 - 3x$ **iii** $12 - \dfrac{x}{2} \geqslant 9$

Just like equations, inequalities can involve brackets and they can often arise from real-life situations.

Example 4

Solve the inequalities.

a $3(a - 6) > 15$ **b** $30 \leqslant 7(b + 2)$

a

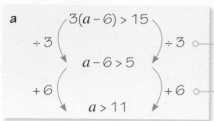

As 3 is a factor of 15 you can divide both sides of the inequality by 3

Then add 6 to each side

b

$$30 \leqslant 7(b + 2)$$
$$30 \leqslant 7b + 14$$
$-14 \qquad -14$
$$16 \leqslant 7b$$
$\div 7 \qquad \div 7$
$$\frac{16}{7} \leqslant b$$
$$2\frac{2}{7} \leqslant b$$

As 7 is not a factor of 30 it is easiest to expand the brackets.

Then subtract 14 from each side.

Then divide both sides by 7

You can convert the improper fraction to a mixed number.

You could also write this solution as $b \geqslant 2\frac{2}{7}$

Example 5

A sales worker is paid a basic wage of £250 a week and £35 for every sale they make. How many sales do they need to make to earn over £800 in a week?

Let the number of sales be n

Total earnings from sales is £$35n$

Total earnings for the week is £$(250 + 35n)$

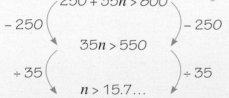

$$250 + 35n > 800$$
$-250 \qquad -250$
$$35n > 550$$
$\div 35 \qquad \div 35$
$$n > 15.7\ldots$$

They need to make at least 16 sales

Assign a letter for the number you want to find.

Use the information in the question to form expressions and set up an inequality.

Solve the inequality in the usual way.

Make sure your answer makes sense in the context of the question. The number of sales must be an integer, so the sales worker needs to make **16** or more sales.

Practice 7.5B

1 Solve the inequalities.

a $4(a - 3) \leqslant 20$ **b** $3(b + 5) > 15$ **c** $5(\frac{c}{2} - 1) \geqslant 11$

2 Ali's science exam is in two parts. He scores 25 marks in the first part. He needs 70 marks overall to pass.
How many marks does he need to score in the second part to pass?

3 The perimeter of a square is more than 50 cm.

  a What can you say about the length of each side of the square?

  b What can you say about the area of the square?

4 **a** The length of a rectangle is 4 cm greater than its width. The perimeter of the rectangle is less than 30 cm.
  Form and solve an inequality to find the range of possible values for the width of the rectangle.

  b How would your answer change if the perimeter of rectangle is instead "no more than 30 cm"?

5 The nth term of a sequence is given by the rule $4n + 5$

  a Find the position in the sequence for the first term that is greater than 100
  What is the value of this term?

  b Find the position in the sequence for the last term that is less than 1000
  What is the value of this term?

6 Bev thinks of a number. 3 more than double her number is more than 50.
Form and solve an inequality to find the range of possible values her number could take.

7 Amina has £300 in the bank. She pays £40 a month into her account.
How long will it be before Amina has over £1000 in her bank account?

What do you think? 💭

1 Explain why $x > 4$ and $x \geqslant 5$ are not the same inequality.

2 **a** $-2 < x \leqslant 4$ is a short way of writing "$-2 < x$ and $x \leqslant 4$".
  List the integer values x can take.

  b State the possible integer values b can take if $-3 \leqslant b < 2$

  c $15 \leqslant c < 25$. Find the possible values of c if

    i c is a square number   **ii** c is prime

    iii c is a factor of 60   **iv** c is a multiple of 3

3 **a** Why is it impossible to form a triangle of sides 8 cm, 2 cm and 3 cm?

  b If the sides of a triangle are a, b and c, write three inequalities in terms of a, b and c that must be satisfied.

Consolidate – do you need more?

1 Write a value of x that satisfies each inequality.

a $x > 12$ **b** $8 > x$ **c** $-3 > x$ **d** $x < 9$

2 Which of these inequalities does $x = 5$ satisfy?

| $x > 4$ | $5 \leqslant x$ | $4 < x$ | $5 > x$ | $x \leqslant 5$ | $6 < x$ |

3 Solve the inequalities.

a $a - 4 < 7$ **b** $10 > b + 6$ **c** $3c \geqslant 27$ **d** $\dfrac{d}{4} < 12$

e $30 \leqslant \dfrac{e}{6}$ **f** $5f - 1 > 29$ **g** $2g - 3 \leqslant 12$ **h** $15 > \dfrac{h}{2} - 7$

4 Solve the inequalities.

a $4(p + 3) > 16$ **b** $5(2 + q) \leqslant 14$

c $3 > 5(x - 2)$ **d** $6(y + 2) \geqslant 1$

5 Bobbie is buying books. Each book costs £12. She has £100 to spend.
What is the maximum number of books she can buy?

Stretch – can you deepen your learning?

1 If $a < b$, then $a^2 < b^2$

Use examples and counterexamples to show Marta's statement is only sometimes true.

2 By thinking of $10 < 2a + 3 < 15$ as two separate inequalities

a find the range of values between which a can lie

b state the possible integer values of a

3 Solve the inequality $\dfrac{1}{2}x + \dfrac{1}{3}(x - 9) \geqslant 17$

Reflect

What's the same and what's different about solving equations and solving inequalities?

Small steps

- Solve equations and inequalities with unknowns on both sides ⊕
- Form and solve equations and inequalities with unknowns on both sides ⊕

Key words

Unknown – a variable (letter), whose value is not yet known

Check – find out if you are correct

Are you ready?

1 Solve each equation.

   a $3x = 930$ **b** $204 = \frac{y}{5}$ **c** $p + 7 = 2$ **d** $12 = 30 - q$

2 Expand the brackets.

   a $7(a - 3)$ **b** $8(5 + b)$ **c** $3(7 - c)$ **d** $4(2d - 5)$

3 Solve each equation.

   a $2p + 7 = 20$ **b** $14 = \frac{x}{2} - 2$ **c** $7 - 3y = 1$ **d** $3(2w + 1) = 5$

4 I think of a number, double it, then add 3 and double it again. The result is 34
Form and solve an equation to find the number I was first thinking of.

Models and representations

Scales and bar models are very useful representations for equations with letters on both sides of the equals sign.

Both of these models show $3x + 2 = x + 6$

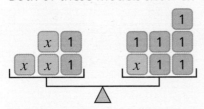

x	x	x	2
x	6		

Since they are balanced, the scales show equality, meaning the "weights" on both sides are equal.

The bar model also shows equality, as the bar representing $3x + 2$ is the same length as the bar representing $x + 6$

When you start solving an equation, you don't know the value of the letter, so it is called an **unknown**. An equation of the form $3x + 2 = x + 6$, with a term in x on the left-hand side and on the right-hand side, is called an "equation with unknowns on both sides".

You could also use algebra tiles or cups and counters if you like.

Example 1

Solve the equation $3x + 2 = x + 6$

Method A

Think of the equation as a balance.

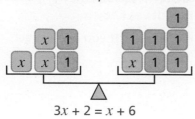

$$3x + 2 = x + 6$$

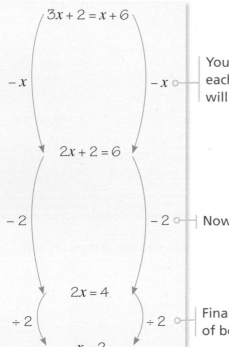

$$3x + 2 = x + 6$$

$-x$ $-x$ ○—

$$2x + 2 = 6$$

-2 -2 ○—

$$2x = 4$$

$÷2$ $÷2$ ○—

$$x = 2$$

You can take one "x" off each side and the scales will still be balanced.

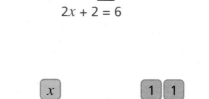

$$2x + 2 = 6$$

Now take 2 off each side.

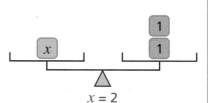

$$2x = 4$$

Finally divide the contents of both sides by 2

$$x = 2$$

You can **check** your answer by substituting into both sides of the equation

Left-hand side (LHS) = $3x + 2 = 3 \times 2 + 2 = 6 + 2 = 8$

Right-hand side (RHS) = $x + 6 = 2 + 6 = 8$

You can see how representing the solution with objects or pictures matches the solution using symbols.

Method B

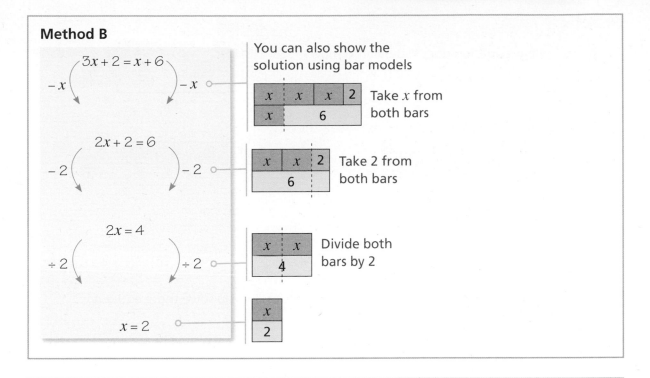

$3x + 2 = x + 6$

$-x$ $-x$

x	x	x	2

x	6

Take x from both bars

$2x + 2 = 6$

-2 -2

x	x	2

6

Take 2 from both bars

$2x = 4$

$\div 2$ $\div 2$

x	x

4

Divide both bars by 2

$x = 2$

x
2

You can also show the solution using bar models

Example 2

Solve the equations.

a $2m + 12 = 5m$ **b** $6y - 3 = 3y + 7$

a

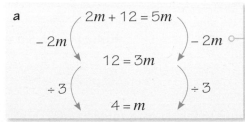

$2m + 12 = 5m$

$-2m$ $-2m$

$12 = 3m$

$\div 3$ $\div 3$

$4 = m$

You start by subtracting $2m$ from both sides as this will leave unknowns on one side only. As a bar model, the equation looks like this:

m	m	12

m	m	m	m	m

You could also show these steps on a balance – try it!

b

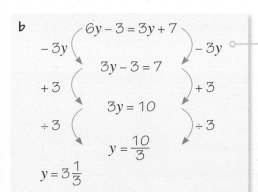

$6y - 3 = 3y + 7$

$-3y$ $-3y$

$3y - 3 = 7$

$+3$ $+3$

$3y = 10$

$\div 3$ $\div 3$

$y = \dfrac{10}{3}$

$y = 3\frac{1}{3}$

You start by subtracting $3y$ from both sides as this will leave unknowns on one side only.

It is difficult to represent negatives on a balance, but you could show this as a bar model.

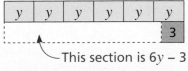

y	y	y	y	y	y

	3

This section is $6y - 3$

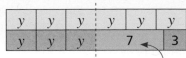

y	y	y	y	y	y

y	y	y	7	3

We have put this equal to $3y + 7$

Taking off $3y$ from the top and bottom bar leaves you with $3y = 10$, as with the method just using symbols.

Practice 7.6A

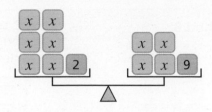

1 **a** What equation does this balance show?

b Solve the equation.

2 **a** What equations do these bar models show?

i

5x	
3x	18

ii

5y	12
7y	2

b Solve the equations.

3 **a** Solve the equations on the cards.

i $3x + 4 = 10$ **ii** $3x + 4 = 7x$ **iii** $3x + 4 = x + 10$

 **b** What's the same and what's different about how you solve these equations?

4 Solve the equations.

a $a + 26 = 5a + 10$ **b** $8b + 7 = 4b + 9$ **c** $11c = 7c + 10$

5 Find the values of the letters.

a $w - 3 = 4w + 6$ **b** $6x - 19 = 2x - 7$ **c** $6y - 7 = 2y - 19$

 **6** **a** Compare solving $4x + 2 = x + 11$ with solving $4(x + 2) = x + 11$

b Solve the equations.

i $3(p + 7) = 8p - 19$ **ii** $3(q + 4) = 2(q - 1)$ **iii** $4(2m - 3) = 5m + 3$

What do you think?

1 Seb thinks of a number. He adds 4 and then multiplies by 5. His result is 7 times the number he started with.

a Calling Seb's number n, which of these equations represents Seb's number puzzle?

$n + 20 = 7n$ $5n + 4 = 7n$ $5(n + 4) = 7n$ $5n + 4 = 7 + n$

b Solve the equation to find Seb's number.

2

> 10 more than 4 times my number is the same as 18 more than double my number.

Form and solve an equation to find the number Faith is thinking of.

3

> I know how to solve equations with unknowns on both sides, but what about inequalities?

How would you advise Ed? Compare solving $5x + 4 = 8x - 20$ with solving $5x + 4 < 8x - 20$

④ Find the smallest integer value of x that satisfies these inequalities.

a $3x + 5 > x + 2$ **b** $5x - 7 > 6 + 2x$ **c** $2x - 7 < 10x - 8$

Remember: when solving equations and inequalities where the unknown has a negative coefficient, it is easier to start by adding to get rid of the negative term.

Example 3

a Solve the equation $10 - 3p = 5p + 6$ **b** Solve the inequality $10 - 3p < 5p + 6$

c Solve the inequality $13 - 2q \geqslant 22 - 5q$

a

$$10 - 3p = 5p + 6$$
$+3p$ ⟍ ⟋ $+3p$
$$10 = 8p + 6$$
-6 ⟍ ⟋ -6
$$4 = 8p$$
$\div 8$ ⟍ ⟋ $\div 8$
$$\frac{1}{2} = p$$

As there is a negative term in p, you can add $3p$ to both sides to simplify the equation.

Now you have an "ordinary" two-step equation.

b

$$10 - 3p < 5p + 6$$
$+3p$ ⟍ ⟋ $+3p$
$$10 < 8p + 6$$
-6 ⟍ ⟋ -6
$$4 < 8p$$
$\div 8$ ⟍ ⟋ $\div 8$
$$\frac{1}{2} < p$$
$$p > \frac{1}{2}$$

You can approach this inequality in exactly the same way.

c

$$13 - 2q \geqslant 22 - 5q$$
$+5q$ ⟍ ⟋ $+5q$
$$13 + 3q \geqslant 22$$
-13 ⟍ ⟋ -13
$$3q \geqslant 9$$
$\div 3$ ⟍ ⟋ $\div 3$
$$q \geqslant 3$$

The coefficients of both terms in q are negative. Adding $2q$ would still leave a negative q term on the right-hand side, so it is best to add $5q$ to both sides.

Practice 7.6B

① Solve the equations and inequalities.

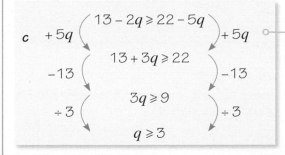

a $11 - 2x = 2x - 5$ **b** $8 - 4y = 6y - 2$ **c** $5 - 15z = 21 - 7z$

d $6 - a < 3 + a$ **e** $5b - 3 < 2 - 5b$ **f** $12 - 4c \geqslant 8 - 2c$

2 Here is a rectangle.

    a Form and solve an equation to find the value of x

    b Hence find the area and perimeter of the rectangle.

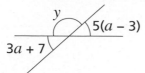

3 The diagram shows two straight lines crossing.

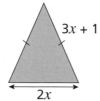

    a Find the value of a

    b Find the value of y

4 The perimeter of the triangle is the same as the perimeter of the rectangle.

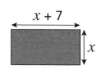

Work out the area of the rectangle.

5 **a** Eight years ago Tiff was half her age now. How old is Tiff now?

    b Two years ago Tiff's father was three times his age 28 years ago. How old is Tiff's father now?

What do you think? 💭

1 Jackson is trying to solve the equation $6x + 5 = x$

> I'm stuck! If I take x from both sides, there's just a blank: $5x + 5 =$

Help Jackson solve the equation.

2 How would you go about solving equations of the form $\frac{3x - 1}{4} = 2x - 9$?

3 Design an equation of the type in question 2 with a solution of

    a $x = 5$ **b** $x = -2$

    Can you involve brackets as well?

Consolidate – do you need more?

1 **a** What equation does this balance show?

    b Solve the equation.

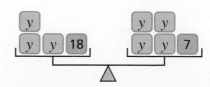

2 a What equations do these bar models show?

i

6x	8

4x	18

ii

5y	9

7y

b Solve the equations.

3 Solve the equations and inequalities.

a $3x + 5 = x + 19$　　　**b** $3x - 5 = x + 19$　　　**c** $3x - 5 < x - 19$

d $6y = 2y + 40$　　　**e** $6y + 10 > 2y + 40$　　　**f** $6y - 10 = 2y + 40$

g $4(n + 5) = 9n$　　　**h** $8n = 5(n + 6)$　　　**i** $8n = 5(n - 6)$

j $3(p + 4) < 4p + 2$　　　**k** $7(p - 2) = 2(p + 8)$　　　**l** $6(5 + p) = 2(7 - p)$

4 I think of a number. Three less than twice my number is 8 more than my number. Work out my number.

5 The area of the rectangle is double the area of the triangle. Find both their areas.

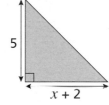
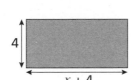

Triangle: height 5, base $x + 2$. Rectangle: height 4, width $x + 4$.

Stretch – can you deepen your learning?

1 a Why is it impossible to solve the equation $a + b = 20$ even though there are unknowns only on one side of the equals sign?

b You are also told $2a + b = 25$. By drawing bar models of the two equations, work out the values of a and b

2 The bar models show $4x + y = 19$ and $2x + y = 14$.

x	x	x	x	y

19

x	x	y

14

Work out the values of x and y.

3 a The sum of two numbers is 20. The difference between the two numbers is 6. Find the two numbers.

b The sum of two numbers is 40. The difference between the two numbers is 17. Find the two numbers.

Reflect

1 Explain the steps you would take to solve an equation or an inequality with unknowns on both sides.

2 How can you check that your answer to an equation with unknowns on both sides is correct? Is the method the same for an inequality?

Small steps

- Identify and use formulae, expressions, identities and equations

Key words

Variable – a numerical quantity that might change, often denoted by a letter, for example x or t

Formula – a rule connecting variables written with mathematical symbols

Identity – a statement that is true no matter what the values of the variables are

Equation – a statement with an equals sign, which states that two expressions are equal in value

Are you ready?

1 Find the value of each expression when $t = 12$

   a $3t + 7$ **b** t^2 **c** $\dfrac{84}{t}$

2 Expand the brackets.

   a $7(a - 3)$ **b** $7(3 - a)$ **c** $10(2t - 4b)$

3 Write down the formulae for

   a   the perimeter of a rectangle length a and width b

   b   the area of a rectangle length a and width b

4 Use the formula $A = 2bc$ to work out A when $b = 7$ and $c = 10$

Models and representations

In this chapter you will use algebraic symbols to represent different types of relationships.

$3x + 5$ is an **expression**. It is a collection of terms. You can work out the value of the expression from different values of x

$3(x + 5) = 36$ is an **equation**. You can solve this to find the value of x

$3(x + 5) \equiv 3x + 15$ is an **identity**. It is true for all values of the letter x

$A = bh$ is a **formula**. It is the formula for the area of a parallelogram of base b and perpendicular height h. You can substitute values for b and h to work out the value of A. You can also use it to form equations to work out b or h if you are given the values of the other letters.

Example 1

a Find the area of the parallelogram.

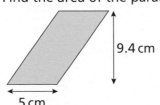

9.4 cm

5 cm

b Find the height of the parallelogram.

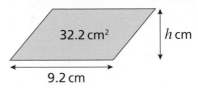

32.2 cm² — h cm

9.2 cm

a $A = bh$ ——| Start by writing down the formula.

$A = 5 \times 9.4$ ——| Substitute in the values of b and h

$A = 47\,cm^2$ ——| Work out the calculation.

b $A = bh$ ——| Start by writing down the formula.

$32.2 = 9.2 \times h$ ——| Substitute in the values you know

$32.2 \div 9.2 = h$ | (A and b) to give you an equation.

$3.5 = h$ ——| Solve the equation.

The height is 3.5 cm

Example 2

A hexagon is formed by cutting a smaller rectangle from the corner of a larger rectangle.

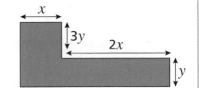

x — $3y$ — $2x$ — y

a Write an expression for the perimeter of the hexagon.

b Write a formula for the perimeter, P, of the hexagon.

c Use your formula to work out the perimeter of the hexagon when $x = 10\,cm$ and $y = 5\,cm$

d Use your formula to work out the value of x if the perimeter of the hexagon is 53 cm and $y = 4\,cm$

a

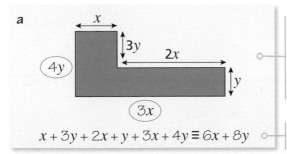

Start by working out the lengths of the unknown sides in terms of x and y. You can do this because you know the original shape was a rectangle and the opposite sides of a rectangle are equal in length.

$x + 3y + 2x + y + 3x + 4y \equiv 6x + 8y$ ——| Add up the sides to find the perimeter. Simplify your answer.

b $P = 6x + 8y$ ——| Put your expression equal to P to give you a formula.

c $P = 6x + 8y$ ——| Start by writing down the formula.

$P = 6 \times 10 + 8 \times 5$ ——| Substitute in the values of x and y

$P = 60 + 40$ ——| Work out the calculations.

$P = 100\,cm$

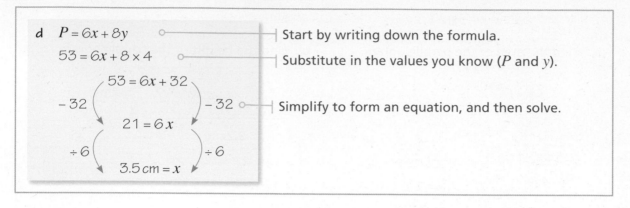

d $P = 6x + 8y$ — Start by writing down the formula.

$53 = 6x + 8 \times 4$ — Substitute in the values you know (P and y).

$53 = 6x + 32$

-32 ⟍ ⟋ -32 — Simplify to form an equation, and then solve.

$21 = 6x$

$\div 6$ ⟍ ⟋ $\div 6$

$3.5\,cm = x$

Practice 7.7A

1 Which of these are formulae and which are equations? Explain how you know.

A $\quad P = \dfrac{F}{A}$

B $\quad A = lw$

C $\quad \pi d = 30$

D $\quad 42 = 16 + 2w$

E $\quad d = st$

F $\quad S = 2\pi r + \pi r^2$

2 Copy and complete the identities.

a $3a + \boxed{} \equiv 10a$

b $4(b + 2c) \equiv 4b + \boxed{}$

c $12x + 9y \equiv 3(\boxed{})$

d $5p + 9p \equiv 20p - \boxed{}$

e $12p \times \boxed{} \equiv 48pq$

3 The perimeter of a rectangle of length l and width w is given by the formula $P = 2(l + w)$

Use the formula

a to find P when $l = 6$ and $w = 20$

b to find w when $P = 19$ and $l = 5$

4 The area of a triangle is given by the formula $A = \dfrac{1}{2}bh$

Use the formula

a to find A when $b = 20$ and $h = 9$

b to find b when $A = 20$ and $h = 5$

5 **a** Find a formula for the perimeter, P, of the hexagon in terms of a and b

b Work out P when $a = 7$ and $b = 3$

c Work out a when $b = 2$ and $P = 21$

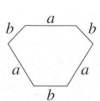

6 **a** Explain why the length of the line with a question mark is $3p$

b Find an expression for the perimeter of the shape.

c Find a formula for the perimeter, P, of the shape.

d Why are parts **b** and **c** different?

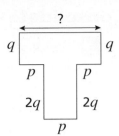

7 Given that $6(p + 2q) + 3(4p + bq) \equiv ap + 33q$, find the values of a and b

What do you think?

1    $y = 2x + 3$ is called the "equation of a straight line" but I think it's a formula because you can use it to work out the value of y when you know x

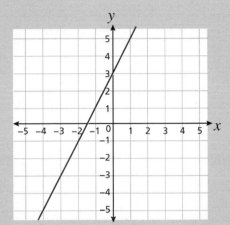

x	-2	-1	0	1
$y = 2x + 3$	-1	1	3	5

Do you agree with Zach?

2 Ali says the perimeter of the shape is $4x + 8y$

Mario says the perimeter of the shape is $2(2x + 4y)$

Lida says the perimeter of the shape is $4(x + 2y)$

Who do you agree with? Why?

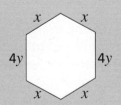

3 The formula for the time period of a pendulum is $T = 2\pi\sqrt{\dfrac{l}{g}}$, where l is the length of the string and g is the acceleration due to gravity, which is $9.8\,\text{m/s}^2$

a Find the time period for a pendulum of length 2 m.

b Darius thinks the time period of a pendulum of length 4 m will be double that of a pendulum of length 2 m. Use calculations to show that Darius is wrong.

Consolidate – do you need more?

1 Which of these are expressions, which are formulae and which are equations? Explain how you know.

A $\boxed{2c + 30 = 212}$ B $\boxed{b^2 + c^2}$ C $\boxed{V = IR}$

D $\boxed{F = \dfrac{9}{5}C + 32}$ E $\boxed{4x - 7}$ F $\boxed{48 = 6x}$

G $\boxed{2\pi r}$ H $\boxed{D = \dfrac{M}{V}}$ I $\boxed{v = u + at}$

2 Copy and complete the identities.

a $8a - \boxed{} \equiv a$

b $2(3b - 5) \equiv \boxed{} - \boxed{}$

c $10x + 5y \equiv 5(\boxed{} + \boxed{})$

d $5p \times 9p \equiv \boxed{}$

e $3(m + 4n) + 2(m - 6n) \equiv \boxed{}$

3 The side length of a square is l

a Write a formula to find the perimeter, P, of the square.

b Use the formula to find P when $l = 23$

c Use the formula to form and solve an equation to find l when $P = 23$

4 A car salesman has a basic salary of £800 a month. For every car he sells, he earns a commission of £150

a Write a formula to find his wage, W, in terms of the number, n, of cars he sells.

b Use the formula to find his wage in a month when he sells 12 cars.

c Use the formula to form and solve an equation to find the number of cars he sells in a month when he earns £1850

5 Sketch a scalene triangle, an isosceles triangle and an equilateral triangle, using letters to represent the lengths of each side.
What's the same and what's different about the formulae for the perimeter of each type of triangle?

Stretch – can you deepen your learning?

1 The formula for the area of a rectangle with length l and width w is $A = lw$

a A rectangle of width 5 cm has an area of 90 cm². Work out the length of the rectangle.

b Write a formula to find the width of a rectangle if you know its length and its area.

c Write a formula to find the length of a rectangle if you know its width and its area.

2 Write a formula to find

    a the diameter of a circle if you know its radius

    b the radius of a circle if you know its diameter

    c the diameter of a circle if you know its circumference

    d the radius of a circle if you know its circumference.

> You may remember this from Book 1. You will explore this in Block 14 of this book.

3 The formula for the area of a trapezium is $A = \frac{1}{2}(a + b)h$

    a Find the value of A given that $a = 7$, $b = 8$ and $h = 10$

    b Find the value of h given that $A = 72$, $a = 6$ and $b = 10$

    c Find the value of b given that $A = 78$, $a = 7$ and $h = 12$

4 The formulae for the sum of the first n positive integers is $S = \frac{1}{2}n(n + 1)$

    a Find the sum of the integers from 1 to 100

    b Find the sum of the integers from 1 to 200

    c Find the sum of the integers from 101 to 200

    d Describe the connection between the formula and the triangular numbers.

Reflect

What's the same and what's different about expressions, formulae, equations and identities? Give examples to support your reasoning.

I have become **fluent in…**	I have developed my **reasoning** skills by…	I have been **problem-solving** through…
■ forming and using algebraic expressions	■ deciding how best to represent a situation	■ modelling problems by writing them as equations and inequalities
■ expanding single brackets	■ knowing when and how to simplify expressions	■ solving multi-step problems by representing them as equations and inequalities
■ expanding a pair of binomials Ⓗ	■ making connections between relationships and their algebraic representations	
■ forming and solving linear equations	■ identifying variables and expressing relations between variables algebraically	■ selecting appropriate methods and techniques to apply to problems
■ forming and solving linear inequalities		
■ identifying formulae, expressions, identities and equations.	■ recognising when to use formulae, expressions, identities and equations.	■ representing equations and inequalities in a wide variety of forms.

Check my understanding

1 **a** n is a number. Write an expression for the number that is

    i 3 more than n **ii** 5 less than triple n **iii** half the square of n

  b Find the values of your expressions when $n = -4$

2 **a** Copy and complete the identity $3(2a - \square + 5) \equiv 6a - 12b + \square$

  b Work out 83×99

  c Expand and simplify $5(2b - 3c) - 2(3c - 4b)$

  d Expand and simplify $(3 - x)(x + 8)$ Ⓗ

3 **a** **i** Factorise $10x + 12y$ **ii** Factorise $10x + xy$ **iii** Factorise fully $10x^2 + 12xy$

  b Why does **a iii** say "Factorise fully"?

4 **a** Ali thinks of a number. Three more than double Ali's number is 40. Find Ali's number.

  b Lida thinks of a number. She adds 3 and then doubles her answer. Her answer is also 40. Find Lida's number.

5 **a** Solve the inequalities.   **i** $6x > 408$ **ii** $\frac{y}{3} - 7 \geqslant 50$ **iii** $10 \leqslant 4(z + 5)$

  b What is the smallest integer solution of each inequality?

6 Solve the equation $10 - 5x = 4(x - 2)$ Ⓗ

7 Which of these is an expression, which is a formula and which is an equation?

  A $\boxed{F = ma}$ B $\boxed{\pi r^2 h}$ C $\boxed{10 = 16 + 8a}$

8 Sequences

In this block, I will learn...

how to generate a sequence given in words

Start at 6 and double every time

$6 \quad 12 \quad 24 \quad 48 \quad \cdots$
$\quad \times 2 \quad \times 2 \quad \times 2 \quad \times 2$

how to describe a sequence

$7 \quad 10 \quad 13 \quad 16 \quad \cdots$
$\quad +3 \quad +3 \quad +3 \quad +3$

- The sequence is linear
- It starts at 7
- The term-to-term rule is "add 3 to the previous term"

how to generate a sequence using a simple algebraic rule

$4n - 11$

1st term: $n = 1$
$\quad 4 \times 1 - 11 = 4 - 11 = -7$

2nd term: $n = 2$
$\quad 4 \times 2 - 11 = 8 - 11 = -3$

3rd term: $n = 3$
$\quad 4 \times 3 - 11 = 12 - 11 = 1$

how to find out whether a given number is in a sequence

Is 80 a term in the sequence given by the rule $3n - 7$?

$3n - 7 = 80$
$+7 \qquad\qquad +7$
$\quad 3n = 87$
$\div 3 \qquad\qquad \div 3$
$\quad n = 29$

80 is the 29th term in the sequence

how to generate a sequence using a complex algebraic rule

$6 - 2n^2$

1st term: $n = 1$
$\quad 6 - 2 \times 1^2 = 6 - 2 = 4$

2nd term: $n = 2$
$\quad 6 - 2 \times 2^2 = 6 - 8 = -2$

3rd term: $n = 3$
$\quad 6 - 2 \times 3^2 = 6 - 18 = -12$

how to find the rule for the nth term of a sequence Ⓗ

$-4 \quad -1 \quad 2 \quad 5 \quad \cdots$
$\quad +3 \quad +3 \quad +3$

$3n: \quad 3 \quad 6 \quad 9 \quad 12$
$\qquad -7 \quad -7 \quad -7 \quad -7$
$\qquad -4 \quad -1 \quad 2 \quad 5$

The nth term is given by $3n - 7$

Small steps

- Generate sequences given a rule in words
- Generate sequences given a simple algebraic rule

Key words

Sequence – a list of items in a given order, usually following a rule

Term-to-term rule – a rule that describes how you get from one term of a sequence to the next

Linear sequence – a sequence whose terms are increasing or decreasing by a constant difference

Geometric sequence – a sequence is geometric if the value of each successive term is found by multiplying or dividing the previous term by the same number

Are you ready?

1 A sequence starts 12, 15, 18, 21, 24…

   a What is the third term of the sequence?

   b Describe how the sequence changes.

   c Predict the next term of the sequence.

2 Draw the next term in the sequence.

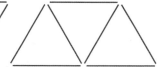

3 Which of these sequences are linear and which are non-linear?

     A   18, 14, 10, 6…

     B   18, 30, 42, 54…

     C   18, 20, 24, 30…

     D   18, 9, 4.5, 2.25…

4 What's the same and what's different about the sequence of odd numbers and the sequence of even numbers?

Models and representations

You can model a **sequence** of shapes by drawing, or by using solid objects, such as cubes, to represent each term.

You can write the sequence as a list: 1, 3, 5, 7...

or in a table:

Position	1	2	3	4
Term	1	3	5	7

You can also describe a sequence in words:

> The sequence starts at 1. The term-to-term rule is "add 2 to the previous term".

You can use descriptions in words to generate a sequence.

Example 1

a   Three sequences all have the first term 10. Find the fourth term of each sequence if the term-to-term rule is:

   i   add 7 to the previous term

   ii   subtract 7 from the previous term

   iii   double the previous term.

b   Are the sequences linear, geometric or neither?

a **i** 10, 17, 24, 31
     The fourth term is 31

   ii 10, 3, −4, −11
     The fourth term is −11

   iii 10, 20, 40, 80
     The fourth term is 80

Start with **10** each time, and apply the **term-to-term rule** until you get to the fourth term.

b **i** Linear

There is a constant difference between the terms, so the sequence is linear.

   ii Linear

Linear sequences can be descending as well as ascending.

   iii Geometric

A **geometric sequence** has a constant multiplier between the terms.

Example 2

The rule to find the next term in a Fibonacci sequence is "add the previous two terms together".

Find the first five terms of the Fibonacci sequence that starts 7, 10...

1st term = 7, 2nd term = 10

3rd term = 7 + 10 = 17

4th term = 10 + 17 = 27

5th term = 17 + 27 = 44

The first five terms are 7, 10, 17, 27, 44

These are the terms you are given.

You find the third term by adding together the first two terms.

You find the fourth term by adding together the second and third, and so on.

Practice 8.1A

1 The first term in a sequence is 100. Find the fourth term if the term-to-term rule is:

   a add 10 to the previous term

   b subtract 10 from the previous term

   c multiply the previous term by 10

   d divide the previous term by 10

2 Here are the first three terms of a sequence of white and black rectangles.

   a Draw the next pattern in the sequence.

   b Describe the term-to-term rule for the number of white rectangles in the pattern.

   c Describe the term-to-term rule for the number of black rectangles in the pattern.

   d Describe the term-to-term rule for the total number of rectangles in the pattern.

3 The sequence shown in the table is linear.

Position	1	2	3	4
Term	18	15	12	9

   a Write the term-to-term rule for the sequence.

   b Find the fifth term in the sequence.

4 **a** Emily is describing a sequence. She says, "The term-to-term rule is add 7 every time." Explain why this is not enough information for someone else to be able to write down the sequence.

   b Fully describe these sequences.

     i 50, 100, 150, 200, 250...

     ii 50, 100, 200, 400...

     iii 50, 30, 10, −10...

5 The sequence shown in the table is geometric.

Position	1	2	3	4
Term	2	20	200	2000

   a What is the term-to-term rule for the sequence?

   b What is the fifth term in the sequence?

6 Here is a flowchart for generating a sequence.

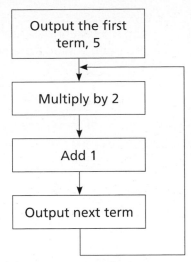

a The first two terms, output by the flowchart are 5 and 11. Find the next three terms.

b Instead of the first term being 5, it is altered to 6. Find the first five terms of the sequence.

c The instructions in the second and third boxes of the flowchart are swapped around. Find the first five terms if the starting number is

i 5

ii 6

7 The term-to-term rule for a sequence is "multiply the previous term by 3 and subtract 2". Find the first five terms of the sequence if the first term is

a 2

b 0

c 1

8 The term-to-term rule for a Fibonacci sequence is "add the previous two terms together".

a A Fibonacci sequence starts with 7 and 8. Find the next five terms.

b The fourth and fifth terms of a Fibonacci sequence are 15 and 24.

i Find the first three terms.

ii What do the first five terms in the sequence have in common?
Will this be true as the sequence continues?

What do you think? 💭

1 Investigate a variation of the Fibonacci sequence. In this new sequence, instead of adding the previous two terms together, you add the previous three terms. Try these starting points:

A ⟦ 1, 1, 1... ⟧   B ⟦ 1, 1, 2... ⟧   C ⟦ 1, 1, 3... ⟧   D ⟦ 1, 2, 3... ⟧

2 Investigate another variation of the Fibonacci sequence. Now, instead of adding the previous two terms together, you find the difference between the previous two terms. Try these starting points:

A ⟦ 1, 1... ⟧   B ⟦ 1, 2... ⟧   C ⟦ 1, 3... ⟧   D ⟦ 2, 4... ⟧

3 a Find the term-to-term rules for these sequences.

i 10, 11, 13, 16, 20...

ii 1, 2, 6, 24, 120...

iii 8000, 2000, 500, 125...

b Are the sequences linear, geometric or neither?

You can also represent sequences using algebraic rules.

⟦ $3n + 1$ ⟧   ⟦ $4n - 2$ ⟧   ⟦ $10 - 6n$ ⟧

These rules tell us the value of the nth term of a sequence; this is the term that is in the nth position.

Rules of the form $an + b$ produce linear sequences.

Example 3

The nth term of a sequence is given by the rule $6n + 3$

a Work out the 8th term of the sequence.

b Find the first term of the sequence that is greater than 100

c Describe the sequence in words.

a When $n = 8$, $6n + 3 = 6 \times 8 + 3$
$$= 51$$

The 8th term means $n = 8$, so substitute $n = 8$ into the expression for the nth term.

b
$$6n + 3 > 100$$
$-3 \quad \quad -3$
$$6n > 97$$
$\div 6 \quad \quad \div 6$
$$n > 16.166...$$

The 17th term will be greater than 100

When $n = 17$, $6n + 3 = 6 \times 17 + 3$
$$= 105$$

Set up an inequality.

Solve this the way you learned in Block 7

n must be an integer.

Substitute into the expression for the sequence to find the required term and check that it is greater than 100

c   When $n = 1, 6n + 3 = 6 \times 1 + 3$
$= 9$

Find the first two terms by substituting $n = 1$ and $n = 2$

When $n = 2, 6n + 3 = 6 \times 2 + 3$
$= 15$

The difference between the first two terms is $15 - 9 = 6$

You know that the sequence is linear as $6n + 3$ is of the form $an + b$

The sequence starts at 9 and goes up 6 every time

You can therefore describe the sequence by giving the first term and the difference.

Later in your studies, you will learn more about the connection between the numbers in the rule for a linear sequence and the way the sequence behaves.

Practice 8.1B

1 a   Work out the first five terms of the sequence given by these algebraic rules.

i $3n + 1$ ii $5n - 2$ iii $4n$ iv $6n + 5$

b   Describe the sequences in words.

c   What connection can you see between the sequences and their algebraic rules?

2 Find the 100th term of each of the sequences in question 1

3 A sequence is given by the rule $10 - 3n$

a   Work out the first five terms of the sequence.

b   Is the sequence linear, geometric or neither?

c   Find

i   the 50th term of the sequence   ii   the 200th term of the sequence.

4 a   Form and solve an equation to find which term of the sequence given by the rule $3n - 1$ is 128

b   Show that 200 is a term in the sequence but 400 isn't.

5 a   Find the value of the first term in the sequence given by the rule $6n - 5$ that is greater than 250

b   i   Find the first three terms of the sequence given by the rule $3n - 70$

ii   Find the last term in the sequence that is negative.

6 a   What do you notice about the last digits of the numbers in the sequence given by the rule $5n + 2$?

b   Deduce the last digits of the numbers in the sequence given by these rules.

i $5n + 3$ ii $5n + a$

c   Find a rule for a sequence so that the last digits of the numbers are always 1

What do you think? 💭

1 Find an algebraic rule for a sequence so that every term in the sequence will

  a   be an even number

  b   be an odd number

  c   end in a 6

2 **a** Find the first five terms and the 10th term of the sequences given by these rules.

    i $4n + 1$ **ii** $3n$ **iii** $6n + 1$ **iv** $8n$

  b In which of the sequences is the 10th term double the 5th term? Why is this true for these sequences but not the other sequences?

  c Write a rule for a sequence where the 20th term is

    i double the 10th term

    ii more than the double the 10th term

    iii less than double the 10th term

Consolidate – do you need more?

1 A sequence has a first term of 20. Find the next three terms of the sequence if the term-to-term rule is:

  a   subtract 3 from the previous term

  b   double the previous term

  c   halve the previous term

  d   add 30 to the previous term.

2 The term-to-term rule for a sequence is "add 5 to the previous term and then double the answer". Find the first five terms of the sequence if the first term is

  a 1   **b** 0   **c** 10

3 The table shows the first two terms of a sequence.

Position	1	2	3	4
Term	2	8		

  a Find the next two terms of the sequence if it is

    i linear   **ii** geometric   **iii** Fibonacci.

  b Repeat part **a** with first two terms 8 and 2

4 **a** Work out the first five terms of the sequences given by these algebraic rules.

    i $2n + 3$ **ii** $3n$ **iii** $4n - 7$ **iv** $8 - 3n$

  b Describe the sequences in words.

5 The nth term of a sequence is given by the rule $8n + 4$

    a Which position in the sequence is the term with value 284?

    b Find the first term in the sequence that is greater than 1000

    c Show that 200 is not a term in the sequence.

Stretch – can you deepen your learning?

1 Seb is investigating the Fibonacci sequence that starts 1, 2, 3…

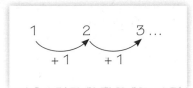

Seb thinks that the sequence is linear as well as Fibonacci. Explain Seb's mistake.

2 Find the missing terms in these linear sequences.

    a 6, ___, 12

    b 6, ___, ___, 12

    c 6, ___, ___, ___, 12

3 **a** Here is part of a Fibonacci sequence.

       5, ___, ___, 17, ___

       i If the second term of the sequence is x, find an expression for the third term in the sequence.

       ii Form and solve an equation to find the value of x

       iii Deduce the other missing terms in the sequence.

    b Find the missing terms in these Fibonacci sequences.

       i 3, ___, ___, 33

       ii 2, ___, ___, ___, 28

       iii ___, 7, ___, ___, 17

4 Here are some sequences you might recognise. Find the next term in each sequence, giving a reason for your answer.

    a J, F, M, A…   **b** M, T, W…

    c 2, 3, 5, 7…   **d** 1, 4, 9, 16…

    e O, T, T, F, F…

Reflect

1 How much information do you need to describe a sequence?

2 Make a list of all the words you know connected to the topic "Sequences", explaining the meaning of each word.

Small steps

- Generate sequences given a complex algebraic rule

- Find the rule for the nth term of a linear sequence Ⓗ

Are you ready?

1 $x = 5$. Work out the value of

    a x^2 **b** $6(x + 1)$ **c** $\dfrac{1}{x}$ **d** $30 - x^2$

2 $y = 10$. Work out the value of

    a $y^2 - 5$ **b** $(y - 5)^2$ **c** $(5 - y)^2$

3 Work out the calculations.

    a $3 + 4^2$ **b** $(3 + 4)^2$ **c** $2 - 3 \times 4^2$

4 **a** Write the coefficient of n in each of these expressions.

      i $3n + 5$ **ii** $2n - 1$ **iii** $6n$

    b Find the first four terms of the sequence given by the rule $6n$

    c What is the connection between your answer to **b** and the 6 times table?

Models and representations

You can use drawings or objects to represent sequences.

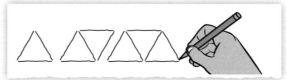

You can use cubes to make sequences, including those that feature squares.

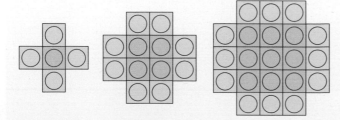

You will now explore sequences given by more complex algebraic rules.

Example 1

a Find the first three terms of the sequences given by these rules.

i $n^2 + 7$

ii $n(n + 1)$

iii $10 - 2n^2$

b State whether each of the sequences in part **a** is linear, geometric or neither.

a i When $n = 1$, $n^2 + 7 = 1^2 + 7$ Substitute $n = 1, 2$ and 3 in turn into the rule and
$$= 1 + 7$$
$$= 8$$
When $n = 2$, $n^2 + 7 = 2^2 + 7$
$$= 4 + 7$$
$$= 11$$
When $n = 3$, $n^2 + 7 = 3^2 + 7$
$$= 9 + 7$$
$$= 16$$

Substitute $n = 1, 2$ and 3 in turn into the rule and evaluate the expression.

ii When $n = 1$, $n(n + 1) = 1(1 + 1)$
$$= 1 \times 2$$
$$= 2$$
When $n = 2$, $n(n + 1) = 2(2 + 1)$
$$= 2 \times 3$$
$$= 6$$
When $n = 3$, $n(n + 1) = 3(3 + 1)$
$$= 3 \times 4$$
$$= 12$$

Substitute $n = 1, 2$ and 3 in turn into the rule and evaluate the expression. Remember you work out what is in the brackets first and then multiply.

iii When $n = 1$, $10 - 2n^2 = 10 - 2 \times 1^2$
$$= 10 - 2 \times 1$$
$$= 10 - 2$$
$$= 8$$
When $n = 2$, $10 - 2n^2 = 10 - 2 \times 2^2$
$$= 10 - 2 \times 4$$
$$= 10 - 8$$
$$= 2$$
When $n = 3$, $10 - 2n^2 = 10 - 2 \times 3^2$
$$= 10 - 2 \times 9$$
$$= 10 - 18$$
$$= -8$$

Remember: $2n^2$ means $2 \times n^2$, not $2 \times n$ and then square.

b

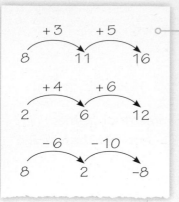

Work out the differences between successive terms of each sequence to see if these are linear.

The differences between terms change, so the sequences are not linear.

State your conclusion giving a reason.

Work out the multipliers between successive terms of each sequence, to see if the multipliers are all the same. You can use decimals to help with the first one. $\frac{11}{8} = 1.375$, $\frac{16}{11} = 1.4545...$

$\times \frac{11}{8}$ $\times \frac{16}{11}$

8   11   16

$\times 3$ $\times 2$

2   6   12

$\times \frac{1}{4}$ $\times -4$

8   2   -8

The multipliers between terms change so the sequences aren't geometric either.

State your conclusion giving a reason.

Practice 8.2A

1 Find the first five terms of the sequences given by these rules.

a n^2 **b** $n^2 + 3n$ **c** $n^2 - 3n$ **d** $7(n + 3)$

e $n(n + 3)$ **f** $(n + 4)(n + 3)$ **g** $(n - 4)(n + 3)$ **h** n^3

2 Find the difference between the 20th terms of the two sequences given by each pair of rules.

a n^3 and n^2 **b** $n^3 - 10$ and $n^2 - 10$ **c** $n^3 - 10n$ and $n^2 - 10n$

3 Here are the first four triangular numbers.

    a Draw the next triangular number.

    b Verify that the nth triangular number is given by the formula $\frac{1}{2}n(n + 1)$

    c Work out the 99th triangular number.

    d Explain why none of the triangular numbers are prime.

4 Compare the first five terms of the two sequences given by each pair of rules.

    a $3n^2$ and $(3n)^2$ **b** $(n + 2)^2$ and $n^2 + 2$

5 Which of these rules will generate linear sequences? How do you know?

A $\boxed{n + 5}$ B $\boxed{n^2 + 5}$ C $\boxed{5 - 3n}$ D $\boxed{n(n + 5)}$ E $\boxed{5(n - 5)}$

F $\boxed{n^5 + 1}$ G $\boxed{(n + 1)^5}$ H $\boxed{500 - 5n}$ I $\boxed{\frac{1}{n}}$

6 Find the first five terms of the sequences given by these rules.

    a 2^n **b** $2^n - 1$ **c** 2^{n-1}

What's the same and what's different?

7 **a** What's the same and what's different about the sequences given by the rules $\frac{1}{n}$ and $\frac{1}{n + 1}$?

    b Will the sequences in part **a** ever contain the term 0?

What do you think?

1 **a** Find the first five terms of the sequences given by these rules.

    i $\frac{n}{n + 1}$ **ii** $\frac{n}{2n + 1}$

    b What do you think will happen to the sequences as n gets very large? Check using a spreadsheet.

    c Predict how these sequences will behave when n gets very large.

    i $\frac{n}{3n + 1}$ **ii** $\frac{n}{an + 1}$

    d Investigate how the behaviour of the sequences changes if you change the constant, for example: $\frac{n}{n + 2}$, $\frac{n}{n + 3}$, and so on.

2 Explore the sequences.

    a $\frac{n^2}{n + 1}$ **b** $\frac{n^2 - 1}{n + 1}$ **c** $\frac{n^2}{n - 1}$ for $n \geqslant 2$

Why can the sequence in part **c** only be explored for $n \geqslant 2$?

Remember the rules for linear sequences are of the form $an + b$, for some numbers a and b.

In the next exercise, you will explore how to find a and b for any given linear sequence.

Practice 8.2B

1 **a** Find the first five terms of the sequences given by these rules.

i $3n + 1$ **ii** $3n - 2$ **iii** $3n - 3$

b Fully describe the sequences in words, including the term-to-term rules.

c What's the same about all the rules and all the descriptions?

2 **a** Find the first five terms of the sequences given by these rules.

i $2n + 5$ **ii** $6n - 2$ **iii** $7n - 5$

b What are the term-to-term rules for each sequence?

c What is the connection between the rule for the nth term of a linear sequence and the term-to-term rule for the sequence?

> You might want to use the word "coefficient" in your answer.

3 Chloe is comparing the sequences given by the rules $4n$, $4n + 1$ and $4n - 1$

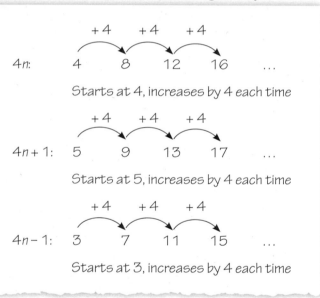

a What's the same and what's different about the sequences?

b Without writing out the sequences, predict how you would describe each of these sequences.

i $4n + 2$ **ii** $4n + 3$ **iii** $4n - 3$ **iv** $4n + 10$

v $4n + a$ **vi** $4n - a$

4 Here is another linear sequence.

5, 8, 11, 14…

a Write the next term of the sequence.

b Write the term-to-term rule for the sequence.

c Find the coefficient of n in the rule for the nth term of the sequence.

d Compare the sequence to the 3 times table.

5, 8, 11, 14…   3, 6, 9, 12…

Will the rule for the sequence be of the form $3n +$ __ or $3n -$ __?

e Find the rule for the nth term of the sequence.

5 **a** Find the rule for the nth term of each linear sequence.

i 2, 4, 6, 8…   **ii** 5, 7, 9, 11…   **iii** 6, 12, 18, 24…

iv 4, 10, 16, 22…   **v** 7, 12, 17, 22…   **vi** 8, 11, 14, 17…

vii 7, 17, 27, 37…   **viii** –4, 3, 10, 17…

b Why can't you use the same method to find the nth term of the sequence 1, 3, 6, 10…?

6 Rob thinks the rule for the nth term of the sequence that starts 8, 9, 10, 11 … will be $n + 1$

a Show that Rob is wrong.

b Find the correct rule for the nth term of the sequence.

7 **a** Here is a pattern of triangles made from pencils.

i Copy the sequence and draw the next term.

ii Find a rule for the total number of matchsticks used to make the nth shape in the pattern.

iii Find a rule for the number of matchsticks used on the perimeter of the nth shape in the pattern.

b Repeat part **a** for this pattern of squares made from pencils.

What do you think? 💭

1 **a** Find the rule for the nth term of the linear sequence that starts 9, 13, 17, 21, 25…

b Explain why the coefficient of n in the nth term rule for the sequence 25, 21, 17, 13, 9… will not be the same as that for the sequence 9, 13, 17, 21, 25…

c Find the rule for the nth term of these linear sequences.

i 25, 21, 17, 13, 9…   **ii** 10, 9, 8, 7…

iii 6, –2, –10, –18…   **iv** 100, 85, 70, 55…

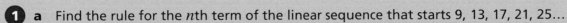

2 **a** I've worked out lots of terms of the sequence $5 + 4n - 1$ and it looks linear. But it's not in the form $an + b$

How would you explain to Marta why the sequence is linear?

b Will the sequence given by the rule $4(n - 1)$ be linear?

c What's the same and what's different about the sequences given by the rules $4(n - 1)$ and $4(1 - n)$?

3 The differences between the terms of the sequence given by n^2 are not the same, but the second differences (the differences between the differences) are.

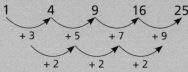

First differences $+ 3$ $+ 5$ $+ 7$ $+ 9$

Second differences $+ 2$ $+ 2$ $+ 2$

Explore the first and second differences of the sequences given by these rules.

a $n(n + 2)$

b $n^2 + 10$

c $2n^2$

d $(2n)^2$

Consolidate – do you need more?

1 Find the first five terms of the sequences given by these rules.

a $n^2 + 4$ **b** $n^2 - n$ **c** $5(n - 4)$

d $n^2 + 3n + 1$ **e** $n(n - 3)$ **f** $(n + 2)(n + 3)$

g $(n - 2)(n - 3)$ **h** $5n^2$

2 Find the sum of the 10th terms of the two sequences given by each pair of rules.

a n^3 and n^2

b $(n - 1)^2$ and $(n - 2)^2$

c $n(n + 1)(n + 2)$ and $n^3 - n^2$

3 Here are the first four rectangle numbers.

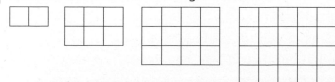

a Draw the next rectangular number.

b Verify that the nth rectangular number is given by the formula $n(n + 1)$

c Work out the 99th rectangular number.

4 **a** The difference between successive terms of a linear sequence is 5.
What is the coefficient of n in the rule for the nth term of the sequence?

b Find the rule for the nth term of these linear sequences.

i 6, 11, 16, 21…   **ii** 13, 16, 19, 22…   **iii** 0, 2, 4, 6…

iv 9, 13, 17, 21…   **v** 7, 8, 9, 10…   **vi** 12, 20, 28, 36…

vii 19, 20, 21, 22…   **viii** 3, 11, 19, 27…

5 Work with a partner to make up some linear sequences and challenge each other to find the rules.

Stretch – can you deepen your learning?

1 The rule n^2 gives the sequence 1, 4, 9, 16…

a

> I can see the rule for the sequence 2, 5, 10, 17… by comparing it to n^2

Compare the two sequences and write down the rule for Ed's sequence.

b By comparing them to the sequence for n^2, work out the rules for finding the nth term of these sequences.

i 2, 8, 18, 32…   **ii** 0, 3, 8, 15…   **iii** 11, 14, 19, 26…

iv 99, 96, 91, 84…   **v** 4, 9, 16, 25…

2 Here is a sequence of fractions.
$$\frac{3}{4}, \frac{5}{7}, \frac{7}{10}, \frac{9}{13} \dots$$

a Find a rule for the sequence formed by the numerators of the fractions.

b Find a rule for the sequence formed by the denominators of the fractions.

c Use your answers to **a** and **b** to find a rule for the nth term of the sequence of the fractions.

d Find the rules for the nth term of these sequences.

i $\frac{4}{5}, \frac{5}{9}, \frac{6}{13}, \frac{7}{17} \dots$

ii $\frac{3}{9}, \frac{5}{11}, \frac{7}{13}, \frac{9}{15} \dots$

iii $\frac{1}{2}, \frac{3}{5}, \frac{5}{10}, \frac{7}{17} \dots$

Reflect

1 Explain how you can tell whether a sequence is linear or not by looking at the rule.

2 Explain how you find an algebraic rule for the nth term of a linear sequence.

3 What other sequences do you think you can "spot" the rule for?

8 Sequences
Chapters 8.1–8.2

I have become fluent in…	I have developed my reasoning skills by…	I have been problem-solving through…
■ finding the next term in sequences ■ using a rule in words to generate sequences ■ using algebraic rules to generate sequences ■ describing sequences in words.	■ making connections between information given in a wide variety of forms ■ making and testing ideas about patterns and relationships ■ identifying how a sequence changes and finding term-to term rules ■ determining whether a number is or isn't in a sequence ■ finding the rule for the nth term of a linear sequence Ⓗ.	■ spotting and describing patterns ■ solving multi-step problems involving sequences ■ representing sequences in a variety of forms ■ working out missing terms in sequences.

Check my understanding

1 a Fully describe the sequences.

    i 40, 80, 120, 160…   **ii** 40, 80, 160, 320…

    iii 40, 20, 10, 5…   **iv** 40, 20, 60, 80…

  b For each sequence, find the sixth term.

2 The term-to-term rule for a sequence is "multiply the previous term by 5 and add 1". The third term of the sequence is 325

  a Find the fourth and fifth terms of the sequence.

  b Find the second and first terms of the sequence.

3 Find the difference between the 10th and 20th terms of the sequences given by these rules.

  a $5n + 1$ **b** n^2 **c** $n(2n - 1)$

  d $1000 - 2n^2$ **e** $4 - 39n$

4 Which of these sequences contain the term 1000?

  A $\boxed{4n}$ B $\boxed{6n + 2}$ C $\boxed{12n - 8}$

5 Find the first term in the sequence given by the rule $3n + 87$ that exceeds 2000

6 Find the rule for the nth term of each sequence. Ⓗ

  a 7, 15, 23, 31…   **b** 10, 20, 30, 40…   **c** 20, 19, 18, 17…

9 Indices

In this block, I will learn...

how to add and subtract expressions with indices

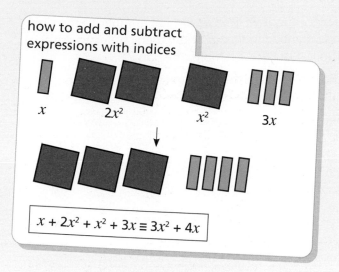

x $2x^2$ x^2 $3x$

$x + 2x^2 + x^2 + 3x \equiv 3x^2 + 4x$

how to use the addition law of indices

$h^5 \times h^3 \times h = h \times h \times h \times h \times h \times h \times h \times h \times h = h^9$

how to simplify powers of powers

$(a^5)^3 = a^5 \times a^5 \times a^5$

$a^5 \times a^5 \times a^5 = a^{15}$

$(a^5)^3 = a^{5 \times 3} = a^{15}$

how to use the subtraction law of indices

$y^8 \div y^5$

$\dfrac{y^8}{y^5} = \dfrac{y \times y \times y \times y \times y \times y \times y \times y}{y \times y \times y \times y \times y}$

$= \dfrac{y \times y \times y \times \cancel{y} \times \cancel{y} \times \cancel{y} \times \cancel{y} \times \cancel{y}}{\cancel{y} \times \cancel{y} \times \cancel{y} \times \cancel{y} \times \cancel{y}}$

$= y \times y \times y$

$y^8 \div y^5 = y^3$

Small steps

- Add and subtract expressions with indices

Key words

Expression – a collection of terms involving mathematical operations

Power (or **exponent**) – this is written as a small number to the right and above the base number, indicating how many times to use the number in a multiplication. For example, the 5 in 2^5

Index (plural: **indices**) – an index number (or power) tells you how many times to multiply a number by itself

Base – the number that gets multiplied when using a power/index

Simplify – rewrite in a simpler form, for example rewrite $8 \times h$ as $8h$

Are you ready?

1 Work out

   a $2 - 3$ **b** $-3 - 2$ **c** $-2 + 3$

2 Which of these contain like terms?

   A $4x$ and $3x$ B $7y$ and y C $2a$ and $5a$ D 6 and $6c$

3 Simplify each of these by collecting like terms.

   a $x + x + x + x$ **b** $x + x + y + y$ **c** $3x + 5x - 2x$

   d $4y - 3x + y$ **e** $15a - 3b - 2b$ **f** $4b - 5b + 10$

Models and representations

Algebra tiles

Algebra tiles are a good way of representing expressions that involve **indices**.

This shows one x^2 tile, three x tiles and two "one" tiles.

This represents the expression $x^2 + 3x + 2$

In this chapter, you will add and subtract **expressions** involving indices and use the identity symbol $\equiv$. It means that the left-hand side is equivalent to the right-hand side. For example $3y + y \equiv 4y$.

Another word for an index is a **power** or an **exponent**.

Example 1

Simplify each expression by collecting like terms.

a $5x^2 + 2x^2$ **b** $4x + 2x^2 + 3x^2$ **c** $7x^3 - 3x^2 - 10x^3$

a $5x^2 + 2x^2 \equiv 7x^2$

You can represent $5x^2 + 2x^2$ using algebra tiles.

In total this is $7x^2$

b $4x + 2x^2 + 3x^2 \equiv 4x + 5x^2$

You can't **simplify** this further because $4x$ and $5x^2$ are not like terms.

You can see this clearly by representing the expression using algebra tiles.

c $7x^3 - 3x^2 - 10x^3 \equiv -3x^3 - 3x^2$

Be careful when working with negative numbers.

Practice 9.1A

1 The area of each tile is shown.

 $= x^2$ $= x$ $= 1$

a Work out each total area.

    i   **ii**   **iii**

b Find an expression for each total area.

    i   **ii**   **iii**

c Find an expression for each total area.

    i   **ii**

    iii

d What's the same and what's different about parts **a**, **b** and **c**?

2 Zach writes $5x^2 + 5x + 3x^2 \equiv 13x^2$. Explain why Zach is incorrect.

3 Write an expression for each total area.

a

b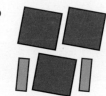

c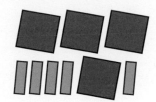

4 Explain how the diagram shows that $x^2 + x^2 + x^2 + x^2 + x^2 \equiv 5x^2$

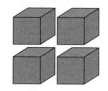

5 Simplify these expressions.

a $x^2 + x^2 + x^2$

b $2x^2 + x^2 + x^2$

c $2x^2 + x^2 - x^2$

d $8x^2 + 5x^2 - 3x^2$

e $y^2 + 4y^2 - 3y^2$

f $5p^2 - 2p^2 + 4p^2$

g $6t^2 + 4t^2 - 9t^2$

h $17m^2 + 19m^2 - m^2$

6 The volume of each cube is x^3. Find an expression for each total volume.

a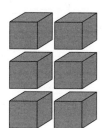

b

c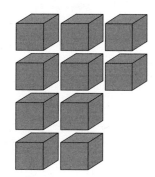

7 Simplify these expressions.

a $5x^3 + 9x^3$

b $7y^3 - 4y^3$

c $10p^3 - 2p^3 - 3p^3$

d $27x^3 - 26x^3$

e $11y^3 + 2y^3 - y^3$

f $16t^3 + 15t^3 - 7t^3$

g $100k^3 - 64k^3$

h $k^3 - 0.64k^3$

8 a Explain how the diagram shows that $5x^3 + 2x^2 \neq 7x^5$

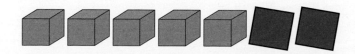

b Explain why $5x^3 + 2x^2$ cannot be simplified.

9 Simplify each expression.

    a $12x^3 + 2x^2 - 4x^3$

    b $15x^2 - 9x^2 + 10x^3$

    c $18x^3 + 2x^2 - 17x^2$

    d $5c^2 + 6c^2 - 2c^2$

    e $k^3 + 12k^2 - 11k^2$

    f $3h^3 + 21h^2 - 2h^3$

10 Simplify these expressions.

    a $12a^5 + 6a^2 - 4a + 15a^2$

    b $8p^2 - 9p + 15p^5 - 6p$

    c $x^3 + 7x^4 - 6x^5 + x^4$

    d $y - 7y^5 + 2y^3 - 6y^5$

    e $18j^4 - 9j^3 + 3j^4 + 9j^3$

    f $q^4 - 10q^3 - 5q^4 + q^2$

11 Simplify these expressions.

    a $p^3 + p^2 + p^2 + m^3 + m^2 + m^3$

    b $25j^4 - 16k^3 - 11j^4 + 6k^3 - 15j^3$

    c $9a^3 + 7a^2 + 12b^3 - 2a^2 + 15b^3$

    d $17x^3 - 15y^2 + 24x - 5x^3 + 8y^2$

What do you think? 💭

1 Sort the following expressions into three sets of like terms.

x^2	x^3	$5x$	$-3x^2$
$-2x$	$-7.5x$	$5x^3$	$-7x^2$
$\dfrac{x^2}{3}$	$2x^2$	$3x$	$-4x^3$

Explain your method to your partner.

2 Marta says, "x^2 and y^2 are like terms so you can add them together."

    a   Why do you think Marta thinks x^2 and y^2 are like terms?

    b   Explain why you cannot add x^2 and y^2 together.

3 Here are some facts about an algebraic expression.

    ■ The expression has got four terms.

    ■ The expression simplifies to $6p^3 - 2p$

    ■ There are three terms involving p^3

Write down three possibilities for the original expression.

Consolidate – do you need more?

1 Simplify each of these expressions.

   a $x^2 + x^2 + x^2$ **b** $a^2 + a^2$ **c** $y^2 + y^2 + y^2 + y^2$

2 Simplify these expressions by collecting like terms.

   a $2x^2 + 3x^2$ **b** $7y^2 + 5y^2$ **c** $m^2 + 4m^2 - 3m^2$

3 Write each expression in its simplest form.

   a $c^2 + 3d^2 + 2c^2$ **b** $10p^2 - 7q + p^2$ **c** $6x^2 + 1 - 6x^2$

4 Simplify each expression.

   a $12y^3 + 2x^2 - 5y^3$ **b** $5x^2 - 9x^2 + 4x^3$ **c** $8x^3 + 2x^2 - 7x^2$

   d $c^2 + 6c^2 - 2c^3$ **e** $4m^3 + 12m^2 - 11m^2$ **f** $20h^3 + 21h^2 - h^3$

Stretch – can you deepen your learning?

1 **a** Copy and complete these expressions.

     i $7x^3 - \boxed{} \equiv 5x^3$ **ii** $7x^3 - \boxed{} \equiv -2x^3$

     iii $7x^3 - \boxed{} \equiv 0$ **iv** $7x^3 - \boxed{} \equiv 7x^3 - x^2$

   b Is it possible to find an expression that will complete this subtraction correctly?

     $7x^3 - \boxed{} \equiv 7x^2$

2 **a** Is the following statement correct?

     $\frac{1}{2}x^3 + \frac{1}{3}x^3 = \frac{x^3}{2} + \frac{x^3}{3}$

     Explain your reasoning.

   b Simplify fully

     $\frac{1}{2}x^3 + \frac{1}{3}x^3$

3 Simplify these expressions.

   a $5f^2g^3 + 7f^2g^3$ **b** $13x^2y + 12xy^2 - 6xy^2 + 2xy^2$

   c $2m^5 + 7m^3 - 8m^5 + 2m^2 + 3m^3 - 6m^2 + m - 1$

Reflect

1 **a** Give an example of three terms that are like terms.

   b Give an example of three terms that are unlike terms.

2 In your own words, explain how to simplify an expression by collecting like terms.

9.2 Multiplying expressions

Small steps

- Simplify algebraic expressions by multiplying indices
- Use the addition law for indices

Key words

Expression – a collection of terms involving mathematical operations

Power (or **exponent**) – this is written as a small number to the right and above the base number, indicating how many times to use the number in a multiplication. For example, the 5 in 2^5

Simplify – rewrite in a simpler form, for example rewrite $8 \times h$ as $8h$

Are you ready?

1 Write each of these using powers.

  a $4 \times 4 \times 4$

  c $7 \times 7 \times 7 \times 7 \times 7$

  b $g \times g \times g \times g \times g \times g$

  d $M \times M \times M \times M \times M \times M \times M \times M$

2 Simplify these expressions.

  a $3 \times g$ **b** $a \times b$ **c** $4 \times g \times h$ **d** $2y \times y$

Models and representations

Expanded form

2^4 in expanded form is $2 \times 2 \times 2 \times 2$

y^3 in expanded form is $y \times y \times y$

 Being able to represent numbers and letters involving **powers** in expanded form is essential for understanding the addition law for indices.

In this chapter you will learn how to multiply algebraic **expressions** involving indices.

Example 1

Simplify $h^5 \times h^3 \times h$

$h^5 \times h^3 \times h$
$= h \times h \times h \times h \times h \times h \times h \times h \times h$
$= h^9$

Rewrite the calculation in expanded form.

Now that the calculation is written in expanded form, you can see that you are finding the result of "h" multiplied by itself nine times, which can be simplified to h^9

The base stays the same and the powers are added.

Practice 9.2A

1 **a** Explain each step in this working.

$$p^2 \times p^5 = (p \times p) \times (p \times p \times p \times p \times p)$$
$$= p \times p \times p \times p \times p \times p \times p$$
$$= p^7$$

b Show that

   i $m^3 \times m^4 = m^7$ **ii** $q^3 \times q \times q^2 = q^6$

c What do you notice about the powers?

2 Simplify each of these expressions.

  a **i** $c^3 \times c^3$ **ii** $c^3 \times c^4$ **iii** $c^3 \times c^5$ **iv** $c^4 \times c^6$

  b **i** $y^3 \times y$ **ii** $y^4 \times y$ **iii** $y^5 \times y$ **iv** $y^{10} \times y$

What method did you use?

3 Write each of these using a single power.

  a $x^2 \times x^5$ **b** $y^3 \times y^4$ **c** $a^5 \times a^6$

  d $4^3 \times 4^5$ **e** $5^3 \times 5$ **f** $b^3 \times b^2 \times b^4$

  g $e^3 \times e^4 \times e^6$ **h** $b^4 \times b \times b^2$ **i** $n^2 \times n^5 \times n^2 \times n^3$

  j $x^4 \times x^{18}$ **k** $p^{35} \times p^{42}$

4 Seb is trying to simplify this expression:

$2q^4 \times 4q^5$

The answer is $6q^9$

Seb has made a mistake. The correct answer should be $8q^9$

  a Why might Seb think the answer is $6q^9$?

  b Explain why $2q^4 \times 4q^5 = 8q^9$

5 Simplify each of these expressions.

  a **i** $x^4 \times x^2$ **ii** $-x^4 \times x^2$ **iii** $-x^4 \times -x^2$

  b **i** $2m^3 \times m^2$ **ii** $m^3 \times 2m^2$ **iii** $3m^3 \times 2m^2$ **iv** $5m^3 \times 4m^2$

6 Work out the missing powers.

  a $x^3 \times x^{\boxed{}} = x^5$ **b** $4^{\boxed{}} \times 4^2 = 4^6$

  c $y^5 \times y^{\boxed{}} = y^{17}$ **d** $h^{\boxed{}} \times h^4 = h^5$

  e $3^4 \times 3^{\boxed{}} = 3^9$ **f** $k^{\boxed{}} \times k^5 = k^{19}$

  g $10^4 \times 10^{\boxed{}} \times 10 = 10^8$

7 Write each of these expressions as a single term.

  a $2^x \times 2^{3x}$ **b** $5^{3x-1} \times 5^{7x}$ **c** $r^{2y+3} \times r^{y-2}$

8 Explain why you cannot add the powers to simplify this expression.

$x^2 \times y^3$

What do you think?

1 a Jackson is simplifying this expression

$e^4 \times e^3$

What mistake has Jackson made?

> You multiply the powers so the answer is e^{12}

b Ed is simplifying this expression

$w^4 \times w$

What mistake has Ed made?

> The answer is w^4 as you add the powers. There is a power 0 on the second w

2 Here are some cards.

| a | a^2 | a^3 | a^4 | a^5 | a^6 | a^7 | a^8 | a^9 | a^{10} |

a Find two cards that multiply to make

  i a^7 **ii** a^{10}

b Find three cards that multiply to make a^{10}

3 Write each of the following as a single power.

a $6 \times 6 \times 6 \times 36$ **b** $5 \times 5 \times 5 \times 25$

(Hint: Remember that $6 \times 6 = 36$)

c $9 \times 3 \times 3$ **d** $4 \times 64 \times 4 \times 4$

What method did you use?

Consolidate – do you need more?

1 Simplify

a $q^7 \times q^3$ **b** $r^5 \times r^2$ **c** $n^3 \times n$ **d** $b^3 \times b^2 \times b^4$

e $2g^6 \times g^2$ **f** $3x^7 \times 2x^2$ **g** $3t^5 \times 2t \times 5t^3$

2 Find the missing powers.

a $a^5 \times a^{\square} = a^7$ **b** $y^{\square} \times y^3 = y^9$ **c** $m^4 \times m^{\square} \times m = m^{12}$

Stretch – can you deepen your learning?

1 **a** Darius says that it is not possible to write 8×4 as a single power.

Beth says that you can think of each number of a power of 2.

Complete Beth's method.

$4 = 2^{\square}$

$8 = 2^{\square}$

$4 \times 8 = 2^{\square} \times 2^{\square} = 2^{\square}$

b Write each of these as single power.

    i 9×27 **ii** $16 \times 8 \times 8$ **iii** $5 \times 25 \times 125$

    iv $100 \times 1000 \times 10\,000 \times 100$ **v** $64 \times 128 \times 32 \times 8$

2 Simplify

    a $f^3 \times h^3 \times f^4 \times h^4$ **b** $3m^2n \times m^8$ **c** $a^2b^3 \times a^3b^2$

    d $x^5y^2 \times y^4x^3$ **e** $m^2n^3 \times mn^4$ **f** $3c^{12}d^5 \times c^4d^5$

    g $6gh^4 \times 2h^3$ **h** $7a^3b^3c^2 \times b^5c^6$ **i** $x^4 \times y^3$

    j $u^4v^7 \times u^8 \times v^6$ **k** $(x + 3)^2 \times (x + 3)^9$ **l** $(b + c) \times (b + c)^6$

3 Find the value of x in each of these identities.

    a $3^x \times 3^x \equiv 3^{10}$ **b** $5^{x+1} \times 5^{2x} \equiv 5^{13}$

    c $a^x \times a^{4x} \times a^2 \equiv a^{32}$ **d** $7^{x-5} \times 49 \equiv 7^6$

    e $m^{2x} \times m^{x-2} \equiv m^x \times m^4$

4 Expand each set of brackets.

    a $3x^5(2x^7 + 11)$ **b** $4a^3(3b^{10} - 12a^8)$ **c** $6x^2y(5xy^5 + 3x^7 - y)$

    d $(x^3 + 4)(x^5 - 7)$ **e** $(2y^4 + 1)(3y^5 - 4)$ **f** $(2x^3 + 4y)(5y^4 - x)$

Reflect

Explain why $t^a \times t^b = t^{a+b}$

Small steps

- Simplify algebraic expressions by dividing indices
- Use the addition and subtraction law for indices

Key words

Expression – a collection of terms involving mathematical operations

Power (or **exponent**) – this is written as a small number to the right and above the base number, indicating how many times to use the number in a multiplication. For example, the 5 in 2^5

Simplify – rewrite in a simpler form, for example rewrite $8 \times h$ as $8h$

Are you ready?

1 Simplify each of these expressions.

   a $p \times p \times p \times p$ **b** $g^2 \times g^3$ **c** $n^5 \times n^4$ **d** $m^3 \times m$ **e** $t^3 \times t^4 \times t^5$

2 Work out

   a $\dfrac{12}{6}$ **b** $\dfrac{14}{4}$ **c** $20 \div 8$

3 Simplify these fractions.

   a $\dfrac{2}{8}$ **b** $\dfrac{6}{9}$ **c** $\dfrac{15}{20}$ **d** $\dfrac{28}{35}$ **e** $\dfrac{8}{12}$ **f** $\dfrac{100}{1000}$

Models and representations

Expanded form

2^4 in expanded form is $2 \times 2 \times 2 \times 2$

y^3 in expanded form is $y \times y \times y$

Being able to represent numbers and letters with **powers** in expanded form is essential for understanding the subtraction law for indices.

In this chapter, you will learn how to divide **expressions** involving indices.

Example 1

a Simplify $7^5 \div 7^2$ **b** Simplify $y^8 \div y^5$

a $7^5 \div 7^2$

$\dfrac{7^5}{7^2} = \dfrac{7 \times 7 \times 7 \times 7 \times 7}{7 \times 7}$

$= \dfrac{7 \times 7 \times 7 \times \cancel{7} \times \cancel{7}}{\cancel{7} \times \cancel{7}}$

$= 7 \times 7 \times 7$

$7^5 \div 7^2 = 7^3$

You can write the calculation as a fraction.

Then you can write both the numerator and the denominator in expanded form.

You can now cancel common factors.

You are now left with the product of three sevens which is 7^3 so $7^5 \div 7^2 = 7^3$

The base stays the same and the powers are subtracted.

b $y^8 \div y^5$

$\dfrac{y^8}{y^5} = \dfrac{y \times y \times y \times y \times y \times y \times y \times y}{y \times y \times y \times y \times y}$ ○———

You can write the calculation as a fraction.

Then you can write both the numerator and the denominator in expanded form.

$= \dfrac{y \times y \times y \times \cancel{y} \times \cancel{y} \times \cancel{y} \times \cancel{y} \times \cancel{y}}{\cancel{y} \times \cancel{y} \times \cancel{y} \times \cancel{y} \times \cancel{y}}$ ○———

You can now cancel common factors.

$= y \times y \times y$

$y^8 \div y^5 = y^3$ ○———

You are now left with the product of three ys which is y^3 so $y^8 \div y^5 = y^3$

The base stays the same and the powers are subtracted.

Practice 9.3A

1 **a** Flo is simplifying $\dfrac{e^5}{e^3}$

Here is Flo's correct working:

$\dfrac{e \times e \times \cancel{e} \times \cancel{e} \times \cancel{e}}{\cancel{e} \times \cancel{e} \times \cancel{e}} \equiv e^2$

Explain why Flo's method works.

b Use Flo's method to show that

i $\dfrac{m^8}{m^3} \equiv m^5$ **ii** $\dfrac{y^3}{y^2} \equiv y$ **iii** $6^7 \div 6^4 = 6^3$

c What do you notice about the powers?

2 Beca is simplifying $\dfrac{u^8}{u^2}$

a Explain the mistake that Beca has made.

b Find the correct answer, showing your working.

The answer is u^4 as 8 divided by 2 is 4

3 Explain why these two cards give the same answer.

A $\dfrac{g^8}{g^4}$ B $g^8 \div g^4$

4 Simplify each expression.

a **i** $d^6 \div d^3$ **ii** $d^6 \div d^4$ **iii** $d^6 \div d^5$

b **i** $\dfrac{y^3}{y}$ **ii** $\dfrac{y^4}{y}$ **iii** $\dfrac{y^5}{y}$

Explain your method.

5 Write each of these with a single power.

a $n^6 \div n^2$ **b** $x^7 \div x^3$ **c** $7^5 \div 7^3$ **d** $q^4 \div q^3$

e $5^4 \div 5$ **f** $y^{50} \div y^2$ **g** $e^4 \div e$ **h** $a^{14} \div a^5$

i $\dfrac{x^6}{x^2}$ **j** $\dfrac{n^7}{n}$ **k** $\dfrac{7^9}{7^5}$ **l** $\dfrac{r^{30}}{r^{10}}$

6　Simplify each of the following.

a　**i** $x^8 \div x^2$ 　　　　　　**ii** $-x^8 \div x^2$ 　　　　　　**iii** $-x^8 \div -x^2$

b　**i** $10m^6 \div m^2$ 　　　　**ii** $10m^6 \div 2$ 　　　　　**iii** $10m^6 \div 2m^2$

7　Simplify these pairs of expressions.

a 　**i** $\dfrac{g^8}{g^2}$ 　　**ii** $\dfrac{g^2}{g^8}$ 　　　　　　**b** 　**i** $\dfrac{8k^3}{2k}$ 　　**ii** $\dfrac{8k}{2k^3}$

c 　**i** $\dfrac{12p^7}{6p^2}$ 　　**ii** $\dfrac{6p^2}{12p^7}$ 　　　　**d** 　**i** $\dfrac{12x^5}{9x^4}$ 　　**ii** $\dfrac{9x^4}{12x^5}$

8　Work out the missing powers.

a 　$x^8 \div x^{\boxed{}} \equiv x^5$ 　　　　　　　**b** 　$4^{\boxed{}} \div 4^2 \equiv 4^6$

c 　$y^{10} \div y^{\boxed{}} \equiv y^{10}$ 　　　　　　**d** 　$q^{\boxed{}} \div q \equiv q^5$

9　Simplify these expressions.

a 　$\dfrac{x^3 \times x^6}{x^4}$ 　　　　**b** 　$\dfrac{y^3 \times y^2}{y^2}$ 　　　　**c** 　$\dfrac{4^5 \times 4^3}{4^2}$

d 　$\dfrac{a^5 \times a^2}{a}$ 　　　　**e** 　$\dfrac{p^2 \times p^3}{p^7}$ 　　　　**f** 　$\dfrac{t^2}{t^4 \times t^2}$

g 　$\dfrac{3^4 \times 3^2 \times 3^2}{3^4}$ 　　**h** 　$\dfrac{y^4 \times y^3}{y^2 \times y}$ 　　**i** 　$\dfrac{2^2 \times 2^3 \times 2^2}{2^7 \times 2^3}$

What do you think? ☁

1　Explain why this expression cannot be simplified further.

$\dfrac{g^7}{m^4}$

2　Here are some expression cards.

| $12y^3$ | $18y^6$ | $10y$ | 12 | $10y^4$ | $6y^8$ | $5y^2$ | $2y^4$ |

a　Rhys picks two cards and then divides the expressions. He gets the answer y^3
　Which two cards does he pick? Find two different answers.

b　Huda picks two cards and then divides the expressions. She gets the answer $\dfrac{3}{y^2}$
　Which two cards does she pick?

3　Emily says, "$8^{17} \div 8^{15} = 1^2$"
Explain why Emily is incorrect.

4　**a**　Explain why $\dfrac{g^3}{g^3} = 1$

　b　Using your answer to part **a**, explain why $h^0 = 1$

5　| $2^9 = 512$ | $2^{11} = 2048$ | $2^{20} = 1\,048\,576$ |

Use this information to work out the answers to these calculations.

a　$1\,048\,576 \div 2048$ 　　　　　**b**　$1\,048\,576 \div 512$

Explain your method to a partner.

Consolidate – do you need more?

1 Simplify

   a $d^{10} \div d^2$ **b** $y^5 \div y^2$ **c** $w^6 \div w^5$

   d $u^6 \div u$ **e** $\dfrac{h^7}{h^5}$ **f** $\dfrac{l^7}{l^4}$

   g $m^3 \div m^3$ **h** $\dfrac{16y^7}{2y^3}$ **i** $\dfrac{24p^{19}}{4p}$

2 Write each of these expressions as a single power.

   a $\dfrac{d^4 \times d^6}{d^2}$ **b** $\dfrac{y^5}{y \times y}$ **c** $\dfrac{w \times w^5}{w^5}$

   d $\dfrac{u^3 \times u^3}{u}$ **e** $\dfrac{h^6 \times h}{h^2 \times h^3}$ **f** $\dfrac{l^{10} \times l^{-3}}{l^2 \times l^2}$

3 Find five different expressions which simplify to h^{10}

Stretch – can you deepen your learning?

1 **a** x and y are positive integers. Find three possible pairs of values for x and y

      $10^x \div 10^y \equiv 10^3$

   b Here is an expression:

      $\dfrac{w^x}{w^y} = w$

      i What is the value of $x - y$?   **ii** What is the value of $y - x$?

2 Simplify these expressions.

   a $f^8g^3 \div f^2g^2$ **b** $f^8g^2 \div f^2g^3$ **c** $f^2g^3 \div f^8g^2$ **d** $f^2g^2 \div f^8g^3$

3 Simplify these expressions.

   a $\dfrac{10m^7n^5}{2m^2n^4}$ **b** $\dfrac{12a^4b^5}{8ab^3}$ **c** $\dfrac{10h^8g^3}{25h^4g^3}$ **d** $20x^2y^3 \div 2xy^5$

4 Simplify each of the following.

   a $7^{5a} \div 7^{2a}$ **b** $y^{5x+3} \div y^{2x}$ **c** $y^{2m+3} \div y^{m+5}$

5 Simplify

   a $x^a \times x^b$ **b** $x^a \div x^b$

Reflect

In your own words, explain how to use the addition and subtraction laws for indices. Give examples to support your reasoning.

Small steps

- Explore powers of powers ⓗ

Key words

Expression – a collection of terms involving mathematical operations

Power (or **exponent**) – this is written as a small number to the right and above the base number, indicating how many times to use the number in a multiplication. For example, the 5 in 2^5

Simplify – rewrite in a simpler form, for example rewrite $8 \times h$ as $8h$

Are you ready?

1 Write each of these as a single power.

   a $6 \times 6 \times 6 \times 6 \times 6$

   b $k \times k \times k \times k \times k \times k \times k \times k$

   c $y \times y \times y$

   d $h \times h \times h \times h \times h \times h \times h \times h \times h \times h$

2 Simplify each of these expressions.

   a $7^3 \times 7^5$ **b** $y^2 \times y^8$ **c** $p^2 \times p^2 \times p^2$ **d** $5m^2 \times m$ **e** $12r^3 \times 2r^3$

3 Expand

   a $s^2(s + s^4)$

   b $x^5(x^3 - x^7)$

   c $2u^3(3u^2 + u^4)$

4 Simplify

   a $12^6 \div 12^2$

   b $y^5 \div y$

   c $\dfrac{g^6}{g^5}$

Models and representations

Expanded form

$a^3 \equiv a \times a \times a$

so $(ab)^3 \equiv ab \times ab \times ab$

Also, $(17ab)^3 \equiv 17ab \times 17ab \times 17ab$

and $(125a^4b^2)^3 \equiv 125a^4b^2 \times 125a^4b^2 \times 125a^4b^2$

In this chapter, you will learn how to **simplify powers** of powers. Writing calculations in expanded form will help you to understand the content of this chapter.

Example 1

a Simplify $(a^5)^3$ **b** Simplify $(2b^7)^4$

a $(a^5)^3 \equiv a^5 \times a^5 \times a^5$

$a^5 \times a^5 \times a^5 = a^{15}$

$(a^5)^3 \equiv a^{5 \times 3} = a^{15}$

You can rewrite the calculation in expanded form.

Use the addition law for indices to simplify the **expression**. You need to add the powers.

Notice the connection between the powers in the question and the power in the answer.

When finding powers of powers, you find the product of the powers: $5 \times 3 = 15$

b $(2b^7)^4 \equiv 2 \times b^7 \times 2 \times b^7 \times 2 \times b^7 \times 2 \times b^7$

$(2b^7)^4 \equiv (2 \times 2 \times 2 \times 2) \times (b^7 \times b^7 \times b^7 \times b^7)$

$(2b^7)^4 \equiv 2^4 \times b^{28} \equiv 16b^{28}$

Simplify the numerical part of the expression by multiplying the numbers.

Use the addition law for indices to simplify the algebraic part of the expression.

$(2b^7)^4 \equiv 16b^{28}$

You raise the integer to the power outside the bracket: $2^4 = 16$

When finding powers of powers, you find the product of the powers: $7 \times 4 = 28$

Practice 9.4A

1 Copy and complete.

a $(4^2)^3 = 4^2 \times 4^2 \times 4^2 = \boxed{}$

b $(5^3)^4 = 5^3 \times \boxed{} \times \boxed{} \times \boxed{} = \boxed{}$

c $(d^4)^3 \equiv \boxed{} \times \boxed{} \times \boxed{} \equiv \boxed{}$

d $(v^5)^6 \equiv \boxed{} \times \boxed{} \times \boxed{} \times \boxed{} \times \boxed{} \times \boxed{} \equiv \boxed{}$

2 Write each of these as a single power.

a   **i** $(5^3)^2$ **ii** $(5^3)^3$ **iii** $(5^3)^4$ **iv** $(5^3)^5$ **v** $(5^3)^6$

b   **i** $(e^2)^5$ **ii** $(e^3)^5$ **iii** $(e^4)^5$ **iv** $(e^5)^5$ **v** $(e^6)^5$

c Explain your method.

d What did you notice about the powers and the power of the final answer?

3 Simplify

a $(x^7)^4$ **b** $(p^4)^2$ **c** $(v^8)^3$ **d** $(d^6)^3$

e $(t^4)^{10}$ **f** $(x^2)^5$ **g** $(p^6)^5$ **h** $(u^3)^4$

i $(4^2)^4$ **j** $(5^3)^9$ **k** $(22^3)^5$ **l** $(176.4^{20})^{50}$

④ Without simplifying, explain why these expressions are equivalent.

$(r^{273})^{95}$

$(r^{95})^{273}$

⑤ Work out the missing numbers.

a $(2^3)^{\square} = 2^{12}$ **b** $(a^{\square})^6 \equiv a^{42}$ **c** $(g^4)^{\square} \equiv g^{40}$

⑥ Faith is trying to simplify $(2x^4)^3$

Faith says the answer is $6x^{12}$

a What mistake has Faith made?

b Show that the answer to the question is $8x^{12}$

⑦ Simplify these expressions.

a $(3x^5)^2$ **b** $(2x^3)^4$ **c** $(5p^3)^3$ **d** $(\frac{1}{2}g^4)^3$

⑧ Simplify

a $5 \times (5^5)^5$ **b** $(m^2)^5 \times m^4$ **c** $(x^3 \times x^4)^2$

d $(n^3)^7 \div n^5$ **e** $(m^3)^5 \times (m^4)^2$ **f** $\dfrac{g^{10}}{(g^2)^3}$

What do you think? 💭

① x and y are positive integers.

$(6^x)^y = 6^{24}$

What could the values of x and y be? Find as many possible pairs of values as you can.

② Find the missing values.

$(10y^{\square})^{\square} = 100\,000y^{30}$

Explain your method.

③ Each card on the left has an equivalent card on the right.

Match the cards.

$g^3(g^2 + g^5)$	g^{10}
$g^3(g^2 \times g^5)$	g^{21}
$(g^2)^3 + (g^5)^3$	$g^5 + g^8$
$(g^2)^3 \times (g^5)^3$	$g^6 + g^{15}$

Consolidate – do you need more?

1 Write each of these expressions in expanded form. The first one has been done for you.

a $(a^4)^5 \equiv a^4 \times a^4 \times a^4 \times a^4 \times a^4$ **b** $(h^3)^4$

c $(s^9)^5$ **d** $(p^{100})^2$

e $(6^9)^3$ **f** $(21^5)^8$

g $(17^3)^4$ **h** $(101^{99})^5$

2 Write each of these as a single power.

a $(a^4)^5$ **b** $(h^3)^4$ **c** $(s^9)^5$ **d** $(p^{100})^2$

e $(6^9)^3$ **f** $(21^5)^8$ **g** $(17^3)^4$ **h** $(101^{99})^5$

3 Write these in expanded form. The first one has been done for you.

a $(2a^4)^5 \equiv 2a^4 \times 2a^4 \times 2a^4 \times 2a^4 \times 2a^4$ **b** $(3h^3)^4$

c $(4s^9)^5$ **d** $(12p^{100})^2$

4 Simplify these expressions involving powers of powers.

a $(2a^4)^5$ **b** $(3h^3)^4$ **c** $(4s^9)^5$ **d** $(12p^{100})^2$

5 Find five different expressions involving powers of powers that simplify to p^{30}

Stretch – can you deepen your learning?

1 Simplify each expression.

a $(a^b)^c$ **b** $(Aa^b)^c$

2 **a** Complete the following.

$(x^3y^2)^3 \equiv x^3y^2 \times \boxed{} \times \boxed{} \equiv \boxed{}$

b Simplify these.

   i $(a^4b^3)^2$ **ii** $(f^3g^6)^4$ **iii** $(tm^6)^3$

   iv $(2p^5q^2)^4$ **v** $(3m^{10}n^6)^3$ **vi** $(a^3b^5c^7)^5$

3 Look at Abdullah's method for writing 8^6 as a power of 2

$8^6 = (2^3)^6 = 2^{18}$

a Explain Abdullah's method to a friend.

b Find the missing value in each of these expressions.

   i $9^4 = 3^{\boxed{}}$ **ii** $100^5 = 10^{\boxed{}}$ **iii** $1000^3 = 10^{\boxed{}}$ **iv** $125^4 = 5^{\boxed{}}$

c Solve this equation.

$16^x = 2^{60}$

4 a Simplify

i $(p^{\frac{1}{9}})^{\frac{2}{5}}$ ii $(h^{\frac{3}{4}})^{\frac{9}{10}}$ iii $(x^{0.1}y^{\frac{5}{6}})^{0.125}$

b Solve to find x

$(m^{\frac{3}{11}})^x \equiv m^{\frac{5}{9}}$

5 Simplify

a $(p^x)^{2x-1}$ b $(h^{3y})^{5-9y}$ c $(m^{5w^2})^{3w^3+2}$ d $(y^{x+1})^{x-5}$ e $(t^{3a^2+1})^{4a^3-a}$

Reflect

In your own words, explain how to find powers of powers.

I have become **fluent in…**

- adding and subtracting expressions with indices
- simplifying algebraic expressions by multiplying indices
- simplifying algebraic expressions by dividing indices
- exploring powers of powers. **H**

I have developed my **reasoning skills by…**

- explaining why the addition law for indices works
- explaining why the subtraction law for indices works
- addressing common misconceptions involving indices
- choosing when to add, subtract or multiply with powers.

I have been **problem-solving through…**

- finding missing powers in expressions
- calculating with indices in different contexts
- using fraction and decimal conversions whilst working with indices
- working with powers of numbers and letters.

Check my understanding

1 Simplify each expression by collecting like terms.

   a $4x^2 + 5x + 3x^2$ **b** $7y - 2y^3 + 3y$ **c** $8p^3 + 2p^2 - 7p^3 + p^2$

   d $10a^3 - 12a^4 - 12a^3$ **e** $6b^4 - 5b^3 + 2b^4 - b^3$ **f** $3h^3 - 2h^2 + 5h^2 - 15h^3$

2 Write each of these as a single power.

   a $a^4 \times a^5$ **b** $s^2 \times s^8$ **c** $5^6 \times 5^2$ **d** $h^2 \times h \times h^{10}$

   e $2^3 \times 2^{17}$ **f** $y^{106} \times y^{94}$ **g** $8^{15} \times 8$ **h** $d \times d^{15} \times d^{100}$

3 Write each of these as a single power.

   a $h^6 \div h^3$ **b** $p^{11} \div p^6$ **c** $17^{10} \div 17^5$ **d** $w^{99} \div w^9$

   e $\dfrac{p^{18}}{p^6}$ **f** $\dfrac{16^{10}}{16^2}$ **g** $\dfrac{q^{17}}{q^{17}}$ **h** $\dfrac{125^{500}}{125^{50}}$

4 Write each of these as a single power.

   a $\dfrac{h^5 \times h^7}{h^4}$ **b** $\dfrac{x^{11} \times x^9}{x^{10}}$ **c** $\dfrac{2^9 \times 2^3}{2^5}$ **d** $\dfrac{y^{100}}{y^{14} \times y^6}$ **e** $\dfrac{a^{15} \times a^9}{a^5 \times a}$

5 Simplify each expression.

   a $3a^5 \times 2a^7$ **b** $7x^9 \times 4x^5$ **c** $\dfrac{28y^6}{4y^3}$ **d** $\dfrac{15x^4 \times 4x}{20x^4}$ **e** $\dfrac{8p^{20} \times 10p^5}{16p^7 \div 5p^2}$

6 Write each of these as a single power. **H**

   a $(x^4)^5$ **b** $(y^6)^3$ **c** $\dfrac{(t^3)^5}{t^7}$ **d** $(h^9)^{10} \times h^8$ **e** $(2k^5)^3$

10 Fractions and percentages

In this block, I will learn...

how to convert between fractions, decimals and percentages

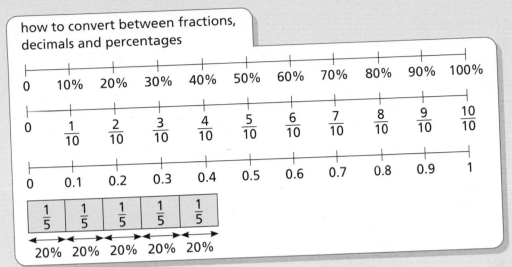

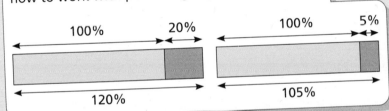

how to work with percentages greater than 100%

| 100% | 20% |
| 100% | 5% |

120%   105%

how to increase and decrease by a given percentage

100% + 18% = 118%

118% as a decimal is 1.18

So £45 increased by 18% is... 1.18 × £45 = £53.10

100% − 18% = 82%

82% as a decimal is 0.82

So £45 decreased by 18% is... 0.82 × £45 = £36.90

how to express one number as a fraction or a percentage of another

36 out of 80 = $\frac{36}{80}$ = $\frac{9}{20}$ = $\frac{45}{100}$ = 45%

45 out of 80 = $\frac{45}{80}$ = 45 ÷ 80 = 0.5625 = 56.25% = 56% to the nearest 1%

how to work out a percentage change

Cost price £160

Selling price £200

Profit = £200 − £160 = £40

Percentage profit = $\frac{£40}{£160}$ × 100% = 25%

how to find the original number given the result of a percentage change **H**

Price with VAT at 20% = £99

100% + 20% = 120%

120% = £99

1% = £0.825

100% = £82.50

Small steps

- Convert fluently between key fractions, decimals and percentages Ⓡ

- Calculate key fractions, decimals and percentages of an amount without a calculator Ⓡ

- Calculate fractions, decimals and percentages of an amount using calculator methods Ⓡ

Key words

Convert – change from one form to another, for example a percentage to a decimal

Equivalent – numbers or expressions that are written differently but are always equal in value

Are you ready?

1 Multiply these numbers by 100

   a 0.12   **b** 0.2   **c** 0.02   **d** 0.125

2 Divide these numbers by 100

   a 40   **b** 44   **c** 4   **d** 2.4

3 Write "three-tenths" as

   a a decimal   **b** a fraction.

4 Write "seventeen-hundredths" as

   a a decimal   **b** a fraction.

5 Which of these statements are true and which are false?

$$10\% = \frac{1}{10}$$ $$20\% = \frac{1}{20}$$ $$50\% = \frac{1}{5}$$ $$25\% = \frac{1}{4}$$ $$30\% = \frac{1}{3}$$

Models and representations

Hundred square

Bar model

| $\frac{1}{10}$ | $\frac{1}{10}$ | $\frac{1}{10}$ | $\frac{1}{10}$ | $\frac{1}{10}$ | $\frac{1}{10}$ | $\frac{1}{10}$ | $\frac{1}{10}$ | $\frac{1}{10}$ | $\frac{1}{10}$ |

| 10% | 10% | 10% | 10% | 10% | 10% | 10% | 10% | 10% | 10% |

Number lines

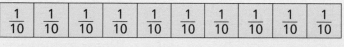

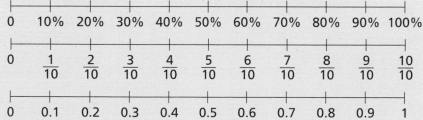

In this chapter, you will revise your learning of fractions, decimals and percentages from Book 1

Example 1

Express these fractions as percentages.

a $\dfrac{3}{10}$ **b** $\dfrac{4}{25}$ **c** $\dfrac{2}{7}$

a $\dfrac{3}{10} = \dfrac{30}{100} = 30\%$

$\dfrac{3}{10} \overset{\times 10}{=} \dfrac{30}{100} = 30\%$ ($\times 10$)

Remember: "per cent" means "out of a hundred". You can change $\dfrac{3}{10}$ to a fraction out of 100 by multiplying the numerator and the denominator by 10

$\dfrac{30}{100}$ means "30 out of 100", so it is the same as 30%

Alternatively, you could have written $\dfrac{3}{10}$ as a decimal, 0.3, and multiplied the decimal by 100%

$0.3 \times 100\% = 30\%$

b $\dfrac{4}{25} = \dfrac{16}{100} = 16\%$

$\dfrac{4}{25} \overset{\times 4}{=} \dfrac{16}{100} = 16\%$ ($\times 4$)

This time you need to multiply the numerator and the denominator by 4, because $100 \div 25 = 4$

Alternatively, you could have **converted** $\dfrac{4}{25}$ to a decimal by working out $4 \div 25 = 0.16$, and multiplied the decimal by 100%

$0.16 \times 100\% = 16\%$

c $\dfrac{2}{7} = 2 \div 7 = 0.285714\ldots$

$0.285714\ldots \times 100\% = 28.5714\ldots\%$

$= 28.6\%\ (3\ \text{sf})$

7 is not a factor of 100, so you could convert $\dfrac{2}{7}$ to a decimal first.

You can then multiply the decimal by 100% and round the percentage to a sensible degree of accuracy.

Example 2

Without using a calculator, work out

a 75% of 36   **b** 15% of £3000   **c** $33\frac{1}{3}\%$ of 90 kg

a $75\% = \dfrac{3}{4}$

$\dfrac{1}{4}$ of $36 = 36 \div 4 = 9$

So $\dfrac{3}{4}$ of $36 = 9 \times 3 = 27$

Remember that $25\% = \dfrac{1}{4}$, so $75\% = 3 \times 25\% = \dfrac{3}{4}$

Find $\dfrac{1}{4}$ by dividing by 4

Find $\dfrac{3}{4}$ by multiplying $\dfrac{1}{4}$ by 3

You can show this using a bar model:

36

| 9 | 9 | 9 | 9 |

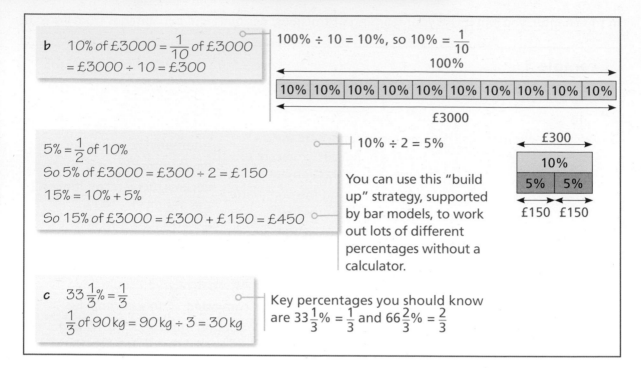

b   10% of £3000 = $\frac{1}{10}$ of £3000
    = £3000 ÷ 10 = £300

100% ÷ 10 = 10%, so 10% = $\frac{1}{10}$

$5\% = \frac{1}{2}$ of 10%
So 5% of £3000 = £300 ÷ 2 = £150
15% = 10% + 5%
So 15% of £3000 = £300 + £150 = £450

10% ÷ 2 = 5%

You can use this "build up" strategy, supported by bar models, to work out lots of different percentages without a calculator.

c $33\frac{1}{3}\% = \frac{1}{3}$
    $\frac{1}{3}$ of 90 kg = 90 kg ÷ 3 = 30 kg

Key percentages you should know
are $33\frac{1}{3}\% = \frac{1}{3}$ and $66\frac{2}{3}\% = \frac{2}{3}$

Practice 10.1A

1 What fraction and percentage equivalents do you know? How can you use them to work out others?

Record your answers and a diagram like this.

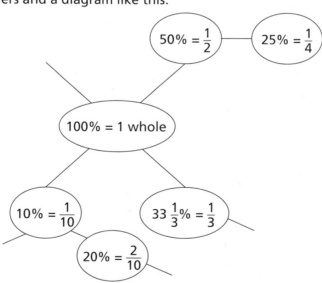

2 **a** What's the connection between percentages and decimals? How can you convert between them?

b Convert these percentages to decimals.

   i 65%   **ii** 80%   **iii** 8%   **iv** 12%   **v** 1%   **vi** 1.5%   **vii** 17.5%

c Convert these decimals to percentages.

   i 0.72   **ii** 0.06   **iii** 0.3   **iv** 0.67   **v** 0.675   **vi** 0.002   **vii** 0.345

3 Seb is converting 45% to a fraction in its simplest form.

   a Explain why Seb is correct.

   b Complete Seb's working.

> I know the fraction will be just under a half.

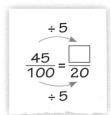

$$\frac{45}{100} = \frac{\square}{20}$$

with $\div 5$ above and $\div 5$ below.

   c How do you know that your answer is less than a half?

4 Match the equivalent fractions and percentages.

65% $\frac{1}{25}$ 74% $\frac{18}{25}$ 4% $\frac{7}{20}$ 72% $\frac{37}{50}$ 35% $\frac{13}{20}$

5 Use your knowledge of fraction and percentage equivalents to work out

   a 50% of 42   **b** 25% of 30   **c** 75% of 84

   d $33\frac{1}{3}\%$ of 150   **e** $66\frac{2}{3}\%$ of 18

6 a Find

     i 10% of 800   **ii** 5% of 800

   b Use your answers to part **a** to work out

     i 20% of 800   **ii** 15% of 800   **iii** 35% of 800

   c

> To find 95% of 800, multiply 10% of 800 by 9 and add on 5% of 800

   Describe a simpler way of working out 95% of 800

7 Benji got $\frac{11}{20}$ of the marks in a test. Abdullah scored 58% in the test. As a decimal, Marta got 0.57 of the test correct. Put the students in order of their test marks, starting with the lowest.

8 Write each set of numbers in order of size, starting with the smallest.

   a $0.3, \frac{1}{3}, 23\%, \frac{1}{4}, 0.225$ **b** $72\%, \frac{3}{4}, \frac{7}{10}, 0.8, 0.77$

What do you think? 💭

1 In how many ways can you use "key" percentages to find 45% of 60? Which is the most efficient?

2 **a** Use the fact that 25% = $\frac{1}{4}$ to convert $\frac{1}{8}$ to a percentage.

   What other multiples of $\frac{1}{8}$ can you find?

   b What fraction is the same as

   i $2\frac{1}{2}$%   **ii** $7\frac{1}{2}$%?

3 30% of a number is 24

   a What is 10% of the number?   **b** What is the number?

   c What other percentages of the number can you work out easily?

4 **a** Work out 1% of

   i 300   **ii** 450   **iii** 27

   b Describe a simpler way of working out 26% of 400 without using a calculator.

   > To find 26% of 400, multiply 1% of 400 by 26

Sometimes it is more efficient to use a calculator to work out fractions or percentages.

Example 3

Work out

a $\frac{7}{8}$ of £2100   **b** 60% of £710   **c** 43% of £602   **d** 2.3% of £168

a $\frac{7}{8}$ of £2100 = £1837.50 ○

You could work this out by hand, but it is quicker on a calculator. If your calculator has a fraction key, enter $\frac{7}{8}$ × 2100

Notice that the calculator gives an answer of £1837.5. You need to interpret this as meaning £1837.50

Or you could enter 2100 ÷ 8 × 7

b 0.6 × £710 = £426 ○─┤ Remember: 60% = $\frac{60}{100}$ = 0.6

c 0.43 × £602 = £258.86 ○─┤ 43% as a decimal is 0.43

d 0.023 × £168 = £3.86 ○─┤ 2.3% as a decimal is 0.023

The calculator shows 3.864. As this is an amount of money, you should give the answer to the nearest penny, which is to 2 decimal places.

Practice 10.1B

1 Use a calculator to work out each of each of these amounts.

   a 37% of 150 cm   **b** 68% of £350

   c 9% of 140 g   **d** 12% of £324

   e 24% of 6.7 km   **f** 61% of 5 m

   g 3% of 52 kg   **h** 89% of £137

2 **a** Which of these decimals is equivalent to $17\frac{1}{2}\%$?

| 1.75 | 0.175 | 0.17.5 | 17.5 | 0.0175 |

   b Use a calculator to find $17\frac{1}{2}\%$ of £1600

   c Check your answer by finding 10%, 5% and $2\frac{1}{2}\%$ of £1600 and adding up the three amounts.

3 **a** Jakub thinks that 2.7% is equivalent to 0.27. Explain why Jakub is wrong.

   b Find 2.7% of 300 kg   **c** Find 1.8% of £40

4 Use a calculator to work out these amounts. Give your answers in correct form and to the nearest penny, if necessary.

   a 7.3% of £200   **b** 12.8% of £65

   c 5.8% of £13   **d** 16% of £3412

5 All prices are reduced by 35% in a sale. Calculate the discount on each of these articles.

   a A TV costing £399   **b** A tablet costing £95

   c A screen protector costing £17.99

6 **a** Find 4% of 12 km. Give your answer in metres.

   b Find 6% of 3 m. Give your answer in centimetres.

   c Find 2.3% of 9 kg. Give your answer in grams.

What do you think? 💭

1 Which is greater, 45% of 63 or 42% of 68? Justify your answer.

2 The pie charts show the percentages of carbohydrate, protein, fats and other things in a protein bar and a chocolate bar.

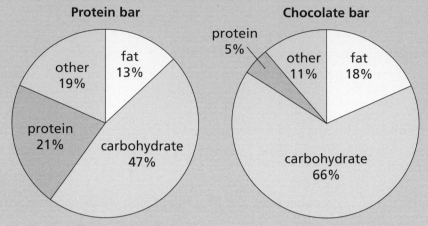

Protein bar

fat 13%
other 19%
protein 21%
carbohydrate 47%

Chocolate bar

protein 5%
other 11%
fat 18%
carbohydrate 66%

The protein bar has mass 34 g and the chocolate bar has mass 40 g.
Compare the amounts of the different food types in each bar.

Consolidate – do you need more?

1 Write these fractions as percentages.

a $\frac{1}{4}$ **b** $\frac{1}{10}$ **c** $\frac{3}{10}$ **d** $\frac{3}{5}$ **e** $\frac{11}{50}$ **f** $\frac{11}{20}$

2 Convert these percentages to decimals

a 45%   **b** 38%   **c** 7%   **d** 12%   **e** 12.5%

3 Write these percentages as fractions in their simplest form.

a 66%   **b** 80%   **c** 15%   **d** 76%   **e** 85%

4 Work these out without using a calculator.

a 25% of 48   **b** 75% of 48   **c** 10% of 180   **d** 5% of 180   **e** 65% of 180

f 45% of 90   **g** 10% of 6000   **h** 1% of 6000   **i** 3% of 6000   **j** 32% of 400

💬 Compare your methods with a partner.

5 Use a calculator to work out these percentages.

a 19% of £65   **b** 27% of £140   **c** 6% of £313

d 6.1% of £313   **e** 82.6% of £35

6 Describe three different ways of working out 35% of 700

Stretch – can you deepen your learning?

1 Can you convert percentages greater than 100% to decimals? Why or why not?

2 a Work out

    i 35% of 80   **ii** 80% of 35

  b What do you notice about your answers?

  c Do you agree with Jackson?
     Can you prove your conclusion?

$a\%$ of $b = b\%$ of a

  d Without using a calculator, work out

    i 36% of 25   **ii** 17% of 50

    iii 64% of 12.5   **iv** 90% of $66\frac{2}{3}$

3 Complete these calculations in as many ways as you can.

  ☐% of ☐ = 30   ☐% of ☐ = 27

Reflect

1 How do you work out a percentage of an amount

  a with a calculator   **b** without a calculator?

2 When is it better to use a calculator method than a non-calculator method?

Small steps

■ Percentage decrease with a multiplier

■ Convert between decimals and percentages greater than 100%

■ Calculate percentage increase and decrease using a multiplier

Key words

Multiplier – a number you multiply by

Decrease – to make something smaller

Reduce – to make something smaller

Increase – to make something larger

Are you ready?

1 Write these percentages as decimals.

    a 80%   **b** 8%   **c** 35%   **d** 23%   **e** 17%

2 60% of a class enjoy watching football. What percentage of the class do not enjoy watching football?

3 18% of the world's population live in China. What percentage of the world's population do not live in China?

4 $\frac{1}{10}$ of a year group were absent one day. What percentage were present?

5 Divide each of these numbers by 100

    a 100   **b** 160   **c** 106   **d** 126

Models and representations

Bar models

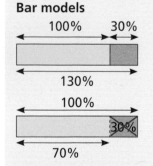

This bar model shows a percentage **increase**.

This bar model shows a percentage **decrease**.

Number lines

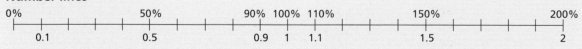

In this chapter, you will explore increasing and decreasing quantities by a given percentage. You can do this with or without a calculator.

Example 1

A shop has a sale. All the prices are reduced by 15%. Work out the sale price of an item that normally costs £280

Method A – Calculation and subtraction

10% of £280 = £280 ÷ 10 = £28
5% of £280 = 10% of £280 ÷ 2 = £28 ÷ 2 = £14
15% of £280 = 10% of £280 + 5% of £280 = £28 + £14 = £42

You can work out the **reduction** in price using the non-calculator methods you looked at in 10.1, or by entering 0.15 × £280 on a calculator.

Sale price = £280 – £42 = £238

You can then find the sale price by subtracting the reduction from the normal price.

Method B – The multiplier method

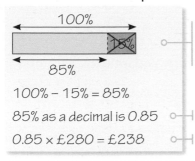

If you decrease an amount by 15%, there will be 85% of the original amount remaining.

Use your knowledge of converting between percentages and decimals.

100% – 15% = 85%
85% as a decimal is 0.85

To find 85%, you can multiply by 0.85

0.85 × £280 = £238

0.85 is the decimal **multiplier** used to find a 15% reduction.

Practice 10.2A

1 A pair of trainers normally cost £90. In a sale, the price of the trainers is reduced by 20%

- **a** What percentage of the original price of the trainers is the sale price?
- **b** Write the decimal equivalent of your answer to part **a**.
- **c** Use your decimal equivalent from part **b** to find the sale price of the trainers.

2 In one year, the value of a car decreases by 18%

- **a** What percentage of the original value is the value of the car after one year?
- **b** Write the decimal equivalent of your answer to part **a**.
- **c** A car originally costs £18 000. Use your decimal equivalent from part **b** to find the value of this car after one year.

3 A family sells their house for £268 000. The estate agent takes 2% of the price as commission.

- **a** What percentage of the price do the family receive?
- **b** Write the decimal equivalent of your answer to part **a**.
- **c** Use your decimal equivalent from part **b** to find the amount of money received by the family.

4 A travel agency reduces the prices of their holidays by 30%

   a What percentage of the original price is left after 30% is taken off?

   b Calculate the new price of a holiday which originally costs £2800

5 A jacket costs £90. The price of the jacket is reduced by 40% in a sale.

 The sale price of the jacket is 0.40 × £90 = £36

   a Explain Ed's mistake.

   b Find the sale price of the jacket.

6 Write down the percentage remaining after each of these reductions.

a 10%	**b** 15%	**c** 35%
d 52%	**e** 60%	**f** 6%
g 75%	**h** 90%	**i** 17%

7 Write down the decimal multiplier you would use to reduce a quantity by each of these percentages.

a 10%	**b** 15%	**c** 35%
d 52%	**e** 60%	**f** 6%
g 75%	**h** 90%	**i** 17%

What do you think?

1 a Which of these words or phrases mean that a quantity will be decreased?

   i a discount of 12%   **ii** a service charge of 17%

   iii a saving of 30%   **iv** a no-claims discount of 11%

   v a surcharge of 9%   **vi** shrink by 4%

  b Write the decimal multiplier for each of the decreases in part **a**.

2 Write down the decimal multiplier you would use to reduce a quantity by each of these percentages.

   a 5% **b** 2.5% **c** 1.5% **d** 0.5% **e** 0.1% **f** 1.1% **g** 10.1%

You can also **increase** quantities by a given percentage. This will mean that the new percentage is greater than 100%

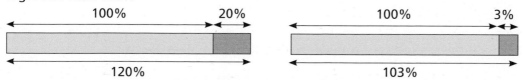

Example 2

Write the decimal equivalents of

a 105%

a $105\% = \dfrac{105}{100} = 1.05$ ○─┤

b 120%

You could also think of 100% = 1 and 5% = 0.05, so 105% = 100% + 5% = 1 + 0.05 = 1.05

c 163%

Think of per cent as "out of 100"

105 ÷ 100 = 1.05

b $120\% = \dfrac{120}{100} = 1.2$ ○─┤ Remember: 1.2 and 1.20 are the same number.

c $163\% = \dfrac{163}{100} = 1.63$

Example 3

A service charge of 18% is added to a restaurant bill of £45. What is the total cost of the bill?

Method A – *Calculation and addition*

18% of 45 = 0.18 × £45 = £8.10 ○─┤ Your calculator display will show 8.1 and this means £8.10

£45 + £8.10 = £53.10 ○─┤ You can find the total bill by adding the service charge to the cost of the food.

Method B – *The multiplier method*

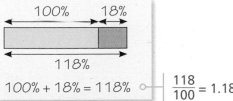

100% + 18% = 118% ○─┤ $\dfrac{118}{100} = 1.18$

118% as a decimal is 1.18 ○─┤ To find 118%, you can multiply by 1.18

1.18 is the decimal multiplier used to find an 18% increase.

1.18 × £45 = £53.10 ○─┤ Your calculator display will show 53.1 and this means £53.10

Practice 10.2B

1 A football team increases its prices for season tickets by 18%

a What percentage of the original prices are the new prices?

b Write the decimal equivalent of your answer to part **a**.

c Use your decimal equivalent from part **b** to find new the prices of

i a standard ticket costing £650

ii an executive ticket costing £950

2 All the employees at company get a 7% pay rise.

    a What percentage of the original salaries are the new salaries of the employees?

    b Write the decimal equivalent of your answer to part **a**.

    c Use your decimal equivalent from part **b** to find new the salaries of

        i a clerk earning £24 000   **ii** a manager earning £38 000

3 The cost of rewiring a house is £820 plus VAT at 20%. Work out the total cost of rewiring the house.

4 What percentage of the original value is the new value after each of these percentage increases?

    a 10%   **b** 15%   **c** 35%

    d 52%   **e** 60%   **f** 6%

    g 75%   **h** 90%   **i** 17%

5 Write down the decimal multiplier you would use to increase a quantity by each of these percentages.

    a 10%   **b** 15%   **c** 35%

    d 52%   **e** 60%   **f** 6%

    g 75%   **h** 90%   **i** 17%

6

The decimal multiplier 1.04 works out an increase of 4%

The decimal multiplier 0.94 works out a decrease of 6%

What is the effect of each of these decimal multipliers?

    a 1.23   **b** 1.08   **c** 0.85   **d** 1.01   **e** 0.62   **f** 1.92   **g** 0.18

7 **a** A laptop costs £1200. A shopkeeper increases the price by 15%. Work out the new price of the laptop.

    b The shopkeeper realises that the price of the laptop was supposed to be decreased by 15% and corrects the error. What is the price of the laptop now?

What do you think? 🗨

1 **a** Which of these words or phrases mean that a quantity will be increased?

     i a discount of 12%   **ii** a service charge of 17%   **iii** a saving of 30%

     iv a no-claims discount of 11%   **v** a surcharge of 9%   **vi** shrink by 4%

    b Write the decimal multiplier for each of the increases in part **a**.

2 A quantity is increased by 20% twice.

    a Explain why this will result in an overall increase of more than 40%

    b Work out the single multiplier that can be used to increase a quantity by 20% twice.

3 The length of a rectangle is increased by 25% and the width of the rectangle is increased by 20%.

    a Find the percentage increase in the area of the rectangle.

    b Why can't you work out the percentage increase in the perimeter of the rectangle?

4 Two investors each buy £3000 of shares. After a year, one investor has made a 12% profit and the other has made a 15% loss.

Explain how Flo can do this.

> I can find the difference between their investments at the end of year by multiplying £3000 by a single number.

Consolidate – do you need more?

1 What percentage of the original value is the new value after each of these percentage changes?

    a an increase of 20%   **b** a decrease of 20%

    c an increase of 2%   **d** a decrease of 2%

    e an increase of 17%   **f** a decrease of 17%

    g an increase of 65%   **h** a decrease of 65%

2 Write down the decimal multiplier you would use for each of the changes described in question 1

3 **a** Work out 38% of £120   **b** Increase £120 by 38%

    c Decrease £120 by 38%

4 In 1980, the population of a village was 1200. By 2020, the population had fallen by 45%. Calculate the population of the village in 2020

5 An arena has a capacity of 2800. An extension increases the capacity of the arena by 35%. Find the new capacity of the arena.

6 Which is greater, 40 decreased by 25% or 25 increased by 40%? How much greater is it?

Stretch – can you deepen your learning?

1 **a** A number is increased by 10% and then the answer is decreased 10%. Will the final answer be greater than, less than, or the same as the original number? Justify your answer.

b How would your answer to **a** change if the decrease happens before the increase? Why?

c A number is increased by 25%. What percentage decrease must be used to get back to the original number?

2 **a** Is it possible to decrease a quantity by more than 100%? Explain why or why not.

b Over a month, the height of a plant increases by 150%

i Explain why the decimal multiplier to find the plant's new height is not 1.5

ii Write the correct decimal multiplier to increase a quantity by 1.5

c A quantity is multiplied by 4.2. By what percentage has the quantity increased?

3 After a 10% pay rise, Sam earns £29 150 a year. How much did Sam earn before the pay rise?

You will explore problems like this in more detail in Chapter 10.5

Reflect

1 What's the same and what's different about using multipliers to find percentage increases and percentage decreases?

2 When might you use different methods for working out percentage increases and percentage decreases?

10.3 Expressing as a percentage

Small steps

- Express one number as a fraction or a percentage of another without a calculator

- Express one number as a fraction or a percentage of another using calculator methods

Key words

Express – write, often in a different form

Factor – a positive integer that divides exactly into another positive integer

Are you ready?

1 Write these fractions as percentages.

    a $\frac{1}{2}$ **b** $\frac{1}{10}$ **c** $\frac{7}{10}$ **d** $\frac{1}{3}$ **e** $\frac{3}{4}$ **f** $\frac{19}{50}$ **g** $\frac{6}{25}$

2 Write these decimals as percentages.

    a 0.7   **b** 0.71   **c** 0.17   **d** 0.07

3 Write these decimals as percentages. Give your answers to the nearest 1%

    a 0.123   **b** 0.345   **c** 0.678   **d** 0.901   **e** 0.082

4 Which is greater, $\frac{4}{5}$ or 75%? How do you know?

Models and representations

Fraction wall

1												

(fraction wall diagram showing rows for $\frac{1}{2}$, $\frac{1}{3}$, $\frac{1}{4}$, $\frac{1}{5}$, $\frac{1}{6}$, $\frac{1}{7}$, $\frac{1}{8}$, $\frac{1}{9}$, $\frac{1}{10}$, $\frac{1}{12}$)

331

Hundred square

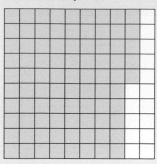

Bar model

$\frac{1}{5}$	$\frac{1}{5}$	$\frac{1}{5}$	$\frac{1}{5}$	$\frac{1}{5}$
20%	20%	20%	20%	20%

Pictures of fractions

Example 1

a   What fraction of the grid is shaded?

b   What percentage of the grid is shaded?

a $\frac{17}{25}$ ○─┤ There are 25 squares altogether and 17 of them are shaded.

b $\frac{17}{25} = \frac{68}{100} = 68\%$ (×4 numerator and denominator)

To **express** the fraction as a percentage, you need to find an equivalent fraction out of 100

25 is a **factor** of 100 as 25 × 4 = 100, so you multiply the numerator and the denominator by 4

Example 2

In a board game, there are 200 cards. 80 are red, 50 are green and the rest are yellow.

a   What percentage of the counters are yellow?

b   20 of the red cards are "Good luck" cards.

　　i   What percentage of the red cards are "Good luck" cards?

　　ii   What percentage of all the cards are the red "Good luck" cards?

a $200 - (80 + 50) = 70$

Fraction of cards that are yellow $= \frac{70}{200}$

Percentage of cards that are yellow $= \frac{70}{200} = \frac{35}{100} = 35\%$ (÷2)

First, find how many yellow cards there are.

Write this as a fraction of the total number of cards.

Convert the fraction to a percentage.

You need to divide the numerator and the denominator by 2 to change the denominator to 100

b i $\quad \div 20$

$$\frac{20}{80} = \frac{1}{4} = 25\%$$

$\div 20$

20 of the 80 red cards are "Good luck" cards.

Write this as a fraction, $\frac{20}{80}$. 80 is not a factor of 100, but you can simplify $\frac{20}{80}$ to $\frac{1}{4}$ and you know that $\frac{1}{4}$ is the same as 25%

b ii $\quad \div 2$

$$\frac{20}{200} = \frac{10}{100} = 10\%$$

$\div 2$

20 of the 200 cards are red "Good luck" cards so the fraction is $\frac{20}{200}$

Practice 10.3A

1 Look at these grids.

a

b

c

d

e
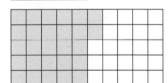

i What fraction of each grid is shaded?

ii What percentage of each grid is shaded?

2
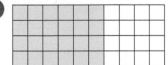

40 isn't a factor of 100 so I won't able to write the shaded fraction as an exact percentage.

a Write the fraction of the grid that is shaded. Give your answer in its simplest form.

b Convert your simplified answer to a percentage to show that Zach is wrong.

3 There are 20 counters in a bag. 14 of the counters are blue and the rest are red. One counter is picked out the bag without looking.

What is the probability that the counter is

a blue   **b** red?

Give each answer as a fraction, a decimal and a percentage.

4 In Year 8 at Oakwood High, 144 of the 240 students are girls.

   a What fraction of the students in Year 8 are girls? Give your answer in its simplest form.

   b What percentage of the students in Year 8 are girls?

5 **a** Write 60 cm as a fraction of 4 m   **b** Write 60 cm as percentage of 4 m

   c What percentage of 2 kg is 600 g?

6 Every week, Marta spends £3 of her £12 pocket money and saves the rest.

   a What fraction of her pocket money does she save?

   b What percentage of her pocket money does she save?

7 Flo is 12 years old. She has been going to school for the last 8 years.
For what percentage of her life has Flo been going to school?

8 Abdullah has written 350 words of a 500-word essay. What percentage of the essay does he still need to write?

What do you think? 💭

1 Explain why you can write "20 out of 40" as a fraction and a percentage without any working. Discuss what other fractions and percentages you might be able to recognise without working.

2 I have some 5p coins and some 10p coins in my pocket. What percentage of the coins are 5p coins if

   a the ratio of 5p coins to 10p coins is 2 : 3   **b** the ratio of 5p coins to 10p coins is 3 : 2

   c the ratio of 5p coins to 10p coins is 3 : 7   **d** the ratio of 5p coins to 10p coins is 1 : 1

   e the ratio of 5p coins to 10p coins is 1 : 3   **f** the ratio of 5p coins to 10p coins is 2 : 1?

3 Express a as a percentage of b when

   a $a = 60$, $b = 120$ **b** $a = 60$, $b = 150$ **c** $a = 60$, $b = 1200$ **d** $a = 60$, $b = 60$

   e $a = 60$, $b = 30$ **f** $a = 60$, $b = 10$ **g** $a = 60$, $b = 40$ **h** $a = 60$, $b = 45$

You can also use a calculator to express one number as a percentage of another. You may need to round your answers to the nearest 1% or another degree of accuracy.

You will explore rounding in detail in Block 12 of this book.

Example 3

There are 30 students in Class 8X. 11 of them walk to school. What percentage of Class 8X walk to school? Give your answer to the nearest 1%

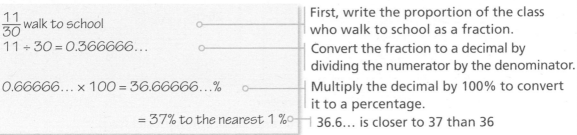

$\frac{11}{30}$ walk to school ○———— First, write the proportion of the class who walk to school as a fraction.

$11 \div 30 = 0.366666\ldots$ ○———— Convert the fraction to a decimal by dividing the numerator by the denominator.

$0.66666\ldots \times 100 = 36.66666\ldots\%$ ○———— Multiply the decimal by 100% to convert it to a percentage.

$= 37\%$ to the nearest 1 % ○—— 36.6… is closer to 37 than 36

Practice 10.3B

1. Amina scores 55 marks out of 80 in a test. Write her mark as

   a a fraction   **b** a decimal

   c a percentage, giving your answer to the nearest 1%

2. 7 out of a class of 31 students are absent. What percentage of the class are absent? Give your answer to the nearest 1%

3. At a college, 87 boys and 95 girls study Maths. What percentage of all those who study Maths are

   a boys   **b** girls?

   Give your answers to the nearest 1%

4. Write 25 as a percentage of 80

5. In a 2.5 kg cake, there is 270 g of fruit. What percentage of the cake is fruit? Give your answer to the nearest 1%

6. 160° of a pie chart is green. What percentage of the pie chart is green? Give your answer to the nearest 1%

7. The school library has 3100 books. 815 of the books are fiction. What percentage of the books in the library are fiction? Give your answer to the nearest 1%

8. Here are Jakub's marks in his end-of-term tests.

Maths	65 out of 80
English	52 out of 75
Science	98 out of 140
French	47 out of 65
History	54 out of 70

   a By writing each mark as a percentage, put the subjects in order of how well Jakub did.

   b What percentage of all the marks available did Jakub get?

9 A standard flag is a rectangle, 114 cm long by 56 cm wide.

i

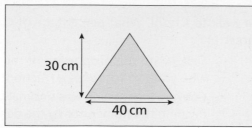

30 cm

40 cm

ii

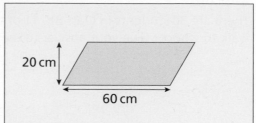

20 cm

60 cm

a What percentage of each flag is yellow?

b Investigate the percentages of different colours on national flags.

What do you think? 💭

1 Beca, Flo and Seb play a game of darts. Beca hits the target 17 times out of 27 attempts, Flo hits it 13 out of 19 times and Seb hits it 20 out of 31 times. Who was best at hitting the target? Justify your answer.

2 Mr Singh is buying a car. The car costs £17 000. Mr Singh pays a £3000 deposit. What percentage of the price of the car has he still to pay?

3 In each part, **a** to **h**, express a as a percentage of b. Give your answer to the nearest 1% where appropriate.

a $a = 12, b = 600$ **b** $a = 12, b = 120$ **c** $a = 12, b = 17$ **d** $a = 12, b = 7$

e $a = 7, b = 12$ **f** $a = 7, b = 700$ **g** $a = 7, b = 3$ **h** $a = 7, b = 37$

Which of these did you work out easily without a calculator?

Consolidate – do you need more?

1 Look at these grids.

a

b

c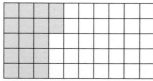

d

e

i What fraction of each grid is shaded?

ii What percentage of each grid is shaded?

2 Chloe scored 48 out of 80 in a test.

    a Write Chloe's mark as a fraction in its simplest form.

    b Write Chloe's mark as a percentage.

3 Express each of these as a percentage.

    a 130 out of 200   **b** 300 out of 500

    c 24 out of 30   **d** 40 out of 64

4 Express 90 cm out of 4 m as a percentage.

5 In a science experiment, 40 g of Chemical X is mixed with 85 g of Chemical Y.

    a What is the total mass of the mixture?

    b What percentage of the mixture is Chemical Y?

6 Abdullah does a two-part test. He scores 19 out 30 in Section A and 45 out of 55 in Section B. Write, giving your answers to the nearest 1%

    a Abdullah's mark in Section A as a percentage

    b Abdullah's mark in Section B as a percentage

    c Abdullah's overall percentage score.

Stretch – can you deepen your learning?

1 The diagram shows two identical overlapping squares. 15% of each square is shaded.

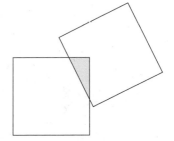

    a What percentage of the whole diagram is shaded?

    b Design a pair of overlapping shapes so that 15% of the whole diagram is shaded. Generalise for other percentages.

    You might want to start by looking at easier percentages.

Reflect

When would you use a calculator to express one quantity as a percentage of another, and when would you not? Describe your method(s) in each case.

Small steps

- Work with percentage change
- Choose appropriate methods to solve percentage problems

Key words

Profit – if you buy something and then sell it for a higher amount,
profit = amount received – amount paid

Loss – if you buy something and then sell it for a smaller amount,
loss = amount paid – amount received

Are you ready?

1 Write these fractions as percentages.

a $\frac{13}{50}$ b $\frac{21}{25}$ c $\frac{9}{20}$ d $\frac{140}{200}$

2 Convert these decimals to percentages.

a 0.46   b 0.37   c 0.17   d 0.24   e 0.8   f 0.06

3 Round these numbers to the nearest whole number.

a 9.1   b 9.14   c 27.5   d 27.45

4 Round these percentages to the nearest 1%

a 18.72%   b 18.27%   c 41.6%   d 41.06%   e 1.9%

Models and representations

Bar models

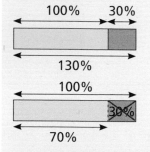

This bar model shows a percentage increase.

This bar model shows a percentage decrease.

Number lines

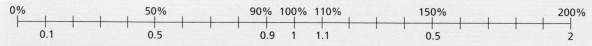

| 0% | | 50% | | 90% 100% 110% | | 150% | | 200% |

0.1   0.5   0.9 1 1.1   0.5   2

Formula

Percentage change = $\frac{\text{change}}{\text{original}} \times 100$

A percentage change is found by comparing the change to an original value.

Examples where you commonly find changes expressed as a percentage included **profit** and **loss** when buying and selling goods, changes in population and changes in measures such as mass, length and temperature.

Example 1

a   A collector buys an antique for £250 and sells it for £300. What is their percentage profit?

b   A trader invests £8000 but after a year their investment has fallen in value to £6300. What is their percentage loss?

a   Profit = £300 – £250 = £50

$$\text{Percentage profit} = \frac{\text{profit}}{\text{original}} \times 100$$

$$= \frac{50}{250} \times 100$$

$$= 20\%$$

First work out the actual profit.
Then express the profit as a percentage of the original value.

You could simplify $\frac{50}{250}$ to $\frac{1}{5}$, and then write this as a percentage, as you know that $\frac{1}{5} = 20\%$

b   Loss = £8000 – £6300 = £1700

$$\text{Percentage loss} = \frac{\text{loss}}{\text{original}} \times 100$$

$$= \frac{1700}{8000} \times 100$$

$$= 21.25\%$$

First work out the actual loss.
Then express the loss as a percentage of the original value. This is the exact answer, but if you were asked to give it to the nearest 1%, then your answer would be 21%

You can enter this on a calculator by using the fraction key.

Example 2

In 1980, the population of a city was 1.7 million people. In 2020, the population was 3.8 million. Find the percentage increase in the population of the city between 1980 and 2020. Give your answer to the nearest 1%

Increase = 3.8 million – 1.7 million = 2.1 million

$$\text{Percentage increase} = \frac{\text{increase}}{\text{original}} \times 100$$

$$= \frac{2.1}{1.7} \times 100$$

$$= 124\%$$

You don't need to write the numbers out in full, you can just work in millions throughout the question.

Your calculator shows 123.5294... This is closer to 124 than 123, so the answer is 124% to the nearest 1%

Notice that the answer is greater than 100%

Practice 10.4A

 1   Flo buys a damaged bike for £80. She repairs it and sells it for £100

a   How much profit does Flo make?

b   What percentage profit does Flo make?

2. Benji buys a games console for £300. Later, when a new model is available, he sells the games console for £120

   a What loss does Benji make on the games console?

   b What percentage loss does Benji make?

3. A shop buys trainers from a manufacturer for £50 a pair. The shop then sells each pair of trainers for £90

   a What percentage profit does the shop make on each pair of trainers?

   b At the end of the season, the shop reduces the price of the trainers to £60 a pair. What percentage profit does the shop make on each pair of trainers after this reduction?

   c The last few pairs of trainers are reduced again to £30 a pair. What percentage loss does the shop make on each pair of trainers after this reduction?

4. Seb buys a book for £12 and sells it for £15. He works out his percentage profit.

   Profit = £15 – £12 = £3

   Percentage profit = $\dfrac{£3}{£15} \times 100\% = \dfrac{1}{5} \times 100 = 20\%$

   Explain Seb's mistake.

5. A plant is 1.6 m tall. One month later, the plant is 2 m tall. Find the percentage increase in the height of the plant.

6. The population of a village is 1200. 40 years ago, the population of the village was 2000. Find the percentage decrease in the population of the village in the last 40 years.

7. Use a calculator to find the percentage profit or loss made on each of these items, giving your answers to the nearest 1%

	Price paid	Selling price
a	£18	£25
b	£260	£180
c	£17	£32
d	£33	£28
e	£3.50	£4.99
f	£19	£55

8. Chloe is running a tuck shop. She buys a pack of 12 chocolate bars for £3.50

   a Chloe sells all the chocolate bars for 60p each. Work out her percentage profit.

   b Work out Chloe's percentage profit if she sells one third of the chocolate bars for 60p each and the rest for 30p each.

What do you think?

1

I use a different method to work out percentage change.

Original price = £200, Selling price = £246

$\dfrac{\text{selling price}}{\text{original price}} = \dfrac{£246}{£200} = 1.23$

1.23 is the decimal multiplier for 123%, so the percentage profit is 23%

Why does Faith's method work? Check Faith's method by using it to work out the percentage changes you found in the practice exercise. Which method do you prefer?

2 In Worked Example 2, the percentage increase in population was greater than 100%

a Can a percentage decrease in population be greater than 100%? Why or why not?

b Think of some situations where percentage increases and decreases of over 100% are possible and also some where they are not possible.

In the last few chapters, you have studied several different types of problems involving percentages. In this exercise, the questions are mixed and you will need to think carefully about what type of calculation you need to carry out to solve each problem.

Example 3

a Express 36 as a percentage of 80

b Work out 36% of 80

a

36	

80

$\dfrac{36}{80} = 36 \div 80 = 0.45 = 45\%$

80 is the whole and 36 is part of the whole.

First write the answer as a fraction and then convert it to a decimal and a percentage.

You can do this by simplifying.

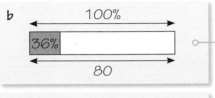

$$\dfrac{36}{80} = \dfrac{9}{20} = \dfrac{45}{100} = 0.45$$

b

100%

36%	

80

The whole is 100%. You want to work out 36%

36% of 80 = 0.36 × 80 = 28.8

You could use a non-calculator method (for example, find 10%, 5%, 1%, and so on) but for 36%, converting to a decimal and multiplying on a calculator is more efficient.

Example 4

A school buys pencils at £8 for a box of 200 and sells them for 10p each.

a If all the pencils are sold, what percentage profit does the school make?

b What percentage profit does the school make if only 60% of the pencils are sold?

c What percentage of the pencils needs to be sold so that the school does not make a loss?

a $200 \times 10p = 2000p = £20$

First find the total amount of money the school gets for selling the pencils.

Profit = £20 – £8 = £12

Percentage profit = $\dfrac{\text{profit}}{\text{original}} \times 100$

$= \dfrac{12}{8} \times 100$

$= 150\%$

You could work this out mentally as $\dfrac{12}{8} = \dfrac{3}{2} = 1.5$

b 10% of 200 = 200 ÷ 10 = 10 pencils

So 60% of 200 = 6 × 20 = 120 pencils

120 × 10p = 1200p = £12

Profit = £12 – £8 = £4

Percentage profit = $\dfrac{\text{profit}}{\text{original}} \times 100$

$= \dfrac{4}{8} \times 100$

$= 50\%$

First you need to find how many pencils were sold.

You could do this as shown, or by working out 0.6 × 200.

Again, you could do this in your head as you know that $\dfrac{4}{8} = \dfrac{1}{2} = 50\%$

c £8 = 800p

800p ÷ 10p = 80 pencils

$\dfrac{80}{200} \overset{\div 2}{\underset{\div 2}{=}} \dfrac{40}{100} = 40\%$

For the school to not make a loss, they need to make at least £8 selling pencils.

Divide 800p by 10p to find out how many pencils they need to sell.

You now need to express this as a percentage of 200

Start by writing the fraction of the pencils sold and then convert this to a percentage.

You studied this in Chapter 10.3

Remember: you can convert the fraction to a decimal using a calculator and then write this as a percentage.

Practice 10.4B

1 **a** Work out 60% of 300

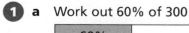

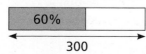
300

b Express 60 as a percentage of 300

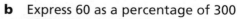

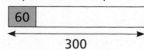
300

2 a A new car costs £15 000. After 3 years, the car has lost 35% of its value. How much is the car worth after 3 years?

b A different car costs £18 000 when new and is worth £14 000 after 3 years. What is the percentage decrease in the value of the car?

3 A rugby player has mass 105 kg. His doctor advises him to lose 15% of his mass.

a Work out the mass of the rugby player if he succeeds in losing 15% of his mass.

b The rugby player actually loses 12 kg. What percentage of his mass has he lost?

4 Copy and complete the table.

Discuss with a partner what degree of accuracy to give your answers to.

	Price paid	Selling price	Percentage profit or loss
a	£30		45% profit
b	£80		15% loss
c	£16	£20	
d	£20	£16	
e	£5.99	£7.49	
f	£80		80% loss

5 In 2019, a driver paid £650 for their car insurance. In 2020, the cost of the insurance had risen by 8%. The driver was entitled to an 8% no-claims discount.

a Explain why the driver did not still pay £650 for their car insurance in 2020

b Work out how much the driver paid for their car insurance in 2020

c Express their 2020 payment as a percentage of their 2019 payment.

6 In this question, give any percentages to the nearest 1%

The table shows the number of boys and girls at a school who are left-handed and right-handed.

	Left-handed	Right-handed	Total
Boys	122		582
Girls			
Total		979	1185

a Copy and complete the table.

b What percentage of the boys are left-handed?

c What percentage of the left-handed students are boys?

d What percentage of the right-handed students are girls?

e What percentage of the students are girls?

f What percentage of the whole school is left-handed?

g Express the number of right-handed boys as a percentage of the number of right-handed girls.

h 22.5% of the 120 staff at the school are left-handed. How many teachers is this?

7 Zach is buying a new phone. The same phone is on sale at two shops.

a From which shop is it cheaper for Zach to buy the phone?

b Express the cost of the phone in Green Phones as a percentage of the cost of the phone in Blue Phones.

c Express the cost of the phone in Blue Phones as a percentage of the cost of the phone in Green Phones.

8 The population of a town grows from 85000 to 120000 over 10 years. If the population continues to grow at the same percentage rate, what will the population of the town be after another 10 years?

What do you think? 💭

1 a A professor earns £58000 a year. She pays no tax on the first £12500 of her earnings, she pays 20% tax on earnings from £12501 to £37500 and 40% tax on earnings over £37500. Express the total amount of tax she pays as a percentage of her earnings.

b The professor gets a pay rise of 4%. How does this change the amount of tax she pays as a percentage of her earnings?

2 Another professor works for three days a week.

a He earns 0.6 of the salary of the professor in question **1**. Why do you think this is?

b Using the same tax limits as in question **1**, express the amount of tax he pays as a percentage of his earnings

i before a 4% pay rise   **ii** after a 4% pay rise.

Consolidate – do you need more?

1 Last year, a tennis club had 150 members. This year, the tennis club has only 126 members.

a How many fewer members does the club have than last year?

b Work out the percentage reduction in the number of members.

2 Last year, a book club had 60 members. This year, the book club has 96 members.

a How many more members does the club have this year than last year?

b Work out the percentage increase in the number of members.

3 Find the percentage profit or loss made on each of these items. Give your answers to the nearest 1%, if necessary.

	Price paid	Selling price
a	£12	£15
b	£15	£12
c	£2500	£1750
d	£300	£354
e	£1750	£2000
f	£97 000	£85 000

4 Kath buys 30 packs of crisps for £8.50. She sells each pack of crisps for 40p. Work out Kath's percentage profit.

5 A pizza shop offers 15% off all orders over £10. Benji buys two pizzas that normally cost £8.50 each. How much does he pay?

6 Sven buys a tablet computer for £650. How much does he sell the tablet for if

    a he makes a profit of 20%   **b** he makes a loss of 30%?

7 When water freezes, it increases in volume by 9%. Calculate the volume of 2 litres of water when it is frozen. Give your answer in cm^3, using the fact that 1 litre = 1000 cm^3

Stretch – can you deepen your learning?

1 A shop sells a laptop for £600. Find the original price of the laptop if

    a the shop makes a 20% profit

    b the shop makes a 20% loss.

> Problems of this kind will be looked at in Chapter 10.5

2 A pizza shop owner is not making a profit. He decides to either decrease the size of the pizzas by 10% or increase the price of pizzas by 10%. Assuming that the number of sales is not affected, which strategy would give him a better profit?

Reflect

How do you decide what approach to take when solving a problem involving percentages? When does the solution involve multiplication and when does it involve division?

Small steps

- Find the original amount given a percentage less than 100% ⊕
- Find the original amount given a percentage greater than 100% ⊕
- Choose appropriate methods to solve complex percentage problems ⊕

Key words

Original value – a value before a change takes place

Reverse percentage – a problem where you work out the original value

Are you ready?

1 Write these percentages as decimals.

   a 45%   **b** 145%   **c** 9%   **d** 109%

2 What decimal multiplier would you use to

   a increase a quantity by 35%   **b** decrease a quantity by 35%

   c decrease a quantity by 65%   **d** increase a quantity by 65%

   e find 35% of a quantity   **f** find 65% of a quantity?

3 Why are some of the answers to the parts of question **2** the same?

Models and representations

Bar models

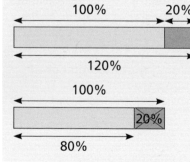

When a value has been increased by 20%, the new value is 120% of the **original value.**

When a value has been decreased by 20%, the new value is 80% of the original value.

Example 1

The price of a shirt is reduced by 20% to £24. What was the price of the shirt before the reduction?

Method A – Using multiples and factors

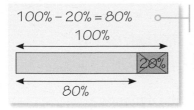

$100\% - 20\% = 80\%$

Work out what percentage of the original price is left after the reduction.

Percentage left = 100% − percentage reduction

80% of the original price is £24

So 10% of the original price is £24 ÷ 8 = £3

80% = 8 × 10%, so 10% = 80% ÷ 10

The original price was 10 × £3 = £30

The original price is 100%, which is 10 × 10%

You could also have worked this out by finding 1%

$$80\% = £24$$
÷ 80 $\qquad$ ÷ 80
$$1\% = 0.30$$
× 100 $\qquad$ × 100
$$100\% = £30$$

Method B – Using an equation

$100\% - 20\% = 80\%$

Work out what percentage of the original price is left after the reduction.

Let the original price be £x

Then $x \times 0.8 = 24$

To find 80% of a quantity, you multiply by 0.8
Set up an equation to show the change that has taken place.

$x = 24 \div 0.8$

$x = 30$

The original price was £30

Rearrange the equation to find the original price.

You could also have worked this out by finding 1%

$$106\% = £24$$
÷ 106 $\qquad$ ÷ 106
$$1\% = £270$$
× 100 $\qquad$ × 100
$$100\% = £27\,000$$

Problems like this where you find the original after a change are often called "**reverse percentage** problems".

Example 2

After a 6% pay rise, an employee earns £28 620 a year. What did the employee earn before the pay rise?

100% + 6% = 106% ○——| Work out the new percentage after the increase.

New percentage = 100% + percentage increase.

Let the original price be £x ○——| To find 106% of a quantity, you multiply by 1.06

Then x × 1.06 = 28 620 ○——| Set up an equation to show the change that has taken place.

x = 28 620 ÷ 1.06 ○————————| Rearrange the equation
x = 27 000   to find the original wage.

Before the pay rise, the employee earned £27 000

Practice 10.5A

1 30% of a number is 72. Work out

   a   10% of the number   **b**   5% of the number   **c**   40% of the number

   d   60% of the number   **e**   100% of the number.

2 150 children visited a museum. This was 20% of the total number of visitors.

   a   How many visitors were there in total?

   b   How many adults visited the museum?

3 Work out the missing numbers.

   a   40% of ☐ = 24   **b**   60% of ☐ = 24   **c**   3% of ☐ = 24

   d   12% of ☐ = 24   **e**   120% of ☐ = 24   **f**   150% of ☐ = 24

4 A pair of jeans cost £59.50 in a "15% off everything" sale. How much did the jeans cost before the sale?

5 A car dealer makes a 25% profit when she sells a car for £4000. How much did she pay for the car?

6 Rhys is solving this problem.

> After a 12% pay rise, a worker earns £168 a day.
> What did the worker earn before the pay rise?

He writes:

> 100% – 12% = 88%.
> £168 × 0.88 = £147.84

Explain what Rhys has done wrong and find the correct answer to the problem.

7 When a liquid freezes, it increases in volume by 8%. What volume of the liquid becomes 2.7 litres when frozen?

8 Junaid is 6% taller this year than he was last year. This year he is 1.43 m tall. How tall was Junaid last year? Give your answer to the nearest centimetre.

9 The value of an investor's shares dropped by 18% to £30 750 in a financial crash. What was the value of the shares before the crash?

What do you think? 💭

1 A car loses 15% of its value in the first year after purchase and then loses 10% of its value every following year. What was the original price of a car that has a value of £12 240 two years after purchase?

2 The height of a plant increases by 10% a year. The plant is now 2 m tall. How tall was the plant three years ago? Give your answer to the nearest centimetre.

3 A shopkeeper buys 20 kg of carrots. He sells them in bags containing 250 g of carrots for £1 each. He makes a 25% profit. How much did he pay for the carrots?

In the next exercise, some of the questions are about finding original values, some are about finding new values and some are about expressing values as percentages. Work with a partner to decide which type of calculation you need to do.

Practice 10.5B

1 A meal at a restaurant costs £94.08 including a service charge of 12%. Work out the cost of the meal without the service charge.

2 A gas bill is liable for VAT at 5%. Find the total cost of a gas bill if £326 of gas has been used.

3 I pay £204 for my car insurance after a 15% no-claims discount. What would my car insurance cost without the no-claims discount?

4 Copy and complete the table. Round your answers to a sensible degree of accuracy, if necessary.

	Price paid	Selling price	Percentage profit or loss
a	£30		20% loss
b		£30	20% profit
c	£65	£75	
d		£100	25% profit
e		£100	50% loss
f	£85		18% loss
g	£700	£665	
h		£649	18% profit

(5) Bev is testing a dice to find out whether it is biased. She rolls the dice and it lands on six 35% of the time. The dice lands on six 147 times. How many times did Bev roll the dice?

(6) Ms Harris sold a house for £462 000, making a 65% profit on the amount she paid for the house. How much profit did she make?

(7) Darius has been saving for 6 months to buy a laptop. The laptop originally cost £800 but has increased in price by 25%. Darius has saved £900. What percentage of the price of the laptop does Darius still need to save?

What do you think? 💭

(1) In 2019, 11.91% of the population of the USA lived in California and 0.32% lived in Rhode Island. 39 512 233 people lived in California. How many people lived in Rhode Island?

Compare methods for solving this problem by both working out the total population of the USA and by not working out the total population of the USA. Is your answer likely to be exact? Why or why not?

(2) A number is increased by 50%. By what percentage must the new number be decreased to return to the original number?

How is this connected to multiplying and dividing fractions?

Consolidate – do you need more?

1 40% of a number is 84
Work out

  a 10% of the number   **b** 5% of the number

  c 80% of the number   **d** 30% of the number

  e 100% of the number.

2 The prices of these items include VAT at 20%. How much did each item cost before VAT was added?

a

£720

b

£2400

c

£1020

3 Ali spends an hour and a half on his homework every night.
This is a 50% increase on last year.

   a How long did Ali spend on his homework last year?

   b Ali intends to spend 30% longer on his homework every night next year.
How long will he spend on his homework every night next year?

4 Copy and complete the table. Round your answers to a sensible degree of accuracy, if necessary.

	Price paid	Selling price	Percentage profit or loss
a	£40	£50	
b	£50	£40	
c	£60		25% profit
d		£80	25% profit
e		£80	20% loss
f	£200		95% loss
g	£75		30% profit
h		£350	30% loss

Stretch – can you deepen your learning?

1 When would you perform each of these operations for solving a percentage problem?
What mistakes might happen?

× 1.05	÷ 1.05	× 0.95	÷ 0.95

2 **a** The length of a rectangle is increased by 25%. What must happen to the width of the rectangle if the area of the rectangle is to remain the same?

   b Describe other pairs of percentage changes to the length and width of a rectangle so that its area remains the same.

   c Describe pairs of percentage changes to the length and width of a rectangle so that its area

     i increases by 50%   **ii** decreases by 10%

   d Investigate other overall percentage changes resulting from a combination of an increase and a decrease.

   e Would your answer to part **a** be the same or different if the shape were a triangle instead of a rectangle? Why or why not?

Reflect

1 What's the same and what's different about finding an original value after a percentage increase and after a percentage decrease?

2 How can you tell when you need to find an original value or a new value when solving a percentage problem?

I have become **fluent** in…	I have developed my **reasoning** skills by…	I have been **problem-solving** through…
■ converting between fractions, decimals and percentages ■ working out percentages of amounts ■ working out percentage increases and decreases ■ expressing one number as a percentage of another ■ working out percentages without a calculator.	■ interpreting the structure of a mathematical problem ■ interpreting relationships multiplicatively ■ understanding percentages greater than 100% ■ knowing when to calculate a percentage and when to express a value as a percentage ■ understanding and working with percentage changes	■ interpreting and solving financial mathematics problems ■ choosing appropriate methods to solve percentage problems ■ solving multi-step problems ■ finding the original value before a percentage change. Ⓗ

Check my understanding

1 Explain how you know that $20\% = \frac{1}{5}$

2 Explain how you could find 45% of 120

   a using a calculator   **b** without using a calculator.

3 Put these calculations in order of size, starting with the smallest.

A $\frac{2}{5}$ of 80   B   65% of 50   C   0.4 × 90   D $33\frac{1}{3}\%$ of 93

4 Work out 262% of £350

5 Last year, a company made £360 000 profit. This year the company's profits fell by 65%. How much profit did the company make this year?

6 2300 people in a town with a population of 15 700 are vegetarian.

   a What percentage of the town's population is vegetarian? Give your answer to the nearest 1%

   b 32% of the vegetarians in the town also eat no dairy produce. How many people is this?

7 In a sale, the normal price of a printer is reduced by 40%. The sale price of the printer is £168. Work out the normal price of the printer. Ⓗ

11 Standard index form

In this block, I will learn...

how to work with large numbers in standard form

$423\,000\,000 = 4.23 \times 10^8$

how to work with decimal numbers in standard form

$0.000\,027 = 2.7 \times 10^{-5}$

how to compare numbers in standard form

$3 \times 10^5 = 3 \times 10 \times 10 \times 10 \times 10 \times 10 = 300\,000$

$7 \times 10^4 = 7 \times 10 \times 10 \times 10 \times 10 = 70\,000$

$300\,000$ is greater than $70\,000$ so 3×10^5 is greater than 7×10^4

how to use standard form in real-life contexts with and without a calculator

Planet	Earth	Saturn	Neptune
Distance to Sun (miles)	9.3×10^7	8.9×10^8	2.8×10^9

the meaning of negative and fractional powers **H**

Power	Fraction
2^{-1}	$\frac{1}{2}$
2^{-2}	$\frac{1}{2^2} = \frac{1}{4}$
2^{-3}	$\frac{1}{2^3} = \frac{1}{8}$
2^{-4}	$\frac{1}{2^4} = \frac{1}{16}$
$2^{\frac{1}{2}}$	$\sqrt{2}$

White
Rose
Maths

Small steps

- Investigate positive powers of 10

- Work with numbers greater than 1 in standard form

Key words

Base – the number that gets multiplied when using a power/index

Index (plural: **indices**) – an index number (or power) tells you how many times to multiply a number by itself

Power (or **exponent**) – this is written as a small number to the right and above the base number, indicating how many times to use the number in a multiplication. For example, the 5 in 2^5

Standard form – a number written in the form $A \times 10^n$ where A is at least 1 and less than 10, and n is an integer

Are you ready?

1 Write each of these as a power of 10.

   a $10 \times 10 \times 10$ **b** $10 \times 10 \times 10 \times 10 \times 10 \times 10$

   c 10×10 **d** 100

   e 1000 **f** $10\,000$

2 Work these out.

   a **i** 6×10 **ii** 6×100 **iii** 6×1000

   b **i** 5.2×10 **ii** 5.2×100 **iii** 5.2×1000

   c **i** 1.65×10 **ii** 1.65×100 **iii** 1.65×1000

3 Work these out.

   a 20×300 **b** 7000×20 **c** 90^2

   d 50×200 **e** 500×800 **f** $700 \times 200 \times 200$

Models and representations

To compare numbers, you could write them in a **place value grid**.

Ten thousands	Thousands	Hundreds	Tens	Ones
	9	0	0	0
5	0	0	0	0

In your Science lessons, you sometimes have to deal with very large numbers or very small numbers.

Here are two very large numbers:

700 000 000 000 000

60 000 000 000 000

It can be easy to make a mistake when reading or comparing numbers like these.

To avoid these errors, very large numbers are often written in **standard form** or standard index form.

> In standard form a number is written as $A \times 10^n$ where A is at least 1 and less than 10, and n is an integer of any size. 10 is the **base** that is being raised to the **power** n. n is called the **index**.

Example 1

Write 800 000 in the form $A \times 10^n$

800 000 is equal to $8 \times 100 000$	Rewrite 800 000 as a number that is at least 1 and less than 10 multiplied by a power of 10
$100 000 = 10 \times 10 \times 10 \times 10 \times 10 = 10^5$	Write the power of 10 in index form.
$800 000 = 8 \times 10^5$	Write the answer in the form $A \times 10^n$ $A = 8$ and $n = 5$

Example 2

Write 423 000 000 in the form $A \times 10^n$

$423 000 000 = 4.23 \times 100 000 000$	All the non-zero digits need to form your number between 1 and 10 In this case A is 4.23
$100 000 000 = 10 \times 10 \times 10 \times 10 \times 10 \times 10 \times 10 \times 10 = 10^8$	Write the power of 10 in index form.
$423 000 000 = 4.23 \times 10^8$	There are 8 digits after the digit 4, so you multiply 4.23 by 10^8

Example 3

Write these numbers in ordinary form.

a 3×10^5 **b** 7.51×10^4

a $3 \times 10^5 = 300\,000$ ○——| $10^5 = 10 \times 10 \times 10 \times 10 \times 10 = 100\,000$

b $7.51 \times 10^4 = 7.51 \times 10\,000$ ○——| A common mistake is to write four zeros in the
$\qquad\qquad = 75\,100$ | answer to a problem like this. This is incorrect.

Practice 11.1A

1 Copy and complete the table.

10^2	10×10	100
10^3	$10 \times 10 \times 10$	1000
10^4		
10^5		
10^6		

2 Write each number as a power of 10

a $10\,000\,000$

b $100\,000\,000$

c 1 million

d $10\,000\,000\,000$

e $100\,000\,000\,000\,000$

f 1 billion

3

10^{21} is the number 1 followed by 21 zeros.

Do you agree with Beca?

4 Here are two definitions.

Googol: the number 1 followed by 100 zeros

Googolplex: the number 1 followed by 1000 zeros

Write these as powers of 10

a a googol

b a googolplex

5 Write these numbers in the form $A \times 10^n$

    a **i** 700   **ii** 7000   **iii** 70 000   **iv** 700 000

    b **i** 90 000   **ii** 20 000   **iii** 50 000   **iv** 60 000

    c **i** 3 000 000 **ii** 5000   **iii** 60 000 000 **iv** 400 000 000

6 Zach is writing these numbers in standard form.

| 20 000 | 35 000 | 43 200 | 51 170 |

    What will be the same about each of his answers? What will be different?

7 Write these numbers in standard form.

    a 26 000   **b** 670   **c** 219 000

    d 63 000 000   **e** 7 540 000   **f** 35

    g 718 000 000   **h** 90 300   **i** 66 120 000 000

    j 1.2 million   **k** 7155   **l** 730 000 025 000

8 **a** Write each set of numbers in ordinary form.

        Set A: 1.4×10^5, 6.32×10^5, 7.182×10^5, 3.0261×10^5

        Set B: 1.32×10^3, 1.32×10^4, 1.32×10^5, 1.32×10^6

    b What patterns do you notice in each set?

9 Write these numbers in ordinary form.

    a 2×10^4 **b** 1.5×10^3 **c** 3.82×10^5

    d 1.79×10^4 **e** 9.91×10^6 **f** 6.5×10^2

    g 2.274×10^5 **h** 6.5823×10^8 **i** 8×10^5

What do you think? 🗨

1

> $2.3 \times 10^4 = 230\,000$

Jackson is incorrect.

a Why do you think Jackson thinks this? What mistake has he made?

b What number is 2.3×10^4 equal to?

c Explain to a partner how you worked out the answer to part **b**.

2 a Write these numbers in standard form.

    i 12600000000   **ii** 12 600 000 000

b Which number was easier to write in standard form? Explain why?

3 Are these numbers written in standard form?

  a 56.3×10^5 **b** 0.62×10^7 **c** $2.7 \times 10^{3.1}$

4

> 1.3×10^6 is 100 times bigger than 1.3×10^4

Do you agree with this statement?

Consolidate – do you need more?

1 Write these numbers in standard form.

  a 7000   **b** 90 000   **c** 600 000 000

  d 800   **e** 10 000 000 000

2 Write these numbers in standard form.

  a 25 000   **b** 110 000   **c** 920

  d 451 000   **e** 76 500 000   **f** 6.3 million

3 Write these numbers in ordinary form.

  a 2×10^5 **b** 3.2×10^4 **c** 6.2×10^3

  d 7.91×10^4 **e** 7.91×10^6 **f** 8.032×10^6

Stretch – can you deepen your learning?

1 Write these numbers in standard form.

  a 12 000 000 000 000 000 000 000 000 000   **b** 67 500 000 000 000 000 000 000

  c 5 300 000 000 000 000 000   **d** 176 350 000 000 000 000 000 000 000

  Explain your method.

2 In a survey in 2018, the population of China was recorded as 1.393×10^9

Do you think that this statistic is exact? Explain your reason.

3 A factory makes jellybeans. Yesterday the factory made 370 918 064 jellybeans.

a Write the number 370 918 064 in standard form.

b Explain why it is unlikely that you would write this number in standard form.

c What could you do with the number before writing it in standard form?

4

My number is 3.6×10^8

Marta

My number is 100 times larger than yours.

Emily

Write Emily's number in standard form. Explain your method.

5 Here are four number cards.

A 1.6×10^4 B 1.6×10^5 C 1.7×10^4 D 1.7×10^6

a Which number is 10 times greater than A?

b Which number is 100 times smaller than D?

c E is a thousand times greater than A. Write E in standard form.

Reflect

1 Why do you write a large number in standard form?

2 When might you write a large number in standard form?

3 Explain to a friend how you write a number in standard form.

Small steps

- Investigate negative powers of 10

- Work with numbers between 0 and 1 in standard form

Key words

Base – the number that gets multiplied when using a power/ index

Index (plural: **indices**) – an index number (or power) tells you how many times to multiply a number by itself

Power (or **exponent**) – this is written as a small number to the right and above the base number, indicating how many times to use the number in a multiplication. For example, the 5 in 2^5

Standard form – a number written in the form $A \times 10^n$ where A is at least 1 and less than 10, and n is an integer

Are you ready?

1. Write each of these fractions as a decimal.

   a $\frac{1}{10}$ b $\frac{1}{100}$ c $\frac{1}{1000}$ d $\frac{1}{10\,000}$

2. Work these out.

   a i 6×1 ii 6×10 iii 6×100 iv 6×1000

   b i 6×0.1 ii 6×0.01 iii 6×0.001

3. Use your power button on your calculator to work these out.

   a 2^6 b 3^4 c 5^3 d 10^6

Models and representations

Place value grid

You can compare numbers by writing them in a place value grid. In this chapter, you will use numbers less than 1.

Ones	tenths	hundredths	thousandths	ten thousandths
0	0	0	1	

In this chapter, you will investigate how to write numbers between 0 and 1 in **standard form** using negative **powers**.

A number between 0 and 1 is written with a negative power of 10

A negative power of 10 represents 1 divided by the corresponding positive power of 10

10^{-1} is equal to $\frac{1}{10}$ which is one-tenth = 0.1

10^{-2} is equal to $\frac{1}{10 \times 10}$ or $\frac{1}{100}$ which is one-hundredth = 0.01

10^{-3} is equal to $\frac{1}{10 \times 10 \times 10}$ or $\frac{1}{1000}$ which is one-thousandth = 0.001

10^{-4} is equal to $\frac{1}{10 \times 10 \times 10 \times 10}$ or $\frac{1}{10\,000}$ which is one ten-thousandth = 0.0001

Example 1

a Write 0.000 02 in standard form.   **b** Write 0.000 027 in standard form.

c Write 3.05×10^{-4} as a decimal.

a $0.000\,02 = 2 \times 10^{-5}$ ○————— Write the number as a decimal between 1 and 10

2 is between 1 and 10

To adjust this to become 0.000 02
you need to multiply by 0.000 01

b $0.000\,027 = 2.7 \times 10^{-5}$ ○————— Write the number as a decimal between 1 and 10

2.7 is between 1 and 10

To adjust this to become 0.000 027
you need to multiply by 0.000 01

c $10^{-4} = 0.0001$ ○————— The power is −4 so you know that
the number will be less than 1

So $3.05 \times 10^{-4} = 3.05 \times 0.0001$

$\qquad\qquad\qquad = 0.000\,305$

Practice 11.2A

1 **a** Copy and complete the table. You can use your calculator to help you.

Power of 10	10^5	10^4	10^3	10^2	10^1	10^0	10^{-1}	10^{-2}	10^{-3}	10^{-4}	10^{-5}
Ordinary number	100 000	10 000			10	1	0.1	0.01			

   b What patterns do you notice?

   c Which powers of 10 give positive answers?

   d Which powers of 10 give negative answers?

2 Write these as a power of 10

   a 0.001   **b** 0.0001   **c** 0.000 001

   d 0.000 0001   **e** 0.000 000 001

3 Chloe is investigating negative powers of 10

> I noticed something about the number of zeros before the 1 and the power of 10

What do you think Chloe noticed?

4 Seb is trying to write 0.005 in standard form. He writes

$$0.005 = 5 \times 0.001 = 5 \times 10^{-3}$$

Use Seb's method to write each of these sets of numbers in standard form.

a **i** 0.0003   **ii** 0.0006   **iii** 0.0007

b **i** 0.2   **ii** 0.02   **iii** 0.002

What patterns do you notice?

5 Write each of these numbers in the form $A \times 10^n$, where A and n are integers.

a 0.000005   **b** 0.007   **c** 0.0000003   **d** 0.000002

6 **a** Write these numbers in standard form.

     i 0.12   **ii** 0.012   **iii** 0.0012   **iv** 0.00012

     b What's the same and what's different about your answers?

7 **a** Write these numbers in standard form.

     i 0.00003   **ii** 0.000032   **iii** 0.0000317

     b What's the same and what's different about your answers?

8 Faith thinks that 0.72×10^{-4} is a number written in standard form.
Explain why Faith is incorrect.

9 Write these numbers in standard form.

a 0.0014   **b** 0.00078   **c** 0.0527

d 0.0000032   **e** 0.005032   **f** 0.0000568

g 0.0096   **h** 0.01056   **i** 0.35

j 0.000000071

10 Write each set of numbers in ordinary form.

Set A: 1.4×10^{-1}, 1.4×10^{-2}, 1.4×10^{-3}, 1.4×10^{-4}

Set B: 3×10^{-3}, 3.6×10^{-3}, 3.65×10^{-3}, 3.651×10^{-3}

What patterns do you notice in each set?

11 Write these numbers in ordinary form.

a 2.4×10^{-3} **b** 1.56×10^{-4} **c** 8.1×10^{-2}

d 7.685×10^{-6} **e** 1.35×10^{-7} **f** 7.6×10^{-1}

g 5.564×10^{-4} **h** 2.367×10^{-6} **i** 6×10^{-3}

What do you think?

1 Here are three number cards.

| 6×10^{-3} | 6.1×10^{-3} | 6.12×10^{-3} |

Without writing each number in ordinary form, state what will be the same about each of them.

2 a Abdullah is trying to write $0.000\,000\,002\,15$ in standard form.

How do I do this?

It is 2.15×10^{-9}

Abdullah   You did that really quickly.

Flo

How do you think Flo got the answer so quickly?

b Flo now writes the number $42\,000$ in standard form.

It is 4.2×10^{-4}

What mistake has Flo made?

c When does a number written in standard form have a negative power?

3 Copy the table and write these numbers in the correct columns.

| 0.42×10^{-2} | 3.61×10^{-4} | 7×10^{4} | 0.7×10^{4} |

| 12.6×10^{5} | 0.003 | 3.276×10^{3} | $2.3 \times 10^{-3.5}$ |

Numbers in standard form	Numbers not in standard form

Compare your answers with a partner. Discuss any that you didn't agree on.

4 a Which number is 10 times smaller than 4.2×10^{-3}?

A 4.2×10^{-4} B 4.2×10^{-2}

Explain why you chose your answer.

b Which number is 10 times greater than 5×10^{-6}?

A 5×10^{-7} B 5×10^{-5}

Consolidate – do you need more?

1 Write these numbers in standard form.

   a 0.007   **b** 0.09   **c** 0.000006   **d** 0.00008   **e** 0.000000005

2 Write these numbers in standard form.

   a 0.00418   **b** 0.00073   **c** 0.04132

   d 0.005381   **e** 0.023   **f** One thousandth

3 Write these numbers in ordinary form.

   a 2×10^{-3} **b** 4.2×10^{-3} **c** 7.56×10^{-2}

   d 6.35×10^{-5} **e** 7.91×10^{-4} **f** 4.308×10^{-6}

Stretch – can you deepen your learning?

1 Write these numbers in standard form.

   a 0.000000000000000402   **b** 0.00000000000318

   c 0.0000000000000000073184   **d** 0.00000000000900041

   e One millionth   **f** One billionth

2 Each of these numbers may look like they are written in standard form, but they are not. Write the numbers in standard form.

   a 0.372×10^{-3} **b** 0.00748×10^{-2} **c** 50×10^{-6}

3 Copy and complete the table. Write all your answers in standard form.

Number	8.2×10^{-4}	1.39×10^{-1}	
10 times greater	8.2×10^{-3}		
100 times greater			
1000 times greater			6.04
10 times smaller	8.2×10^{-5}		
100 times smaller			
1000 times smaller			

Reflect

1 What type of number will have a negative power when written in standard form? Which numbers will have positive powers?

2 Explain how you write a number between 0 and 1 in standard form.

White Rose Maths

Small steps

■ Compare and order numbers in standard form

■ Mentally calculate with numbers in standard form

Key words

Index (plural: **indices**) – an index number (or power) tells you how many times to multiply a number by itself

Are you ready?

1 Write these numbers in standard form.

    a 20 000   **b** 9600   **c** 3 560 000

    d 0.005   **e** 0.0306   **f** 0.000 008 215

2 Write these numbers in ordinary form.

    a 6×10^3 **b** 3.2×10^4 **c** 4.15×10^5 **d** 3.05×10^3

    e 6×10^{-4} **f** 7.1×10^{-2} **g** 1.842×10^{-5}

3 Write <, > or = between the two numbers to make each statement correct.

    a 720 ◯ 1380   **b** 950 000 ◯ 906 398   **c** 1 350 062 ◯ 2 million

    d 0.34 ◯ 1.7   **e** 0.003 ◯ 0.03   **f** 7 tenths ◯ 0.7

4 Work these out in your head.

    a **i** 2000 + 5000   **ii** 2000 + 500   **iii** 2000 + 50   **iv** 2000 + 5

    b **i** 300 × 2   **ii** 300 × 20   **iii** 300 × 200   **iv** 300 × 2000

    c **i** 1 − 0.3   **ii** 1 − 0.4   **iii** 1 − 0.03   **iv** 1 − 0.04

Models and representations

Place value chart

Place value charts are useful for comparing large or small numbers

Millions			Thousands			Ones		
H	T	O	H	T	O	H	T	O
	2	6	9	1	1	1	0	3

The hundreds, tens and ones columns are repeated below each place value.

This shows 26 millions, 911 thousands and 103 ones.

In this chapter, you will compare numbers written in standard form and perform calculations without using a calculator.

Example 1

Which of these numbers is greater?

3×10^5 or 7×10^4

$3 \times 10^5 = 3 \times 10 \times 10 \times 10 \times 10 \times 10 = 300\,000$

$7 \times 10^4 = 7 \times 10 \times 10 \times 10 \times 10 = 70\,000$

$300\,000$ is greater than $70\,000$

so 3×10^5 is the greater number.

3×10^5 is greater than 7×10^4 as it has the greater power of 10

One method for answering this is to write each number in ordinary form. Another method is to compare the **indices**.

10^5 is greater than 10^4

The numbers must be in standard form to make this comparison.

Practice 11.3A

1. Ed and Beca are comparing numbers in standard form.

   7×10^4 5×10^3

   Ed says that he will write each of the numbers in ordinary form and then compare them.

   a. Write each of the numbers in ordinary form.

   b. Which is the greater number?

   c. Beca says that she can compare the numbers without writing them in ordinary form. Explain how Beca can do this.

2. Which of these are greater than 2.6×10^5?

   A 2.7×10^5 B 9.95×10^3 C 3×10^4

   D 6.61×10^6 E 1.8×10^{-6} F 1.86×10^5

3. Which of these are smaller than 7×10^2?

   A 7×10^3 B 6.5×10^2 C 1.8×10^4

   D $30\,000$ E 3.85×10^{-1} F 8.04×10^2

4 Copy the table and write each of these numbers in the correct column.

| 6×10^4 | 6×10^5 | 6×10^6 |

| 1.8×10^5 | 1.9×10^5 | 1.2×10^5 |

| 2.356×10^7 | 1.6×10^{11} | 4.8×10^{-7} |

Numbers greater than 1 million	Numbers less than 1 million

5 Compare these numbers using <, > or =

a 4.5×10^3 ◯ 4.2×10^3

b 7×10^4 ◯ $55\,000$

c 8.95×10^5 ◯ 2.8×10^6

d 1.03×10^{-2} ◯ 1.5×10^{-2}

e 1.03×10^{-2} ◯ 1.5×10^{-3}

f 1.03×10^{-2} ◯ 1.5×10^{-1}

6 Write down the greatest number from each set.

Set A: 2.6×10^4, 2.8×10^4, 3×10^2

Set B: 3.92×10^4, 3.92×10^3, 3.92×10^2

Set C: 1.46×10^5, 8.2×10^4, 1.87×10^4

Set D: 1.651×10^5, 1.605×10^6, 1.65×10^5

7 Jakub has these number cards.

| 1.3×10^{-2} | 1.3×10^{-4} | 1.3×10^{-1} |

Jakub says that the greatest number is 1.3×10^{-4} as 4 is the greatest number.

What mistake has Jakub made? Which is the greatest number?

8 Put the following numbers in ascending order.

4.2 million 3.97×10^5 $73 \times 100\,000$

9 Here are the masses of three planets in our Solar System.

Planet	Mercury	Venus	Earth
Mass (kg)	3.301×10^{23}	4.867×10^{24}	5.972×10^{24}

a Which is heavier, Mercury or Venus? How do you know?

b Which is heavier, Venus or Earth? How do you know?

10 Write down three numbers that lie between 6.5×10^4 and 6.9×10^5.
Check your answers with a partner.

What do you think?

1 Benji thinks that 7×10^4 is greater than 2×10^6

   a Why might Benji think this?   **b** What mistake has he made?

   c Explain why 2×10^6 is greater than 7×10^4

2 Zach and Faith are discussing numbers in standard form.

> I know that 3.6×10^5 is greater than 1.5×10^5 because 3.6 is greater than 1.5

Zach

> I know that 3.6×10^5 is greater than 1.5×10^6 because 3.6 is greater than 1.5

Faith

Do you agree with Zach and Faith?

3 What numbers could replace x to make this statement correct?

$$6.3 \times 10^5 > 7.6 \times 10^x$$

Explain your answer.

Example 2

Work out $6 \times 10^3 \times 7$

Give your answer in standard form.

$6 \times 10^3 \times 7 = (6 \times 7) \times 10^3$ ○──── This calculation can be rearranged and performed mentally.

         $= 42 \times 10^3$ ○──── Your answer is no longer in standard form because 42 is greater than 10.

$42 \times 10^3 = 4.2 \times 10 \times 10^3$ ○──── 4.2 is between 1 and 10.

         $= 4.2 \times 10^1 \times 10^3$ ○──── You can use the addition law for indices and the fact that $42 = 4.2 \times 10$ to help you rewrite your answer in standard form.

         $= 4.2 \times 10^{(1+3)}$

         $= 4.2 \times 10^4$ ○──── $42 \times 10^3 = 4.2 \times 10^4$

Practice 11.3B

1 Complete the calculations. Give your answers in standard form.

a **i** $7 \times 10^8 \times 10$ **ii** $7 \times 10^8 \times 100$

    iii $7 \times 10^8 \times 1000$ **iv** $7 \times 10^8 \times 10\,000$

b **i** $10 \times 5 \times 10^3$ **ii** $100 \times 5 \times 10^3$

    iii $1000 \times 5 \times 10^3$ **iv** $10\,000 \times 5 \times 10^3$

c **i** $7.9 \times 10^{-6} \times 10$ **ii** $7.9 \times 10^{-6} \times 100$

    iii $7.9 \times 10^{-6} \times 1000$ **iv** $7.9 \times 10^{-6} \times 10\,000$

d **i** $80\,000 \times 10$ **ii** $100 \times 200\,000$

    iii $10 \times 720\,000\,000$ **iv** $1\,000\,000 \times 32\,600$

2 Sort these numbers into a copy of the table.

| 0.2×10^5 | 2×10^7 | 20×10^{-3} | 2.54×10^2 |

| 64×10^{11} | 256×10^{-9} | 5.3×10^3 | 16.4×10^{305} |

| $20\,000 \times 10^{19}$ | 0.756×10^{-1} | 208×10^{10} | 10×10^3 |

Numbers written in standard form	Numbers not written in standard form

3 Rewrite each number in standard form.

a **i** 30×10^3 **ii** 30×10^4 **iii** 30×10^5 **iv** 30×10^{27}

b **i** 80×10^7 **ii** 800×10^7 **iii** 8000×10^7 **iv** $800\,000 \times 10^7$

c **i** 20×10^3 **ii** 12×10^3 **iii** 23×10^3 **iv** 105×10^3

d **i** 80×10^{14} **ii** 97×10^{14} **iii** 230×10^{14} **iv** 10.5×10^{14}

e **i** 1250×10^{-3} **ii** 125×10^{-3} **iii** 12.5×10^{-3} **iv** 0.125×10^{-3}

4 Work these out. Give your answers in standard form.

a **i** $2 \times 10^3 \times 2$ **ii** $2 \times 10^3 \times 3$

    iii $2 \times 10^3 \times 4$ **iv** $2 \times 10^3 \times 5$

b **i** $6 \times 10^3 \div 2$ **ii** $6 \times 10^3 \div 3$

    iii $6 \times 10^3 \div 4$ **iv** $6 \times 10^3 \div 5$

What do you think? ☻

1 Here are some standard form calculations.

A $\quad 3.6 \times 10^3 \times 2.1 \times 10^3$
B $\quad 5 \times 10^5 \times 4 \times 10^3$
C $\quad 3.1 \times 10^{-3} \div 1.4 \times 10^{-2}$

D $\quad 3 \times 10^4 \times 2$
E $\quad 8 \times 10^4 \div 2$

Discuss these questions with a partner.

a Which of them are easy to do in your head? **b** Which ones are harder? Why?

c What mistakes do you think people make?

2 Find two possible pairs of missing numbers to make each statement correct.

a $\boxed{} \times 10^4 \times \boxed{} = 6 \times 10^5$ **b** $\boxed{} \times 10^4 \div \boxed{} = 1.5 \times 10^4$

Consolidate – do you need more?

1 Write <, > or = to make each statement correct.

a $5 \times 10^6 \bigcirc 7 \times 10^4$

b $7.1 \times 10^3 \bigcirc 4.5 \times 10^3$

c $6.9 \times 10^3 \bigcirc 8 \times 10^5$

d $1.35 \times 10^{-2} \bigcirc 1.4 \times 10^{-2}$

e $7 \times 10^{-3} \bigcirc 3 \times 10^{-1}$

f $7.25 \times 10^4 \bigcirc 3$ million

g $1200 \bigcirc 2.8 \times 10^2$

2 Which of these are greater than 3×10^4?

A $\boxed{5 \times 10^6}$ B $\boxed{2 \times 10^4}$ C $\boxed{2 \times 10^5}$

D $\boxed{3.1 \times 10^4}$ E $\boxed{7.356 \times 10^3}$ F $\boxed{8 \times 10^{-5}}$

3 Which of these are less than 2×10^{-2}?

A $\boxed{2 \times 10^{-3}}$ B $\boxed{4.1 \times 10^{-2}}$ C $\boxed{1.5 \times 10^{-2}}$

D $\boxed{2.5 \times 10^{-2}}$ E $\boxed{9.8 \times 10^{-3}}$ F $\boxed{4 \times 10^{-1}}$

4 Work these out in your head.

a $2 \times 10^3 \times 3$ **b** $5 \times 10^7 \times 3.4$ **c** $3.2 \times 10^3 \times 2$ **d** $9 \times 10^{-3} \div 2$

Stretch – can you deepen your learning?

1 a $P \times 10^x > P \times 10^y$

Which of these is true?

A $\boxed{x > y}$ B $\boxed{y > x}$ C $\boxed{x = y}$

Explain your answer.

b $M \times 10^x > N \times 10^x$

Which of these is true?

A $\boxed{M > N}$ B $\boxed{N > M}$ C $\boxed{M = N}$

Explain your answer.

2 Ed has written two numbers in standard form. The first part of each number is missing.

A $\times 10^5$ B $\times 10^7$

a Which number is greater? How can you tell if you know only part of the number?

b If the numbers were not written in standard form, could you tell which number was the greater?

3 Each square represents one digit. Suggest what the missing digits could be.

a i $1.\boxed{} \times 10^4 > 1.6 \times 10^4$ **ii** $1.\boxed{} \times 10^4 > 1.6 \times 10^3$

b i $6.35 \times 10^3 < 6.\boxed{} \times 10^3$ **ii** $6.35 \times 10^3 < 6.3\boxed{} \times 10^3$

4 Find the set of values of x that makes each statement correct.

a $6 \times 10^4 > 6 \times 10^x$ **b** $6 \times 10^4 > 5 \times 10^x$ **c** $6 \times 10^4 > 7 \times 10^x$

Reflect

1 How do you compare numbers in standard form? What different methods do you know?

2 How can the laws of indices help you to rewrite numbers in standard form?

Small steps

- Add and subtract numbers in standard form
- Multiply and divide numbers in standard form
- Use a calculator to work with numbers in standard form

Key words

Index (plural: **indices**) – an index number (or power) tells you how many times to multiply a number by itself

Commutative – when an operation can be in any order

Are you ready?

1 Write these numbers in standard form.

  a 40 000   **b** 36 000   **c** 175 000   **d** 2 million

  e 287 100 000   **f** 0.0006   **g** 0.062   **h** 0.000 1934

2 Write these numbers in ordinary form.

  a 1.6×10^3 **b** 7.52×10^3 **c** 2.85×10^5 **d** 1.9×10^7

  e 7×10^{-4} **f** 8.1×10^{-3} **g** 3.656×10^{-2} **h** 1.8×10^{-1}

3 Write these as a single power of 10

  a $10^5 \times 10^2$ **b** $10^4 \times 10^3$ **c** $10^6 \times 10^{-4}$ **d** $10^2 \times 10^5 \times 10^{-1}$

  e $10^6 \div 10^2$ **f** $10^{25} \div 10^3$ **g** $10^4 \div 10^{-3}$ **h** $10^{100} \div 10$

Models and representations

A calculator can be used to work with numbers in standard form.

There is often a "$\boxed{\times 10^x}$" button on a calculator that can be used to work with standard form.

You know from the previous chapter that $3 \times 10^9 \times 7 = 21 \times 10^9$

You can use a calculator to give this answer in correct standard form.

If you type in $\boxed{2}\boxed{1}\boxed{\times 10^x}\boxed{9}\boxed{=}$ your calculator will rewrite this in standard form.

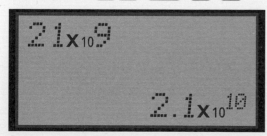

Example 1

Work out each calculation, giving your answers in standard form.

a $7 \times 10^3 + 4 \times 10^3$ **b** $4 \times 10^3 + 6 \times 10^5$

Method A

a $7000 + 4000 = 11\,000$ ○——┤ Change each number from standard
      $11\,000 = 11 \times 10^3$ form to an ordinary number.
             $= 1.1 \times 10^4$ ○——┤ Convert your answer back to standard form.

b $4000 + 600\,000 = 604\,000$
      $604\,000 = 6.04 \times 10^5$

Method B

a $(7 + 4) \times 10^3 = 11 \times 10^3$ ○——┤ The powers are the same, so you can re-order
             $= 1.1 \times 10^4$ ○——┤ the calculation and simply add the numbers.
                                        Make sure that your final answer is in standard form.

b $4 \times 10^3 = 0.04 \times 10^5$ ○———┤ Adjust one of the numbers so
      $0.04 \times 10^5 + 6 \times 10^5 = (0.04 + 6) \times 10^5$ that the powers are the same.
                          $= 6.04 \times 10^5$ ○——┤ Then use the same method as in part **a**.

Practice 11.4A

1 Write these numbers in standard form.

   a **i** 12×10^4 **ii** 12×10^5 **iii** 12×10^6

   b **i** 26×10^3 **ii** 260×10^3 **iii** 2600×10^3

   What do you notice?

2 Write these numbers in standard form.

   a **i** 0.3×10^2 **ii** 0.3×10^3 **iii** 0.3×10^4

   b **i** 0.07×10^2 **ii** 0.52×10^2 **iii** 0.0056×10^2

3 Jakub and Faith are adding the following numbers

   $2.3 \times 10^4 + 5.6 \times 10^3$

I will write each number in ordinary form and then add them together.

$2.3 \times 10^4 = 23\,000$
$5.6 \times 10^3 = 5600$
$23\,000 + 5600 = 28\,600$

I will write each number so that they have the same power of 10

$5.6 \times 10^3 = 0.56 \times 10^4$
so $2.3 \times 10^4 + 0.56 \times 10^4 = 2.86 \times 10^4$

Discuss the methods with your partner. Whose method do you prefer?

4 Work these out. Give your answers in standard form.

    a $6 \times 10^3 + 2 \times 10^3$ **b** $6 \times 10^3 + 2 \times 10^4$ **c** $6 \times 10^3 + 2 \times 10^5$

5 Work out these additions. Give your answers in standard form.

    a $3 \times 10^4 + 5 \times 10^3$ **b** $1.6 \times 10^4 + 7 \times 10^2$

    c $2.8 \times 10^6 + 3.5 \times 10^5$ **d** $5.9 \times 10^5 + 3.78 \times 10^5$

    e $7.45 \times 10^4 + 3.6 \times 10^2$ **f** $1.578 \times 10^4 + 4.9 \times 10^3$

    g $5.13 \times 10^4 + 2.8 \times 10^5$ **h** $7.508 \times 10^3 + 4.75 \times 10^2$

    i $3 \times 10^{-3} + 4.1 \times 10^{-2}$ **j** $6.8 \times 10^{-4} + 2.3 \times 10^{-1}$

    k $6.64 \times 10^3 + 3.5 \times 10^4 + 1.8 \times 10^5$

    Check your answers using a calculator.

6 Work out these subtractions. Give your answers in standard form.

    a $8 \times 10^3 - 3 \times 10^2$ **b** $7 \times 10^5 - 7 \times 10^3$

    c $2.9 \times 10^5 - 3 \times 10^4$ **d** $1.892 \times 10^4 - 1.7 \times 10^4$

    e $3.855 \times 10^6 - 1.31 \times 10^4$ **f** $7.83 \times 10^4 - 6.84 \times 10^3$

    g $4 \times 10^{-2} - 3 \times 10^{-3}$ **h** $1.78 \times 10^{-1} - 2.5 \times 10^{-2}$

    i $9.8 \times 10^{-3} - 1.65 \times 10^{-5}$

    Check your answers using a calculator.

7 The diagram shows a rectangle.

    The length of the rectangle is 6.3×10^4 cm

    The width of the rectangle is 1.65×10^3 cm

    Find the perimeter of the rectangle. Give
    your answers in standard form.

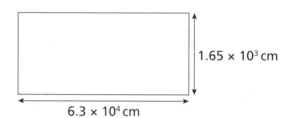

1.65×10^3 cm

6.3×10^4 cm

8 The table below shows the distances of three of the planets from the Sun at a
    particular time.

Planet	Earth	Saturn	Neptune
Distance to Sun (miles)	9.3×10^7	8.9×10^8	2.8×10^9

    Assume that the Sun and all three planets lie in a straight line at this particular time.
    Use the table to find the distance between

    a Saturn and Earth   **b** Neptune and Earth.

    Give your answers in standard form.

9 The perimeter of this triangle is 1.6×10^5 mm

    Find the length of the side labelled x
    Give your answer in standard form.

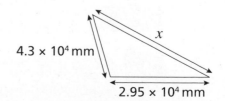

4.3×10^4 mm

x

2.95×10^4 mm

What do you think? 💭

1 Abdullah is performing these calculations with numbers in standard form.

> $2.5 \times 10^5 + 3 \times 10^7$
>
> $2\,500\,000 + 30\,000\,000 = 32\,500\,000 = 3.25 \times 10^7$

What mistake has Abdullah made?

2 Here are three calculations.

> **a** $2.1 \times 10^4 + 5 \times 10^4$ **b** $3.6 \times 10^4 + 2.8 \times 10^5$ **c** $3 \times 10^3 + 1.8 \times 10^6$

Find the answer to each calculation.

Which method did you use?

Did you use the same method for each calculation?

3 Seb is working out $1.7 \times 10^{24} + 2.3 \times 10^{23}$

a Explain why you would not advise Seb to write each number in ordinary form first.

b How could Seb work out the answer without using a calculator?

4 Chloe is working out $7.2 \times 10^4 - 4.3 \times 10^7$

Chloe says that she knows that the answer is negative. Explain how Chloe knows this.

Example 2

Work out

a $(3.2 \times 10^2) \times (2 \times 10^2)$ **b** $\dfrac{4 \times 10^5}{2 \times 10^6}$

a $(3.2 \times 10^2) \times (2 \times 10^2) = (3.2 \times 2) \times (10^2 \times 10^2)$ — Multiplication is **commutative**, so the calculation can be rearranged to make it easier.

$3.2 \times 2 = 6.4$

$10^2 \times 10^2 = 10^4$

So $(3.2 \times 10^2) \times (2 \times 10^2) = 6.4 \times 10^4$ — Use your knowledge of **index** laws: when multiplying powers of 10 you add the powers.

b $\dfrac{4 \times 10^5}{2 \times 10^6}$

$4 \div 2 = 2$

$10^5 \div 10^6 = 10^{-1}$ — Use your knowledge of index laws: when dividing powers of 10 you subtract the powers.

So $\dfrac{4 \times 10^5}{2 \times 10^6} = 2 \times 10^{-1}$

Practice 11.4B

1 Work these out.

a $(3 \times 10^2) \times (2 \times 10^3)$ **b** $(3 \times 10^2) \times (2 \times 10^4)$ **c** $(3 \times 10^2) \times (2 \times 10^5)$

d $(3 \times 10^2) \times (2 \times 10^7)$ **e** $(3 \times 10^2) \times (2 \times 10^{-5})$

What do you notice?

Explain your method to a partner.

2 Jackson is working out $3 \times 10^4 \times 4 \times 10^2$

This is what he writes

$$3 \times 10^4 \times 4 \times 10^2 = 3 \times 4 \times 10^4 \times 10^2$$
$$= 12 \times 10^6$$

Jackson's teacher wants him to write his answer in standard form. His teacher says that his final answer is not quite right.

How can Jackson improve his answer?

3 Calculate the answers to these multiplications. Write your answers in standard form.

Do not use a calculator for any of these questions.

a $(6 \times 10^3) \times (4 \times 10^4)$ **b** $(7 \times 10^2) \times (3 \times 10^6)$

c $(4 \times 10^4) \times (5 \times 10^2)$ **d** $(1.4 \times 10^3) \times (2 \times 10^5)$

e $(2.7 \times 10^4) \times (3 \times 10^4)$ **f** $(6.8 \times 10^5) \times (2 \times 10^4)$

g $(3.45 \times 10^3) \times (5 \times 10^6)$ **h** $(7.6 \times 10^8) \times (4 \times 10^3)$

i $(7 \times 10^5) \times (1.2 \times 10^{-2})$ **j** $(3.9 \times 10^6) \times (5 \times 10^{-3})$

k $(8.6 \times 10^8) \times (6 \times 10^{-5})$ **l** $(3 \times 10^{-4}) \times (3.9 \times 10^7)$

m $(1.86 \times 10^8) \times (3 \times 10^4)$ **n** $(4.95 \times 10^5) \times (4 \times 10^{-1})$

4 Work out these divisions.

a $(8 \times 10^7) \div (2 \times 10^2)$ **b** $(8 \times 10^7) \div (2 \times 10^3)$ **c** $(8 \times 10^7) \div (2 \times 10^4)$

d $\dfrac{8 \times 10^7}{2 \times 10^5}$ **e** $(8 \times 10^7) \div (2 \times 10^{-4})$

What do you notice? Explain your method.

5 Work out the missing numbers.

a $(12 \times 10^9) \div (2 \times 10^{\square}) = 6 \times 10^3$ **b** $(6 \times 10^{\square}) \div (4 \times 10^3) = 1.5 \times 10^5$

c $(2.4 \times 10^2) \div (2 \times 10^{\square}) = 1.2 \times 10^{-3}$ **d** $(2.4 \times 10^{\square}) \div (2 \times 10^2) = 1.2 \times 10^{-3}$

6 **a** Here is a division calculation in standard form.

$$\frac{4 \times 10^8}{8 \times 10^5}$$

i Explain why the answer is 0.5×10^3

ii Explain why this is the same as 5×10^2

b Here is another division calculation in standard form.

$$\frac{4 \times 10^5}{8 \times 10^8}$$

Explain why the answer to this question is 5×10^{-4}

c Use the answers to parts **a** and **b** to work out these divisions.

i $\dfrac{4 \times 10^{12}}{8 \times 10^3}$ **ii** $\dfrac{4 \times 10^5}{8 \times 10^7}$ **iii** $\dfrac{4 \times 10^5}{8 \times 10^{-8}}$

7 Calculate the answers to these divisions. Write your answers in standard form.

Do not use a calculator for any of these questions.

a $(8 \times 10^7) \div (4 \times 10^4)$ **b** $(8 \times 10^5) \div (2 \times 10^3)$

c $(2.4 \times 10^6) \div (2 \times 10^4)$ **d** $(7.5 \times 10^5) \div (5 \times 10^2)$

e $(6.6 \times 10^7) \div (6 \times 10^6)$ **f** $(3.6 \times 10^4) \div (6 \times 10^2)$

g $(2.4 \times 10^6) \div (8 \times 10^2)$ **h** $(1.84 \times 10^8) \div (2 \times 10^5)$

i $(12 \times 10^5) \div (8 \times 10^3)$ **j** $(6 \times 10^9) \div (8 \times 10^5)$

k $(8 \times 10^{-7}) \div (5 \times 10^4)$ **l** $(2.8 \times 10^5) \div (2 \times 10^7)$

m $(8 \times 10^{-3}) \div (4 \times 10^{-8})$ **n** $(6.9 \times 10^5) \div (3 \times 10^{-3})$

What do you think? 💭

1 Ed is working out $(4 \times 10^5) \div (5 \times 10^2)$

The answer is 0.8×10^3

This is 800 in ordinary form.

This is 8×10^2

Explain how Ed could convert his first answer into standard form without having to convert it to ordinary form first.

2 Find integer values for x and y that will make these calculations correct.

a $(2.5 \times 10^x) \times (3 \times 10^y) = 7.5 \times 10^6$ **b** $(3.8 \times 10^x) \div (2 \times 10^y) = 1.9 \times 10^6$

c $(x \times 10^6) \times (y \times 10^{12}) = 6 \times 10^{18}$ **d** $(x \times 10^4) \times (y \times 10^5) = 1.2 \times 10^{10}$

Consolidate – do you need more?

1 Work out

a $5 \times 10^4 + 3 \times 10^4$

b $6 \times 10^5 + 4 \times 10^4$

c $6 \times 10^5 + 4 \times 10^3$

d $1.5 \times 10^4 + 2.3 \times 10^3$

e $7.8 \times 10^8 + 8.5 \times 10^5$

f $3.9 \times 10^{-3} + 4.2 \times 10^{-2}$

g $3.9 \times 10^{-3} + 4.2 \times 10^{-4}$

h $7 \times 10^8 - 2 \times 10^5$

i $6.8 \times 10^5 - 4.8 \times 10^2$

2 Work out

a $(3 \times 10^5) \times (2 \times 10^6)$

b $(9 \times 10^2) \times (2 \times 10^5)$

c $(4.3 \times 10^4) \times (2 \times 10^{15})$

d $(2.4 \times 10^7) \times (3 \times 10^{-3})$

e $(8 \times 10^4) \times (3 \times 10^6)$

f $(7.9 \times 10^8) \times (2 \times 10^{-2})$

g $(9 \times 10^7) \div (3 \times 10^4)$

h $(8 \times 10^9) \div (2 \times 10^{12})$

i $(4.8 \times 10^{15}) \div (4 \times 10^3)$

j $(1.5 \times 10^7) \div (3 \times 10^{10})$

k $(6.6 \times 10^{-2}) \div (6 \times 10^6)$

l $(1.8 \times 10^8) \div (9 \times 10^{-3})$

Stretch – can you deepen your learning?

1 A plank of wood is 1.26×10^4 mm long.

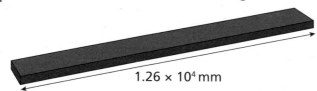

1.26×10^4 mm

Two pieces, each of length 4.5 m, are cut from the plank. The remaining piece of wood is then cut into two pieces.

Find the length of one of these pieces.

2 Look at these four number cards.

| 5×10^4 | 1.6×10^3 | 4.85×10^5 | 3.9×10^2 |

a Which two cards have the greatest total? Find this total.

b Which two cards have the smallest difference? What is this difference?

c Which two cards have the greatest difference? What is this difference?

3 Find the area of this trapezium.

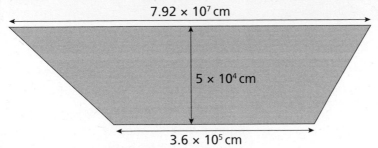

7.92 × 10⁷ cm

5 × 10⁴ cm

3.6 × 10⁵ cm

> Look up the formula for the area of a trapezium. You will cover this in detail in Block 14

Reflect

1 What rules have you learned for adding and subtracting numbers in standard form? Which is the most efficient?

2 What methods have you learned for multiplying and dividing numbers in standard form? How are these linked to the rules of indices?

ⓗ 11.5 Negative and fractional powers

Small steps

■ Understand and use negative indices ⓗ

■ Understand and use fractional indices ⓗ

Are you ready?

1 Work these out without a calculator.

    a 2^3 **b** 2^4 **c** 2^5 **d** 2^6

    e 3^2 **f** 3^3 **g** 5^2 **h** 5^3

2 Work these out using a calculator if you need to.

    a 5^5 **b** 6^3 **c** 11^4 **d** 0.6^3

3 Write each of these as a power of 2

    a 32   **b** 8   **c** 128   **d** 256

4 Which of these numbers can be written as powers of 5?

| 75 | 5 | 5000 | 625 |

| 25 | 20 | 125 | 1 |

5 Work these out using a calculator if you need to.

    a $\sqrt{100}$ **b** $\sqrt{49}$ **c** $\sqrt{4}$ **d** $\sqrt{0.36}$

    e $\sqrt[3]{27}$ **f** $\sqrt[3]{125}$ **g** $\sqrt[4]{10\,000}$ **h** $\sqrt[5]{32}$

Models and representations

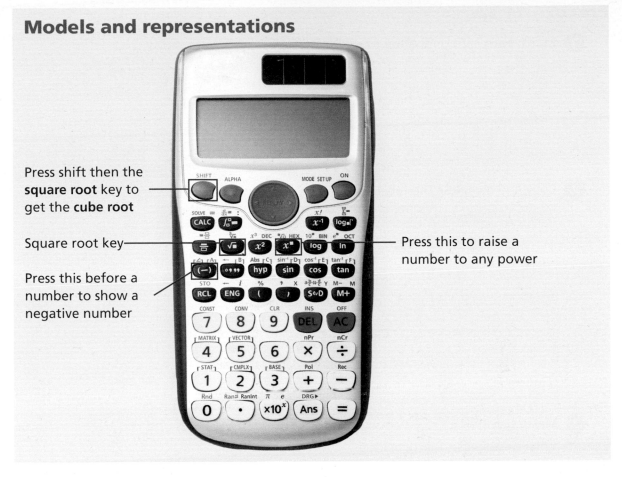

Press shift then the **square root** key to get the **cube root**

Square root key

Press this before a number to show a negative number

Press this to raise a number to any power

In this section, you will look at the connection between negative powers and **reciprocals**.

Example 1

Write 3^{-2} as a simplified fraction

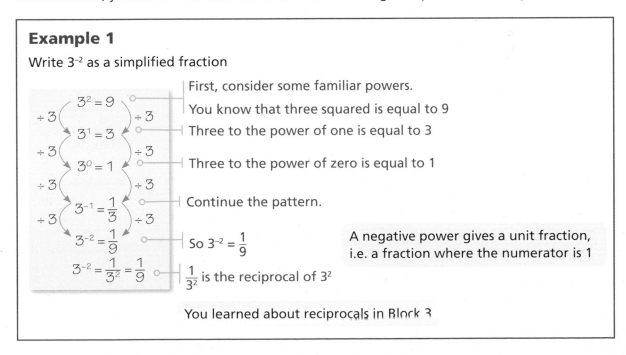

First, consider some familiar powers.

You know that three squared is equal to 9

Three to the power of one is equal to 3

Three to the power of zero is equal to 1

Continue the pattern.

So $3^{-2} = \frac{1}{9}$

A negative power gives a unit fraction, i.e. a fraction where the numerator is 1

$\frac{1}{3^2}$ is the reciprocal of 3^2

You learned about reciprocals in Block 3

$3^2 = 9$

$3^1 = 3$

$3^0 = 1$

$3^{-1} = \frac{1}{3}$

$3^{-2} = \frac{1}{9}$

$3^{-2} = \frac{1}{3^2} = \frac{1}{9}$

Practice 11.5A

1 Match each card on the top row with a card from the bottom row.

| 2^{-1} | 3^{-1} | 4^{-1} | 5^{-1} | 8^{-1} | 10^{-1} |

| $\frac{1}{3}$ | $\frac{1}{5}$ | 0.5 | 0.1 | $\frac{1}{4}$ | $\frac{1}{8}$ |

2 Copy and complete the table.

Power	Fraction
2^{-1}	$\frac{1}{2}$
2^{-2}	$\frac{1}{2^2} = \frac{1}{4}$
2^{-3}	
2^{-4}	
2^{-5}	

3 Marta says that $6^{-1} = \frac{-1}{6}$

What mistake has Marta made?

4 Copy and complete the following.

a $3^{-2} = \frac{1}{3^2} = \frac{1}{\square}$

b $10^{-2} = \frac{1}{10^2} = \frac{1}{\square}$

c $5^{-2} = \frac{1}{\square^2} = \frac{1}{\square}$

d $6^{-2} = \frac{1}{\square^{\square}} = \frac{1}{\square}$

5 Write each of these as a simplified fraction.

a 3^{-3} **b** 2^{-6} **c** 5^{-3} **d** 6^{-2}

e 10^{-3} **f** 50^{-1} **g** 11^{-3} **h** 0.5^{-3}

i $(-2)^{-1}$ **j** $(-2)^{-2}$ **k** $(-2)^{-3}$

6 Do these cards have the same value? Explain your answer.

| $(-5)^{-2}$ | -5^{-2} |

7 Write each of these as a decimal.

a 2^{-3} **b** 4^{-1} **c** 10^{-2}

d 25^{-1} **e** 5^{-2} **f** $(-10)^{-2}$

8 Write each pair of numbers as powers of the given base number.

a 8 and $\frac{1}{8}$ as powers of 2   **b** 16 and $\frac{1}{16}$ as powers of 4

c 16 and $\frac{1}{16}$ as powers of 2   **d** 25 and $\frac{1}{25}$ as powers of 5

e 125 and $\frac{1}{125}$ as powers of 5   **f** 625 and $\frac{1}{625}$ as powers of 5

g 1000 000 and $\frac{1}{1\,000\,000}$ as powers of 10

What do you notice?

What do you think? 🧠

1 A teacher asks her class to write down the value of 2^{-3}

Here are some of her students' answers.

A $\boxed{-8}$ B $\boxed{\frac{1}{6}}$ C $\boxed{0.125}$ D $\boxed{\frac{1}{8}}$

a Which of the answers are correct?

b What mistakes have the other students made?

2 Copy and complete these sentences.

a **i** If I know that $2^{10} = 1024$, then 2^{-10} is equal to $\boxed{}$

     ii If I know that $3^{-5} = \frac{1}{243}$ then 3^5 is equal to $\boxed{}$

b Explain your method to your partner.

3 Work out the value of each of these cards.

You may use your calculator.

What do you notice about each set?

Set A   **i** $\boxed{\left(\frac{1}{2}\right)^{-1}}$ **ii** $\boxed{\left(\frac{1}{3}\right)^{-1}}$ **iii** $\boxed{\left(\frac{1}{4}\right)^{-1}}$ **iv** $\boxed{\left(\frac{1}{5}\right)^{-1}}$

Set B   **i** $\boxed{\left(\frac{2}{3}\right)^{-1}}$ **ii** $\boxed{\left(\frac{3}{4}\right)^{-1}}$ **iii** $\boxed{\left(\frac{5}{2}\right)^{-1}}$ **iv** $\boxed{\left(\frac{2}{5}\right)^{-1}}$

Set C   **i** $\boxed{\left(\frac{1}{2}\right)^{-1}}$ **ii** $\boxed{\left(\frac{1}{2}\right)^{-2}}$ **iii** $\boxed{\left(\frac{1}{2}\right)^{-3}}$ **iv** $\boxed{\left(\frac{1}{2}\right)^{-4}}$

In this section, you will investigate the meaning of fractional indices.

Example 2

Find

a $16^{\frac{1}{2}}$ **b** $27^{\frac{1}{3}}$

a $16^{\frac{1}{2}} \times 16^{\frac{1}{2}} = 16^1$ Use the laws of indices that you already know.
$16^{\frac{1}{2}} \times 16^{\frac{1}{2}} = 16$

$4 \times 4 = 16$ Think about: "What number do you multiply by itself to get 16?"
$16^{\frac{1}{2}} = 4$ 4 multiplied by itself gives 16 so 16 to the power of one half is equal to 4

$16^{\frac{1}{2}} = \sqrt{16} = 4$ So using inverse operations, the power of one half must mean the same as finding the square root of the base number.

b $27^{\frac{1}{3}} \times 27^{\frac{1}{3}} \times 27^{\frac{1}{3}} = 27^1$ Use the laws of indices that you already know.
$27^{\frac{1}{3}} \times 27^{\frac{1}{3}} \times 27^{\frac{1}{3}} = 27$

   Think about: "What number do you multiply by itself twice to get 27?"

$3 \times 3 \times 3 = 27$ 3 multiplied by itself twice gives 27

$27^{\frac{1}{3}} = 3$ So 27 to the power of one-third is equal to 3

$27^{\frac{1}{3}} = \sqrt[3]{27} = 3$ So using inverse operations, the power of one-third must mean the same as finding the cube root of the base number.

Practice 11.5B

① Copy and complete the table.

Expression with powers	Working
$16^{\frac{1}{2}}$	$\sqrt{16} = \boxed{}$
$25^{\frac{1}{2}}$	$\sqrt{\boxed{}} = \boxed{}$
$36^{\frac{1}{2}}$	$\sqrt{\boxed{}} = \boxed{}$
$64^{\frac{1}{2}}$	$\sqrt{\boxed{}} = \boxed{}$
$100^{\frac{1}{2}}$	$\sqrt{\boxed{}} = \boxed{}$

What do you notice?

② Zach is working out $27^{\frac{1}{3}}$

The answer is 9, because $\frac{1}{3}$ of 27 is 9

Explain the mistake Zach has made. Why do you think he might have made this mistake?

③ Evaluate these.

a $8^{\frac{1}{3}}$ **b** $16^{\frac{1}{4}}$ **c** $32^{\frac{1}{5}}$ **d** $100^{\frac{1}{2}}$

e $1000^{\frac{1}{3}}$ **f** $343^{\frac{1}{3}}$ **g** $125^{\frac{1}{3}}$ **h** $625^{\frac{1}{4}}$

i $0.25^{\frac{1}{2}}$ **j** $(-125)^{\frac{1}{3}}$ **k** $(-32)^{\frac{1}{5}}$

④ Match each expression with a power to its correct value.

$64^{\frac{1}{2}}$	$64^{\frac{1}{3}}$	$64^{\frac{1}{6}}$	64^{0}

4	8	1	2

⑤ Work out the missing powers.

a $49^{\boxed{}} = 7$ **b** $1000^{\boxed{}} = 10$

c $125^{\boxed{}} = 5$ **d** $81^{\boxed{}} = 9$

e $81^{\boxed{}} = 3$ **f** $128^{\boxed{}} = 2$

g $100\,000^{\boxed{}} = 10$ **h** $(-8)^{\boxed{}} = -2$

⑥ Work out each of these, giving your answers to 5 decimal places.

a $7^{\frac{1}{2}}$ **b** $7^{\frac{1}{3}}$ **c** $100^{\frac{1}{3}}$

d $72^{\frac{1}{5}}$ **e** $0.76^{\frac{1}{2}}$ **f** $(-2)^{\frac{1}{3}}$

What do you think? 🗨

1 **a** Use the laws of indices to explain why $x^{\frac{1}{2}} \times x^{\frac{1}{2}} = x$

b Explain why a number to the power of $\frac{1}{2}$ is the same as its square root.

2 **a** Jackson is trying to work out $(-25)^{\frac{1}{2}}$. He thinks that the answer is -5
Explain why Jackson might think this.

b Jackson uses his calculator to work out the answer.

$$(-25)^{\frac{1}{2}}$$

His calculator shows an error message.

ERROR

Why is it not possible to find $(-25)^{\frac{1}{2}}$?

c Instead, Jackson types this into his calculator

$$-25^{\frac{1}{2}}$$

His calculator then says

$$-5$$

Explain why the calculator now shows -5

d Explain why this example shows the importance of using brackets.

3 **a** Sort these cards into two groups: those that have an answer that can be calculated and those that don't.

| $(-20)^{\frac{1}{2}}$ | $(-20)^{\frac{1}{3}}$ | $(-20)^{\frac{1}{4}}$ | $(-20)^{\frac{1}{5}}$ |

| $(-7)^{\frac{1}{2}}$ | $(-10)^{\frac{1}{3}}$ | $(-17)^{\frac{1}{4}}$ | $(-20)^{\frac{1}{5}}$ |

Can calculate the answer	Can't calculate the answer

b What patterns do you notice?

c How can you determine when it is possible to calculate a power of $\frac{1}{2}$ and when it isn't? What about other fractional powers?

Consolidate – do you need more?

1 Write each of these as a simplified fraction.

a 2^{-3} **b** 3^{-2} **c** 6^{-1} **d** 7^{-2}

e 5^{-3} **f** $(-3)^{-2}$ **g** 2^{-5} **h** $(-7)^{-3}$

2 Evaluate

a $9^{\frac{1}{2}}$ **b** $25^{\frac{1}{2}}$ **c** $8^{\frac{1}{3}}$ **d** $125^{\frac{1}{3}}$

e $81^{\frac{1}{4}}$ **f** $32^{\frac{1}{5}}$ **g** $(-100\,000)^{\frac{1}{5}}$ **h** $(1.44)^{\frac{1}{2}}$

Stretch – can you deepen your learning?

1 Work these out without using a calculator.

a $3^2 \times 2^{-1}$ **b** $2^{-3} \times 25^{\frac{1}{2}}$ **c** $2^{-1} + 3^{-1}$

d $2^{-1} - 3^{-1}$ **e** $\dfrac{125^{\frac{1}{3}}}{32^{\frac{1}{5}}}$ **f** 3×2^{-3}

2 Emily is working out $36^{0.5}$

> I think the answer is $\sqrt{36}$, which is equal to 6

Do you agree with Emily? Explain your answer.

3 Work these out without using a calculator.

a $81^{0.5}$ **b** $144^{0.5}$ **c** $16^{0.25}$ **d** $32^{0.2}$

e $625^{0.25}$ **f** $0.125^{0.\dot{3}}$ **g** $(-27)^{0.\dot{3}}$ **h** $-27^{0.\dot{3}}$

4 a Choose the correct answer to $16^{-\frac{1}{2}}$

A $\boxed{8}$ B $\boxed{4}$ C $\boxed{-4}$ D $\boxed{\dfrac{1}{4}}$ E $\boxed{\dfrac{1}{4}}$

b Explain your reasoning.

5 Work these out without using a calculator.

a $25^{-\frac{1}{2}}$ **b** $36^{-\frac{1}{2}}$ **c** $64^{-\frac{1}{3}}$ **d** $81^{-\frac{1}{4}}$

e $32^{-\frac{1}{5}}$ **f** $49^{-0.5}$ **g** $\left(\frac{1}{5}\right)^{-2}$ **h** $\left(\frac{2}{5}\right)^{-2}$

i $(1.5)^{-2}$ **j** $\left(\frac{9}{4}\right)^{-\frac{1}{2}}$

Reflect

What is the difference between a negative power and a fractional power?

11 Standard index form
Chapters 11.1–11.5

White Rose Maths

I have become fluent in...

- working with numbers greater than 1 in standard form
- working with numbers between 0 and 1 in standard form
- calculating with numbers in standard form.

I have developed my reasoning skills by...

- investigating powers of 10
- explaining mental methods of calculating with standard form
- explaining why a number is or isn't in standard form.

I have been problem-solving through...

- using standard form in different contexts
- finding missing values in standard form calculations
- exploring fractional and negative indices. **H**

Check my understanding

1 Write each number in standard form.

  a 20 000   **b** 300 000   **c** 5000   **d** 70 000 000

  e 410 000   **f** 19 000   **g** 650   **h** 8 703 000 000

2 Write these as ordinary numbers.

  a 3×10^4 **b** 7×10^8 **c** 4.1×10^2 **d** 6.05×10^{11}

3 Write each number in standard form.

  a 0.002   **b** 0.000 0009   **c** 0.000 335

  d 0.78   **e** 0.000 008 01   **f** 0.008 200 59

4 Write these as ordinary numbers.

  a 5×10^{-4} **b** 6.7×10^{-10} **c** 1.99×10^{-7} **d** 2.04×10^{-2}

5 Write 37×10^5 in standard form.

6 Work out these additions.

  a $4 \times 10^4 + 5 \times 10^3$ **b** $1.8 \times 10^4 + 4 \times 10^2$ **c** $3.2 \times 10^6 + 5.5 \times 10^5$

7 Work out these multiplications.

  a $(4 \times 10^2) \times (3 \times 10^3)$ **b** $(5 \times 10^2) \times (7 \times 10^4)$ **c** $(2 \times 10^4) \times (9 \times 10^5)$

8 Write each of these as a fraction in its simplest form. **H**

  a 2^{-3} **b** 2^{-4} **c** 5^{-3}

9 Work out **H**

  a $27^{\frac{1}{3}}$ **b** $16^{\frac{1}{4}}$ **c** $100^{\frac{1}{2}}$

12 Number sense

White Rose Maths

In this block, I will learn...

how to round numbers to a given number of significant figures

$864|25 = 86\,400$ to 3 s.f.

$0.072|6 = 0.073$ to 2 s.f.

the difference between significant figures and decimal places

$0.072|6 = 0.073$ to 2 s.f.

$0.07|26 = 0.07$ to 2 d.p.

how to solve problems with money

			Units used	Price of each unit	Amount £
Current reading	Previous reading			16.5p	_____
6108	4870	Units used	_____		_____
		V.A.T at 5%			_____
		Total charge, including V.A.T.			23.45
		Previous amount owing	_____		_____
		Amount to pay			

how to convert metric units

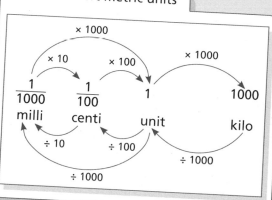

how to perform calculations involving time

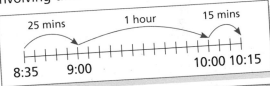

how to convert units of area and volume Ⓗ

how to work with calendars

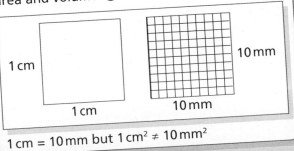

$1\,cm = 10\,mm$ but $1\,cm^2 \neq 10\,mm^2$

12.1 Rounding, estimating and calculating

Small steps

- Round numbers to powers of 10, and to 1 significant figure **R**
- Estimate the answer to a calculation
- Calculate using the order of operations **R**

Key words

Significant figure – the most important digits in a number that give you an idea of its size

Estimate – an approximate answer or to give an approximate answer

Order of operations – the rules that tell you the order in which to perform each part of a calculation

Are you ready?

1 Write the value of 872.1 rounded to

   a the nearest integer

   b the nearest ten

   c the nearest hundred.

2 Which is the most significant figure in each of these numbers?

   a 913   **b** 0.71   **c** 3.008   **d** 0.000 65

3 Round the numbers in question **2** to 1 significant figure.

4 Which is the correct formula for the circumference of a circle?

   $d = \pi C$ $C = \pi r$ $C = \pi d$ $C = \pi r^2$

Models and representations

Number lines are very useful when rounding numbers.

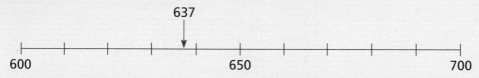

637 is closer to 600 than 700 so 637 = 600 to the nearest hundred.

You can also model rounding like this:

There are 6 hundreds. To the nearest hundred, the number must be 600 or 700

The next digit is in the tens column. There are only 3 tens, which is less than halfway on the number line (halfway is 5 tens), so 637 = 600 to the nearest hundred.

Example 1

In a survey in 2019, the population of Leeds was 792 525

a Write the number 792 525

  i to the nearest thousand

  ii to 1 significant figure

  iii to the nearest hundred.

b Which do you think is the most useful approximation for the population of Leeds in 2019? Why?

a i 792 525 = 793 000 to the nearest thousand

This is the thousands digit.　　5 hundreds means you are past halfway on the number line.

ii 792 525 = 800 000 to 1 significant figure

This is the first **significant figure**.　　9 is greater than 5 so you round to 800 000

iii 792 525 = 792 500 to the nearest hundred

79 25 25

This is the hundreds digit.　　2 tens is less than halfway to the next hundreds, so you round to 792 500

b 793 000, because it is still quite accurate but easy to say and remember

You could round these using number lines if you prefer.

Example 2

A Christmas song was downloaded 613 837 times in December. Estimate the mean number of times the song was downloaded per day in the month.

First, round both numbers to 1 significant figure.

a 613 837 = 600 000 to 1 significant figure

31 = 30 to 1 significant figure

600 000 ÷ 30 = 20 000

6 13 837

1 is less than 5, so you round to 600 000

There are 31 days in December.

The calculation for the exact mean is 613 837 ÷ 31

3 1

1 is less than 5, so you round to 30

Practice 12.1A

1 Round each number to the nearest ten.

    a 76   **b** 376   **c** 3076   **d** 3876

    e 47   **f** 47.2   **g** 47.672   **h** 837

2 Round each number to the nearest thousand.

    a 156 312   **b** 2099   **c** 19 412   **d** 19 518

3 Round each number to 1 significant figure.

    a 870   **b** 87   **c** 0.87   **d** 12 750

    e 0.1089   **f** 0.0189   **g** 5 555 555

4 By rounding each number to 1 significant figure, estimate the answers to these calculations.

    a 58×72 **b** 7.3×9.6 **c** 88^2 **d** $512 \div 4.78$

    e $0.812 \div 98$

Can you tell whether your estimates are overestimates or underestimates? Explain how you decide.

5 Use $\pi = 3$ to estimate the circumference of a circle with

    a diameter 12 cm   **b** radius 12 cm   **c** diameter 28 m   **d** radius 47 cm

6 The circumference of a circle is 32 cm

    a Estimate the diameter of the circle.   **b** Estimate the radius of the circle.

7 £1 is worth 5.62 Polish złoty.

    a Estimate

        i the number of złoty you can buy with £350

        ii the cost, in £, of a flight that costs 300 złoty.

    b Are your estimates overestimates or underestimates? Explain how you know this.

8 A mount for a picture is made by cutting a square from a rectangle of card, as shown.

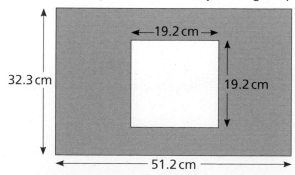

Estimate the area of the mount.

How accurately should you give the answer?

What do you think?

1 **a**

I think that $\sqrt{90}$ is about 9.5

Explain why Beca's estimate is a good one.

b

I think that $\sqrt{150}$ is about 12.5

Explain why Benji's estimate is not as good as Beca's.

2 Here is Flo's working for rounding 8723 to 2 significant figures.

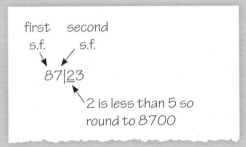

first   second
s.f.   s.f.

87|23

2 is less than 5 so
round to 8700

a Round each number to 2 significant figures.

   i 18.71   **ii** 36 904   **iii** 232.8   **iv** 47 017 863

b

$\pi = 3.141\,592\,653\,59\ldots$

Round π to

   i 1 significant figure   **ii** 2 significant figures

   iii 3 significant figures   **iv** 4 significant figures.

3 The circumference of a tree trunk is 243 cm

$200 \div 3$ $240 \div 3.1$ $240 \div 3$

Jakub   Abdullah   Flo

Compare the students' strategies for estimating the diameter of the tree trunk.

Remember the **order of operations** that you studied in Book 1

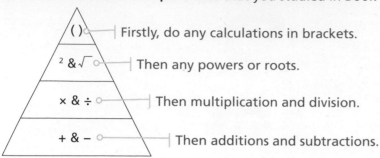

( )	Firstly, do any calculations in brackets.
^{2} & $\sqrt{\ }$	Then any powers or roots.
× & ÷	Then multiplication and division.
+ & −	Then additions and subtractions.

You will need to use this when **estimating** answers to more complex calculations.

Example 3

Estimate the answer to each calculation.

a $19.8 + \dfrac{39.3}{4.3}$ **b** $\dfrac{19.8 + 39.3}{4.3}$ **c** $\dfrac{19.8 + 39.3}{0.43}$

a $19.8 + \dfrac{39.3}{4.3} \approx 20 + \dfrac{40}{4}$ First round each number to 1 significant figure.
Division comes before addition in the order of operations.

$\quad = 20 + 10$ Then do the addition.

$\quad = 30$

b $\dfrac{19.8 + 39.3}{4.3} \approx \dfrac{20 + 40}{4}$ First round each number to 1 significant figure, as before.
The whole of the numerator is to be divided by the denominator, so this time you do the addition first.

$\quad = \dfrac{60}{4}$

Finally do the division.

$\quad = 15$

c $\dfrac{19.8 + 39.3}{0.43} \approx \dfrac{20 + 40}{0.4}$ Start as before.

Here you have a decimal denominator.

$\quad = \dfrac{60 \times 10}{0.4 \times 10}$ Multiplying both the numerator and the denominator by 10 makes the denominator into an integer.

$\quad = \dfrac{600}{4}$ You could do this mentally or using short division.

$\quad = 150$

Practice 12.1B

1 Work out

    a $12 \times 4 + 2$ **b** $12 + 4 \times 2$ **c** $12 \div 4 + 2$ **d** $12 + 4 \div 2$

    e $12 \times 4 + 2^2$ **f** $12 + 4 \times 2^2$ **g** $12 \div 4 + 2^2$ **h** $12 + 4 \div 2^2$

    i $12 \times (4 + 2)$ **j** $(12 + 4)^2 + 2$ **k** $(12 \div 4) + 2^2$ **l** $(12 \times 4) \div 2^2$

2 **a** Match the equivalent calculations.

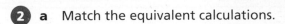

$$\frac{20}{0.5} \qquad \frac{2}{0.4} \qquad \frac{200}{5} \qquad \frac{20}{0.05} \qquad \frac{20}{4} \qquad \frac{200}{0.4} \qquad \frac{2000}{4} \qquad \frac{2000}{5}$$

b Work out

i $\dfrac{6}{0.2}$ **ii** $\dfrac{60}{0.3}$ **iii** $\dfrac{6}{0.03}$

iv $\dfrac{60}{0.4}$ **v** $\dfrac{0.6}{0.02}$ **vi** $\dfrac{600}{0.04}$

3 Estimate the answers to these calculations.

a $\dfrac{286}{0.37}$ **b** $\dfrac{411}{0.18}$ **c** $\dfrac{1.9}{0.051}$

d $\dfrac{0.59}{0.062}$ **e** $\dfrac{1800}{0.23}$ **f** $\dfrac{78}{0.004}$

4 **a** An average raindrop has mass 0.2 g. How many raindrops are there on average in 1 kilogram of rainwater?

b A two-inch nail has mass 0.79 g. A box of nails has mass 400 g.

i Estimate the number of nails in the box.

ii Is your answer an underestimate or an overestimate?

5 Estimate the answer to each calculation.

a $87 + 48 \times 23$ **b** $87 + 48 \div 23$ **c** $(87 + 48) \div 23$ **d** $87 + 48 - 23$

e $\dfrac{87 + 48}{23}$ **f** $\dfrac{87 - 48}{2.3}$ **g** $\dfrac{87 + 4.8}{0.23}$

What do you think?

1 Jackson estimates the answer to $120 \div 0.25$ by rounding both numbers to 1 significant figure. Seb leaves 120 as it is and rounds 0.25 to 1 significant figure. Emily works out the exact answer by thinking of the calculation as $120 \div \dfrac{1}{4}$
Compare the students' answers.

2 Emily scores 38 out of 48 in a test. Estimate her mark as a percentage.

3 Estimate the answers to these calculations. Show each step of your working.

a $\sqrt{5.3 + 8.7 \times 4.8}$ **b** $\dfrac{2.08^3 \times \sqrt{109}}{0.185}$ **c** 6.1% of £27 999

Consolidate – do you need more?

1 Round each number to the nearest hundred.

    a 841   **b** 2841   **c** 16 709   **d** 12 047

    e 19 881   **f** 19 981   **g** 92

2 Round each number to 1 significant figure.

    a 137   **b** 2.167   **c** 23.51   **d** 0.561

    e 0.001 23   **f** 619 904

3 By rounding each number to 1 significant figure, estimate the answers to these calculations.

    a $812 \div 1.8$ **b** 3.8×4.8 **c** 3.8×48

    d 99^2 **e** $32\,910 + 47\,009$

4 Find the value of

    a $8 - 2 \times 3$ **b** $(8 - 2) \times 3$ **c** $8 - 2 \times 3^2$ **d** $8 - (2 \times 3)^2$

    e $8 \div 2 + 3$ **f** $8 \div (2 + 3)$ **g** $8^2 \div 2 + 3$

5 Work out

    a $10 \div 2$ **b** $10 \div 0.2$ **c** $30 \div 0.6$ **d** $4 \div 0.08$

6 Estimate the value of

    a 8.7×0.28 **b** $8.7 \div 0.28$ **c** $87 \times (2.8 + 4.9)$

Stretch – can you deepen your learning?

1 How do you estimate distances? Compare estimating the length of a book, a person's height, the length of your classroom and the length of a field. What could you use to help?

2 The diameter of a 10p coin is 2.45 cm. What other objects will this help you to estimate the size of?

3 In the last century, students learned to estimate square roots using "the square root algorithm". Research the square root algorithm and find out how accurately you can work out square roots without using a calculator.

4 Describe examples of where estimation is used in real-life situations.

Reflect

Explain how you can estimate the answer to a calculation if the numbers are

a integers   **b** decimals.

What's the same and what's different?

Small steps

- Round numbers to a given number of decimal places

- Understand and use error interval notation (H)

Key words

Decimal places – the number of digits to the right of the decimal point in a number

Degree of accuracy – how precise a number is

Error interval – the range of values a number could have taken before being rounded

Are you ready?

1 What is the value of the 3 in each number?

    a 237   **b** 2.37   **c** 23.7   **d** 0.237

2 Round these numbers to the nearest integer.

    a 7.4   **b** 18.8   **c** 2.5

    d 8.47   **e** 5.05

3 Estimate the number that each arrow is pointing to.

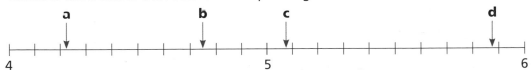

Models and representations

Place value grids

Place value grids can be used to represent decimal numbers.

Tens	Ones	tenths	hundredths	thousands
	6	4		
	5	1	0	7
	3	5	9	

6.4 has only 1 digit after the decimal point. It is given to 1 decimal place.

5.107 is given to 3 decimal places.

3.59 is given to 2 decimal places.

You can use number lines to help you round to any number of **decimal places**.

Example 1

$\pi = 3.14159...$

Round π to

a 1 decimal place **b** 2 decimal places **c** 3 decimal places

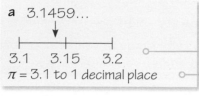

a 3.1459...

3.1   3.15   3.2
$\pi = 3.1$ to 1 decimal place

π is greater than 3.1, so use a number line from 3.1 to 3.2

You can see that π is closer to 3.1 than 3.2

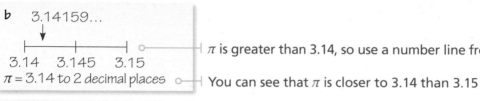

b 3.14159...

3.14   3.145   3.15
$\pi = 3.14$ to 2 decimal places

π is greater than 3.14, so use a number line from 3.14 to 3.15

You can see that π is closer to 3.14 than 3.15

c 3.14159...

3.141   3.1415   3.142
$\pi = 3.142$ to 2 decimal places

π is greater than 3.141, so use a number line from 3.141 to 3.142

You can see that π is closer to 3.142 than 3.141

Example 2

Round each number to the given degree of accuracy.

a 4.68 to 1 decimal place **b** 8.764 to 2 decimal places

c 7.397 to 2 decimal places

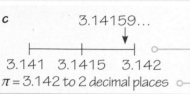

a 4.68

4.6   4.65   4.7
4.68 = 4.7 to 1 decimal place

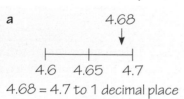

b 8.764

8.76 8.765 8.77
8.764 = 8.76 to 2 decimal places

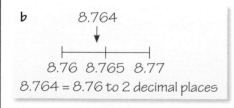

c 7.397

7.39 7.395 7.40
7.397 = 7.40 to 2 decimal places

The next number with two decimal places after 7.39 is 7.40

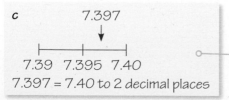

Practice 12.2A

1 Use the number line to round each number to 1 decimal place.

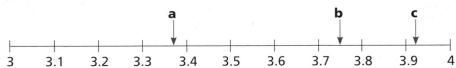

2 **a** **i** Draw a number line from 5.8 to 5.9 and label the number halfway between 5.8 and 5.9

  ii Mark the approximate position of 5.83 on your number line.

  iii Round 5.83 to 1 decimal place.

 b **i** Between which two numbers with 1 decimal place does 7.62 lie?

  ii Round 7.62 to 1 decimal place.

 c Round these numbers to 1 decimal place.

  i 9.14   **ii** 23.67   **iii** 9.58   **iv** 0.72

3 Explain the mistakes that these students have made, and give the correct answers.

a

> 6.2|34   34 is greater than 5, so 6.234 = 6.3 to 1 decimal place

b

> 4.3|2   2 is less than 5, so 4.32 = 4.30 to 1 decimal place

c

> 8.9|7   7 is more than 5, so 8.97 = 8.10 to 1 decimal place

4 Use the number line to round each number to 2 decimal places.

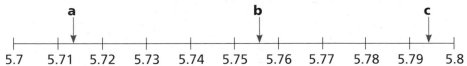

5 **a** **i** Draw a number line from 4.23 to 4.24 and label the number halfway between 4.23 and 4.24.

  ii Mark the approximate position of 4.238 on your number line.

  iii Round 4.238 to 2 decimal places.

 b **i** Between which two numbers with 2 decimal places does 18.367 lie?

  ii Round 18.367 to 2 decimal places.

 c Round these numbers to 2 decimal places.

  i 8.577   **ii** 0.368   **iii** 9.471   **iv** 62.345

6 Use a calculator to work out the answers to these calculations. Give each answer correct to 2 decimal places.

    a 7.86×4.29 **b** $7.86 \div 4.29$

    c $\sqrt{7.86 + 4.29}$ **d** $\sqrt{7.86 - 4.29}$

7 **a** A circle has diameter 7.8 cm. Find its circumference, correct to 2 decimal places.

    b A circle has circumference 1 m. Find, in centimetres correct to 1 decimal place, the diameter of the circle.

8 £1 = €1.14. Convert these amounts of euros to pounds, giving your answers to a sensible degree of accuracy.

    a €35   **b** €100   **c** €2500   **d** €1 million

What do you think?

1 What's the same and what's different about rounding a number to 1 decimal place and rounding a number to the nearest tenth?

2 Use a calculator to work out the answers to these calculations. Give each answer correct to **i** 2 significant figures and **ii** 2 decimal places.

    a 30.7^2 **b** 0.38^2 **c** 12.38×49.57 **d** $0.72 \div 38$

When is it more appropriate to give answers to 2 significant figures, and when to 2 decimal places? Discuss with a partner.

3

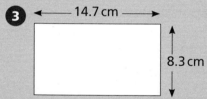

    a Estimate the area of the rectangle by rounding each length to the nearest integer.

    b Use a calculator to find the exact area of the rectangle.

    c Find the difference between your estimate and the actual area of the rectangle.

    d Express this difference as a percentage of the exact area, giving the percentage correct to 1 decimal place.

As you can see from question **3** in "What do you think?", when you work with rounded measurements, your answers are less accurate than working with exact numbers. It is useful to know how far measurements could be from the actual values by finding **error intervals**.

Example 3

A pencil is 12 cm long, measured to the nearest centimetre.

a Which of these measurements could be the length of the pencil?

   11.36 cm   11.52 cm   11.601 cm   12.01 cm   12.49 cm   12.5 cm

b Write the error interval for the length, l, of the pencil in cm.

a 11.52 cm, 11.601 cm, 12.01 cm, 12.49 cm	All of these numbers round to 12 to the nearest integer. Notice that 11.36 would round to 11 and 12.5 would round to 13
b $11.5 \leqslant l < 12.5$	You read this as "l is greater than or equal to 11.5 and less than 12.5" 11.5 is included in the error interval as 11.5 rounds to 12 to the nearest integer, but 12.5 isn't included as 12.5 rounds to 13 to the nearest integer.

Practice 12.2B

1 **a** Which of these numbers round to 7 to the nearest integer?

   7.4   7.403   6.91   7.51   6.51   6.3999   7.06

b Which is the correct error interval for a number x that rounds to 7 to the nearest integer?

   $6.5 < x < 7.5$ $6.5 < x \leqslant 7.5$ $6.5 \leqslant x < 7.5$ $6.5 \leqslant x \leqslant 7.5$

2 The size, a, of an angle is given by the error interval $37.5° \leqslant a < 38.5°$

a Write three numbers, correct to 1 decimal place, that the angle could be.

b What is the size of the angle correct to the nearest integer?

3 **a** A number, y, is given as 6.3 to 1 decimal place. Copy and complete the error interval for y

   $6.25 \leqslant y < \_\_\_\_$

b Write the error intervals for these numbers.

   i $p = 30$ to the nearest integer

   ii $q = 15.4$ to 1 decimal place

   iii $r = 2.72$ to 2 decimal places

c Compare the error intervals for 8, correct to the nearest integer, and 8.0, correct to 1 decimal place.

4 The length, x, of a field is given as 40 m. Write the error interval for x if this measurement is

a correct to the nearest metre

b correct to 1 significant figure.

5 The length of a swimming pool is 50 m, correct to the nearest centimetre.

    a   What is the shortest possible length of the pool?

    b   Write the error interval for the length, l, of the pool.

6 Copy and complete the error interval for each statement.

    a $a = 8000$ to the nearest integer, _____ $\leq a <$ _____

    b $b = 8000$ to the nearest ten, _____ $\leq b <$ _____

    c $c = 8000$ to the nearest hundred, _____ $\leq c <$ _____

    d $d = 8000$ to the nearest thousand, _____ $\leq d <$ _____

7 A number is 6000 to 1 significant figure. What's the same and what's different about the possible values of the number if it represents the area of a field in m² or the number of people at concert?

8 A number n when rounded is 400. Write the error interval for n if it has been rounded to

    a   1 significant figure

    b   2 significant figures

    c   3 significant figures.

What do you think?

1 A rod is 42 cm long, correct to the nearest centimetre. A second rod is 34 cm long, correct to the nearest centimetre. Work out

    a   the greatest possible length of a rod made by putting the two rods together

    b   the shortest possible length of a rod made by putting the two rods together

    c   the greatest possible difference between the lengths of the two rods

    d   the shortest possible difference between the lengths of the two rods.

2 **a**   A number, x, is 20 correct to the nearest 5. Write an error interval for x

    b   A number, y, is 20 correct to the nearest 2. Write an error interval for y

    c   A number, z, is 20 correct to the nearest a. Write an error interval for z

Consolidate – do you need more?

1 Round these numbers to 1 decimal place.

    a 9.25   **b** 41.42   **c** 8.37

    d 2.128   **e** 9.357

2 Round these numbers to 2 decimal places.

    a 45.671   **b** 9.388   **c** 12.816   **d** 48.479 28

3 Use a calculator to work out the answers to these calculations. Give each answer correct to 1 decimal place.

   a The area of a rectangle 8.7 cm long and 3.9 cm wide

   b The area of a triangle with base 8.6 cm and height 9.2 cm

   c The square root of 800

4 $x = 5.12$, $y = 29.4$ and $z = 9.13$. Find, correct to 2 decimal places

   a ab **b** $\dfrac{a}{b}$ **c** $\dfrac{b}{a}$ **d** c^2

   e $\dfrac{ab}{c}$ **f** $\dfrac{a+b}{c}$ **g** $\dfrac{a}{bc}$

5 £1 is worth 1.83 Australian dollars (A\$). Convert these amounts of Australian dollars to pounds, giving your answers to a sensible degree of accuracy.

   a A\$100   **b** A\$2000   **c** A\$50 000

Stretch – can you deepen your learning?

1 The length of a square is given as 8 cm ± 0.5 cm

   a What do you think ± 0.5 means?

   b Write an error interval for

     i the length of the square, x

     ii the perimeter of the square, P

     iii the area of the square, A

   c What's the same and what's different about your error intervals?

2 Faith runs 100 m in 12.5 seconds. Find an error interval for Faith's speed in metres per second if

   a the length is exact and the time is correct to 1 decimal place

   b the length is correct to the nearest metre and the time is exact

   c the length is correct to the nearest metre and the time is correct to 1 decimal place.

   Give the ends of your error intervals correct to 2 decimal places where necessary.

3 Find some examples of the use of decimals in textbooks for other subjects, or in newspapers and magazines. Can you tell which of the numbers are exact and which are rounded?

Reflect

1 What's the same and what's different about rounding to decimal places and rounding to significant figures?

2 Explain how you write an error interval.

12.3 Calculating in context

Small steps

- Convert metric measures of length
- Convert metric units of mass and capacity
- Calculate with money
- Solve problems involving time and the calendar

Key words

centi – one hundredth

milli – one thousandth

kilo – one thousand

Are you ready?

1 How many centimetres are there in a metre?

2 How many millimetres are there in

    a a centimetre?   **b** a metre?

3 How many grams are there in a kilogram?

4 Work out

    a 9.36×10 **b** 9.36×100 **c** 4.7×1000 **d** $8.93 \div 10$

    e $1267 \div 100$ **f** $38 \div 1000$

Models and representations

Place value grids can be used for multiplication and division by powers of 10

Tens	Ones	tenths	hundredths
	5	7	2
× 10	× 10	× 10	
5	7	2	

This shows $5.72 \times 10 = 57.2$. Each digit becomes 10 times greater, so the number becomes 10 times greater.

In the same way $8.7 \div 100 = 0.087$

Tens	Ones	tenths	hundredths	thousands
	8	7		
			÷ 100	÷ 100
	0	0	8	7

You may need to add zeros at the start of the number after a division to show where the digits are in relation to the decimal point.

You can also use **place value counters**.

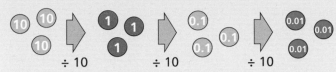

In Book 1, you learned the meaning of **centi-**, **milli-** and **kilo-** when dealing with length, mass and capacity.

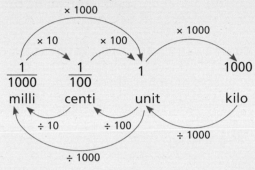

Example 1

A flowerbed is 8.5 m long and 60 cm wide. Find the area of the flowerbed. Give your answer in m².

$60 \, cm = 60 \div 100 \, m = 0.6 \, m$ ○——— There are 100 centimetres in a metre, so to convert from cm to m you divide by 100

area of flowerbed $= 8.5 \times 0.6 = 5.1 \, m^2$ ○——— Once the units are the same, you can find the area.

Example 2

The mass of a golf ball is 45 g. A box full of golf balls has mass 11.5 kg. The box has mass 250 g when empty. How many golf balls are in the box?

In everyday speech, people might say, "The weight of a golf ball is 45 g" but the correct word is "mass".

$11.5 \, kg = 11.5 \times 1000 = 11\,500 \, g$ ○——— The mass of the box full of golf balls is given in kilograms, so first convert this to grams to make the calculations easier. Remember: 1 kg = 1000 g

$11\,500 \, g - 250 \, g = 11\,250 \, g$ ○——— Subtract the mass of the empty box from the total mass to find the mass of the golf balls.

$11\,250 \div 45 = 250$ ○——— Each golf ball has mass 45 g, so divide the total
There are 250 golf balls in the box.   mass of the golf balls by 45 to find how many there are.

Practice 12.3A

1 Convert each of these lengths to centimetres.

a 4 m   **b** 4.5 m   **c** 80 mm   **d** 725 mm

e 1.03 m   **f** 6 mm   **g** 6.5 mm   **h** 6 m

i 3000 mm   **j** 3000 m

2 **a** Which calculation(s) will tell you how many pieces of ribbon 12 cm long can be cut from a 6 m length of ribbon?

$\boxed{6 \div 12}$ $\boxed{60 \div 12}$ $\boxed{12 \div 6}$ $\boxed{1200 \div 6}$ $\boxed{600 \div 12}$ $\boxed{12 \div 600}$ $\boxed{6 \div 0.12}$ $\boxed{1.2 \div 6}$

b Use your chosen calculation to work out how pieces of ribbon there will be.

3 In each pair, decide which value is the greater mass.

a 4 kg, 400 g   **b** 0.3 kg, 250 g   **c** 80 g, 0.07 kg

d 2.5 kg, 6000 g   **e** 125 g, 0.7 kg

4 A can of energy drink contains 50 mg of caffeine.

a How many of the cans would you need to drink to consume 1 g of caffeine?

b Zach drinks two cans of energy drink every day for 30 days. How much caffeine does he consume in total? Give your answer in grams.

5 Convert these measurements to litres.

a 5000 ml   **b** 500 ml   **c** 75 cl   **d** 330 ml

e 2 cl   **f** 8000 cl

6 A bottle contains 2 litres of water.

a How many glasses of 125 ml can be poured from the bottle?

b How many glasses of 175 ml can be poured from the bottle?

c How many glasses of 25 cl can be poured from the bottle?

7 Which is greater, 330 ml or 20 cl? Explain how you know.

8 The recommended dose of a cough medicine is 5 ml. How many doses are there in a 30 cl bottle of the medicine?

What do you think?

1

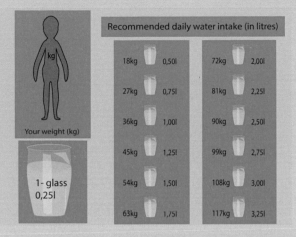

Recommended daily water intake (in litres)

18kg	0,50l
27kg	0,75l
36kg	1,00l
45kg	1,25l
54kg	1,50l
63kg	1,75l
72kg	2,00l
81kg	2,25l
90kg	2,50l
99kg	2,75l
108kg	3,00l
117kg	3,25l

Your weight (kg)

1- glass
0,25l

The diagram shows the recommended daily water intake for different weights.

a Ed weighs 54 kg. How many 0.25 litre glasses of water should he drink a day?

b Ed only has glasses with capacity 300 ml. How many of these should he drink a day?

c Is the amount of water recommended directly proportional to a person's mass? Explain how you know.

2

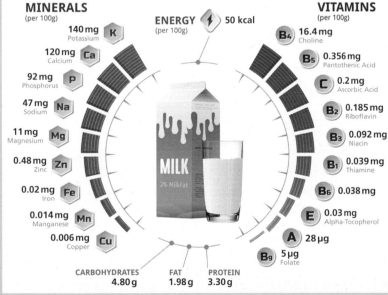

The diagram shows information about the minerals and vitamins in 100 g of milk. A typical glass contains 250 g of milk.

a How much sodium is there in a glass of milk? Give your answer in grams.

b The daily recommended dose of calcium is 0.7 g. How many glasses of milk do you need to drink a day if you get all your calcium from milk?

c The daily recommended dose of Vitamin C is 40 mg. Would it be sensible to get your daily dose of Vitamin C just by drinking milk? Explain why or why not.

£1 = 100p. Converting between pounds and pence involves multiplication and division by 100

Example 3

Pencils costs 12p each.

a How much will 2000 pencils cost?

b A school has 1300 students. How much will it cost the school to buy three pencils for each student if orders over £400 get an 8% discount?

a $12p \times 2000 = 24\,000p$

$24\,000 \div 100 = 240$

The cost is £240

> Multiply the cost of 1 pencil by 2000 to find the total cost.
>
> Divide by 100 to convert pence to pounds.
>
> Alternatively, you could convert 12p to £0.12 and find the answer by working out £0.12 × 2000 = £240

b $1300 \times 3 = 3900$ pencils needed

$3900 \times 12p = 46\,800p$

$46\,800 \div 100 = £468$

$100\% - 8\% = 92\%$

$£468 \times 0.92 = £430.56$

> First work out the total number of pencils needed.
>
> Then find the total cost and convert to pounds. You could have worked out 3900 × £0.12
>
> As the cost is over £400, there is an 8% discount.
>
> You learned how to use multipliers in Block 10

Practice 12.3B

1 A packet of crisps costs 45p and a drink costs £0.85

    a Explain why the total cost of a packet of crisps and a drink is not 45.85p

    b Find the total cost of a packet of crisps and a drink.

    c Ed wants to buy four packets of crisps and three drinks. He has £5.
    Does he have enough money? Show working to justify your answer.

2 A power company charges 14p per unit for electricity and 4.5p per unit for gas.

    a In a month, a family uses 600 units of electricity and 200 units of gas.
    Work out the total cost of their electricity and gas during the month.

    b How much more or less would it cost with a different supplier who charges 13p per unit for electricity and 5.5p per unit for gas?

3 A company pays 65p to post small letters and £1.20 to post large letters.
Which is cheaper, posting 300 small letters and 50 large letters or posting 200 small letters and 130 large letters? How much cheaper is it?

4 A shop sells cartons of orange juice for 82p.

    a How many cartons of orange juice can you buy for £5?

    b What is the minimum number of coins needed to give the change from £5 if you buy the number of cartons you calculated in part **a**? Explain your answer.

5 A tub of 80 marbles costs £4.29. Individual marbles cost 6p. Find the minimum the cost of buying 500 marbles.

6 An adult cinema ticket costs £6.30 and a child ticket costs £4.20. Popcorn costs 85p for a portion and drinks are £1.30 each. Find the total cost of a cinema trip for two adults and five children if they share four packets of popcorn and each have a drink.

7

Meter reading this time	Meter reading previous time		Units used	Price of each unit (in pence)	Amount (£)
4703	3425	Units used	_____	18.5p	_____
		V.A.T at 5%			_____
		Total charge, including V.A.T.			_____
		Previous amount owing			12.46
		Amount to pay			_____

Here is an electricity bill. Work out the missing amounts and find the total amount to pay.

What do you think? 💭

1 Beca has some 20p coins and some 10p coins in her purse. She has four times as many 20p coins as 10p coins. Altogether she has £2.70 in her purse. How many 10p coins does she have?

2 Marta and Jackson share £4 in the ratio 13:3. Jackson gets his share in 5p coins. How many coins does he get?

3 A van hire company charges £18 per hour and 85p per mile travelled. Jakub hires a van for a 120-mile journey and pays a total of £246. For how long did Jakub hire the van?

There are 24 hours in a day and 7 days in a week.

One hour is 60 minutes and one minute is 60 seconds.

Number lines are very useful for working out intervals of time.

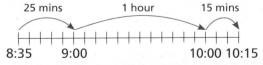

From 8:35 to 10:15 is 25 minutes + 1 hour + 15 minutes = 1 hour and 40 minutes altogether.

You should know the number of days in each month.

Months	Number of days
January, March, May, July, August, October, December	31
April, June, September, November	30
February	28, or 29 in leap years

Example 4

a How many days are there from 12th May to 4th August inclusive?

b Convert your answer from part **a** into

   i hours   **ii** minutes   **iii** seconds.

a Days in May = 31 − 11 = 20 — There are 31 days in May, but as you are starting on the 12th, you do not count May 1st to May 11th.

Days in June = 30

Days in July = 31 — Use known facts for the other months.

Days in August = 4 — You are only including up to August 4th.

Total number of days = 85 — Add to find the total.

b **i** 85 × 24 = 2040 hours — Each day has 24 hours.

  ii 2040 × 60 = 122 400 minutes — Each hour has 60 minutes.

  iii 122 400 × 60 = 7 344 000 seconds — Each minute has 60 seconds.

Practice 12.3C

1 How many hours are there in

  a a week   **b** 12 days   **c** August?

2 The calendar shows the days of the week in November 2021

  a How many Tuesdays are there in November 2021?

  b What day of the week is 31st October 2021?

  c What is the date of the second Friday in December 2021?

  d On what day of the week is 1st January 2022?

NOVEMBER						
S	M	T	W	T	F	S
	1	2	3	4	5	6
7	8	9	10	11	12	13
14	15	16	17	18	19	20
21	22	23	24	25	26	27
28	29	30				

3 How many minutes are there in

  a $\frac{3}{4}$ hour   **b** $\frac{2}{3}$ hour   **c** $4\frac{1}{2}$ hours

  d 1 day   **e** $3\frac{1}{2}$ days?

4 Seb watched four episodes of his favourite TV show without a break. Each episode is 50 minutes long. Seb started watching the first episode at 5:40 p.m. At what time did he finish watching the fourth episode?

5 Ed catches the school bus at 7:40 a.m. The bus arrives at school at 8:22 a.m. How long is Ed's journey?

6 A news report claims that over the last 5 years, the average length of a pop song has decreased by 20 seconds and is now 3 minutes and 30 seconds.

    a What is the percentage decrease in the length of the average pop song in the last 5 years? Give your answer to 2 significant figures.

    b Flo is working out the number of average-length songs that could be played in 2 hours.

> $120 \div 3.3 = 36.3636\ldots$, about 36 songs

    Explain why Flo is wrong, and find the correct answer to the question.

7 A flight leaves Singapore at 18:00. It takes 13 hours and 15 minutes to fly to London. The time in London is 8 hours behind the time in Singapore. At what time (local time) does the flight land in London?

8 How many seconds are there in a leap year? Give your answer in standard form correct to 2 significant figures.

What do you think? 💭

1 There are 52 weeks in a year.

Do you agree with Chloe? Explain your answer.

2 How many days are there from 1st January 2010 to 31st December 2060 inclusive?

3 How many minutes are there between noon today and the start of your next birthday?

Consolidate – do you need more?

1 How many mm are there in

    a 8 cm   **b** 23.6 cm   **c** 1 m   **d** 2.5 m?

2 Find the area of a rectangle 12 cm long and 45 mm wide. Give your answer in cm².

3 Which is heavier, 900 g or 3.5 kg?

4 A can contains 330 ml of juice and costs 89p. Marta wants to drink 2 litres of juice.

    a How many cans must she buy?

    b How much will Marta pay for the juice?

5 A box of 10 pairs of compasses costs £13.50. Pencils cost 11p.

    a How much does it cost to buy a pair of compasses and a pencil for every student in a class of 30 students?

    b How much more does it cost to buy a pair of compasses and a pencil for every student in a class of 32 students?

6 In many countries in the northern hemisphere, December, January and February are called the "winter months".

    a How many days are there in these winter months in a leap year?

    b How many hours of "winter" are there in a non-leap year?

7 In what month are you answering this question? How many minutes are there in this month?

Stretch – can you deepen your learning?

1 A glass and its contents has mass 600 g when half full and mass 0.84 kg when $\frac{4}{5}$ full. What is the mass of the glass when it is empty?

2 Many people still measure height in feet and inches and mass in stones, pounds and ounces. Investigate the relationships between these units. How can you convert them to metric equivalents?

3 5! (called "5 factorial") means $5 \times 4 \times 3 \times 2 \times 1$

    a How many days is 5! hours?

    b How many hours is 6! minutes?

    c Is 7! hours more or less than 6 months? Explain your answer.

4 **a** How can you tell if a year is a leap year?

    b How many leap years will there be in your lifetime if you live to be 85?

Reflect

1 What unit conversions do you know? Explain how you can convert between the units.

2 Why is calculating with time different from calculating with other measures?

Small steps

- Convert metric units of area **H**
- Convert metric units of volume **H**

Key words

Area – the amount of space inside a 2-D shape

Volume – the amount of space taken up by a 3-D shape

Capacity – how much space a container holds

Are you ready?

1 Find the area of each of these shapes. Give your answers in cm²

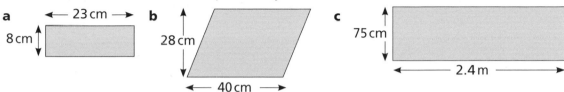

a ← 23 cm → 8 cm

b 28 cm ← 40 cm →

c 75 cm ← 2.4 m →

2 How many mm are there in

a 1 cm **b** 1 m **c** 1 km?

3 How many cubes of side 1 cm are needed to make this cuboid?

4 Work out the volume of a cuboid that measures 12 cm by 20 cm by 40 cm

Models and representations

Area can be represented on squared or dotty paper.

Volume can be represented using cubes.

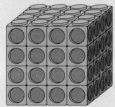

Example 1

Show that $1\,cm^2 = 100\,mm^2$

$1\,cm = 10\,mm$
$1\,cm^2 = 1\,cm \times 1\,cm$
$\quad\ = 10\,mm \times 10\,mm$
$\quad\ = 100\,mm^2$

You can also see this from a diagram.

1 cm

1 cm

10 mm

10 mm

Example 2

a Change $8\,cm^2$ to mm^2

b Change $65\,000\,mm^2$ to cm^2

a $8 \times 100 = 800\,mm^2$ ○── Use the fact that $1\,cm^2 = 100\,mm^2$

b $65\,000 \div 100 = 650\,cm^2$ ○── Every $100\,mm^2$ is $1\,cm^2$, so you need to divide $65\,000$ by 100

Practice 12.4A

1 Use the fact that $1\,cm^2 = 100\,mm^2$ to convert

    a $8\,cm^2$ to mm^2 **b** $90\,cm^2$ to mm^2 **c** $2000\,mm^2$ to cm^2 **d** $40\,mm^2$ to cm^2

2 Find the area of each shape in **i** cm^2 and **ii** mm^2

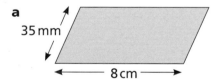

a 35 mm, 8 cm

b 30 mm, 5 cm

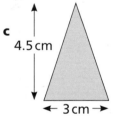

c 4.5 cm, 3 cm

3 **a** Find the area of a square of side 4 m, giving your answer in m^2

    b Find the area of a square of side 4 m, giving your answer in cm^2

    c Copy and complete: $1\,m^2 = $ _____ cm^2

4 Express these areas in cm^2

    a $6\,m^2$ **b** $9.2\,m^2$ **c** $250\,m^2$

5 Express these areas in m^2

    a $8000\,cm^2$ **b** $75\,000\,cm^2$ **c** $1\,000\,000\,cm^2$

6 **a** A postage stamp has area $5\,cm^2$. How many postage stamps would cover an area of $1\,m^2$?

    b The surface of an e-reader has area $400\,cm^2$. How many e-readers would cover an area of $1\,m^2$?

7 How many mm^2 would be the same as $1\,m^2$?

What do you think? 💭

1 A map is drawn to a scale of 1 : 50 000

   a What distance, in kilometres, does 1 cm on the map represent?

   b What area, in km², does a square of side 1 cm on the map represent?

   c What area, in cm², does a square of side 3 mm on the map represent?

2 Which is the same as 1 km²?

   A ⎡100 m²⎤   B ⎡1000 m²⎤   C ⎡100 000 m²⎤   D ⎡1 000 000 m²⎤   E ⎡1 000 000 000 m²⎤

3 A square of land that is 100 m by 100 m covers 1 hectare.

   a Write the area of a hectare in

      i m²   **ii** cm²

   b How many hectares are there in 1 km²?

You can find the volume, V, of a cuboid by multiplying its length, l, width, w, and height, h

$$V = l \times w \times h$$

You will look at volume in detail in Book 3

Example 3

a Find the volume of this cuboid in

   i cm³   **ii** mm³

b Complete: 1 cm³ = ___ mm³

a i $6 \times 2 \times 5 = 60 \, cm^3$ ——○—— Use the formula $V = l \times w \times h$

  ii $60 \times 20 \times 50 = 60\,000 \, mm^3$

b $1 \, cm^3 = 1 \, cm \times 1 \, cm \times 1 \, cm$ ——○—— or, from part **a**, because 60 cm³ = 60 000 mm³,

       $= 10 \, mm \times 10 \, mm \times 10 \, mm$ 1 cm³ = 60 000 ÷ 60

       $= 1000 \, mm^3$ $= 1000 \, mm^3$

Practice 12.4B

1 Use the fact that 1 cm³ = 1000 mm³ to convert

   a 9 cm³ to mm³   **b** 0.2 cm³ to mm³   **c** 2600 mm³ to cm³ **d** 50 mm³ to cm³

2 A cuboid is 6 cm long, 4 cm wide and 20 mm deep.

   a By converting all the lengths to cm, find the volume of the cuboid in cm³

   b By converting all the lengths to mm, find the volume of the cuboid in mm³

   c Use your answers to **a** and **b** to verify that 1 cm³ = 1000 mm³

3 **a** Find the volume, in m³, of a cube of side length 2 m

b By converting 2 m to cm, find the volume of the cube in cm³

c Using your answers to **a** and **b**, copy and complete

1 m³ = _____ cm³

4 Express these volumes in cm³

a 6 m³ **b** $\frac{1}{4}$ m³ **c** 0.025 m³

5 Express these volumes in m³

a 200 000 cm³ **b** 50 000 000 cm³ **c** 3 × 10⁸ cm³

6 **a** The capacity of a hot tub is 0.75 m³. 1 litre = 1000 cm³
How many litres of water does the hot tub hold?

b The capacity of a bath is about 150 litres. Express this in

i cm³ **ii** m³

7 Express 1 m³ in mm³. Give your answer in standard form.

What do you think? 💭

1 A shipping container has volume 32.3 m³

A box measures 50 cm by 25 cm by 40 cm
Approximately how many of these boxes will fit in
one shipping container? Why might your answer not
be exactly what the container can carry?

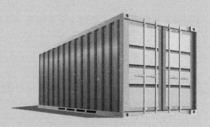

2 A box has volume 1 m³

a How many cubes of side 1 cm are needed to fill the box?

b

> You will need half as many cubes
> if they are of side length 2 cm

Explain why Jakub is wrong.

c Find the correct number of cubes of side 2 cm needed to fill the box.

d How many cubes of side 10 cm are needed to fill the box?

e How many cubes of side 5 cm are needed to fill the box?

3 What's the same and what's different about 1 cm³ and 1 ml?

Consolidate – do you need more?

1 Express these areas in cm²

a 3 m² **b** 0.5 m² **c** 16 m² **d** $\frac{1}{4}$ m²

e 500 mm² **f** 8000 mm² **g** 25 mm²

2 Express these areas in m^2

   a 80 000 cm^2 **b** 5000 cm^2 **c** 300 cm^2 **d** 642 000 cm^2

3 Express these volumes in cm^3

   a 2.5 m^3 **b** $\frac{3}{4}$ m^3 **c** 5000 mm^3

   d 100 mm^3 **e** 62 m^3

4 Express these volumes in m^3

   a 2 000 000 cm^3 **b** 32 000 000 cm^3 **c** 50 000 cm^3 **d** 850 000 cm^3

5 1 litre = 1000 cm^3. Express the following in litres.

   a 45 000 cm^3 **b** 800 cm^3 **c** 1 m^3

   d $\frac{1}{2}$ m^3 **e** 600 000 mm^3

Stretch – can you deepen your learning?

1 In this question, give your answers to 3 significant figures.
1 inch is about 2.54 centimetres.

   a Change 18 inches to centimetres

   b Change 18 $inches^2$ to cm^2

   c Change 18 $inches^2$ to mm^2

   d Change 18 $inches^2$ to m^2

2 Which is the best estimate for the capacity of a bucket?

   A 500 cm^3 **B** 5 litres   **C** 500 ml

   D 5 cl   **E** 5 m^3

3 A jug holds $1\frac{3}{4}$ litres of water. How many jugs are needed to fill this cuboid with water?

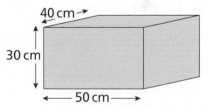

4 The volume of a cylinder is given by the formula $V = \pi r^2 h$, where r is the radius and h is the height.

   a Show that a tank of diameter 1.2 m and height 80 cm can hold more than 900 litres of liquid.

   b Find some possible dimensions of an oil tank with capacity 2000 litres.

Reflect

Explain why 1 $cm^2 \neq 10$ mm^2 and 1 $cm^3 \neq 10$ mm^3

I have become fluent in...

- rounding numbers to a given number of significant figures
- rounding numbers to a given number of decimal places
- converting between metric units of length, mass and capacity
- calculating with time and the calendar.

I have developed my reasoning skills by...

- interpreting the structure of a mathematical problem
- making links between different calculations
- choosing the correct units for a calculation
- understanding when to work with decimal relationships and when not to
- interpreting and creating error intervals. **H**

I have been problem-solving through...

- interpreting and solving financial mathematics problems
- solving problems with time and the calendar
- converting between metric units of area and volume. **H**

Check my understanding

1 Show that $8.02^2 - 1.95^2$ is approximately 60

2 Write these numbers to 2 significant figures.

   a 845 million   **b** 0.0159

3 A rectangle is 9.73 cm long and twice as wide.

Use a calculator to find the area of the rectangle, giving your answer correct to 1 decimal place.

4 Without using a calculator, work out $\dfrac{12}{0.03}$

5 Steve makes components in a factory. He is paid 7p for every component he makes and £12 per day. How much does he earn in a week when he works six days and makes 7500 components?

6 **a** How many days are there from 18th May to 10th August inclusive?

   b A film starts at 10:40 p.m. and finishes at 00:20 a.m. How long is the film?

7 $x = 10.15$ to 2 decimal places. Write the error interval for x **H**

8 How many squares of area 16 cm² can be cut from a larger square of area 4 m²? **H**

13 Angles in parallel lines and polygons

In this block, I will learn...

how to find missing angles on lines and at a point

$a + b = 180°$

$p = q$
$t = w$

$d + e + f = 360°$

the rules for angles between parallel lines and a transversal

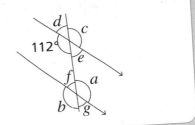

112°

the properties of special quadrilaterals

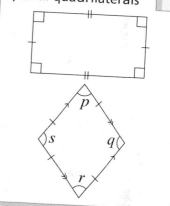

how to solve complex angle problems

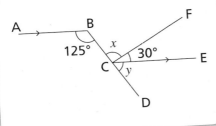

A   B   F
125° x 30°
  C   E
    y
      D

how to prove geometrical facts **H**

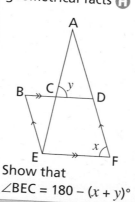

A

B C y D

E x F

Show that
$\angle BEC = 180 - (x + y)°$

how to construct angle bisectors and perpendicular bisectors **H**

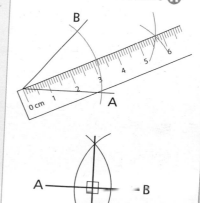

how to work out interior and exterior angles of a polygon

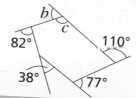

b
c
82°   110°
38°   77°

Small steps

■ Understand and use basic angle rules and notation ®

Key words

Adjacent – next to each other

Vertically opposite angles – angles opposite each other when two lines cross

Are you ready?

1 Classify each of these angles as acute, obtuse or reflex.

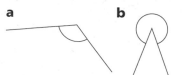

a   **b**   **c**   **d**   **e** 87°   **f** 198°

2 Give the three-letter name for each angle labelled with an arc.

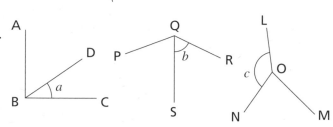

3 Draw angles of size

a 50°   **b** 150°   **c** 250°

 4 Use a ruler and pair of compasses to construct a triangle with sides of length 8 cm, 7 cm and 4 cm. Measure the three angles in your triangle and find their total. Is the result what you expected?

Models and representations

Geoboards

Straws

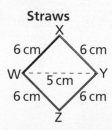

Both these angles are equal as they have a single arc

The hatch marks show these two sides are equal

The double arcs show these two angles are equal

You also use **hatch marks** and **arcs** to show when sides and angles are equal.

In this chapter, you will revise using the facts and rules you studied in Book 1

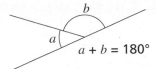

$a + b = 180°$

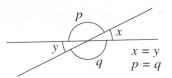

$x = y$
$p = q$

| Angles in a full turn add up to 360° | **Adjacent** angles on a straight line add up to 180° | **Vertically opposite angles** are equal |

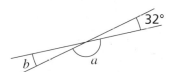

$e + f + g = 180°$

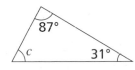

$w + x + y + z = 360°$

Angles in a triangle add up to 180°

Angles in a quadrilateral add up to 360°

Example 1

Find the angles labelled with letters. Give reasons for your answers.

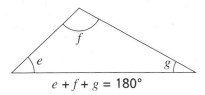

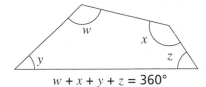

$a = 180° - 32° = 148°$

Angles on a straight line add up to 180°

$b = 32°$

Vertically opposite angles are equal.

a and 32° are adjacent angles on a straight line.

Make sure you give the mathematical reason for your answer. It's not good enough to say things like "because they add up to 180°"

b and 32° are opposite each other and share a common vertex.

$87° + 31° = 118°$

$c = 180° - 118° = 62°$

Angles in a triangle add up to 180°

Find the sum of the angles you already know in the triangle.

Subtract this sum from 180° to find the unknown angle.

Give the mathematical reason.

$d = 180° - 130° = 50°$

Angles on a straight line add up to 180°

$e = 360° - 313° = 47°$

Angles in a full turn add up to 360°

d and 130° are adjacent on a straight line.

e and 313° together make a full turn.

You could also say, "Angles at a point add up to 360°"

$47° + 50° + 135° = 232°$

$f = 360° - 232° = 128°$

Angles in a quadrilateral add up to 360°

Find the sum of the angles you already know in the quadrilateral.

Subtract this sum from 360° to find the unknown angle.

Give the mathematical reason.

Practice 13.1A

1 Work out the size of each angle labelled with a letter. Give reasons for your answers.

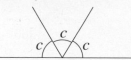

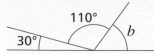

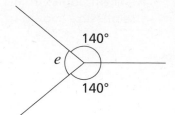

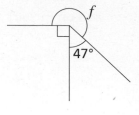

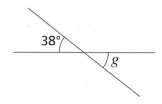

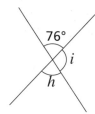

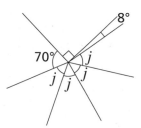

2 Work out the size of each angle labelled with a letter. Give reasons for your answers.

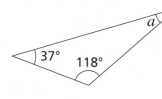

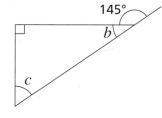

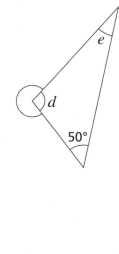

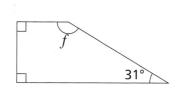

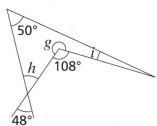

3 In this diagram, explain how you know that PQR is not a straight line.

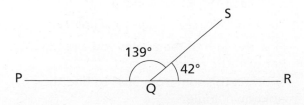

What do you think?

1 Angles a and b are adjacent and together form a straight line. Find the sizes of a and b if

    a a is three times the size of b

    b b is half the size of a

    c a and b are in the ratio $3:5$

2 Name the quadrilaterals which have exactly

    a one pair of equal angles

    b two pairs of equal angles.

3
> The only special quadrilaterals that contain right angles are rectangles and squares.

Do you agree with Flo? Justify your answer.

Sometimes working out unknown angles involves interpreting hatch marks, or forming and solving equations.

Example 2

Work out the size of $\angle ACD$. Give a reason for each step in your working.

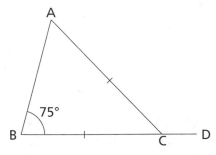

$\angle BAC = 75°$

Base angles in an isosceles triangle are equal. ○———

$\angle ACB = 180° - (75° + 75°) = 30°$

Angles in a triangle add up to $180°$

$\angle ACD = 180° - 30° = 150°$

Angles on a straight line add up to $180°$

The hatch marks on AC and AB tell you that triangle ABC is isosceles.

Example 3

Work out the size of ∠QRS.
Give a reason for each step
in your working.

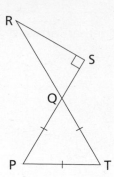

∠PQT = 60°

Angles in an equilateral triangle. ○————— The hatch marks on PQ, QT and PT tell
you that triangle PQT is equilateral. So
∠PQT = 180° ÷ 3 = 60°

∠RQS = 60°

Vertically opposite angles are equal.

90° + 60° = 150° ○————— ∠QSR = 90° as it is marked with a right
angle symbol.

∠QRS = 180° − 150° = 30°

Angles in a triangle add up to 180°

Practice 13.1B

1 Work out the size of each angle labelled with a letter. Give reasons for your answers.

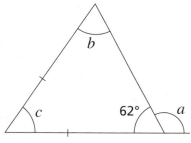

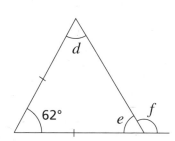

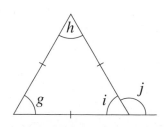

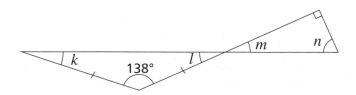

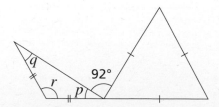

2 Form and solve equations to find the value of each letter. Give reasons for your answers.

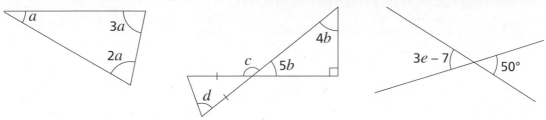

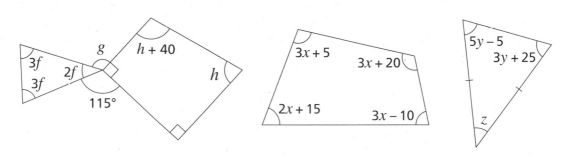

3 Work out the size of ∠BEC. Give a reason for each step in your working.

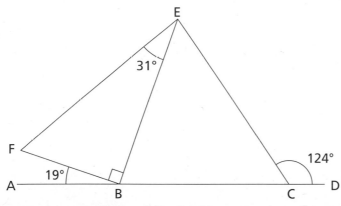

What do you think? 💭

1 Explain why the angles in an equilateral triangle are all 60°

2 **a** One angle in a triangle is known to be 40°. Could the triangle be isosceles? If so, what could the other angles be? Is there more than one possible answer? If it cannot be isosceles, explain why not.

b Repeat part **a** if the known angle is

    i 4°   **ii** 110°   **iii** 90°   **iv** 160°

Consolidate – do you need more?

1 Three angles meet at a point. Two of the angles are each 96°
What is the size of the third angle?

2 Find the size of each unknown angle. Give reasons for your answers.

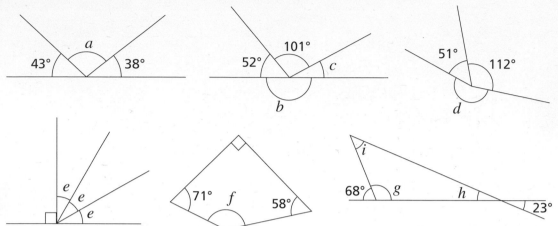

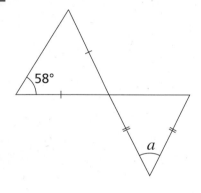

3 One angle in an isosceles triangle is 68°. What might the other two angles be?

4 Work out the value of each letter.

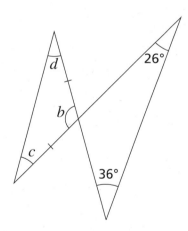

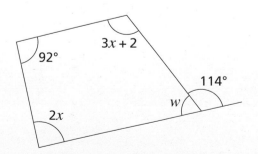

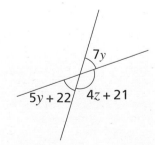

Stretch – can you deepen your learning?

1 Explain why the sum of the angles in a quadrilateral is always 360°

2 The diagram shows part of a regular octagon.

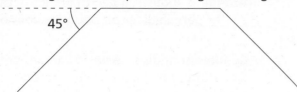

a Calculate the size of one interior angle of a regular octagon.

b Calculate the sum of all the interior angles of an octagon.

3 Explain why a rectangle is not usually a regular quadrilateral.

4 Sketch a quadrilateral with angles in the ratio 1 : 2 : 3 : 4

Reflect

How do you solve a problem with unknown angles? What do you look for?

Small steps

- Investigate angles between parallel lines and the transversal
- Identify and calculate with alternate and corresponding angles
- Identify and calculate with co-interior, alternate and corresponding angles

Key words

Alternate angles – a pair of angles between a pair of parallel lines on opposite sides of a transversal

Co-interior angles – a pair of angles between a pair of parallel lines on the same side of a transversal

Corresponding angles – a pair of angles in matching positions compared with a transversal

Transversal – a line that crosses at least two other lines

Are you ready?

1 Which diagram(s) show a pair of parallel lines?

A

B

C

2 Work out the size of each angle labelled with a letter.

3 Work out the size of each angle labelled with a letter.

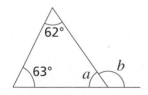

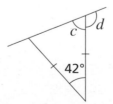

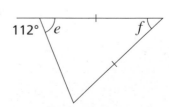

Models and representations

Straws

You can use straws to model parallel lines and a line crossing them, which is called a **transversal**.

Dynamic geometry software

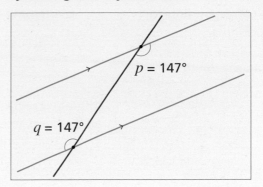

$p = 147°$

$q = 147°$

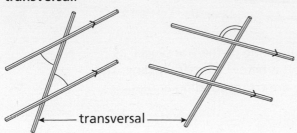

transversal

Use these models along with drawing and measuring to investigate the properties of the angles formed by parallel lines and a transversal.

Look at the angles labelled in each diagram. Notice that they are equal.

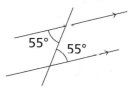

55° 55°

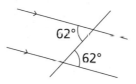

62°
62°

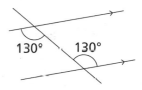

130° 130°

Pairs of angles like this are called **alternate angles**.

They are between the parallel lines but on opposite (alternate) sides, like x and y in this diagram.

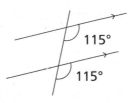

x y

Alternate angles in parallel lines are equal.

Look at the angles labelled in each diagram. Notice that they are equal.

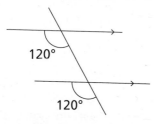

120°

120°

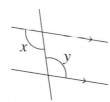

115°

115°

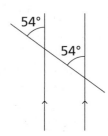

54°

54°

Pairs of angles like this are called **corresponding angles**.

They are in the same corresponding positions compared with the parallel lines and the transversal.

In this diagram, x and y are both to the right of the transversal and below the parallel lines.

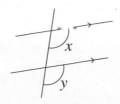

x

y

Corresponding angles in parallel lines are equal.

Example 1

Write an angle that is

a corresponding to x **b** alternate to x

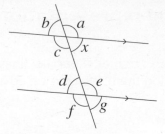

a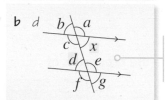

Angle g is on the same side of the transversal as angle x and, like x, it is below the parallel line. This means that x and g are corresponding angles.

b

Angle d is on the opposite side of the transversal to angle x and, like x, it is between the pair of parallel lines. This means that x and d are alternate angles.

Example 2

Find the sizes of the labelled angles, giving reasons for your answers.

a $a = 132°$ *(corresponding angles are equal)*

You need to recognise that a and the 132° angle are corresponding, and so they are equal. They are both above a parallel line and on the left of the transversal.

b $b = 132°$ *(alternate angles are equal)*

Angle b is on the opposite side of the transversal to angle a and, like a, it is between the pair of parallel lines.

Therefore angle b is alternate to a, and so they are equal.

Another reason that angle b is 132° is that it is vertically opposite the given angle of 132°, and vertically opposite angles are equal.

c $c = 132°$ *(corresponding angles are equal)*

You need to recognise that c is corresponding to b, and so these angles are equal. They are both below a parallel line and on the right of the transversal.

Another reason that angle c is 132° is that it is vertically opposite angle a, and vertically opposite angles are equal.

Practice 13.2A

1 In each diagram, identify the angle that is alternate to x

a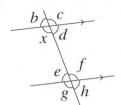

b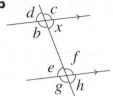

c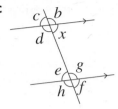

d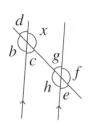

e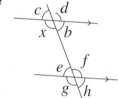

f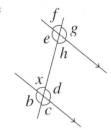

2 In each diagram, identify the angle that is corresponding to x

a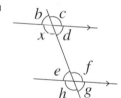

b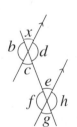

c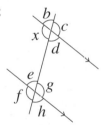

d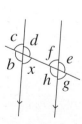

e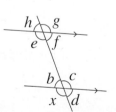

f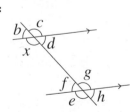

3 **a** Name four pairs of corresponding
angles on the diagram.

b Name two pairs of alternate angles
on the diagram.

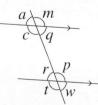

4 **a** Use the rules about corresponding angles, alternate angles and angles on a straight line to work out the seven unknown angles in the diagram.

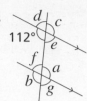

b What other angles rules could you have used?

5 Work out the size of each angle labelled with a letter. Give reasons for your answers.

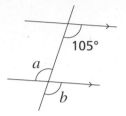

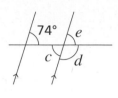

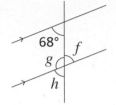

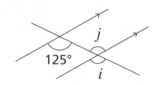

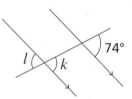

6 Find the size of each angle, giving reasons for your answers.

a ∠QST **b** ∠RSU **c** ∠TSU

7

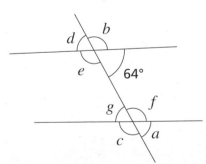

Angle *f* = 64° because corresponding angles are equal.

a Explain why Abdullah is wrong.

b Which angles can you work out in this diagram?

What do you think? 🗨

1 Find the size of each angle, giving reasons for your answers.

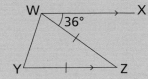

    a ∠WZY   **b** ∠ZWY

2 Find the size of each angle, giving reasons for your answers.

                             a ∠BCF   **b** ∠GFC   **c** ∠FGB

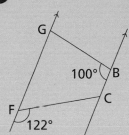

3 Work out the values of x and y. Justify your answers.

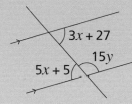

Look at the labelled angles in each diagram. Notice that they add up to 180°

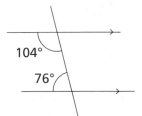

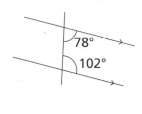

Angles like p and q are called **co-interior angles**.

They are on the same side of the transversal and between the parallel lines.

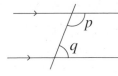

Co-interior angles add up to 180°

Example 3

Find the values of the labelled angles, giving reasons for your answers.

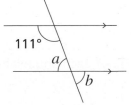

a $a = 180° - 111° = 69°$
(co-interior angles add up to 180°)

You need to recognise that a and the 111° angle are co-interior. They are on the same side of the transversal and between the parallel lines.

b $b = 69°$
(vertically opposite
angles are equal)

○── You need to recognise that a and b are opposite
each other and share a common vertex.

Practice 13.2B

1 Copy the diagrams. In each case, mark an angle b that is co-interior with a.

a

b

c

2 For each diagram, write down the letter of an angle that is

i corresponding to x

ii alternate to x

iii co-interior to x

iv vertically opposite x

a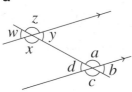

b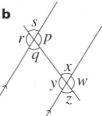

c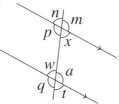

3 **a** Name two pairs of co-interior angles in the diagram.

b Name two pairs of alternate angles in the diagram.

c Name four pairs of corresponding angles in the diagram.

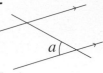

4 Use the rules for co-interior angles, vertically opposite
angles and adjacent angles on a straight line to work
out the sizes of the unknown angles.

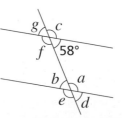

5 Work out the size of each angle labelled with a letter. Give reasons for your answers.

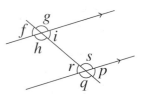

What do you think? 💭

1. Use the rules for corresponding angles and adjacent angles on a straight line to prove that co-interior angles add up to 180°

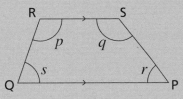

2. PQRS is a trapezium. What relationships do you know about angles p, q, r and s?

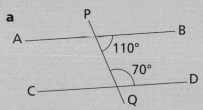

3. In each case, decide if AB is parallel to CD. How do you know?

a

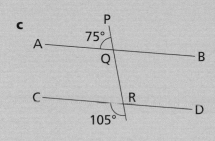

b

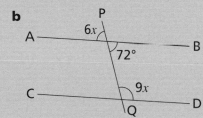

c

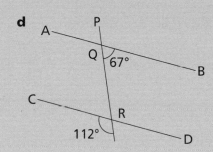

d

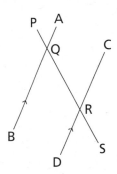

Consolidate – do you need more?

1. Copy and complete these statement

   a ∠BQR is corresponding to _____

   b ∠AQR is alternate to _____

   c ∠BQR is co-interior with _____

   d ∠BQR is alternate to _____

   e ∠CRS is corresponding to _____

2. Work out the size of each angle labelled with a letter, giving reasons for your answers.

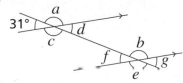

3 Work out the size of each angle labelled with a letter, giving reasons for your answers.

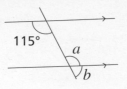

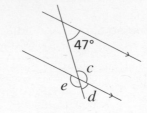

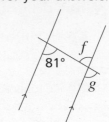

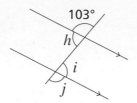

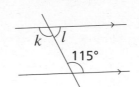

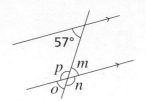

Stretch – can you deepen your learning?

1 Copy the diagram. How many sets of co-interior, corresponding and alternate angles can you find?

2 Work out the size of angle ABC in each diagram. For part **c**, give your answer in terms of x and y.

a

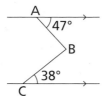

A
47°
B
38°
C

b

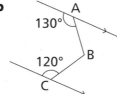

130° A
120° B
C

c

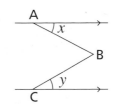

A x
B
C y

3 Work out the size of each unknown angle in these triangles, giving reasons for your answers.

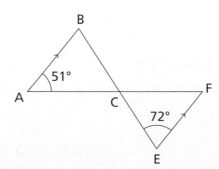

B
51°
A   C   F
72°
E

Reflect

Make a display explaining the rules about angles on parallel lines.

Small steps

- Construct triangles and special quadrilaterals
- Investigate the properties of special quadrilaterals
- Identify and calculate with sides and angles in special quadrilaterals
- Understand and use the properties of diagonals of quadrilaterals **H**

Key words

Adjacent sides/angles – sides or angles that are next to each other

Bisect – to cut in half

Construct – draw accurately using a ruler and compasses

Diagonal – a line that is neither horizontal nor vertical; in quadrilaterals, a line segment that joins two opposite vertices

Isosceles – having two sides the same length

Opposite sides/angles – sides or angles that are not next to each other

Are you ready?

1. Describe the properties of **a** an isosceles triangle, **b** a scalene triangle and **c** an equilateral triangle.

2. Draw these angles accurately. **a** 35° **b** 310° **c** 160°

3. Draw an accurate copy of the diagram.

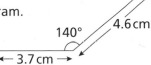

4. How many different types of quadrilateral can you name? How can you tell them apart?

Models and representations

Rods and geoboards (or geoboard apps)

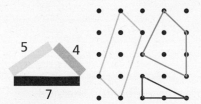

Set squares

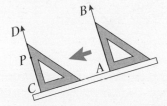

Card and paper fasteners

Geo strips

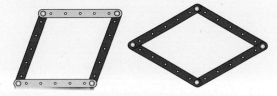

In this chapter, you will revise **constructing** triangles and quadrilaterals from Book 1 and look more closely at the angle properties of quadrilaterals.

Example 1

By constructing two triangles, make an accurate copy of quadrilateral ABCD.

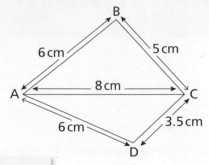

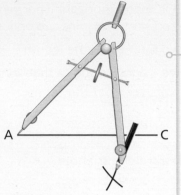

Start by drawing AC 8 cm long. Then use compasses to draw an arc of radius 6 cm long from A and another arc of radius 3.5 cm long from C

Join A and C to the point where the two arcs cross to construct triangle ACD

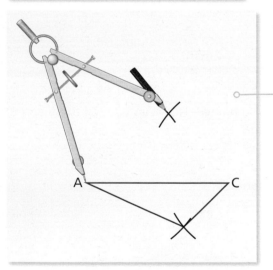

Then construct triangle ABD by drawing an arc of radius 6 cm long from A and another arc of radius 5 cm long from C

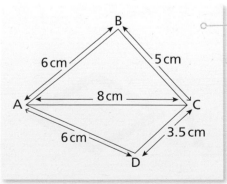

Complete the diagram, labelling the sides.

You could construct triangle ABC and then triangle ADC

Practice 13.3A

1 Use a ruler and a pair of compasses to construct an equilateral triangle with sides of length 5 cm. Use a protractor to verify that all the angles are 60°

2 Follow these steps to construct triangle PQR using a ruler and a protractor.

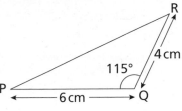

■ Start by drawing PQ 6 cm long.

■ Use a protractor to draw an angle of 115° at Q

■ Mark point R 4 cm from Q

■ Join R and P to complete the triangle.

💬 Why are constructions like this called SAS?

3 Construct triangle XYZ using a ruler and a protractor.

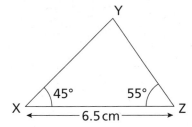

💬 How do you start your construction? Why are constructions like this called ASA?

4 Make an accurate copy of each triangle. They are not drawn accurately.

a

b

c

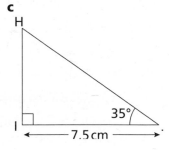

5 a Create parallelogram PQRS by constructing two triangles PQR and PRS.

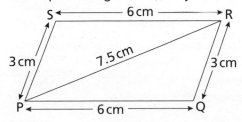

b Measure

i ∠PSR and ∠PQR   **ii** ∠SPQ and ∠SRQ

What do you notice?

6 **a** Create rhombus WXYZ by constructing two isosceles triangles WXY and WXZ

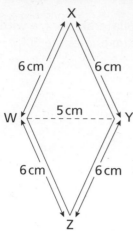

b Measure

i ∠WXY and ∠WZY

ii ∠XWZ and ∠XYZ

What do you notice?

7 **a** Create rhombus ABCD by constructing two isosceles triangles ABC and ADC

b Verify that ABCD is a rhombus by measuring the sides.

c Is ∠ABC equal to ∠ADC?

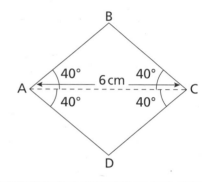

What do you think? 💭

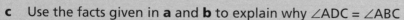

1 Here is a parallelogram, ABCD

a Explain why ∠ABD = ∠BDC

b Explain why ∠ADB = ∠DBC

c Use the facts given in **a** and **b** to explain why ∠ADC = ∠ABC

d Use the fact that the angles in a triangle add up 180° to deduce a fact about ∠DAB and ∠DCB

e Explain why ∠DAB + ∠ADC = 180°

2 **a** Sketch an isosceles trapezium. What can you deduce about its angles?

b Sketch a non-isosceles trapezium. What can you deduce about its angles?

3 Draw an accurate kite, PQRS, by drawing two identical triangles, PQR and PRS. Choose your own dimensions.

Compare your kite with a partner's.

What's the same and what's different?

What can you deduce about the angles in a kite?

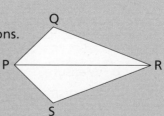

Because co-interior angles add up to 180°, two pairs of angles in a trapezium will add up to 180°

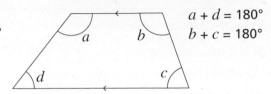

$a + d = 180°$
$b + c = 180°$

In an **isosceles** trapezium, there will also be two pairs of equal angles.

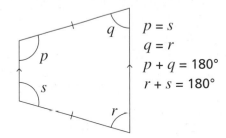

$p = s$
$q = r$
$p + q = 180°$
$r + s = 180°$

In both a parallelogram and a rhombus, **opposite angles** are equal, and **adjacent angles** add up to 180°

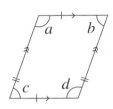

$a + b = 180°$
$b + d = 180°$ etc.
$a = d$
$b = c$

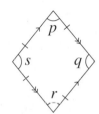

$p + q = 180°$
$s + p = 180°$ etc.
$p = r$
$s = q$

In a kite, one pair of opposite angles are equal.

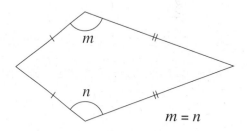

$m = n$

In both a rectangle and a square, opposite and adjacent angles are all equal to 90°

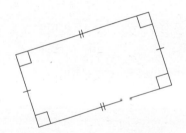

Example 2

EFGH is a parallelogram.

a Work out the sizes of angles a, b and c

b Calculate the size of $\angle$FEH

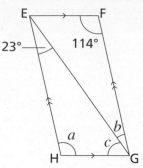

a $a = 114°$ (opposite angles of a parallelogram are equal)

$b = 23°$ (alternate angles are equal)

In triangle EGH,

$114° + 23° + c = 180°$ (angles in a triangle add up to 180°)

So $c = 180° - 114° - 23° = 43°$

b $\angle$FEH $= 180° - 114° = 66°$
(co-interior angles add up to 180°)

Use your knowledge of the properties of angles in quadrilaterals.

You may need to use rules about angles in parallel lines and other rules you have learned in previous chapters.

You could also find this by adding 23° and 43°

Practice 13.3B

1 ABCD is a parallelogram.

Work out the sizes of angles x, y and z

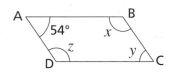

2 PQRS is a rhombus.

Work out the size of each of each angle.

a $\angle$QRS **b** $\angle$QSP **c** $\angle$PQS

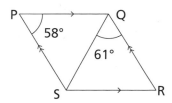

3 Work out the size of each angle labelled with a letter.

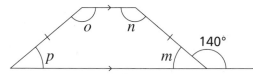

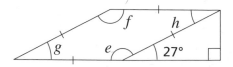

4 a Work out the size of each angle.

 i $\angle$ADB **ii** $\angle$DCB

b What can you say about the lengths of BC and BD? Why?

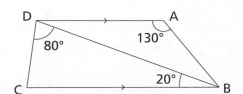

5 Work out the values of the letters in these diagrams.

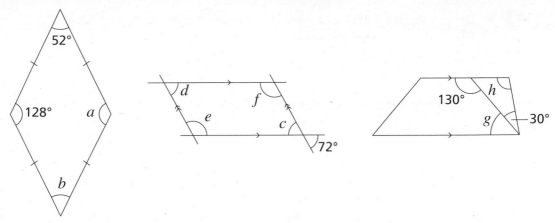

What do you think? 💭

1 How much information do you need to be able to construct a parallelogram?

2 Ed says, "A kite has two pairs of equal sides that each share a common vertex." Do you agree?

3 ABCD is a rectangle. Work out the sizes of angles a and b

Can you work out any of the other angles?

A **diagonal** joins two opposite vertices of a polygon such as a quadrilateral.

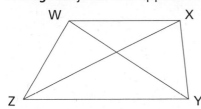

WY and XZ are the diagonals for quadrilateral WXYZ

You will now investigate the diagonals of quadrilaterals. In particular, you will look at whether diagonals **bisect** each other at the point where they meet.

To bisect means to cut in half.

You will also investigate whether diagonals bisect the angles at the vertices where they start and finish.

Practice 13.3C

1 **a** Draw a square ABCD and mark on its diagonals AC and BD, as shown. Label the point where they cross X

b Measure AC and BD. What do you notice?

c Measure AX, XC, BX and XD. What do you notice?

d Name any isosceles triangles in your diagram.

e Deduce the size of angles XAB, XBA and AXB

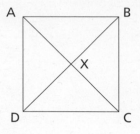

2 Draw a rectangle PQRS and investigate its diagonals. What's the same and what's different about the diagonals of a square and the diagonals of a rectangle?

3 Investigate these claims, and decide whether they are true or false.

a The diagonals of a rhombus are perpendicular to each other.

b The diagonals of a parallelogram are equal in length.

c The diagonals of a kite are perpendicular to each other.

d The diagonals of a rhombus bisect its angles.

e The diagonals of a rhombus are equal in length.

f The diagonals of a parallelogram bisect its angles.

g The diagonals of a trapezium are not equal in length.

4 Find the sizes of a and b in this diagram.

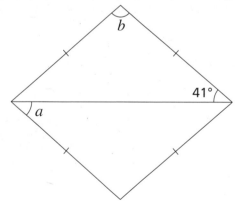

5 Work out the sizes of the angles marked a, b, c, p and q.

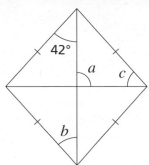

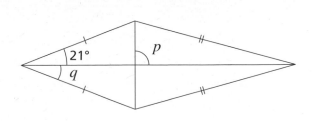

6 a Draw a rhombus with diagonals of length 6 cm and 8 cm.

b Draw a kite with diagonals of length 6 cm and 8 cm.

c Draw a square with diagonals of length 6 cm.

What do you think?

1 All quadrilaterals have exactly two diagonals.

Do you agree with Faith?

2 SQ isn't a diagonal of PQRS because it's horizontal, not diagonal.

Do you agree with Seb?

3 Which quadrilaterals could match the descriptions on each of these cards?

Card A	**Card B**	**Card C**	**Card D**
My diagonals are equal in length.	My diagonals bisect each other.	My diagonals are perpendicular.	My diagonals bisect my interior angles.

Consolidate – do you need more?

1 Use a ruler and compasses to construct these triangles.

a

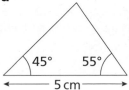

b

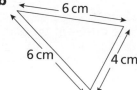

c

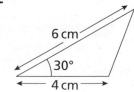

2 Use a ruler and compasses to construct these quadrilaterals.

a

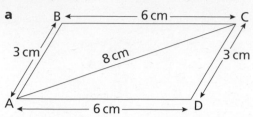

b

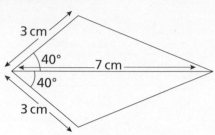

3 Find the size of each angle labelled with a letter.

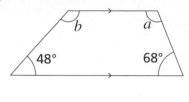

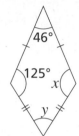

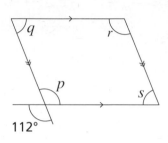

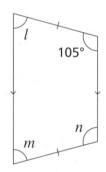

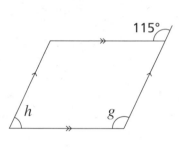

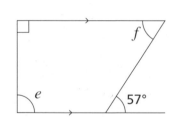

Stretch – can you deepen your learning?

1 **a** Sketch any pentagon. Verify that it has exactly five diagonals.

  b Investigate the number of diagonals for other polygons. Can you find a rule for the number of diagonals of an n-sided polygon?

2 Two shapes are congruent if they are exactly the same shape and size. Investigate whether the diagonals of different quadrilaterals form sets of congruent triangles.

Reflect

1 Describe the properties of the different types of quadrilaterals in terms of their sides and angles.

2 Describe the steps you would take to construct a

  a triangle   **b** parallelogram   **c** kite.

What information would you need in each case?

Small steps

- Understand and use the sum of the exterior angles of any polygon
- Calculate and use the sum of the interior angles in any polygon
- Calculate missing interior angles in regular polygons

Key words

Exterior angle – an angle between the side of a shape and a line extended from the adjacent side

Interior angle – an angle on the inside of a shape

Polygon – a closed 2-D shape with straight sides

Regular polygon – a polygon whose sides are all equal in length and whose angles are all equal in size

Are you ready?

1 Write down the sum of the angles

   a in a triangle   **b** in a quadrilateral   **c** on a straight line.

2 Work out the sizes of the unknown angles.

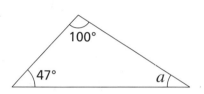
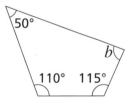
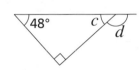

3 Name a polygon with

   a 5 sides   **b** 8 sides   **c** 10 sides.

4 Which of these shapes are hexagons?

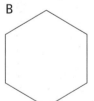
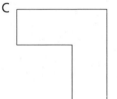
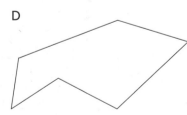

A   B   C   D

Models and representations

Dotty paper

Geoboards/geoboard apps

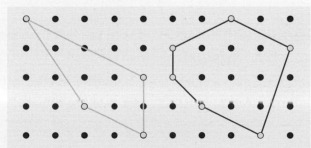

So far you have studied the **interior angles** of **polygons**.

These are the angles at the vertices inside a shape.

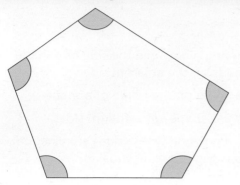

The shaded angles are the interior angles of the pentagon

Now you will explore the **exterior angles** of polygons.

These are formed by extending the sides of the polygon at each vertex.

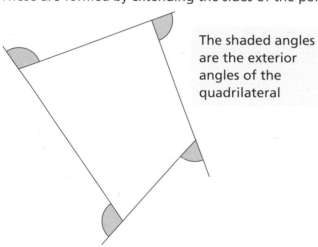

The shaded angles are the exterior angles of the quadrilateral

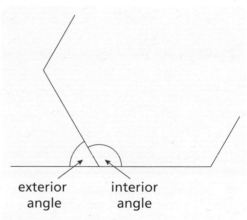

exterior angle   interior angle

Interior and exterior angles of a polygon are adjacent on a straight line, so they sum to 180°

Example 1

Work out the exterior angles of this triangle.

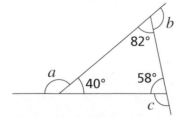

$a = 180° - 40° = 140°$
$b = 180° - 82° = 98°$
$c = 180° - 58° = 122°$

Use the fact that angles on a straight line add up to 180° to work out each of the exterior angles.

Interior angle + exterior angle = 180°

Practice 13.4A

1 Work out the sizes of the unknown angles.

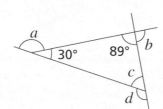

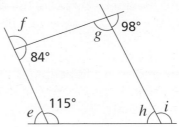

2 a What is the size of each interior angle of an equilateral triangle?

b What is the size of each exterior angle of an equilateral triangle?

c Find the sum of the exterior angles of an equilateral triangle.

3 a What is the size of each interior angle of a rectangle?

b What is the size of each exterior angle of a rectangle?

c Find the sum of the exterior angles of a rectangle.

4 a Work out the sizes of the exterior angles of each of these shapes.

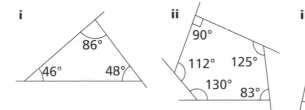

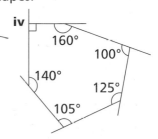

b Find the sum of the exterior angles of each shape.

5

I think that the sum of the exterior angles of any polygon is 360°

Test Zach's conjecture by drawing three polygons, measuring the exterior angles and finding their total.

6 Zach's conjecture is correct. Use it to work out the sizes of the labelled angles in these diagrams.

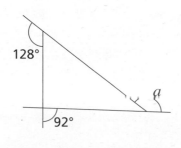

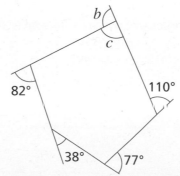

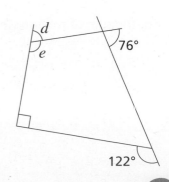

What do you think? 💭

1 Use dynamic geometry software to verify that the sum of the exterior angles of a polygon is 360°

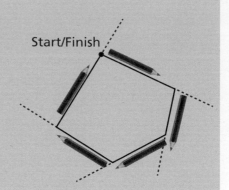

Start/Finish

2 a Explain why turning a pencil through the exterior angles of this pentagon illustrates that the sum of the exterior angles of the pentagon is 360°

b Verify the result by turning a pencil through the exterior angles of other polygons.

3 a Explain why each exterior angle of a regular hexagon is 60°

b Work out the size of each exterior angle of

i a regular pentagon   **ii** a regular decagon   **iii** a regular 30-sided shape.

c What happens to the size of each exterior angle as the number of sides of a regular polygon increases?

d The exterior angles of a regular shape are each 15°. How many sides does the shape have?

You now know that the sum of the exterior angles of any polygon is 360°

You will now explore the sum of the interior angles.

The sum of the interior angles of a triangle is 180°

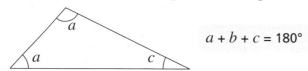

$$a + b + c = 180°$$

Any quadrilateral can be split into two triangles by drawing a diagonal.

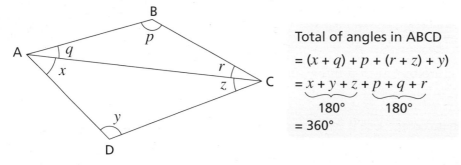

Total of angles in ABCD

$$= (x + q) + p + (r + z) + y$$

$$= \underbrace{x + y + z}_{180°} + \underbrace{p + q + r}_{180°}$$

$$= 360°$$

So, the sum of the interior angles of any quadrilateral is 360°

Practice 13.4B

1 Which of these diagrams would be helpful for finding the sum of the interior angles of a pentagon? Which would not? Explain your answers.

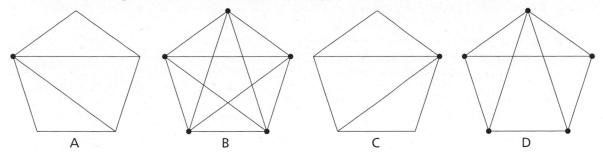

A   B   C   D

2 Use the diagram to explain why the sum of the interior angles of a pentagon is 540°

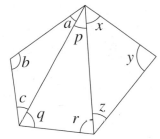

3 a Split a hexagon into triangles by drawing diagonals from one vertex.

b How many triangles are formed inside the hexagon?

c Work out the sum of the interior angles of a hexagon.

4 Repeat the steps of question **3** for a heptagon and an octagon.

5 Copy the table and use your results from questions **3** and **4** to fill in the missing information.

Shape	No. of sides	No. of triangles formed	Sum of interior angles
Triangle	3	1	1 × 180° = 180°
Quadrilateral	4	2	2 × 180° = 360°
Pentagon	5	3	3 × 180° = 540°
Hexagon	6	4	
Heptagon	7		
Octagon	8		

6 a Use your table from question **5** to predict the number of triangles formed by joining one vertex of a nonagon to all the other vertices.

b Work out the sum of the interior angles of a nonagon.

What do you think? 💭

1 What is the connection between the number of sides in a polygon and the number of triangles formed by joining one vertex of the polygon to all the other vertices?

2 a A polygon has n sides. Which expression shows the number of triangles formed by joining one vertex of the polygon to all the other vertices?

| $2n$ | $n + 2$ | $2 - n$ | $n - 2$ | $\dfrac{n}{2}$ |

b Deduce an expression for the sum of the interior angles of any polygon.

You can use the results you've discovered to find missing angles in polygons.

Example 2

Find the size of the unknown angle in this pentagon.

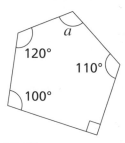

Angle sum of a pentagon = 540°

Total of known angles is:
90° + 100° + 120° + 110° = 420°

a = 540° − 420° = 120°

You know this from question 2 of Practice 13.4B, or you can use the formula described next in Example 3

Example 3

Find the sizes of angles a and b

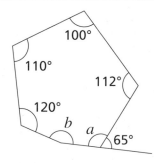

a = 180° − 65° = 115° (angles on a straight line add up to 180°)

Angle sum of a hexagon
= (6 − 2) × 180° = 4 × 180° = 720°

Total of known angles is
115° + 112° + 100° + 110° + 120° = 557°

So b = 720° − 557° = 163°

65° is an exterior angle. Subtract 65° from 180° to find the interior angle at this vertex.

For a hexagon, $n = 6$

The sum of the interior angles of a polygon with n sides is given by $(n - 2) \times 180°$

Practice 13.4C

1 Work out the sum of the interior angles in

   a an octagon   **b** a decagon.

2 Work out the sizes of the missing angles in these shapes.

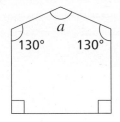
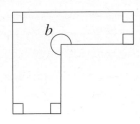
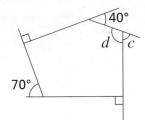

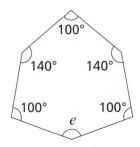
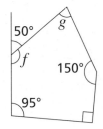
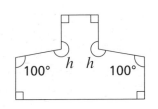

3 Six of the angles in an octagon are 140°. The other two angles are equal.
Work out the size of the equal angles.

4 Jackson and Marta are working out the size of an interior angle of a regular nonagon.

Jackson's method	**Marta's method**
Angle sum of nonagon	Exterior angle of nonagon
$= (9 - 2) \times 180° = 1260°$	$= 360° \div 9 = 40°$
Each interior angle $= 1260 \div 9 = 140°$	Each interior angle $= 180° - 40° = 140°$

Whose method do you prefer?

Use your preferred method to work out the size of an interior angle of

   a a regular pentagon   **b** a regular octagon   **c** a regular 20-sided shape.

5 The diagram shows a regular
hexagon ABCDEF

Work out the size of angle x

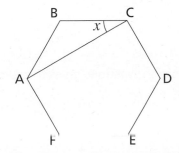

6 Four of the exterior angles of a pentagon are 90°, 60°, 70° and 90°
Work out the interior angles of the pentagon.

7 The angles in a polygon add up to 900°. How many sides does the polygon have?

What do you think?

1 a The exterior angles of a regular polygon are each 20°
How many sides does the polygon have?

b The interior angles of a regular polygon are each 170°
How many sides does the polygon have?

2 a Is it possible for the exterior angle of a regular polygon to be 150°?
Why or why not?

b Is it possible for the interior angle of a regular polygon to be 150°?
Why or why not?

3 The diagram shows two identical squares. Find the size of angle x

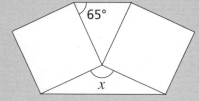

4 Sven positions two regular 12-sided tiles so that they meet edge to edge. A third regular shape is placed so that it touches both of these tiles and makes a full turn. Name the third shape.

Consolidate – do you need more?

1 Which diagram shows an exterior angle?

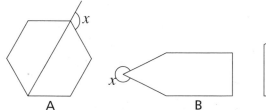

A   B   C   D

2 Which of these expressions gives the sum of the exterior angles of a polygon with n sides?

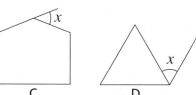

| 180° | 180n° | 360° | 360n° | (n + 2) × 180° |

3 Which of these expressions gives the sum of the interior angles of a polygon with n sides?

$n \times 180°$	$(n - 2) \times 180°$	$(n + 2) \times 180°$

$2n \times 180°$	$(n - 2) \times 360°$	$(n + 2) \times 360°$

4 Find the sizes of the angles labelled with letters.

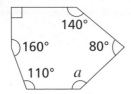

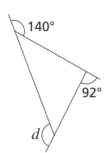

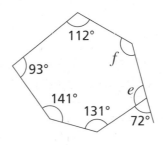

Stretch – can you deepen your learning?

1 **a** Find a formula, in terms of n, for e, the exterior angle of a regular polygon with n sides.

b Find two formulae, in terms of n, for i, the interior angle of a regular polygon with n sides.

c Write an equation connecting i and e

(Hint – you could start with your answer to **a** or with the rule for the sum of the interior angles of a polygon.)

2 **a** A shape is said to "tessellate the plane" if the plane (a flat surface) can be covered with identical copies of the shape without leaving holes and without overlapping. Investigate which regular polygons tessellate the plane. What is the connection between whether a regular polygon will tessellate and the sizes of its exterior angles?

b Investigate whether any irregular polygons tessellate the plane.

Reflect

1 Explain the difference between interior and exterior angles of polygons.

2 Explain how you can find the sizes of the interior and exterior angles of regular polygons.

How is this different from finding the sizes of the interior and exterior angles of irregular polygons?

Small steps

- Solve complex problems with angles in parallel lines
- Prove simple geometric facts (H)

Key words

Give a reason – state the mathematical rule(s) you have used, not just the calculations you have done

Proof – an argument that shows that a statement is true

Are you ready?

1 State three rules about angles in parallel lines.

2 Write down the letter of an angle that is

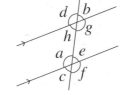

  a corresponding to a **b** alternate to a
  c co-interior with a **d** vertically opposite a

3 Work out the size of each angle labelled with a letter, giving reasons for your answers.

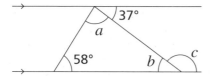

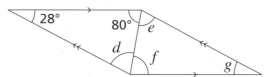

Models and representations

You can use any of the models and representations from the previous chapters in this block to support your working.

Dynamic geometry software is very useful for demonstrating that results are true, but it does not provide a mathematical **proof**.

A proof shows that a claim is true for all cases, not just for some specific examples.

When writing a proof, each statement you make must follow on logically from the previous statement, or must be justified with a reason.

First, here are some multi-step problems based on the angle facts that you have been studying.

Example 1

AB is parallel to CE. Work out the sizes of angles x and y

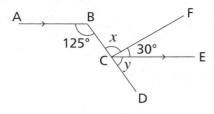

$\angle BCE = 125°$
(Alternate angles are equal)

So $x + 30° = 125°$
$x = 95°$

$y + 30° + x = 180°$ (Angles on a straight line add up to $180°$)

So $y = 180° - 30° - 95° = 55°$

The angle labelled x is not equal to any of the given angles, so you need to start by finding a relevant angle. Since AB is parallel to CE, angles ABC and BCE are equal.

You can see from the diagram that $\angle BCF + \angle FCE = \angle BCE$. Solve the equation to find x

$\angle BCD$ is a straight line.

You can now work out y using your value of x

Example 2

Find the size of $\angle RQT$.

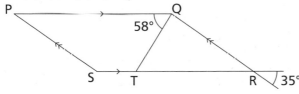

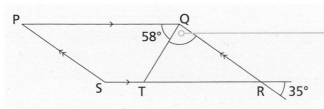

$\angle QRT = 35°$
(Vertically opposite angles are equal)

$\angle PQR = 180° - 35° = 145°$
(Co-interior angles add up to $180°$)

$\angle RQT = 145° - 58° = 87°$

Start by marking the angle you need to work out on your diagram.

You cannot work out $\angle RQT$ directly, so start by using the information on the diagram to work out nearby angles.

As PQ is parallel to SR

Because $\angle PQR = \angle PQT + \angle RQT$

Practice 13.5A

1 Work out the size of each angle labelled with a letter. Give a reason for each step of your working.

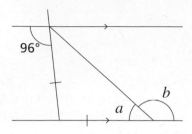

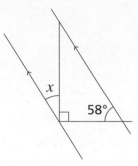

2 Work out the value of each letter.

Hint: You might find it helpful to label the points on the diagram with capital letters so that you can refer to the angles that you need to calculate in your working.

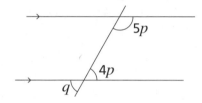

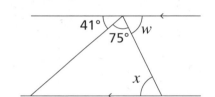

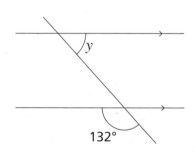

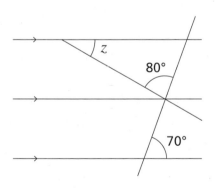

3 Find the sizes of the angles labelled a, b and c

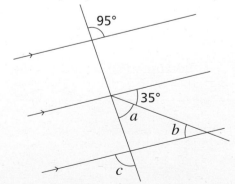

4 Work out the value of x

Give a reason for each step of your working.

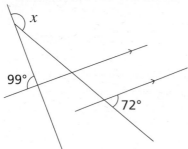

5 Find the sizes of the angles labelled w, x, y and z

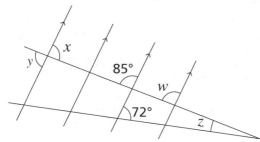

6 Work out the size of $\angle ABC$

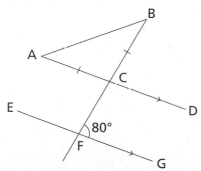

What do you think? 💭

1 AC is parallel to DG

HI is perpendicular to DG

BE = EF

Work out the size of $\angle BFG$

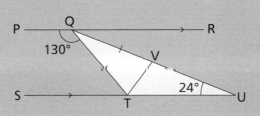

2 PR is parallel to SU

$\angle QTV$ is an isosceles triangle.

Work out

a $\angle QTV$ **b** $\angle TVU$

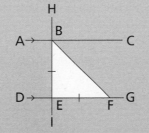

459

Next, you will use the rules you have learned for proving facts about angles and shapes.

Example 3

a Work out the size of angle x

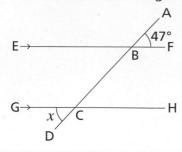

b Prove that angles x and y are equal.

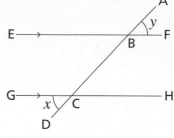

a $\angle EBC = 47°$
(Vertically opposite angles are equal)

$x = \angle EBC$
(Corresponding angles are equal)

So $x = 47°$

> You cannot find x directly, but you can use the given angle to find $\angle EBC$ first.
>
> Now you have an angle that is directly related to x

a $\angle EBC = y$
(Vertically opposite angles are equal)

$x = \angle EBC$
(Corresponding angles are equal)

so $x = y$

> The proof follows the same steps as part **a**, but this time you don't have a specific value.

Example 4

Prove that quadrilateral PQRS is a trapezium.

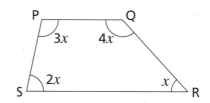

$x + 2x + 3x + 4x = 360°$ (Angles in a quadrilateral add up to 360°)

$10x = 360°$

$x = 36°$

$\angle PSR = 2x = 2 \times 36° = 72°$

$\angle SPQ = 3x = 3 \times 36° = 108°$

$\angle PSR + \angle SPQ = 72° + 108° = 180°$

So $\angle PSR$ and $\angle SPQ$ are co-interior angles because they add up to 180°

So PQ is parallel to SR.

Therefore, PQRS is a trapezium.

> A trapezium is a quadrilateral with one pair of parallel lines.

> To prove that PQRS is a trapezium, you need to show that it has one pair of parallel lines. Working out the angles will help you to do this.

> When writing a proof, justify all your steps giving mathematical reasons.

Practice 13.5B

1 Prove that triangle BEF is a right-angled triangle.

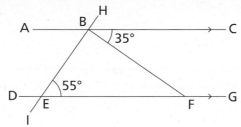

2 **a** Show that ∠BEC = 30°

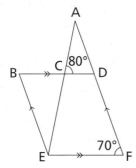

b Show that ∠BEC = 180° − (x + y)

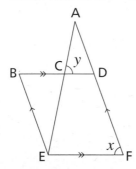

3 **a** What do you know about the sum of the angles labelled p, q and r?

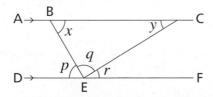

b Which pairs of angles on the diagram are equal?

c Use the diagram to prove that the angle sum of a triangle is 180°

4 **a** In parallelogram ABCD, ∠ABC = 65°

Sketch parallelogram ABCD and use the rules for angles in parallel lines to work out the other three angles in the parallelogram.

b Use the rules for angles in parallel lines to prove that the opposite angles in a parallelogram are equal.

5 In each diagram, PQR is an isosceles triangle and PRS is a straight line.

a Prove that ∠PQR = (2a − 180)°

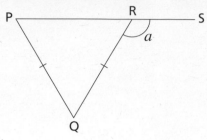

b Prove that ∠QRS = $(90 + \frac{x}{2})°$

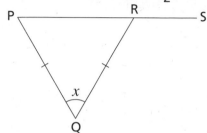

What do you think?

1 Prove that ∠QRS = 2a

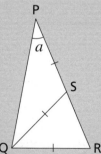

2 Prove that ACD is an isosceles triangle.

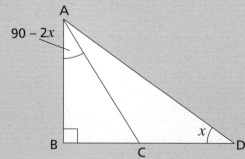

Consolidate – do you need more?

1 Copy the diagram and write on the sizes of all the unmarked angles.

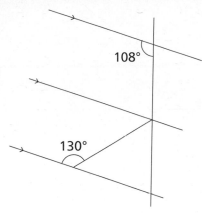

2 Find the sizes of the angles labelled *a*, *b*, *c*, *d* and *e*

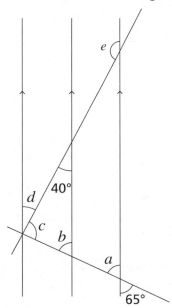

3 Work out

   a ∠ADC   **b** ∠EDC   **c** ∠DEC   **d** ∠ECD

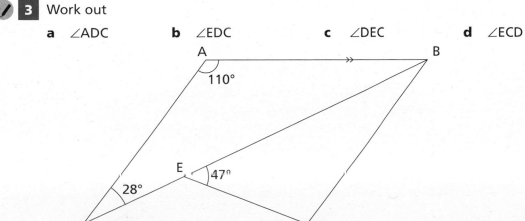

Stretch – can you deepen your learning?

1 **a** Work out ∠WVZ

b XY = 8 cm, VW = 12 cm and YW = 5 cm

Work out the length of ZW

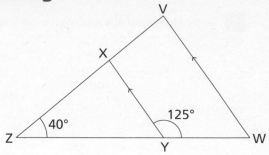

2 Find the values of x and y

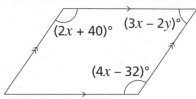

3 AB is parallel to CD. Prove that WY is parallel to XZ

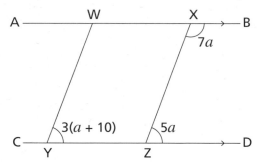

Reflect

1 Explain how you can break down a complicated geometrical problem into smaller steps.
What do you look for?

2 How do you prove a geometrical statement?

Small steps

■ Construct an angle bisector **H**

■ Construct a perpendicular bisector of a line segment **H**

Are you ready?

1 Draw

a a circle with radius 5 cm

b a semicircle with diameter 8 cm

2 Follow the steps to draw a yin-yang symbol.

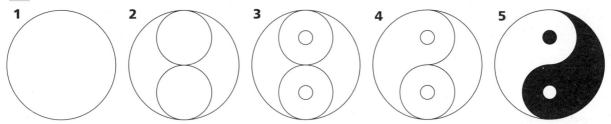

Models and representations

You can use coins or counters to show points that are the same distance from other points or lines.

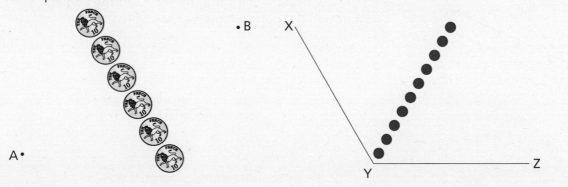

Follow the steps in the worked examples to learn how to construct an angle **bisector** without using a protractor.

An angle bisector is a line that cuts an angle in half.

Example 1

Draw an angle of 50 degrees and construct its angle bisector.

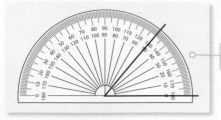

Start by drawing an angle of 50°
Make the arms of the angle about 5 cm long.

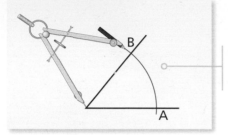

Use a pair of compasses to draw an arc of radius 4 cm long from the vertex of the angle that cuts both arms of the angle. Label these points of intersection A and B

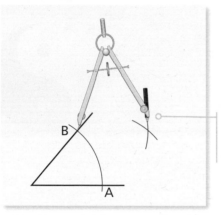

Next, draw arcs of the same length from A and B

Make sure you draw them long enough so that they meet.

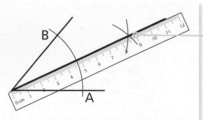

Draw a straight line from the vertex of the angle, through the point of intersection of the two arcs.

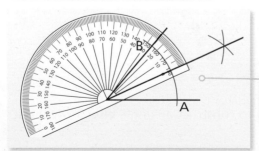

Verify that both angles formed are 25°

You have **constructed** the bisector of the 50° angle.

Practice 13.6A

1 Draw these angles and use the method shown in Example 1 to construct their bisectors.

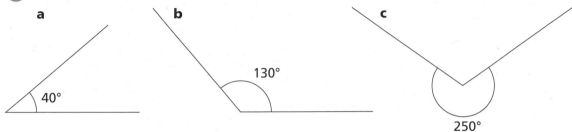

a 80°   **b** 110°   **c** 140°   **d** 30°

What length should the radius of the arcs be? Can you change the radius during the construction? Why or why not?

2 Draw these angles and bisect them using a ruler and a pair of compasses.

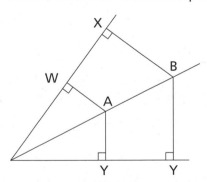

a 40°   **b** 130°   **c** 250°

Is it easier or harder to construct the angle bisector when neither of the arms are horizontal nor vertical? What strategies did you use?

3 The shortest distance from a point to a line is the length of the perpendicular from that point to the line. Construct an angle and its bisector, and measure the shortest distances from some points on the bisector to the arms of the angle.

Measure and compare
AW and AY, BX and BZ

Verify that the angle bisector is equidistant from the arms of the angle.

4 The diagram shows the three villages of Alwick, Broten and Carlton.

Copy the diagram and show the position of a straight road through Broten that is equidistant from Alwick and Carlton.

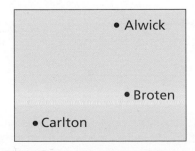

• Alwick

• Broten

• Carlton

5 **a** Construct an accurate copy of triangle ABC

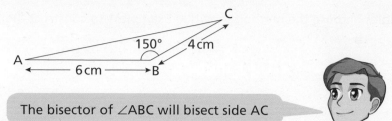

b

The bisector of ∠ABC will bisect side AC

Do you think Ed is correct? Construct the bisector of ∠ABC to test Ed's conjecture.

What do you think?

1 The incircle of a triangle touches all three sides. The centre of an incircle is where the angle bisectors of the triangle meet. Construct an equilateral triangle with sides of length 6 cm. Bisect the angles of your triangle and draw the incircle as shown.

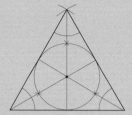

2 Investigate the position of the centre of the incircle for different types of triangle, for example right-angled and obtuse-angled triangles.

Do you need to construct all three angle bisectors? Why or why not?

Follow the steps in the next worked example to learn how to construct the **perpendicular bisector** of a straight line segment.

The perpendicular bisector of a straight line segment cuts it in half and is at right angles to the line.

Example 2

Draw an 8 cm line segment and construct its perpendicular bisector.

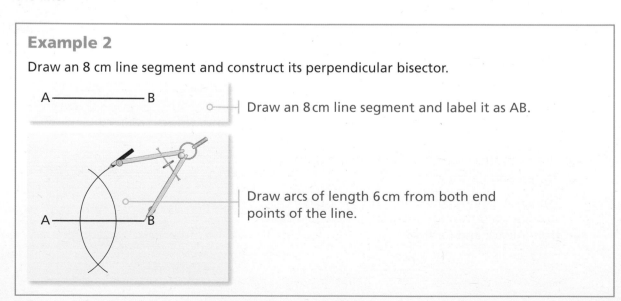

Draw an 8 cm line segment and label it as AB.

Draw arcs of length 6 cm from both end points of the line.

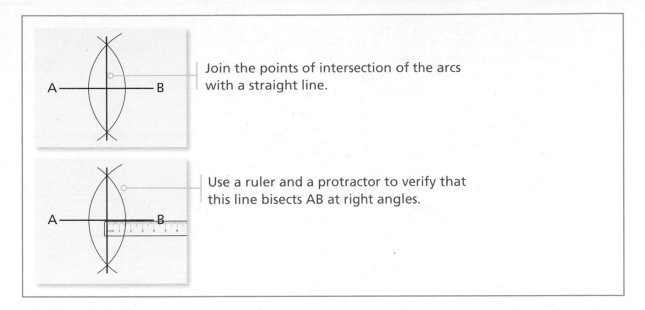

Join the points of intersection of the arcs with a straight line.

Use a ruler and a protractor to verify that this line bisects AB at right angles.

Practice 13.6B

1 Repeat Example 2, but this time draw smaller arcs above and below AB instead of long arcs that go through AB

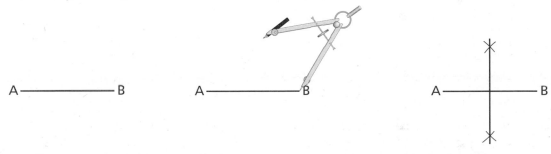

2 Practise drawing perpendicular bisectors using segments of different lengths and orientations.

What must be the radius of the arcs compared with the length of the line you are bisecting? Do both arcs need to have the same radius?

3 **a** Sketch a triangle that looks like ABC. Find the point at which the bisector of ∠BAC meets the perpendicular bisector of the line BC

b Repeat part **a** for different types and sizes of triangle.

4 Draw a line segment AB. Construct a perpendicular bisector of AB, as shown, labelling the points of intersection of the arcs X and Y. Join the points to form quadrilateral AXBY. What do you notice about the quadrilateral? Why is this?

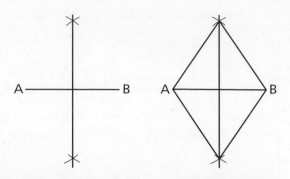

What do you think?

1 The circumcircle of a triangle touches all three vertices. The centre of a circumcircle is where the perpendicular bisectors of the sides meet.
Construct an equilateral triangle with sides of length 6cm. Construct the perpendicular bisectors of two of its sides and draw the circumcircle, as shown.

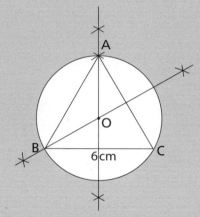

Why do you only need to construct two perpendicular bisectors?

2 Investigate the position of the centre of the circumcircle for different types of triangle, for example right-angled and obtuse-angled triangles.

Consolidate – do you need more?

1 Practise drawing angle bisectors using angles of different sizes and orientations. Include bisecting angles in triangles, quadrilaterals and other shapes.

2 Practise drawing perpendicular bisectors using lines of different lengths and orientations. Include drawing the perpendicular bisectors of sides of triangles, quadrilaterals and other shapes.

3 Repeat Practice 13.6B question 3 with several different triangles to practise drawing angle and perpendicular bisectors.

Stretch – can you deepen your learning?

1

> The points on the circumference of a circle are all equidistant from the centre.

Do you agree with Beca?

2 **a** Investigate quadrilaterals whose diagonals bisect the interior angles.

b Starting with **i** a regular pentagon and **ii** a regular hexagon, investigate whether the diagonals of regular shapes bisect angles and/or bisect each other.

Reflect

Describe the steps you need to take to construct an angle bisector and a perpendicular bisector of a line. What's the same and what's different?

I have become **fluent** in…	I have developed my **reasoning** skills by…	I have been **problem-solving** through…
■ working out missing angles on straight lines and at a point ■ working out missing angles in parallel lines ■ finding missing angles in triangles, quadrilaterals and other polygons ■ identifying special quadrilaterals ■ using mathematical equipment accurately.	■ justifying my answers in angle problems, using precise mathematical language ■ identifying shapes from their properties ■ deducing information from geometrical diagrams ■ investigating relationships between angles ■ writing formal geometric proofs. Ⓗ	■ breaking larger problems down into smaller steps ■ interpreting complex diagrams ■ interpreting problems in a wide variety of forms ■ performing and interpreting geometrical constructions. Ⓗ

Check my understanding

1 Work out the values of a, b and c

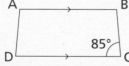

2 ABCD is a trapezium. Explain why ∠ABC must be 95°

3 Work out the size of the angle labelled x
Give a reason for each stage of your working.

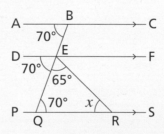

4 The interior angles of a regular polygon are 168° How many sides does the polygon have?

5 **a** Construct a triangle ABC with AB = 9 cm, BC = 7 cm and AC = 6 cm

    b Find the point where the perpendicular bisector of AB meets the angle bisector of ∠ACB Ⓗ

6 Prove that $x = a + b$ Ⓗ

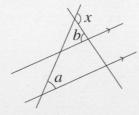

14 Area of trapezia and circles

In this block, I will learn...

how to calculate the area of rectangles, triangles and parallelograms

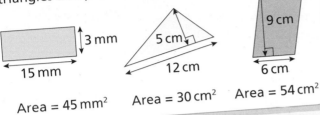

3 mm

15 mm

Area = 45 mm²

5 cm

12 cm

Area = 30 cm²

9 cm

6 cm

Area = 54 cm²

how to calculate the area of a trapezium

3 cm

5 cm

7 cm

Area = 25 cm²

1 cm

6 cm

6 cm

Area = 21 cm²

to find the area of a circle with and without a calculator

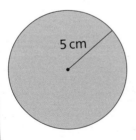

5 cm

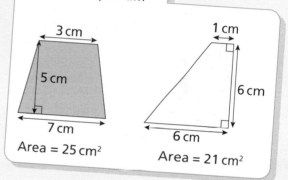

Area = πr^2

= $\pi \times 5^2$

= $\pi \times 25$

= 25π cm² (without a calculator)

= 78.54 cm² (with a calculator)

how to calculate the area and perimeter of a compound shape

3 cm

9 cm

5 cm

4 cm

Area = 51 cm²

Perimeter = 36 cm

Small steps

■ Calculate the area of triangles, rectangles and parallelograms ®

■ Calculate the area of a trapezium

Key words

Area – the space inside a 2-D shape

Formula – a rule connecting variables written with mathematical symbols

Perpendicular height – the height of a shape measured at a right angle to the base

Are you ready?

1 Write down the mathematical name for each shape.

a

b

c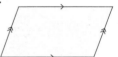

2 These rectangles are drawn on a centimetre squared grid.

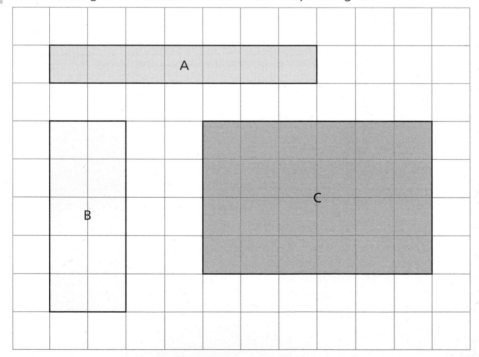

Work out the area of each rectangle.

3 Work out these calculations.

a 4×9

b $\dfrac{3 \times 6}{2}$

c $\dfrac{1}{2} \times 3 \times 6$

d $5 \times \dfrac{1}{2} \times 8$

Models and representations

Geoboards

You can make shapes on a geoboard to see how the dimensions are linked to the area.

In this chapter, you will work out the **areas** of triangles and quadrilaterals.

When working out the areas of triangles and quadrilaterals you should always use the **perpendicular height**. This is always at right angles to one of the sides.

The perpendicular height can be outside the shape...

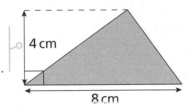

4 cm

8 cm

...or inside the shape.

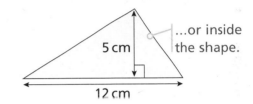

5 cm

12 cm

Example 1

Find the area of each shape.

a

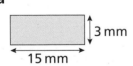

3 mm

15 mm

b

9 cm

6 cm

c

5 cm

12 cm

a   Area = $l \times h$

Area = 3×15

Area = $45\,\text{mm}^2$

The **formula** for the area of a rectangle is

area = length × height. This is also referred to as length × width.

b   Area = $b \times h$

Area = 6×9

Area = $54\,\text{cm}^2$

The formula for the area of a parallelogram is

area = base × perpendicular height.

c   Area = $\dfrac{b \times h}{2}$

Area = $\dfrac{12 \times 5}{2}$

Area = $\dfrac{60}{2}$

Area = $30\,\text{cm}^2$

The formula for the area of a triangle is

area = $\dfrac{\text{base} \times \text{perpendicular height}}{2}$

or area = $\frac{1}{2}$ base × perpendicular height.

Multiplying by $\frac{1}{2}$ is the same as dividing by 2

Practice 14.1A

1 Work out the area of each rectangle, triangle, parallelogram or square.

a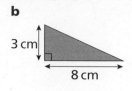
5 cm
2 cm

b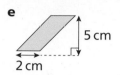
3 cm
8 cm

c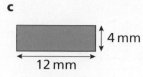
4 mm
12 mm

d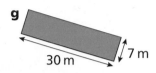
5 mm
4 mm

e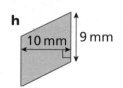
5 cm
2 cm

f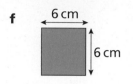
6 cm
6 cm

g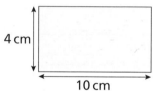
7 m
30 m

h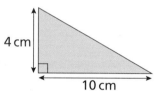
10 mm
9 mm

i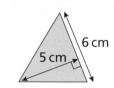
6 cm
5 cm

2 Chloe says, "The areas of the rectangle and the triangle are the same because the dimensions are the same."

4 cm
10 cm

4 cm
10 cm

Do you agree with Chloe? Explain your reasons.

3 **a** Work out the area of each shape.

i
6 cm
20 cm 20 cm
6 cm

ii
12 cm
5 cm
13 cm

iii
8 m
11 m 9 m 11 m
8 m

b What mistakes do you think people might make when working out areas of shapes like those in part **a**?

4 Work out the area of each rectangle.

a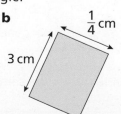
0.5 cm
2 cm

b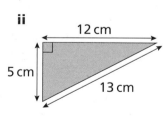
$\frac{1}{4}$ cm
3 cm

c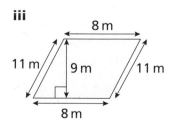
5 cm
x cm

5 a Work out the area of each triangle. Give your answers in cm²

i

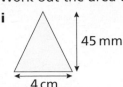

45 mm

4 cm

ii

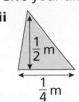

$\frac{1}{2}$ m

$\frac{1}{4}$ m

iii

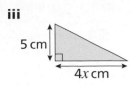

5 cm

4x cm

b Compare your strategy with a partner.

6 Find the missing lengths in these rectangles and parallelograms.

a

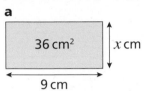

36 cm² x cm

9 cm

b

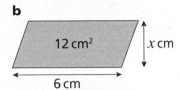

12 cm² x cm

6 cm

c

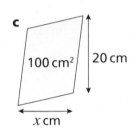

100 cm²   20 cm

x cm

d

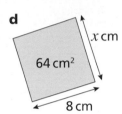

x cm

64 cm²

8 cm

7 Samira is working out the height of this triangle. Here is her working.

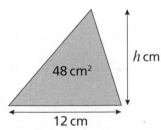

h cm

48 cm²

12 cm

$$A = \frac{b \times h}{2}$$

$$48 = \frac{12 \times h}{2}$$

$$96 = 12 \times h$$

$$8 = h$$

a Explain Samira's method.

b Show that Samira is correct.

8 Find the missing lengths of these triangles.

a

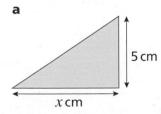

5 cm

x cm

b

x cm

13 cm

c

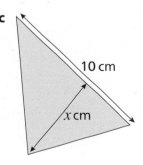

10 cm

x cm

Area = 35 cm²

Area = 13 cm²

Area = 62 cm²

What do you think? 💭

1 Sketch a triangle that has the same area as this parallelogram.

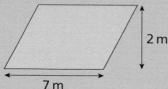

2 m

7 m

Compare your answer with a partner. Are they the same or different?

2 Four identical rectangles are placed together to make one large rectangle.

The area of the large rectangle is 192 cm²

The length of one of the smaller rectangles is 3 times the width.

Work out the length and width of the large rectangle.

The formula for the area of a trapezium is

area $= \frac{1}{2}(a + b)h$

where a and b are the parallel sides and h is the perpendicular height.

You can understand this formula by thinking about the area of a parallelogram.

If you place two congruent trapezia next to each other like this, you get a parallelogram.

The area of a parallelogram is $b \times h$

so you can write this as $(a + b) \times h$

If you halve the shape, you will get back to the original trapezium.

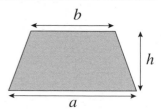

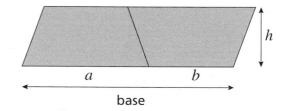

base

Example 2

Find the area of the trapezium.

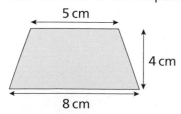

5 cm

4 cm

8 cm

Area $= \frac{1}{2}(a + b)h$ ⊸——⊣ The formula for the area of a trapezium is area $= \frac{1}{2}(a + b)h$

Area $= \frac{1}{2} \times (5 + 8) \times 4$ ⊸—⊣ Substitute the values into the formula.

Area $= \frac{1}{2} \times 13 \times 4$

Area $= 26$ cm² ⊸————⊣ Don't forget the units in your answer.

Practice 14.1B

1 Write down the letters that represent the lengths of the parallel sides in each trapezium.

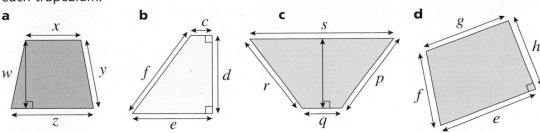

a   **b**   **c**   **d**

2 Work out the area of each trapezium.

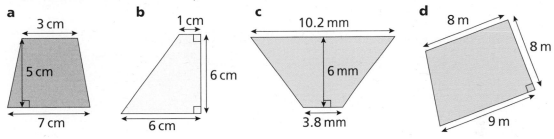

a   **b**   **c**   **d**

3 A trapezium has a perpendicular height of 7 cm and parallel sides of 2 cm and 6 cm. Work out the area of the trapezium.

4 Work out the area of each trapezium.

a

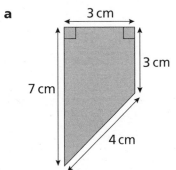

b

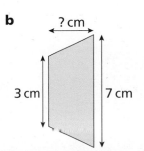

5 The area of each trapezium is 30 cm²

Find each perpendicular height.

a

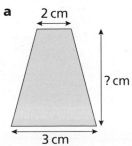

b

What do you think?

1 A trapezium has an area of 80 cm². What could be the dimensions?
How many possibilities can you find? You may find it useful to sketch your answers.

2 Benji works out the area of this trapezium by splitting it into a triangle and a rectangle.

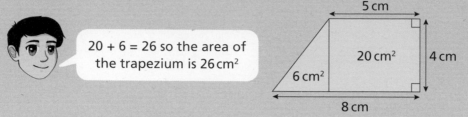

20 + 6 = 26 so the area of the trapezium is 26 cm²

a Use the formula for the area of a trapezium to check that Benji is correct.

b Explain why Benji's method wouldn't work for this example.

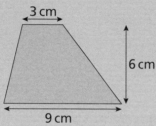

Consolidate – do you need more?

1 Work out the area of each shape.

a
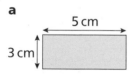

b
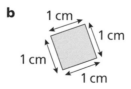

c
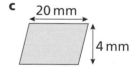

d
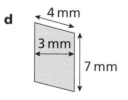

e

f
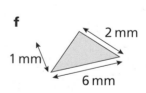

g
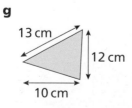

h
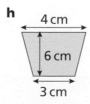

i
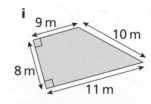

2 Copy and complete the table.

Shape	Base (cm)	Perpendicular height (cm)	Area (cm²)
Rectangle	5	7	
Triangle	5	7	
Triangle	9	4	
Parallelogram	5	7	
Parallelogram	9	4	
Rectangle	20		100
Parallelogram		12	36
Triangle		11	66

Stretch – can you deepen your learning?

1 Work out the length of each unknown side.

a

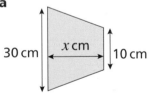

b

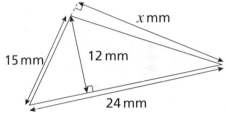

2 The area of each trapezium is given. Work out each missing length.

a

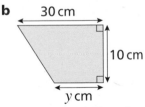

Area = 210 cm²

b

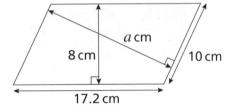

Area = 210 cm²

c

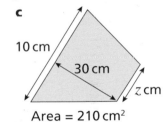

Area = 210 cm²

3

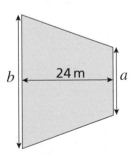

Area = 420 m²

a Suggest two possible pairs of values for a and b.

b What do you know about a and b? Generalise your findings.

4 In a trapezium, one of the parallel sides is 4 cm longer than the other.

The perpendicular height is 2 cm longer than the shorter parallel side.

The length of the shorter parallel side is x cm.

The area of the trapezium is 49 cm^2

a Show that $x^2 + 4x - 45 = 0$

b Ed finds that $(x + 2)^2 = 49$

Use Ed's working to find the value of x

Reflect

1 Explain why the area of a triangle is given by $A = \frac{1}{2}bh$

2 Sketch and label a rectangle, parallelogram, triangle and trapezium where each has an area of 20 cm^2

14.2 Area of a circle

Small steps

- Investigate the area of a circle
- Calculate the area of a circle and parts of a circle without a calculator
- Calculate the area of a circle and parts of a circle with a calculator

Key words

Sector – a part of a circle formed by two radii and a fraction of the circumference

Radius – the distance from the centre of a circle to a point on the circle

Diameter – the distance from one point on a circle to another point on the circle through the centre

Pi – pronounced "pie" and written using the symbol π. It is the ratio of the circumference of a circle to its diameter

Are you ready?

1 Make a copy of the diagram. Use the words to label the parts of a circle.

radius   diameter   circumference   centre

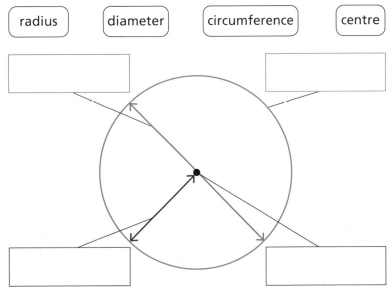

2 Which of these circles show a radius?

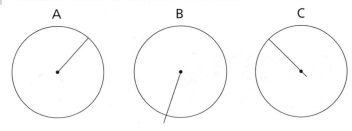

3 Which of these circles show a diameter?

A   B   C

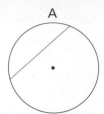 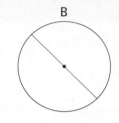 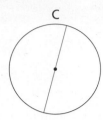

4 Copy and complete these sentences.

a The radius is _____ the diameter.

b The diameter is double the _____.

5 Copy and complete the table.

Radius	Diameter
4 cm	
	100 mm
	25 cm
3.4 m	

6 Round each of these numbers to

  i 1 decimal place   **ii** 2 decimal places.

a 47.28467   **b** 0.949   **c** 107.9995

Models and representations

You can investigate the area of a circle by using the formula for the area of a parallelogram.

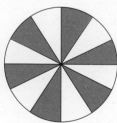

Split the circle into **sectors**.

h

l

Place the sectors side by side so the shape looks like a parallelogram.

The formula for circumference is
$c = \pi d$ or $c = 2\pi r$

The more sectors the circle is split into, the closer it becomes to a parallelogram.

The height represents the **radius** of the circle so it can be labelled r

r

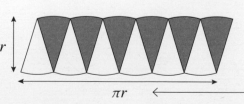

πr

The length represents half of the circumference so it can be labelled πr

The area of a parallelogram is $l \times h$

Therefore the area of a circle is $\pi r \times r$ or πr^2

In this chapter, you will learn how to find the area of a circle. ○ | You will need to use the
formula: area of a circle = πr^2

Example 1

1 Work out the area of each circle. Give your answers in terms of π

a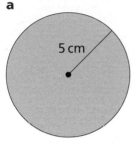

5 cm

b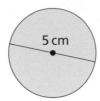

5 cm

a Area = πr^2 ○──── | Remember: with algebra, when a number is
written next to a symbol it means multiply.

Area = $\pi \times 5^2$

Area = $\pi \times 5 \times 5$

Area = $\pi \times 25$ ○ | When writing your answer in terms of π you should
always write the number before the symbol.

Area = **$25\pi\,cm^2$** ○──── | Make sure you include units with your answer.

b Area = πr^2 ○──── | You are given the **diameter** so first you need find the radius.

Area = $\pi \times 2.5^2$ ○──── | 5 ÷ 2 = 2.5, so the radius is 2.5 cm

Area = $\pi \times 2.5 \times 2.5$

Area = $\pi \times 6.25$ | Your answer could also be
written as $6\frac{1}{4}\pi\,cm^2$

Area = **$6.25\pi\,cm^2$**

Practice 14.2A

1 Find the area of each circle. Give your answers in terms of π

a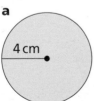

4 cm

b

1 m

c

13 mm

d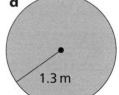

1.3 m

2 Ed says, "The area of the circle is $49\pi\,m^2$."

    a Explain Ed's mistake.

    b In terms of π, what is the area of the circle?
Write your answer as a fraction and as a decimal.

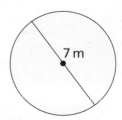

7 m

3 Find the area of each circle. Give your answers in terms of π

a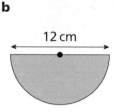
20 mm

b
1 m

c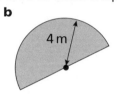
22 cm

d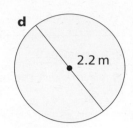
2.2 m

4 Work out the area of each shape. Give your answers in terms of π

a
12 cm

b
12 cm

c
6 cm

Compare your method with a partner.

5 Find, in terms of π, the area of these shapes.

a
4 m

b
4 m

c
0.4 m

6 Here are the areas of some circles.

    i 9π cm²   **ii** 16π cm²   **iii** 100π cm²   **iv** 36π cm²

    a What do you notice about the numbers?

    b Find the radius of each circle.

What do you think?

1 A logo is made up of a blue rectangle and a white circle.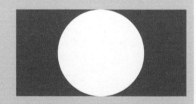

The rectangle has a width of 3 cm and a length of 7 cm

Find, giving your answers in terms of π, the area of the logo that is

a white   **b** blue.

2 The radius of a circle is $7y$ cm

Write the area of the circle in terms of y and π

3 The circumference of a circle is 12π cm. Show that the area of the circle is 36π cm²

Example 2

1 Work out the area of each of these shapes. Give your answers to 1 decimal place.

a

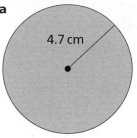

4.7 cm

b

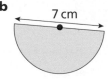

7 cm

a   Area = πr^2

Area = $\pi \times 4.7^2$

Area = $\pi \times 22.09$

Area = $69.397781\ldots$

Area = 69.4 cm² (1 dp)

> You will need a calculator for these questions.
>
> Always write down each step of your working.
>
> Round your answer to the requested degree of accuracy.

7 cm

> To work out the area of a semicircle, first imagine it as a full circle.

b   Area = πr

Area = $\pi \times 3.5^2$

Area = $\pi \times 12.25$

Area = $38.4845100\ldots$

Area of the semicircle = $19.242255\ldots$

Area of the semicircle = 19.2 cm² (1 dp)

> The diameter is given, so first work out the radius.
>
> 7 ÷ 2 = 3.5, so the radius is 3.5
>
> When you have worked out the area of the full circle, halve it to find the area of the semicircle.
>
> Make sure you do not round any answers until the very end of your working.

Practice 14.2B

1 Find the area of each circle. Give your answers to 1 decimal place.

a

23 cm

b

1 m

c

17.2 mm

d

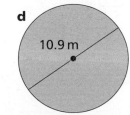

10.9 m

2 Work out the area of each of these semicircles. Give your answers to 1 decimal place.

a
15 m

b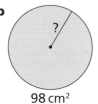
4 cm

c
54.3 mm

3 Four friends share a 12-inch pizza equally.

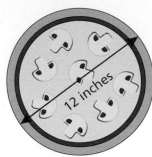

12 inches

What is the area of the piece of pizza each friend gets?
Give your answers to 1 decimal place.

4 Find the radius of each of these circles. Give your answers to 1 decimal place.

a
?
70 cm²

b
?
98 cm²

c
?
0.45 m²

5 Draw a circle with an area of 28.27 cm²

6 A pond has an area of 10 m²

Work out the diameter of the pond.

7 Copy and complete the table.

Radius (cm)	Diameter (cm)	Area in terms of π (cm²)	Area to 1 dp (cm²)
	17		
		49π	
			237.8

What do you think?

1 Determine which deal is better value for money.

Deals

Deal 1   1 × 18 inch pizza
£9.99   2 litres lemonade

Deal 2   2 × 9 inch pizzas
£7.99   2 litres lemonade

2 A circular sticker is placed on a square as shown.

a Show that, when $a = 18$ cm, the sticker covers more than 75% of the square.

b Investigate this for other values of a. Explain your findings.

a cm

3 The centre of a circle has coordinates $(-3, -5)$

$(-3, 2)$ is a point on the circumference of the circle.

Work out the area of the circle.

4 **a** The circumference of a circle is 100 cm. Find the area of the circle.

b The area of a circle is 100 cm². Find the circumference of the circle.

c Can the area and the circumference of a circle have the same numerical value? Justify your answer.

Consolidate – do you need more?

1 Work out the area of each circle. Giving your answers

i in terms of π **ii** to 1 decimal place.

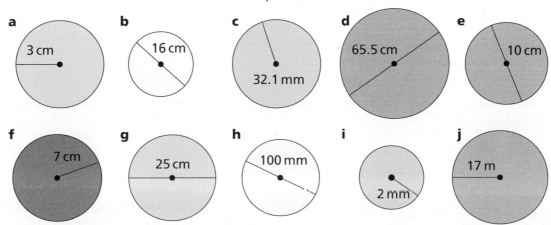

a 3 cm

b 16 cm

c 32.1 mm

d 65.5 cm

e 10 cm

f 7 cm

g 25 cm

h 100 mm

i 2 mm

j 17 m

2 Work out the area of each of these. Give your answers to 1 decimal place.

a

10 m

b

6 cm

c

14.2 m

d

5 m

e

8 cm

f

50 mm

Stretch – can you deepen your learning?

1 The floor in an art gallery is to be painted blue.

The diagram shows the region of the floor to be painted.

A tin of paint covers 6.4 m²

Each tin of paint costs £12.95

Work out how much it costs to paint the floor.

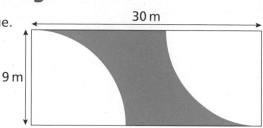

30 m

9 m

2 A shape is made up of two circles that have the same centre.

The diameter of the large circle is 12 cm

The area of the shaded region is 50 cm²

Find the diameter of the smaller circle.

3 The ratio of the area of the rectangle to the area of the circle is $1 : \pi$

The ratio of the length to the width of the rectangle is $4 : 1$

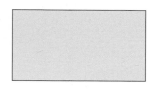

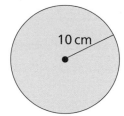

10 cm

a Work out the area of the rectangle.

b Work out the perimeter of the rectangle.

Reflect

1 When finding areas of circles, explain why it is more accurate to leave your answers in terms of π rather than working them out on a calculator.

2 What do you think are the most common mistakes when working out the area of a circle?

14.3 Area of compound shapes

Small steps

■ Calculate the perimeter and area of compound shapes

Key words

Compound shape – also known as a composite shape, this is a shape made up of two or more other shapes

Are you ready?

1 Work out the area of each shape.

a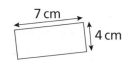
7 cm
4 cm

b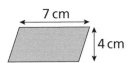
7 cm
4 cm

c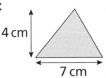
4 cm
7 cm

d
3 cm
4 cm
4 cm

e Which shapes have the same area? Why?

f Which shape can you work out the perimeter of?

2 Work out the area of each shape. Give your answers to 1 decimal place.

a
8 cm

b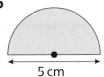
5 cm

c
7.5 cm

3 Find the missing value in each diagram.

a

3	21
?	

b

17	
9	?

c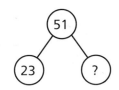
51
23
?

Models and representations

Colour

Colour coding side lengths will help you to see which side lengths are linked and to work out missing lengths before finding the area or perimeter.

The two shorter vertical blue side lengths add together to make the longer vertical blue side.

The two shorter horizontal yellow side lengths add together to make the longer horizontal yellow side.

In this chapter, you will find the area and perimeter of **compound shapes**.

Remember, the perimeter is the distance around the outside of a shape. When dealing with compound shapes, you may need to work out some missing lengths before you can find the perimeter.

Example 1

Work out the area of this compound shape.

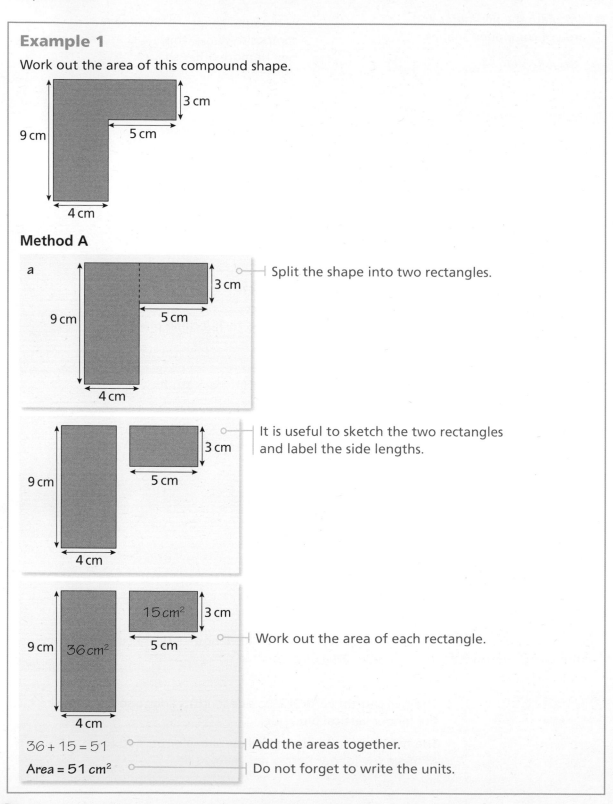

Method A

a   Split the shape into two rectangles.

It is useful to sketch the two rectangles and label the side lengths.

Work out the area of each rectangle.

$36 + 15 = 51$ Add the areas together.

Area = $51 \, cm^2$ Do not forget to write the units.

Method B

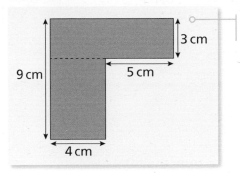

You could have split the shape up horizontally instead of vertically.

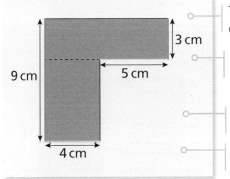

This makes finding the area of each rectangle more difficult as you do not have all the side lengths.

The compound shape has a vertical side length of 9 cm and a shorter vertical side length of 3 cm

The missing vertical side = 9 − 3 = 6 cm
The missing horizontal side = 4 + 5 = 9 cm

As the two shorter sides add together to make the longer side, you can work out the missing length by subtraction.

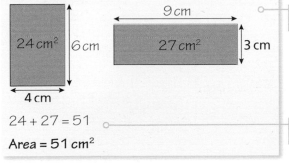

You can then work out the area of each rectangle.

24 + 27 = 51

Area = 51 cm²

Finally add the areas together to get the total area of the compound shape.

Method C

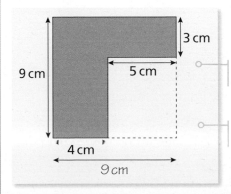

You could think of the compound shape as a complete rectangle and then subtract the area of the blank space.

If it were complete, it would measure 9 cm by 9 cm giving an area of 81 cm²

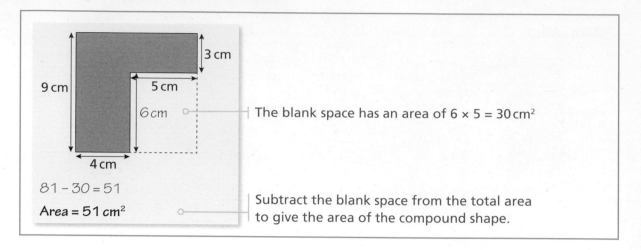

The blank space has an area of 6 × 5 = 30 cm²

81 − 30 = 51

Area = 51 cm²

Subtract the blank space from the total area to give the area of the compound shape.

Practice 14.3A

1 Find the all missing side lengths of each compound shape.

a

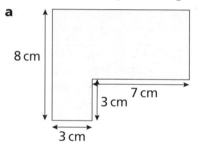

b

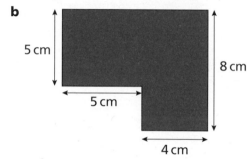

c

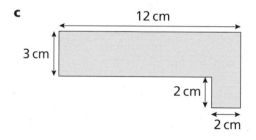

d

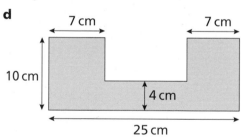

e Find the perimeter of each shape.

2 Find the area of each compound shape.

a

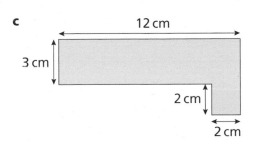

b

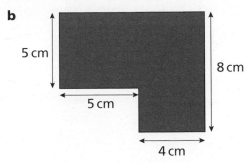

c

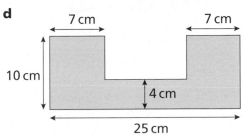

d

Compare your methods with a partner. Did you work them out in the same way?

3 Faith says, "I am going to work out the area by splitting the shape into 3 rectangles, like this."

Chloe says, "I think there is a quicker way."

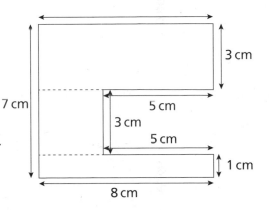

a Which method do you think Chloe is thinking of?

b Work out the area of the compound shape.

c Explain why you could not find the perimeter of the shape by adding the perimeters of the three rectangles.

4 Find the area of each of these shaded regions.

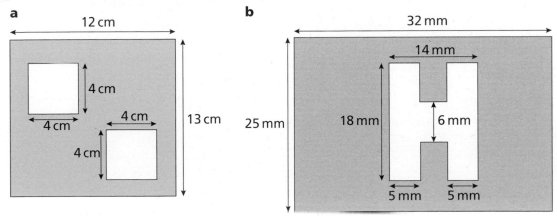

a

b

c What do you think might be meant by the perimeter of each region? Which sides would you include and exclude, and why?

What do you think? 💭

1 Three rectangles, A, B and C, are joined together to form a compound shape.

The length and width of rectangle B are each 3 cm more than the length and width of rectangle A.

Rectangle C is an enlargement of A by scale factor 3

Work out the area of the compound shape.

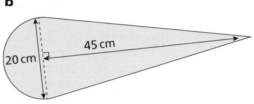

2 Is it possible to work out the perimeter of the compound shape in question **1**? Why or why not?

Example 2

Find the area of each compound shape.

a

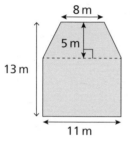

b

This compound shape is made up of a rectangle and a trapezium.

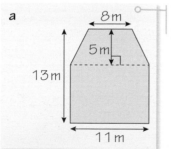

First split up the shape and find any missing lengths.

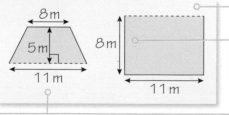

The height of the rectangle must be 8 m because the total height of the shape is 13 m and the height of the trapezium is 5 m: 13 − 5 = 8

The base of the trapezium must be 11 m as it is opposite the base of the rectangle.

Area of trapezium $= \frac{1}{2}(a + b)h$

$= \frac{1}{2}(8 + 11)5$

$= 47.5\,\text{m}^2$

Area of rectangle = length × width

$= 11 × 8$

$= 88\,\text{m}^2$

Total area $= 47.5 + 88 = 135.5\,\text{m}^2$

Work out the area of the trapezium and the rectangle separately and then add them together.

b

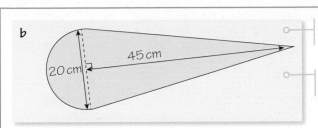

This compound shape is made up of a semicircle and a triangle.

Work out the area of each shape individually and then add them together.

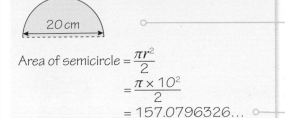

You have been given the diameter so you need to half it to find the radius: $20 \div 2 = 10$

$$\text{Area of semicircle} = \frac{\pi r^2}{2}$$

$$= \frac{\pi \times 10^2}{2}$$

$$= 157.0796326\ldots$$

Do not round your answer here as you are not at the end of the question yet.

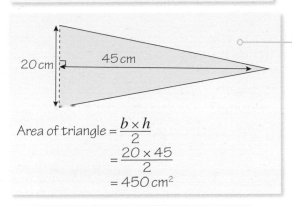

Find the area of the triangle.

$$\text{Area of triangle} = \frac{b \times h}{2}$$

$$= \frac{20 \times 45}{2}$$

$$= 450\,\text{cm}^2$$

Total area $= 157.0796326\ldots + 450$

$$= 607.0796326\ldots$$

$$= 607\,\text{cm}^2 \text{ (3 sf)}$$

Round the final answer to a reasonable degree of accuracy. This is usually to 3 significant figures.

Practice 14.3B

1. Work out the area of each shape. You may find it useful to sketch each diagram.

a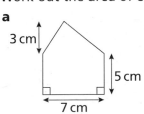

b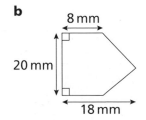

c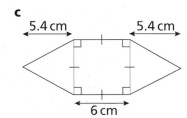

2 Jakub and Marta have split this shape differently to find the area.

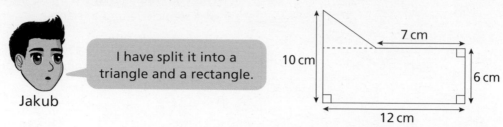

Jakub

I have split it into a triangle and a rectangle.

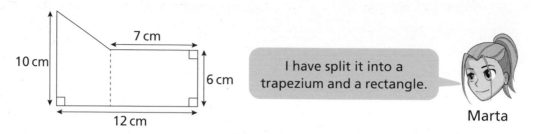

Marta

I have split it into a trapezium and a rectangle.

a Use both Jakub's and Marta's methods to find the area.

b Which method did you prefer? Why?

3 Work out the area of each of these compound shapes.

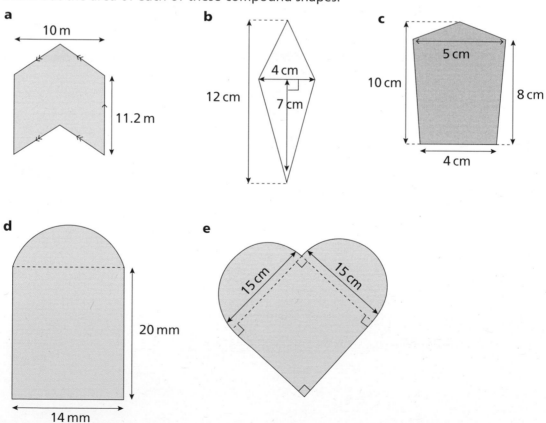

④ Here is the plan view of Mr White's rectangular garden.

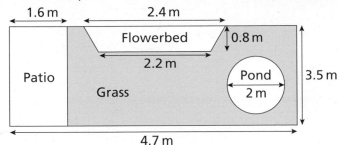

Mr White is going to buy turf to cover the region labelled "Grass". He wants to know the area of the grass so that he knows how much turf to buy.

a Work out the area of the part of the garden labelled "Grass".

b Turf costs £12 per whole square metre. Calculate how much Mr White will pay for the turf.

c Mr White hires a gardener to turf his garden and plant some flowers. The gardener charges £200 for the first day of work and then gives a discount of 15% for any additional days. It takes the gardener 3 days to do the work. Work out how much Mr White will pay the gardener.

What do you think?

① Here is a regular hexagon.

In how many different ways could you work out the area of the hexagon? Compare your answer with a partner.

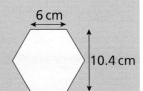

② What percentage of this shape is shaded?

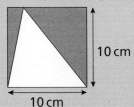

Did you need to work out the area of each shape?

③ The perimeter of a single shape is 100 cm. What might the area of the shape be? Sketch your shapes to justify your answers.

④ The perimeter of a compound shape is 100 cm. What might the area of the shape be? Sketch your shapes to justify your answers.

Consolidate – do you need more?

1 Find the area of each of these compound shapes.

a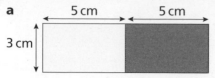

b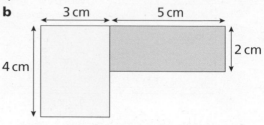

c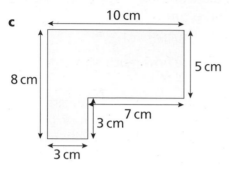

d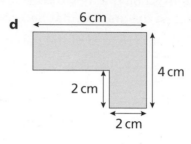

e Find the perimeter of each shape.

2 Work out the area of each of these shapes.

a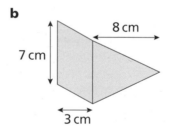

b

3 The diagram shows a pond and a circular path.

a Find the area of the pond.

b Work out the area of the path.

c A fence is built on the inside and outside of the path. Find the total length of fencing needed.

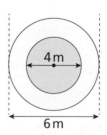

Stretch – can you deepen your learning?

1 What percentage of this shape is shaded? Give your answer to 1 decimal place.

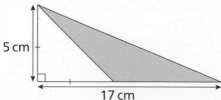

2

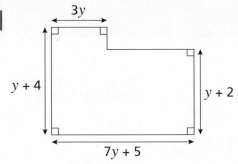

Find a formula for the area of the compound shape in terms of y

Write your formula as simply as possible.

3 A shape consists of a semicircle of radius x cm and a rectangle x cm by $2x$ cm.

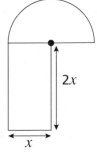

    a Find the area of the shape in terms of x and π

    b Find the perimeter of the shape in terms of x and π

    c Investigate how the area and perimeter of the shape would change if the semicircle and rectangle were in different positions. Here are two examples.

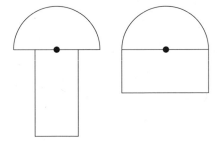

Reflect

Is this statement true or false? "There is only ever one way to split up a compound shape." Give examples to support your answer.

I have become **fluent** in…	I have developed my **reasoning** skills by…	I have been **problem-solving** through…
■ calculating the area of rectangles, triangles and parallelograms ■ calculating the area of trapezia ■ calculating the area of circles ■ calculating the perimeter of compound shapes.	■ finding and explaining how to find missing lengths in compound shapes ■ investigating the area of a circle ■ exploring different ways of splitting compound shapes.	■ finding missing lengths given the area of a shape ■ calculating with area in different mathematical contexts ■ finding shapes that satisfy given information ■ working with both perimeter and area.

Check my understanding

1 Find the area of each shape.

a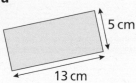

5 cm
13 cm

b

7 cm
6 cm

c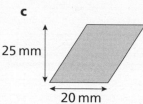

25 mm
20 mm

d

6 m
2.5 m
4 m

e Explain why you can only find the perimeter of one of the shapes.

2 Find the area of each circle. Give your answers

  i in terms of π **ii** to 1 decimal place.

a

7 cm

b

9 cm

3 Work out
  a the area in cm²
  b the perimeter in cm

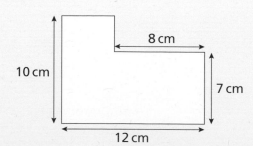

8 cm
10 cm
7 cm
12 cm

15 Line symmetry and reflection

In this block, I will learn...

to recognise lines of symmetry

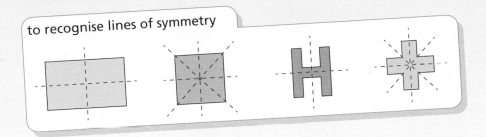

how to reflect shapes in horizontal and vertical lines

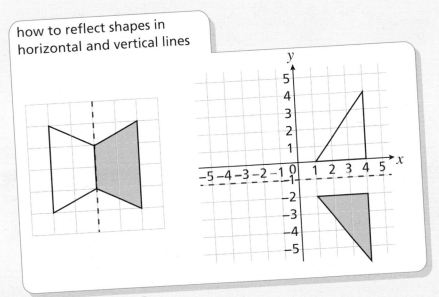

how to reflect shapes in diagonal lines

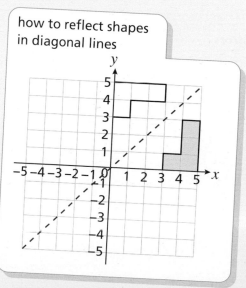

Small steps

■ Recognise line symmetry

Key words

Line of symmetry – a line that cuts a shape exactly in half

Polygon – a closed 2-D shape with straight sides

Regular – a shape that has equal sides and equal angles

Symmetrical – when one half of a shape is the mirror image of the other

Are you ready?

1 **a** Write the mathematical name of each polygon.

i  **ii** **iii** **iv** **v** **vi**

b Which polygons are regular? **c** Which polygons are isosceles?

2 Which butterfly has a symmetrical pattern?

A  B  C

Discuss how you know.

Models and representations

Mirrors

Mirrors are great for helping you see the symmetry of a shape.

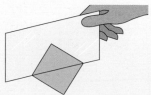

Paper folding

You can cut out a shape and fold it to see if it has one or more **lines of symmetry**.

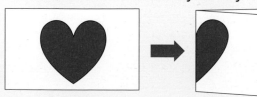

Example 1

How many lines of symmetry does each shape have?

a

b

a Two lines of symmetry
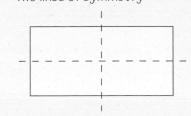

You can check the lines of symmetry of a rectangle by folding a piece of A4 paper.

A common mistake is thinking that a rectangle has four lines of symmetry.

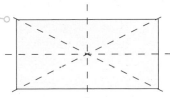

If you fold a rectangular sheet of paper, you will see that the two triangles formed don't fit exactly on top of each other. Therefore, the diagonals are not lines of symmetry.

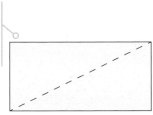
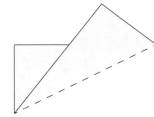

b No lines of symmetry

The triangle has no lines of symmetry. It is not possible to fold it along any one line to create two identical shapes. This is because it is a scalene triangle, all the sides are different lengths.

Practice 15.1A

1 Here are some shapes.

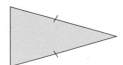

Write the name of the shape with

a exactly one line of symmetry

b exactly four lines of symmetry

c no lines of symmetry

d an infinite number of lines of symmetry.

2 For which of these shapes is the dashed line a line of symmetry?

A   B   C   D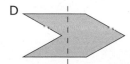

Explain how you know.

3 How many lines of symmetry does each of these shapes have?

a   b   c   d

What do you notice?

4 Beca says, "This hexagon has six sides so it must have six lines of symmetry."

Explain why Beca is wrong.

How many lines of symmetry does the hexagon have?

5 How many lines of symmetry does each of these shapes have?

    a   a semicircle   **b**   a regular octagon   **c**   an isosceles trapezium

You may find it useful to sketch the shapes.

6 A triangle has the following angles.

26°   77°   77°

    a   How many lines of symmetry does the triangle have?

    b   How do you know?

What do you think?

1 Ed says, "All pentagons have five lines of symmetry."

Chloe says, "I disagree. It depends on what type of pentagon it is."

Draw a diagram to show that Chloe is correct.

2 Jakub has drawn a pattern on a square to change the number of lines of symmetry it has.

    a   Do you agree with Jakub?

    b   Draw some patterns on your own squares so that they have

Now my square has only one line of symmetry.

        i   1 line of symmetry   **ii**   2 lines of symmetry.

3 Here are three logos.

Abdullah says, "They all have an infinite number of lines of symmetry because they are circles."

Do you agree with Abdullah?

A   B   C

White Rose Maths

Consolidate – do you need more?

1 Which of these shapes have at least one line of symmetry?

A    B   C   D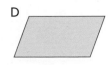

E   F 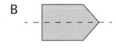   G   H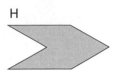

2 For which of these shapes is the dotted line a line of symmetry?

A   B   C   D

E   F    G   H

Compare your answers with a partner.

3 Here are some shapes.

A    B   C   D

E   F   G   H

Copy and complete the table to sort the shapes.

	No lines of symmetry	One line of symmetry	More than one line of symmetry
Triangle			
Quadrilateral			

Stretch – can you deepen your learning?

1 On a squared grid, copy and complete each pattern to make it symmetrical.

a 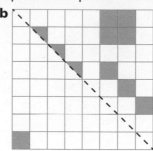 **b** 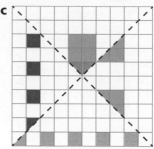 **c**

Create your own patterns on a squared grid that have **i** one, **ii** two, **iii** three, **iv** four lines of symmetry. Are any of these not possible? Explain why.

2 Investigate the lines of symmetry in different types of triangle and summarise your findings.

3 How many lines of symmetry does each of these shapes have?

a a regular 5-sided polygon

b a regular 6-sided polygon

c a regular 10-sided polygon

d a regular n-sided polygon

4 Three of the angles in a quadrilateral are 52°, 129° and 51°. How many lines of symmetry does the quadrilateral have? Explain how you know.

Reflect

1 In your own words, explain what is meant by a "line of symmetry".

2 Draw a shape that has one line of symmetry and another shape with no lines of symmetry.

Small steps

■ Reflect a shape in a horizontal or vertical line

Key words

Congruent – exactly the same size and shape, but possibly in a different orientation

Reflection – a transformation resulting in a mirror image

Vertex (plural: **vertices**) – a point where two line segments meet; a corner of a shape

Are you ready?

1 Which of these diagrams show a correct line of symmetry?

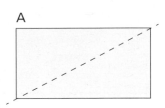

A

B

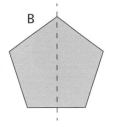

C

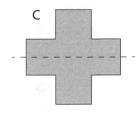

2 Write down the coordinates of A, B, C and D.

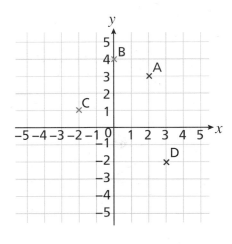

3 Some lines are drawn on the grid.

   a Which lines are vertical?

   b Which line has equation $y = 2$?

   c Which line has equation $x = 2$?

   d Write the equation of line D.

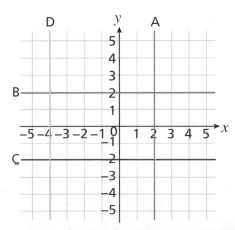

Models and representations

Reflections

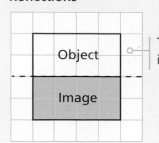

The starting shape is called the object.

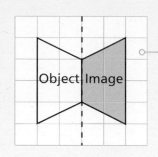

The reflected shape is called the image.

In this chapter, you will learn how to reflect shapes in horizontal and vertical lines.

Example 1

Reflect each shape in the given mirror line.

a

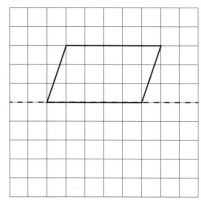

b

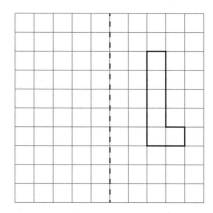

a

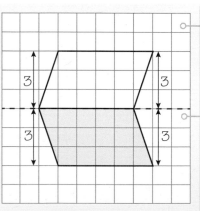

Count the squares vertically from each **vertex** to the mirror line and then count the same number of squares on the other side of the mirror line to find its image.

Join the new vertices to draw the image of the shape.

Two of the vertices of the parallelogram are on the mirror line. The other two are 3 squares away.

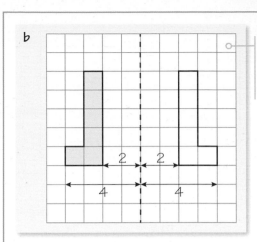

b

The hexagon is positioned two squares away from the mirror line. Therefore, the reflected image must also be two squares away from the mirror line, on the other side.

A reflected image is always **congruent** to the original. This means that it is exactly the same size and shape as the object. It is just in a different orientation.

Practice 15.2A

1 Copy each shape and mirror line onto squared paper and then draw the reflection.

a 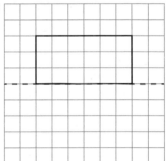   b 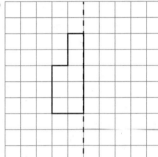   c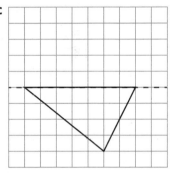

2 Copy each shape and mirror line onto squared paper and then draw the reflection.

a 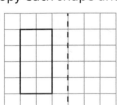   b 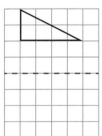   c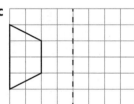

3 Which diagrams show a correct reflection in the mirror line?

A 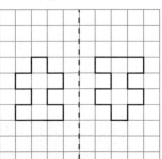   B 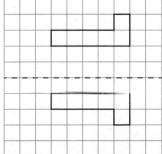   C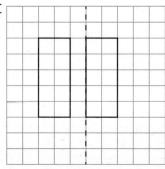

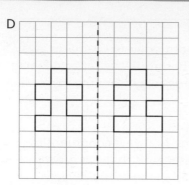

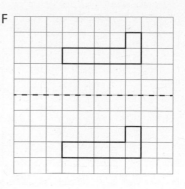

4 a Copy each diagram and reflect the shape in the mirror line.

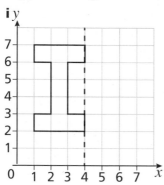

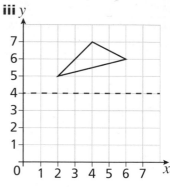

b Write the coordinates of the vertices of each of your reflected shapes.

5 For each diagram, write the coordinates of the vertices of the reflected image.

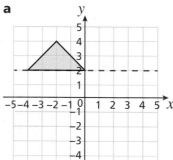

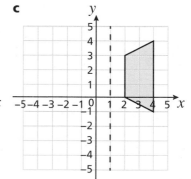

6 Benji says, "The mirror line for this reflection is the y-axis."
Beca says, "I think the mirror line is $x = 0$."

Who do you agree with?

Discuss your thinking with a partner.

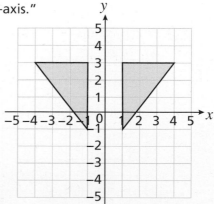

7 Write the equation of each mirror line for each reflection.

a

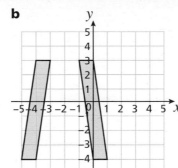

b

c

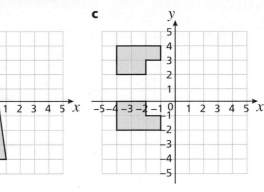

8 Amina has reflected shape A in the mirror line to form image B.
Explain the mistake Amina has made.

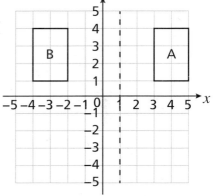

What do you think?

1 Chloe has drawn a rectangle with coordinates (–4, 1), (–4, 4), (0, 1) and (0, 4).

> I'm going to reflect my rectangle in the *x*-axis.

What are the coordinates of her new rectangle?

2 Copy this shape onto squared paper and reflect it in the mirror line.

Why is this reflection different from the ones you have drawn so far in this chapter?

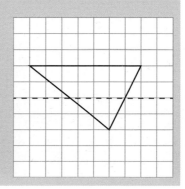

Consolidate – do you need more?

1 **a** Which of these diagrams show a correct reflection in the mirror line?

i 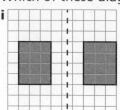 ii 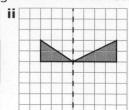 iii 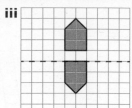 iv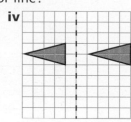

b For any incorrect reflections, explain the mistake that has been made.

2 Copy each pattern and mirror line onto squared paper then shade the reflected image.

a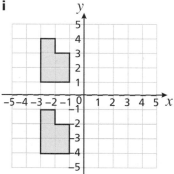

b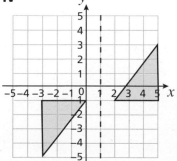

3 **a** Which of these show a correct reflection in the mirror line?

i 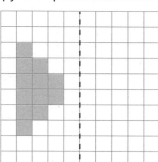 ii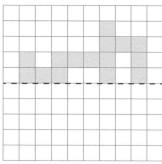

iii 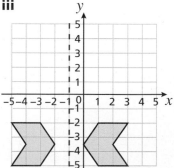 iv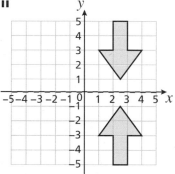

b For any incorrect reflections, explain the mistake that has been made.

Stretch – can you deepen your learning?

1 Filipo reflects this triangle in the line $x = -3$

He says that the coordinates of each vertex have stayed the same.

a Explain why Filipo thinks this.

b Explain why Filipo is mistaken.

c Find the equation of another mirror line such that exactly one vertex of ABC does not move.

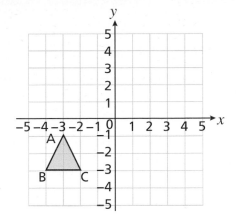

2 A trapezium has vertices at (–5, 3), (–2, 7), (1, 7) and (2, 3).

It is reflected in the line $y = 3$ so that the object and the image join to form a hexagon.

a Find the coordinates of the vertices of the hexagon.

b Work out the area of the hexagon.

3 A rectangle has vertices at (1, –4), (8, –4), (1, 2) and (8, 2). It is reflected in the line $y = 4$

Find the equations of the four lines that border the reflected image.

Reflect

What do you think are the most common mistakes made when reflecting a shape?

Small steps

■ Reflect a shape in a diagonal line

Key words

Diagonal – a line that is neither horizontal nor vertical; in quadrilaterals, a line segment that joins two opposite vertices

Are you ready?

1 Which diagrams show a correct line of symmetry?

A 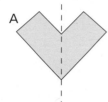   B    C

2 Copy and complete the table of values for the line $y = x$.

x	1	2	3	4	5
y					

3 Which of these points lie on the line $y = x$?

(3, 3)　　(5, −5)　　(−1, 1)　　(3, 4)　　(−6, −6)　　(1.5, 1.5)

Models and representations

Tracing paper

When reflecting shapes, tracing paper can be used as an alternative to counting squares.

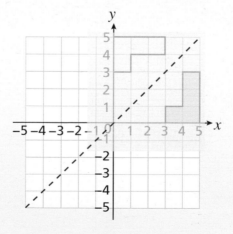

In this chapter, you will learn how to reflect shapes in a **diagonal** mirror line.

Tracing paper can be a useful tool to help you.

Example 1

Reflect the shape in the line $y = x$

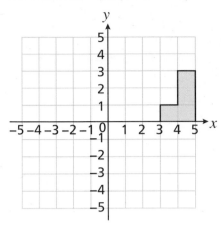

Method A

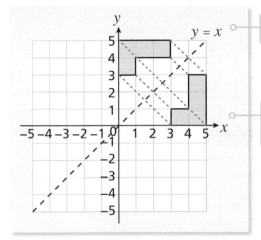

The line $y = x$ is a diagonal line that goes through (0, 0), (1, 1), (2, 2), and so on.

To reflect the shape, count from each vertex to the mirror line and then count the same distance on the other side of the mirror line.

Because the mirror line is diagonal, you should count diagonally through the squares. For example, from (3, 0) to (0, 3) is one and a half diagonals of the squares. The diagonal lines must be perpendicular to the mirror line.

Method B

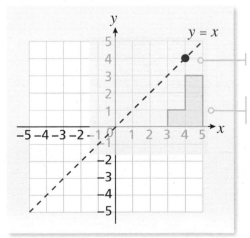

Draw a dot on the mirror line.

Place tracing paper on your diagram and trace the line, the dot and the shape.

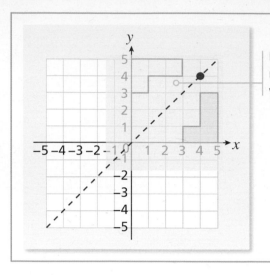

Flip the tracing paper over and line up the mirror line and the dot. The tracing paper will show you where to draw the reflection.

Practice 15.3A

1 Copy each shape and draw its reflection in the mirror line.

a

b

c

2 Copy each shape and draw its reflection in the mirror line.

a

b

c

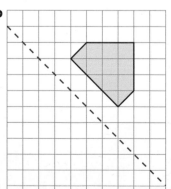

What's the same and what's different about questions **1** and **2**?

Which reflections were harder to draw?

3 Emily has reflected rectangle A in the given mirror line.

   a Explain the mistake she has made.

   b Draw the correct reflection.

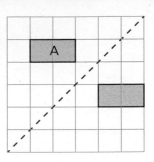

4 Which diagrams show a correct reflection in the mirror line?

For any that are incorrect, draw the correct reflection.

a

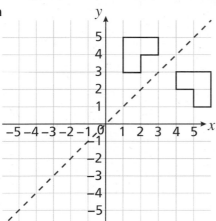

b

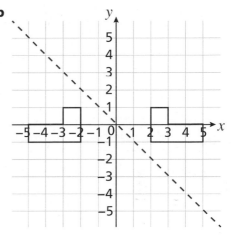

c

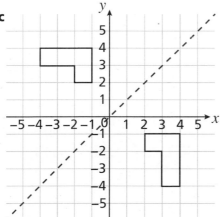

d

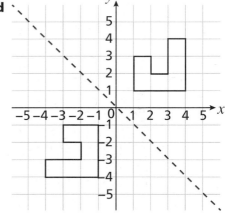

5 **a** Make a copy of this grid. Plot the points (0, 1), (−2, 1) and (−2, 5) and join them to form a triangle.

   b On the same grid, reflect the triangle in the line $y = x$

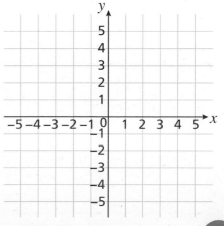

What do you think? 🌐

1 Jackson says, "If I reflect the rectangle in the line, it won't move."

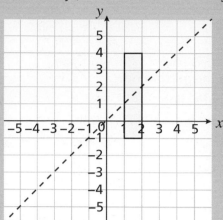

Do you agree with Jackson? Explain your reasons.

2 A rectangle with vertices at the points (–5, –1), (–5, 3), (–3, –1) and (–3, 3) is reflected in the line $y = -x$. Write the coordinates of the vertices of the reflected rectangle.

Consolidate – do you need more?

1 Copy each shape and reflect it in the given mirror line.

a 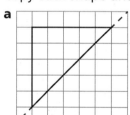 b 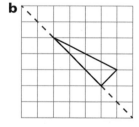 c 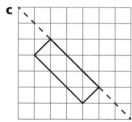 d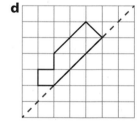

2 Copy each shape and reflect it in the given mirror line.

a 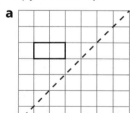 b 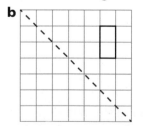 c 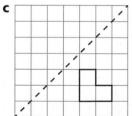 d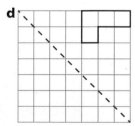

Stretch – can you deepen your learning?

1 Write the equation of the mirror line for each of these reflections.

a

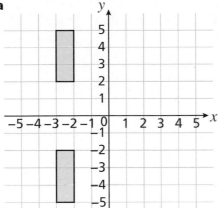

b

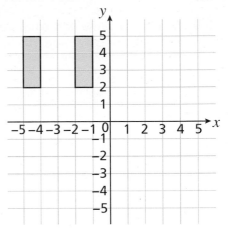

c

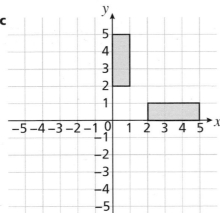

d

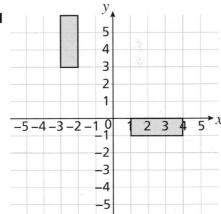

e

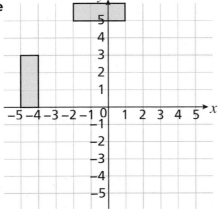

f

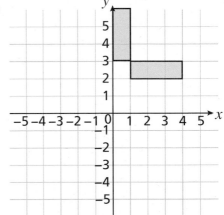

2 The line $y = 4$ is reflected in the line $y = x$.

Write the equation of the reflected line.

Investigate the equations of other lines reflected in the lines $y = x$ and $y = -x$

Reflect

1 In your own words, explain how to reflect a shape in a diagonal line.

2 What's the same and what's different between reflecting a shape in a diagonal line and reflecting a shape in a horizontal or vertical line?

15 Line symmetry and reflection

Chapters 15.1–15.3

White
Rose
Maths

I have become **fluent in...**	I have developed my **reasoning** skills by...	I have been **problem-solving** through...
■ recognising line symmetry ■ reflecting shapes in horizontal and vertical lines ■ reflecting shapes in diagonal lines ■ recognising the mirror line for a reflection.	■ exploring patterns with line symmetry ■ explaining misconceptions related to reflection ■ explaining where a line of reflection must be for given a shape and its reflection.	■ completing a shape, given the line of reflection ■ working with line symmetry and reflection in different contexts ■ finding possible coordinates of shapes given the line of reflection.

Check my understanding

1 In which shape is a line of symmetry drawn correctly?

A

B

C

2 Copy each shape and reflect it in the given mirror line.

a b c d

3 Write the equation of the mirror line for these reflections.

a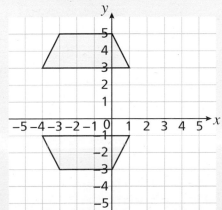

b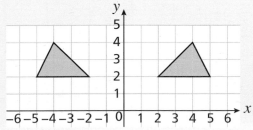

523

In this block, I will learn...

how to set up a statistical enquiry

A hypothesis is an idea that you want to investigate to find out whether it is true or false. Importantly, your hypothesis should be a statement not a question.

how to draw and interpret various charts and graphs

Name	Goals
Flo	
Benji	
Jakub	

Key

 = 2 goals

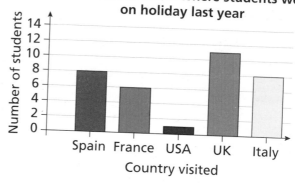

A bar chart to show where students went on holiday last year

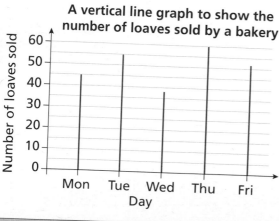

A vertical line graph to show the number of loaves sold by a bakery

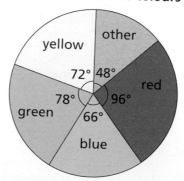

A pie chart to show students' favourite colours

about comparing and interpreting data

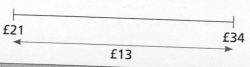

£21   £34

£13

The range is £13

16.1 Collecting data

Small steps

- ■ Set up a statistical enquiry
- ■ Design and criticise questionnaires

Key words

Hypothesis – an idea to investigate that might be true or false

Primary data – data you collect yourself

Questionnaire – a list of questions to gather information

Sample – a selection taken from a larger group

Secondary data – data already collected by someone else

Are you ready?

1 a Find the frequency of each category.

Animal	Tally	Frequency
horse	ⅢⅢ	
dog	ⅢⅢ ⅢⅢ |||	
cat	ⅢⅢ |||	
rabbit	||||	

b Which is the most common animal?   **c** What is the total frequency?

2 Which of these terms best describes each data item?

| Qualitative data | | Continuous quantitative data | | Discrete quantitative data |

a Height   **b** Eye colour

c House number   **d** Distance to school

e Subject at school   **f** Number of cars in a car park

Models and representations

Data collection sheet

Method of travel	Tally	Frequency
walk	ⅢⅢ ||||	9
bike	|||	3
bus	ⅢⅢ ⅢⅢ ||	12
car	ⅢⅢ		6

In this chapter, you will use tally charts to collect your own data. This is called **primary data** because you are collecting it yourself.

In this chapter you will set up your own statistical enquiry. This will involve using primary data (data you have collected yourself) and **secondary data** (data taken from somewhere else) in order to test a **hypothesis**.

A hypothesis is an idea that you want to investigate to see if it is true or false.

Your hypothesis should be a statement, not a question. For example, "Most people travel to school by bus." You would then set up an enquiry to see whether this statement is true or false.

Example 1

Set up a statistical enquiry to find out the most popular flavour of crisps of everyone in your class.

First set a hypothesis.

I think that:

The most popular flavour of crisps in my class is prawn cocktail.

Flavour	Tally	Frequency
salt & vinegar		
ready salted		
cheese & onion		
prawn cocktail		
other		

It doesn't matter if your hypothesis is true or false. The point of the enquiry is to find out. This sort of enquiry might be useful for the school's cafeteria to know which flavours of crisps to stock more of.

An 'other' option is essential so that you are covering all possibilities. It also gives people who don't like crisps an option to choose.

Ask every person in your class and tally the results.

Flavour	Tally	Frequency
salt & vinegar	＃＃ ＃＃ /	11
ready salted	＃＃	5
cheese & onion	＃＃ //	7
prawn cocktail	///	3
other	////	4
	Total:	30

The total frequency should sum to the number of students in your class. The class is the **sample** for this enquiry.

The enquiry could be repeated with a larger sample to check that your hypothesis is still false. The larger the sample, the more accurate the results will be.

The tally chart shows that the most popular flavour of crisps is salt & vinegar, so my hypothesis was false.

Assess whether your hypothesis is true.

Practice 16.1A

1 **a** Set up a statistical enquiry to find out how students in your class travel to school. You should write a hypothesis and use a tally chart to collect your data.

b State the size of your sample.

c Was your hypothesis true or false?

d Did everyone in your class have the same hypothesis?

e Did everyone's tally chart look exactly the same?

2 Hypothesis: More people have a pet dog than any other pet.

a Collect data to test this hypothesis. **b** Was the hypothesis true or false?

3 Amina thinks taller people have bigger feet.

a How could she find out?

b What graph could she draw to display her findings?

4 Rhys wants to find out the most popular sport of students in Year 8.

a Suggest a suitable hypothesis for Rhys to test.

b Does he need to collect information about everyone in Year 8? Why or why not?

What do you think?

1 Abdullah sets the following hypothesis.

"Most people like to play football in their spare time."

He asks his friends what they like to do in their spare time and records the information in this tally chart.

Activity	Tally	Frequency
Play football	ЖЖ ЖЖ	10
Do homework		0
Do jobs around the house		0

Find two things wrong with Abdullah's enquiry.

You will now learn about **questionnaires**. You may have come across these yourself, when people gather information about events or products.

Often companies will send out questionnaires to find out how they can improve their products and services.

Sometimes questionnaires are designed poorly. For example, some questionnaires use **leading questions**. A leading question is one that suggests a particular answer:

"When would you like to join the homework club?"

"Did you enjoy our excellent delivery service?"

Other questionnaires don't contain appropriate categories to choose from. Here is an example.

Example 2

Chloe wants to find out how long people spend reading.

She uses this question.

> **How long do you spend reading?**
>
> 0-1 hours ☐   1-2 hours ☐   3-4 hours ☐

a   What is wrong with Chloe's question?

b   Design a better question for Chloe to use.

a   There are three things wrong with Chloe's question.

   1 She hasn't given a time frame. She could have said either "How long do you spend reading per day?" or "How long do you spend reading per week?"

   2 The option boxes overlap.

   3 There is no option for someone who reads for more than 4 hours.

Not adding a time frame for the question is a common mistake.

If you spend 1 hour reading you wouldn't know whether to tick the first box or the second one.

b

> **How long do you spend reading per week?**
>
> Less than 1 hour ☐
>
> $1 \leqslant h < 2$ ☐
>
> $2 \leqslant h < 3$ ☐
>
> 3 hours or more ☐

Inequality symbols can be used to ensure that all possibilities are covered.

There are now options for people who don't read at all and for people who read for more than 3 hours.

Practice 16.1B

1   A café owner wants to find out if people like their menu.

The café owner asks this question.

> **How would you rate our menu?**
>
> Excellent ☐   Good ☐   Satisfactory ☐

a   Do you think this is a good question?

b   Design a better question for the café owner.

2 Zach wants to know how much people spend on music in a month.

He asks this question.

> **How much do you spend on music?**
>
> £5 to £10 ☐   £10 to £20 ☐   £20 to £30 ☐

a   Write down all the things that are wrong with his question.

b   Design a better question for Zach to use.

3 Flo wants to know how people feel about a new housing development.

She asks this question.

> **Do you agree that the new housing development will ruin the view?**
>
> Yes ☐   Not sure ☐   No ☐

a   What is wrong with Flo's question?

b   Flo asks ten of her friends.

Why might Flo's sample be biased?

c   With a partner, consider the pros and cons of building a new housing development.

4 Design suitable questions to find out

a   how often people go to the cinema per month

b   how much time students spend on their homework per week

c   how much people spend on their food shopping per week.

5 With a partner, consider who you would ask the questions you designed in question **4**. How would you make sure your sample was fair?

What do you think?

1 Jackson is watching a local TV news programme.

During the programme, a reporter claims that "40% of students do not do any exercise out of school".

In order to find out whether or not this is true, Jackson decides to do a survey.

Jackson plays for the school football team and decides to ask his teammates.

a   Give two reasons why this is not a suitable sample to take.

b   Suggest a better sample for Jackson to use.

c   Design a questionnaire and ask everyone in your class to complete it. Does the data from your class support the reporter's hypothesis?

Consolidate – do you need more?

1 Set up a statistical enquiry with your classmates to see whether this hypothesis is true or false.

"The most popular pizza topping combination is chicken and sweetcorn."

2 Emily has just arrived back in the UK from a holiday abroad.

She has been asked to fill out a customer satisfaction survey.

One of the questions on the survey is:

> **How would you rate the overall quality of service provided by our airline staff?**
>
> Good ☐   Poor ☐

a Describe one problem with the response section.

b Suggest better options for the response section of the question.

3 Kevin is carrying out a survey about eating fast food.

Here is one of the questions on his survey.

> **How often do you eat fast food?**
>
> All of the time ☐   Sometimes ☐   Not very much ☐

Write down **one** criticism of

a Kevin's question

b Kevin's response section.

Stretch – can you deepen your learning?

1 Darius is conducting a survey to find out if students use the internet to help them with their homework. Here are two questions that Darius plans to ask his friends.

> **Do you have the internet at home and use it to help you with your homework?**

> **Don't you agree that the internet is great for helping with homework?**

a Write down **one** criticism of each question.

b Suggest more suitable wording for the second question.

Design a response section for your question.

c Darius also wants to find out how long his friends spend online.

Write down a suitable question that Darius could ask.

Design a response section for your question.

2 There are two supermarkets in a town: *Pricewise* and *Freshfoods*.

The manager at *Pricewise* is carrying out a survey.

A member of staff stands at the entrance to their store and stops shoppers at random. The shoppers are asked the following question:

Do you agree that our store provides better value for money than *Freshfoods*?
Strongly agree ☐   Agree ☐   Don't know ☐

Write down **one** criticism of

    a the question

    b the response section

    c the method of sampling.

3 A local council wants to find out about people's opinions of their services. They want to ask 200 people. Three possible methods of surveying are discussed.

Method 1	Method 2	Method 3
Visit the local shopping centre on Monday morning and ask 200 people.	Find an online telephone directory and ring 200 people up at random.	Carry out a door-to-door survey in one particular area of town.

    a Write down a criticism of each method of sampling.

    b Suggest a better way for the town council to get people's opinions of their services.

Reflect

1 Explain what a hypothesis is.

2 Write down three common mistakes that people sometimes make when designing questions for a questionnaire.

Small steps

- Draw and interpret pictograms, bar charts and vertical line charts **ℝ**
- Draw and interpret pie charts **ℝ**
- Choose the most appropriate diagram for a given set of data

Key words

Axis – a reference line on a graph

Key – used to identify the categories present in a graph. A key on a pictogram tells you how many items/people each picture stands for

Are you ready?

1 Name each of these types of chart or graph.

a

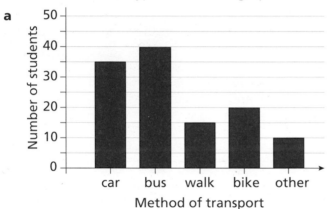

b

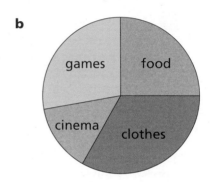

c

Type of tree	Number of trees
elm	🌳🌳🌳🌳🌳
ash	🌳🌳🌳🌳🌳🌳🌳
oak	
beech	🌳🌳

Key: 🌳 = 10 trees

2 State the number of degrees in a full turn.

3 Use a protractor to draw an angle of 48°

Models and representations

Pictograms

Name	Goals
Flo	●●
Benji	●●●◖
Jakub	●●●●●

Key
● = 2 goals

Pictograms always need a key to show what value each picture represents.

Bar charts and line charts

Make sure your **axes** are labelled and your chart has a title.

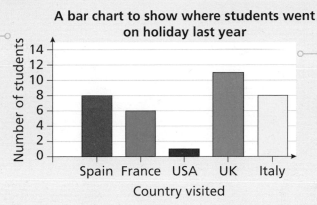

A bar chart to show where students went on holiday last year

There should be a gap between each bar in a bar chart.

Both bar charts and line charts can be drawn vertically or horizontally.

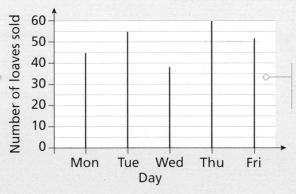

A line chart is like a bar chart but it uses a line rather than a bar.

Pie charts

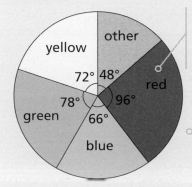

Each sector of a pie chart is measured in degrees. The sum of the angles in all the sectors must be 360° because the angles around a point sum to 360°

Unlike other charts and graphs, it is difficult to determine the sample size from a pie chart. This is usually given in the question.

Example 1

The table shows information about sandwiches sold at a shop one day. Draw a pie chart to represent the information.

Sandwich	Frequency
cheese	55
tuna	27
egg	38

360° ÷ 120 = 3° for each sandwich

Sandwich	Frequency	Working	Angle
cheese	55	55 × 3	165
tuna	27	27 × 3	81
egg	38	38 × 3	114
	120		360

You need to work out the angle to represent each type of sandwich.

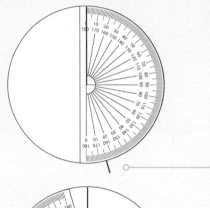

Measure and mark the first angle.

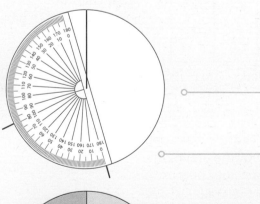

Compete the first sector and mark the next angle.

Line up the 0° on the protractor with the end of the first sector.

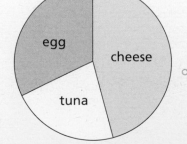

Finish drawing the pie chart.

Remember to label each sector or give a key.

Practice 16.2A

1 Here is a pictogram showing how students travel to school.

How many students

a get the bus to school

b walk to school

c were asked in total?

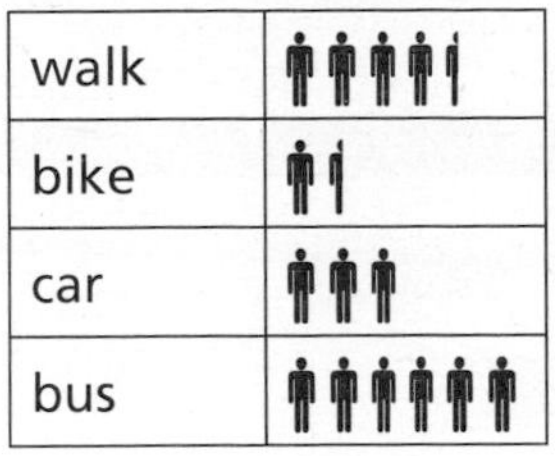

walk	🧍🧍🧍🧍
bike	🧍🧍
car	🧍🧍
bus	🧍🧍🧍🧍🧍

Key: 🧍 = 4 students

2 Junaid has drawn this pictogram.

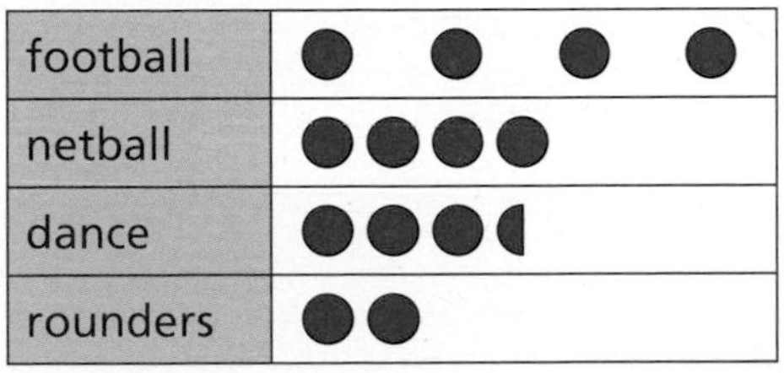

football	● ● ● ●
netball	●●●●
dance	●●●◖
rounders	●●

Find two things wrong with Junaid's pictogram.

3 a Draw a pictogram to represent this data.

	Minutes spent on the bus
Monday	60
Tuesday	20
Wednesday	50
Thursday	50
Friday	80

b Compare your pictogram with a partner. Is it the same? Did you use the same key?

4 The bar charts show the numbers of gold medals won by the USA, Germany and France in the 2016 Summer Olympics and the 2018 Winter Olympics.

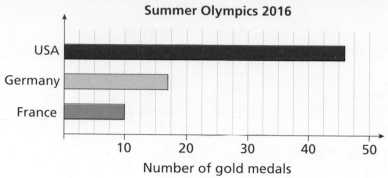

Summer Olympics 2016

Number of gold medals

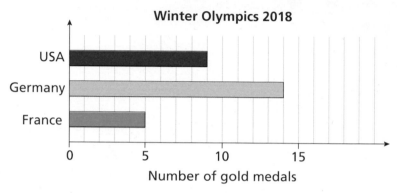

Winter Olympics 2018

Number of gold medals

a How many gold medals did the USA win at the Winter Olympics?

b Ali says, "France won more gold medals at the Winter Olympics than at the Summer Olympics." Do you agree with Ali? Explain your answer.

c Find the total number of gold medals won by the three countries at each Olympics.

5 The bar chart shows the numbers of students who went on holiday to different countries.

a Seven students went to Italy. Copy and complete the bar chart to include this information.

b How many more students went to Spain than the USA?

c How many fewer students went to France than the UK?

Holiday destination

Number of students

Country visited

6 Abdullah asks his class what sport they prefer. He wants to draw a pie chart of the results.

a Copy and complete the table.

Sport	Number of children	Number of degrees
tennis	7	7 × 10 = 70°
netball	8	
football	12	
hockey	5	
rugby	4	
Total	**36**	**36 × 10 = 360°**

b Draw a pie chart to display Abdullah's results. Make sure you label your pie chart.

7 A garage records the types of car that come in for servicing one day.

Fuel type	Frequency	Number of degrees
diesel	11	
petrol	20	
electric battery	8	
hydrogen fuel cell	1	

a Copy and complete the table.

b Draw a pie chart to show the information in the table.

8 60 people were asked to choose their favourite vegetable.

The results are shown in this pie chart.

Favourite vegetable

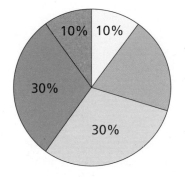

Key	
cabbage	■
carrots	■
cauliflower	□
peas	■
sweetcorn	□

a What percentage of people chose peas?

b How many people chose

  i carrots

  ii cabbage?

What do you think?

1 Which bar chart represents the data in the picture?

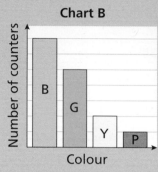

Chart A

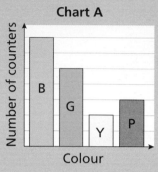

Chart B

Chart C

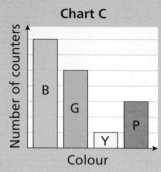

How did you choose? Did you need to count all the colours?

2 Use these clues to work out what pet each bar represents on this bar chart.

- The number of people who own a cat is half of the number of people who own a dog.
- More people own a dog than any other animal.
- More people own a horse than a rabbit.
- One fewer person owns a hamster than a cat.

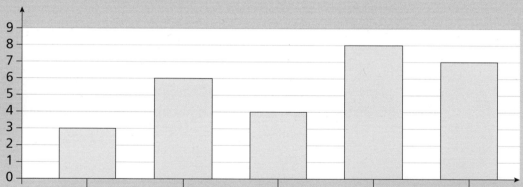

3 Here is a partially completed table showing the meal choices in a restaurant.

a Copy and complete the table.

b Draw a pie chart to represent the meal choices.

Meal choice	Frequency	Number of degrees
meat	20	
fish		96°
vegetarian	13	78°
vegan		
gluten free	4	
Total	**60**	**360°**

Consolidate – do you need more?

1 Match each table to the correct chart.

Class A	
dog	8
cat	2
rabbit	7
snake	12

Class B	
dog	4
cat	1
rabbit	3
snake	4

Class C	
dog	4
cat	1
rabbit	6
snake	8

Class D	
dog	8
cat	2
rabbit	7
snake	3

Chart 1

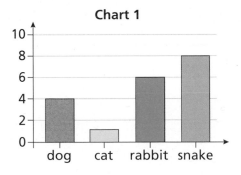

Chart 2

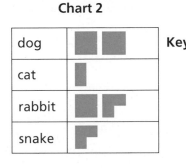

Chart 3

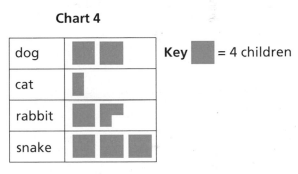

Chart 4

2 The bar chart shows the favourite fruit of each student in Class 8A.

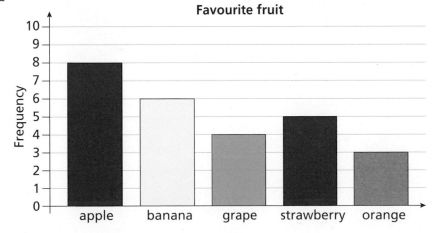

a What is Class 8A's favourite fruit?

b How many students' favourite fruit is grapes?

c How many fewer students' favourite fruit is oranges than bananas?

d How many students are in Class 8A?

3 The table shows the numbers of pets owned by a group of students.

Number of pets	Number of students
0	8
1	14
2	6
3	2

    a Draw each type of diagram to represent the information in the table.

       i Pictogram   **ii** Bar chart   **iii** Pie chart

    b Which diagram did you prefer to draw? Why?

Stretch – can you deepen your learning?

1 The bar chart shows the ages of students in a choir.

    a What is the modal age?

    b Work out the range of the ages.

    c What percentage of the students in the choir are 13 years old?

    d What fraction of the students in the choir are 10 years old?

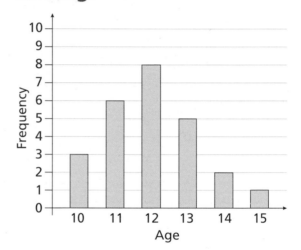

2 The table shows the colours of the cars in a car park.

Colour	red	blue	silver	black	white	grey	other
Frequency	43	71	120	101	142	17	6

    a Explain why a pictogram would not be suitable for this data.

    b Why would a pie chart be difficult to draw for this data? How could you do it?

3 The pie charts show information about the numbers of bedrooms in the houses in two towns.

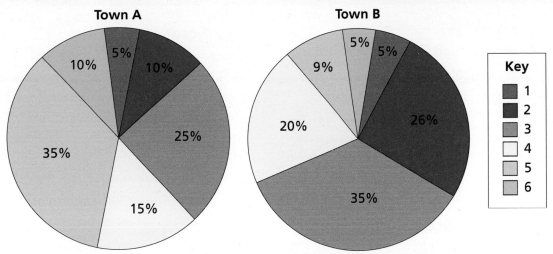

Town A

Town B

Key
1
2
3
4
5
6

In town A there are 300 houses, and in town B there are 1200 houses.

a In which town are there more houses with five bedrooms?

b How many 3-bedroomed houses are there in total in the two towns?

c In town A, 50 more 2-bedroomed houses are built.

Draw a new pie chart for town A that includes the 50 new houses.

d Criticize the scaling of the original pie charts.

Reflect

1 What's the same and what's different about a pictogram and a bar chart?

2 Why might it be harder to draw a pie chart compared to a pictogram or a bar chart?

Small steps

- Draw and interpret multiple bar charts
- Draw and interpret line graphs
- Represent and interpret grouped quantitative data
- Choose the most appropriate diagram for a given set of data

Key words

Compare – to evaluate two or more things to find the differences between them

Line graph – this has connected points and shows how a value changes over time

Multiple bar chart – a way to represent several, related sets of data

Are you ready?

1 The bar chart shows how many people have certain types of pet.

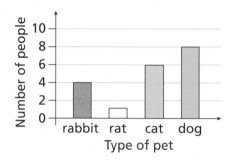

- **a** How many people have a dog?
- **b** How many people have a rat?
- **c** How many more people have a cat than a rabbit?
- **d** How many people were asked altogether?

2 **a** Draw a set of axes, labelled from 0–10 on the x-axis, going up in 1s, and from 0–1000 on the y-axis, going up in 100s.

- **b** Plot these points on your graph.
    - **i** (5, 800)
    - **ii** (3, 400)
    - **iii** (7, 250)

3 Write the meaning of each inequality in words.

- **a** $t > 4$
- **b** $h \geqslant 163$
- **c** $15 \leqslant w \leqslant 20$

Models and representations

Multiple bar charts

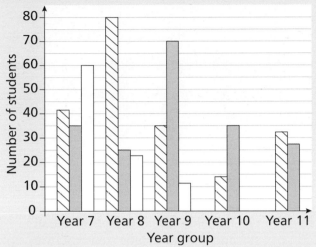

Multiple bar charts can display several pieces of information.

This example shows how many students in Years 7 to 11 study French, Spanish and Arabic.

Key
- ⧄ French
- ▨ Spanish
- ☐ Arabic

A key is used to distinguish between the bars.

Line graphs

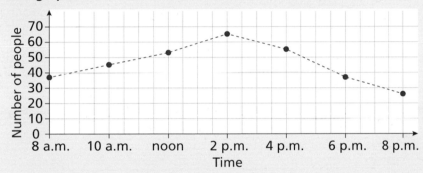

A line graph shows change over time.

The graph above shows the number of people in a shop, recorded at 2-hour intervals.

The points are joined with dotted lines as the number of people between the points is only an estimate. For instance, at 11 a.m. there could have been a sudden drop in the number of people in the shop, but you don't know this because the data was recorded only every 2 hours.

The dotted line shows the trend of the data, in other words, how the data changes. There were fewer people early in the morning and late in the evening, and there was a peak in the middle of the day.

Frequency diagrams

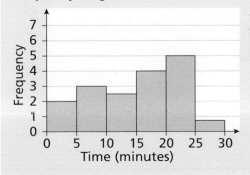

A frequency diagram can be used to display continuous data.

Unlike a bar chart, the bars should touch.

Example 1

The bar chart shows the numbers of students in Years 7, 8 and 9 who play football and rugby.

a   How many students in Year 8 play football?

b   In which year do the same number of students play football and rugby?

c   How many students in Year 7 play rugby?

d   What is the difference between the total number of students who play football and the total number of students who play rugby?

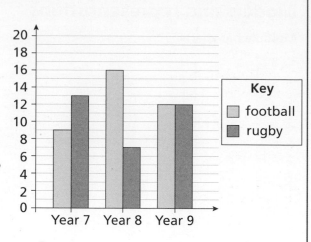

a   16

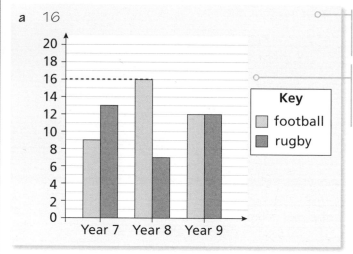

The blue bars represent football and the *x*-axis label tells you the year group

Read across from the top of the blue Year 8 bar to the *y*-axis to find out how many Year 8 students play football.

b   Year 9

For Year 9, the blue bar for football and the purple bar for rugby are the same height, so they represent the same number of students.

c   13

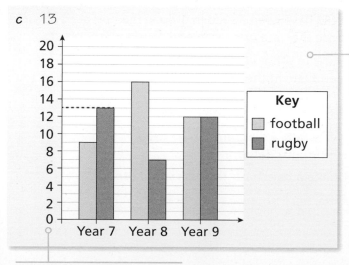

This is harder to read because the top of the bar doesn't sit exactly on a number. Ensure that you check the scale carefully.

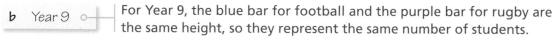

This scale is going up in 2s.

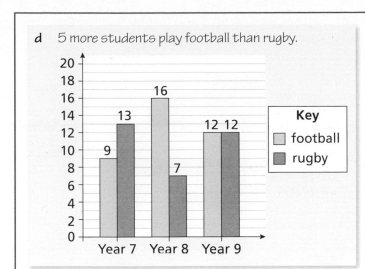

d 5 more students play football than rugby.

Find the number of students represented by each bar.

Add up the total for the bars representing football.

9 + 16 + 12 = 37

Add up the total for the bars representing rugby.

13 + 7 + 12 = 32

Find the difference by subtracting one from the other.

37 − 32 = 5

Practice 16.3A

1 The bar chart shows sales of tea and coffee on five different days at a café.

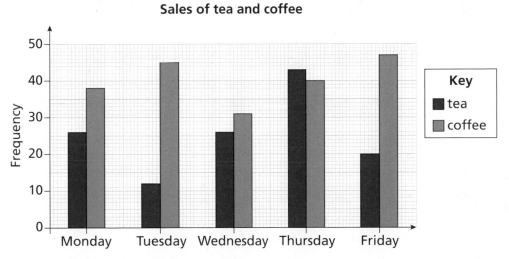

Sales of tea and coffee

a What does each small division on the vertical axis represent?

b How many teas were sold on Friday?

c How many coffees were sold on Tuesday?

d Just by looking at the chart, can you tell whether the café sold more teas or coffees during the week?

e Work out the total sales of tea and coffee to check whether your answer to part **d** was correct.

2 The bar chart shows the numbers of matches won, drawn and lost by two football teams, Trinity United and White Rose Athletic, during this season.

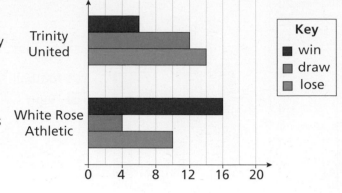

a Which team won more matches?

b How many matches did Trinity United lose?

c How many matches has each team played this season?

d Points are awarded as follows
- 3 points for a win
- 1 point for a draw
- 0 points for a loss

Work out the total number of points scored by each football team.

3 The bar chart shows the numbers of actors in a theatre group each year.

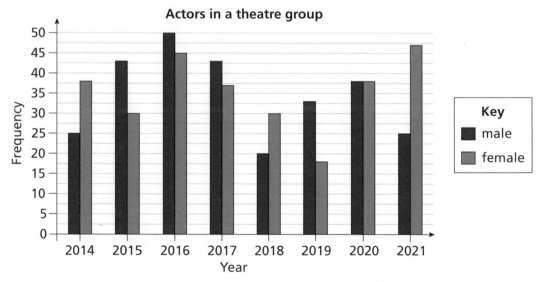

In which year did the theatre group have

a the same number of male and female actors

b the most actors?

c In 2018, the theatre group had 50 actors.

i What percentage of these actors were female?

ii What fraction of these actors were male?

d The theatre group predicts that in 2022 there will be a 20% increase in the number of male actors, whilst the number of female actors will stay the same. Approximately how many actors are the theatre group expecting to have in 2022?

What do you think? 🌐

1 The table shows the numbers of students who have school dinners or packed lunches in a secondary school.

	Year 7	Year 8	Year 9	Year 10	Year 11	Total
School dinner	101			78	55	
Packed lunch	19	34	27			
Total	120	120	120		110	600

a Copy and complete the two-way table.

b Draw a multiple bar chart to represent the information in the table.

c Compare your bar chart with a partner. What's the same and what's different?

Example 2

The table shows information about the temperature between 9 a.m. and 3 p.m. on a day in March.

Time	09:00	10:00	11:00	12:00	13:00	14:00	15:00
Temperature (°C)	14	16	20	26	24	20	18

Draw a line graph to represent the information in the table.

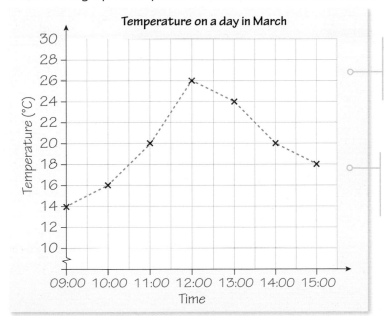

The labels on your axes need to represent each piece of information and do not need to start at zero.

On a line graph, the points are usually joined with a dotted line to show that the points in between are estimates.

Think of the information as coordinates.

(9:00, 14)

Time	09:00	10:00	11:00	12:00	13:00	14:00	15:00
Temperature (°C)	14	16	20	26	24	20	18

Plot each point on the graph.

Practice 16.3B

1 The line graph shows the growth of some grass over 10 days.

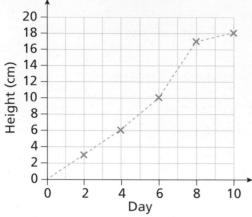

Growth of grass

a How tall was the grass on day 4?

b On what day did the grass reach 10 cm?

c Estimate the height of the grass on day 3

d Estimate when the grass reached a height of 14 cm

2 The table shows the mass of a foal at the end of each week for the first 8 weeks of its life.

Week	1	2	3	4	5	6	7	8
Mass (kg)	45	55	70	80	95	100	100	120

Copy and complete the line graph to show the information in the table.

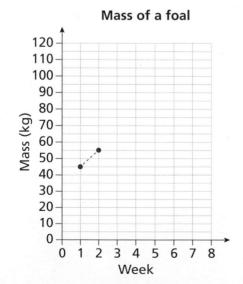

Mass of a foal

3 The line graph shows the average rainfall for the first 8 days in September.

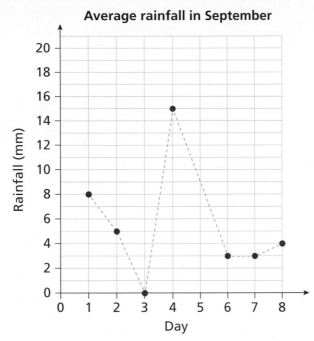

Beca says, "It rained every day." Do you agree with Beca? Explain your answer.

4 The line graph shows the population of a town between 1950 and 2020

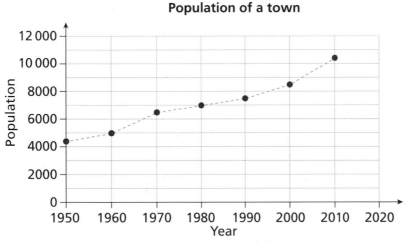

a What was the population in 1980?

b In which year was the population 5000?

c Estimate the increase in population between 1950 and 2020. Why is this an estimate?

What do you think? 💭

1 The line graph and bar chart both show the distance above ground of a bird.

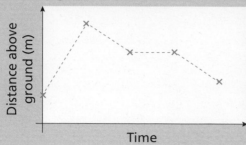

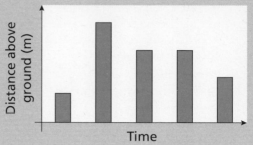

Which representation is more appropriate?

Explain your choice to a partner.

2 What could this line graph represent?

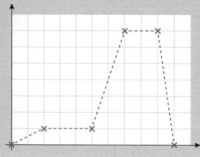

Example 3

The times taken (in minutes) by a group of students to complete a Maths test were recorded.

Here are the results.

| 32 | 24 | 40 | 35 | 28 | 31 | 25 | 27 | 40 | 22 | 37 | 30 |
| 31 | 23 | 39 | 35 | 27 | 26 | 39 | 28 | 35 | 29 | 29 | 31 |

a Put the data into a grouped frequency table.

b Draw a frequency diagram to represent the data.

a

Class interval

Time taken, minutes (m)	Frequency
$20 < m \leqslant 25$	3
$25 < m \leqslant 30$	9
$30 < m \leqslant 35$	7
$35 < m \leqslant 40$	5

The quickest time was 23 minutes and the slowest time was 40 minutes, so it makes sense to have four equal groups of 5 minutes for the class intervals.

b

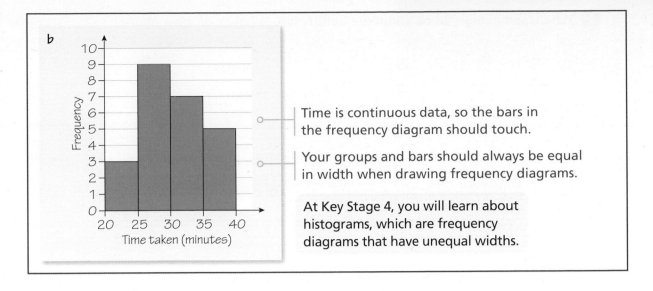

Time is continuous data, so the bars in the frequency diagram should touch.

Your groups and bars should always be equal in width when drawing frequency diagrams.

At Key Stage 4, you will learn about histograms, which are frequency diagrams that have unequal widths.

Practice 16.3C

1 The table shows the times taken (in minutes) by a group of people to complete a puzzle.

Time taken, t (minutes)	Frequency
$2 < t \leqslant 4$	8
$4 < t \leqslant 6$	14
$6 < t \leqslant 8$	11
$8 < t \leqslant 10$	5

On a copy of this grid, draw a frequency diagram to represent the data in the table.

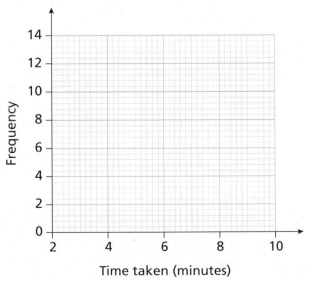

2 Here is some information about the battery life, in hours, of some phones.

12	15	6	22	9	11
20	21	19	15	7	16
24	10	13	7	12	13
23	12	8	18	18	20
21	17	14	9	11	22

a Copy and complete the table to show the battery life of the phones.

Time in hours (h)	Tally	Frequency
$5 < h \leqslant 10$		
$10 < h \leqslant 15$		

b Draw a frequency diagram for the data in your table.

What do you think? 💭

1 What type of data is represented in a grouped frequency table and by a frequency diagram?

2 Explain why the bars in a frequency diagram touch, and the bars in a bar chart do not touch.

3 Ed has created this grouped frequency table for the heights of some trees.

Height in metres (h)	Frequency
$5 \leqslant h \leqslant 10$	12
$10 \leqslant h \leqslant 15$	7
$15 \leqslant h \leqslant 20$	16
$20 \leqslant h \leqslant 25$	2

What mistake has Ed made?

Consolidate – do you need more?

1 The table shows the numbers of girls and boys attending some school clubs.

	Chess	Robot	Drama	Music
Boys	12	11	9	10
Girls	8	15	14	10

Copy and complete the multiple bar chart to represent the information in the table. Include a key.

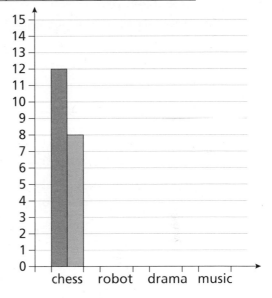

2 The table shows the temperature between 9 a.m. and 3 p.m. on a day in April.

Time	09:00	10:00	11:00	12:00	13:00	14:00	15:00
Temperature (°C)	13	16	21	22	22	19	17

Draw a line graph to represent the information on a copy of this grid.

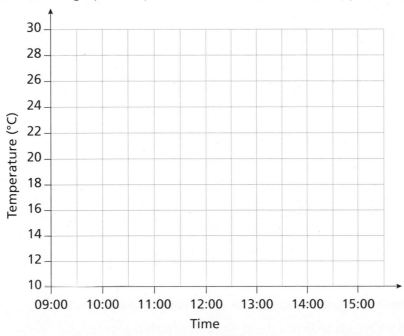

3 The line graph shows the change in the height of a tree over 5 years.

a How tall was the tree when it was planted?

b After how many years was the tree 10 metres tall?

c How much did the tree grow between year 2 and year 3?

d What could have happened in year 4?

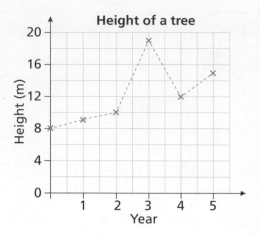

Stretch – can you deepen your learning?

1 There are 250 students in Year 8 at a school.

Here is some information about how the students travel to school.

	Walk	Car
Boys	45	22
Girls	39	16

The rest of the students travel to school by bus.

$\frac{3}{8}$ of the students who travel by bus are girls.

a Draw a multiple bar chart to represent how all the students in Year 8 travel to school.

b In Year 8, how many students are:

i girls   ii boys?

2 Represent the information in the table on a single graph.

	Jan	Feb	Mar	Apr	May	Jun	Jul	Aug	Sep	Oct	Nov	Dec
Sunrise	8:00	7:30	6:30	6:00	5:30	5:00	4:30	5:00	6:00	7:00	7:00	7:30
Sunset	16:00	16:30	17:30	19:30	20:30	21:00	21:30	20:30	19:30	18:30	16:30	16:00

3 Electricity is measured in kilowatt hours (kWh).

The graph shows the amounts of electricity used at different times of the year in one household.

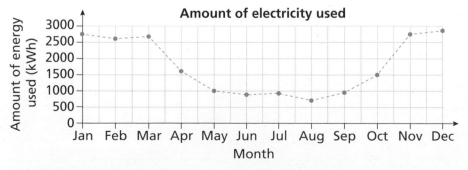

a Describe three things that you can see by looking at the graph.

b Describe three things that you could work out from the graph.

4 The frequency diagram represents the distances walked by some students in a school.

a Draw a grouped frequency table to match the graph.

b Has everyone in your class drawn exactly the same table? How do your answers differ?

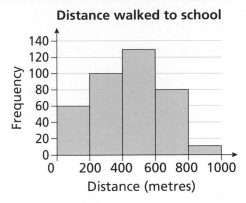

Distance walked to school

Reflect

1 Why can line graphs be useful when interpreting data?

2 What type of graph would you draw for

a discrete data **b** continuous data?

Is there more than one possible graph for each?

Small steps

- Find and interpret the range
- Compare distributions using charts
- Identify misleading graphs

Key words

Compare – to evaluate two or more things to find the differences between them

Distribution – the way in which data is spread out

Misleading graphs – a graph that suggests an incorrect conclusion or assumption

Range – the difference between the greatest value and the smallest value in a set of data

Are you ready?

1 Write down the greatest and the smallest number in each list.

   a 51, 9, 83, 112   **b** 1050, 150, 500, 950   **c** 12.7, 13, 12.4, 12.9

2 Work out

   a 140 – 83   **b** 295 – 273   **c** 58.1 – 24.6

Models and representations

Number lines

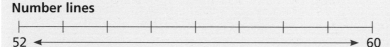

52   60

Number lines are useful for ordering data and working out the difference between a pair of numbers.

Graphs and charts

Pie charts

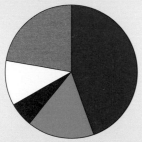

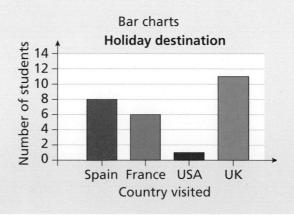
Bar charts

In this chapter, you will find the **range** of a set of numbers. The range is a measure of the spread of a set of data.

You find the range by subtracting the smallest value from the largest value.

Example 1

a Here are the shoe sizes of six people.

6   10   10   5   11   7

Find the range of the shoe sizes.

b Max goes shopping to buy some clothes. The range in the cost of his items is £13. The cheapest item he bought was £21. How much did the most expensive item cost?

a 5, 6, 7, 10, 10, 11

First identify the smallest and largest values. You can do this by putting the data in order.

11 − 5 = 6

Subtract 5 from 11

Therefore, the range of the shoe sizes is 6

b £21   £34

£13

The most expensive item was £34

If the cheapest item is £21 and the range of the costs is £13, the most expensive item must be £21 + £13 = £34

Practice 16.4A

1 Here are the masses of some parcels.

220 g   385 g   148 g   431 g   500 g

Work out the range of the masses.

2 Here are the numbers of miles that Beth drives each day in a week.

31   84   120   43   17   35   102

a Beth says, "The range of the miles I drive is 17 to 102."

Explain the mistake Beth has made.

b Find the range of the number of miles she drives in a week.

3 Here are the temperatures in six cities around the world.

13°C   34°C   8°C   19°C   19°C   27°C

Work out the range of the temperatures.

4 Jackson is working out the range of some data he collected.

He has made a mistake in each calculation. Explain Jackson's mistakes.

a 63 cm, 103 cm, 92 cm, 47 cm, 150 cm   The range = 150 − 63 = 87 cm

b 3 kg, 520 g, 950 g, 730 g, 5 kg   The range = 950 − 3 = 947 g

5 A set of numbers has a range of 15

The largest number is 58

What is the smallest number?

6 The range of the times taken by a group of people to complete a puzzle is 42 seconds.

The fastest time was 2 minutes 34 seconds.

What was the slowest time?

7 The bar chart shows the number of children in each year group of a primary school.

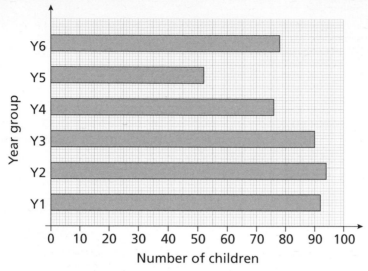

Which year group is

a the smallest

b the largest?

c Find the range of the number of children in the year groups.

8 The pictogram shows the numbers of people who visited the seaside between May and September.

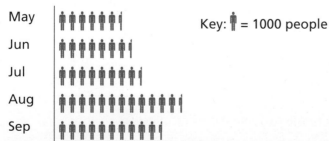

Chloe says, "The number of people ranged from 6500 to 10 500."

Do you agree with Chloe? Explain your answer.

What do you think?

1 Benji says, "The range of these amounts is 497.5."

| 4*l* | 3*l* | 200 m*l* | 2.5*l* | 500 m*l* |

What mistake has Benji made?

Find the range of the amounts on the cards.

2 These are the heights of eight students standing in a line.

| 153 cm | 148 cm | 141 cm | 149 cm | 145 cm | 155 cm | 152 cm | 150 cm |

a Another student joins the line.

The range of the students' heights stays the same.

What can you say about the height of the ninth student?

b Another student joins the line.

The range increases by 2 cm.

What could be the height of the tenth student?

Example 2

The bar chart shows the average cost of shopping at five different supermarkets.

a Explain why the chart is misleading.

b Explain how the chart could be improved.

c Compare the cost of the shopping at each supermarket.

a The scale is too big and the bars are on a blank background so it is hard to tell the difference between them.

b Reduce the size of the scale and draw it on a squared grid.

c The average cost is highest at the Health Store and lowest at Fresh Company. The exact amounts are difficult to tell from the graph.

Practice 16.4B

1 The pie charts show information about the numbers of different coloured cars in a two car parks.

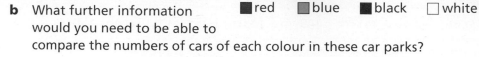

Car park A **Car park B**

a Marta says, "There are far more red cars in car park B than in car park A."

Explain why Marta might not be correct.

b What further information would you need to be able to compare the numbers of cars of each colour in these car parks?

■ red ■ blue ■ black □ white ■ other

c What comparisons can you make about the different-coloured cars in each car park?

2 The bar chart shows the average percentage score in a test for five different classes.

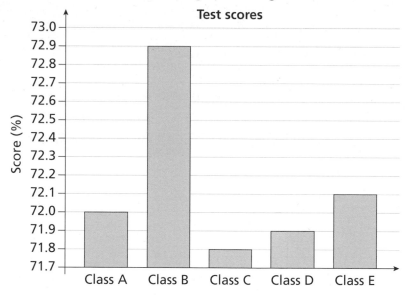

a Faith says, "Class B's average score was significantly higher than any other class."

Explain why Faith thinks this.

b Explain why this chart might seem misleading.

c Compare the average scores of the different classes.

3 The pie charts show information about Emily's and Seb's positions in their last 50 running races.

a Why might these pie charts be considered to be misleading?

b Compare Emily's and Seb's results for their last 50 running races.

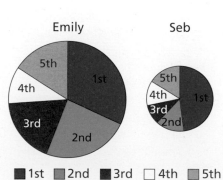

Emily Seb

■ 1st ■ 2nd ■ 3rd □ 4th ■ 5th

What do you think? 🌑

1 The bar chart shows the times taken by a group of students to run 100 m

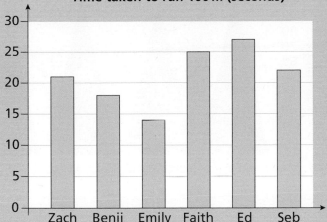

Time taken to run 100 m (seconds)

a Lydia says, "Ed is the best runner because his bar is the tallest."

Explain why Lydia is incorrect.

b Compare the runners' times to run 100 m

c Explain why a bar chart isn't a suitable way of representing this data.

2 Investigate the use of charts and graphs in newspapers and magazines. Are any of them misleading?

Why do you think this might be?

Consolidate – do you need more?

1 Work out the range of each set of data.

a 11, 15, 14, 18

b 40, 70, 20, 45, 80, 55, 90

c £200, £150, £190, £210, £192, £190

d 83 g, 67 g, 62 g, 18 g, 23 g

2 The range of a set of numbers is 25

The smallest number is 17

What is the largest number?

3 The range of a set of numbers is 40

The largest number is 100

What is the smallest number?

Stretch – can you deepen your learning?

1 The chart shows information about the recycling habits of four regions.

a Explain why the graph is misleading.

b Compare the recycling habits of each of the regions.

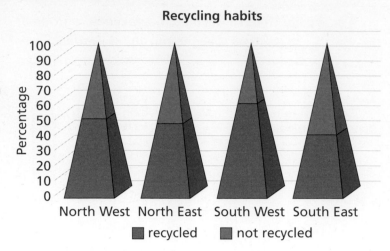

Recycling habits

Percentage

North West North East South West South East

■ recycled   ■ not recycled

2 Given that x is positive and the range of the expressions on these cards is 18, find the value of x

| x | | $2x$ | | $5x$ | | $7x$ |

3 Here are three students making statements about the range.

Marta: The range of a set of numbers could be 0

Benji: The range of a set of numbers could be negative.

Faith: The range of a set of numbers could be a decimal.

Explain whether each student is correct.

Use an example to support your explanation.

4 These sets of numbers have the same range.

A  | 104 | 108 | 103 | x

B  | 1004 | 1008 | y

a Write down five possible pairs of values for x and y

b Discuss your strategy with a partner.

Reflect

1 What do you think are the most common mistakes when finding the range of a set of data?

2 Explain what features of a graph can make it misleading.

I have become **fluent in...**	I have developed my **reasoning skills by...**	I have been **problem-solving through...**
■ drawing charts and graphs ■ representing grouped, quantitative data ■ finding the range.	■ criticising and improving questionnaires ■ interpreting charts and graphs ■ interpreting the range ■ identifying when and how graphs are misleading ■ comparing sets of data.	■ finding missing values in a data set, given the range ■ working out missing values in a chart or graph ■ completing partially completed tables and charts.

Check my understanding

1 Chloe has created a questionnaire about running.

> **How far do you run?**
>
> Not that far ☐   Quite far ☐   Very far ☐

  a State **two** things that are wrong with Chloe's question.

  b Create a better question.

2 The table shows how many points Mario and Filipo each scored in four games.

	Game 1	Game 2	Game 3	Game 4
Mario	18	12	14	17
Filipo	15	20	14	16

  a Draw a multiple bar chart to represent this data.

  b Find the range of points scored by Mario.

3 A group of 60 people were asked to name their favourite colour.

18 people said red, 14 said blue, 9 said green, 4 said pink and the rest said "other".

Draw a pie chart to show this data.

4 The table shows the heights of some trees.

Height of tree, h (m)	$5 < h \leqslant 10$	$10 < h \leqslant 15$	$15 < h \leqslant 20$	$20 < h \leqslant 25$
Number of trees	35	78	68	19

  a How many trees are there in total?

  b How many trees are taller than 20 m?

  c The tallest tree measures 24.7 m and the shortest is 5.09 m

  Find the range in the height of the trees.

17 Measures of location

White Rose Maths

In this block, I will learn...

how to identify the mode of a set of data

1, 1, 2, 2, 2, 2, 3, 3, 4, 4, 5, 5, 5, 5, 6, 6, 7, 7, 7, 8, 8, 9

The modes are 2 and 5

when it is appropriate to use each type of average

£220, £220, £220, £220, £220, £250, £250, £250, £270, £300, £580

Mean? Median? Mode?

how to find averages from frequency tables **H**

Number of goals	Frequency
0	5
1	8
2	6
3	4
4	1
Total	24

Height, h (cm)	Frequency
$150 < h \leqslant 160$	10
$160 < h \leqslant 170$	27
$170 < h \leqslant 180$	29
$180 < h \leqslant 190$	22
$190 < h \leqslant 200$	12
Total	100

how to identify outliers

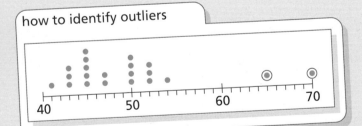

how to compare distributions

	Median time (seconds)	Range of times (seconds)
Teenagers	34	58
Over 65s	47	12

Small steps

- Understand and use the mean, median and mode
- Choose the most appropriate average

Key words

Average – a number representing the typical value of a set of data

Mean – the result of sharing the total of a set of data equally between them

Median – the middle number in an ordered list

Mode – the item which appears most often in a set of data

Modal class – the class in a set of grouped data that contains the highest frequency

Are you ready?

1 Find the mean of each set of data.

   a   9   12   20   32   **b**   7   7   9   15   19

2 Find the median of each set of data.

   a   7   10   15   19   20   22   30   **b**   6   2   10   15   7

3 Bobbie has four cats. Their masses are 6 kg, 5.8 kg, 7 kg and 4.9 kg

   Find

   a   the mean mass of the cats   **b**   the median mass of the cats.

Models and representations

Number line

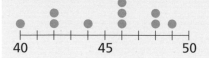

Division

$$\begin{array}{r} 4\ 7\cdot 2 \\ 5\overline{\smash{)}2\,^2 3\,^3 6\cdot{}^1 0} \end{array}$$

Vertical line chart

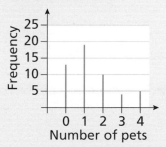

Number of pets

A number line is useful for putting items of data in order and finding the **median**.

You may need to use the formal method of division to find the **mean**.

Bar charts, or other graphs such as vertical line charts, can be useful for identifying the **mode**.

The mode is the most common item in a set of data.

> You learned how to find the mean and the median in Book 1

Example 1

Here are the marks of five students in a test.   25   30   32   36   36

Find the mean, median and mode of the marks.

$$\text{Mean} = \frac{\text{total}}{\text{number of items}} = \frac{159}{5}$$

$$= 31.8$$

To find the mean of a set of data, you find the total of the data and divide this by the number of data items.

In this case, the mean isn't an integer, and isn't one of the items of data in the list either.

25   30   (32)   36   36

Median = 32

The median is the middle number when the data is arranged in order.

In this case, the scores are already in order so you don't need to rearrange them.

Mode = 36

The mode is the one that occurs most frequently.

36 appears twice, and all the other numbers occur only once, so 36 is the mode.

Example 2

The ages of six infants at a playgroup are 19 months, 20 months, 17 months, 14 months, 16 months and 21 months. Find the mean, median and mode of the ages of the infants.

$$\text{Mean} = \frac{\text{total}}{\text{number of items}} = \frac{107}{6}$$

$$= 17.8 \text{ months (1 d.p.)}$$

Use the formula for finding the mean.

There is an even number of items in the data set, so there is no single middle number.

14   16   (17   19)   20   21

$$\text{Median} = \frac{17 + 19}{2} = 18$$

You find the median by finding the mean of the middle pair of numbers.

There is no mode.

No item occurs more often than any of the others, so there is no mode.

Sometimes a data set can have more than one mode. For example, in the set 2, 2, 2, 3, 3, 4, 4, 4, 5, 5, 6, 7 the modes are 2 and 4

Practice 17.1A

1 Find the mode of each of these sets of numbers.

a 5   7   7   9   10   12   12   12   14   15

b 3   6   3   7   2   8   3   4   5   7

c 1   4   5   5   6   6   6   7   8   10

d 10   9   4   9   4   10   4   9   11   9

For which sets is it easier to find the mode? Why?

2 Here are the scores obtained when a dice is rolled 7 times.

5   3   3   5   4   6   3

Find the mean, median and mode of the scores on the dice.
Give your answers to 1 decimal place, if necessary.

3 The masses, in kilograms, of the players in an ice hockey team are

| 95.2 | 89.2 | 98.8 | 93.4 | 89.7 | 94.2 |

   a Explain why the data has no mode.

   b Find the mean mass of the ice hockey players.

   c Find the median mass.

     Hint: Remember to put the masses in order first.

   d Which, if any, of the averages will change if the heaviest player is replaced by a player of mass 94 kg?

4 **a** What's the same and what's different about finding the median of these three sets of data?

     i 6 6 8 10 10   **ii** 6 7 8 10   **iii** 6 7 7 10

   b What's the same and what's different about finding the mode of the three sets of data in part **a**?

   c Find the mean of each set of data in part **a**.

5 **a** Do you agree with Ed? Why or why not?

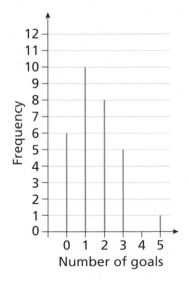

The mean doesn't have to be a member of the data set, but the mode always is.

   b Does the median always have to be a member of the data set? Why or why not?

6 The vertical line chart shows the numbers of goals scored by a team over a season.

Which average can you tell just from looking at the graph?

7 A car dealer records the number of sales she makes each day of one week.

    1   0   1   1   2   4   2

a The dealer says, "The modal value of the number of sales this week was 1." What do you think she means by the phrase "modal value"? Is she correct that the modal value is 1?

b Which of the three averages for this set of data is the greatest: the mean, the median or the mode?

8 The heights of 10 plants (in centimetres) are shown on the number line.

a What is the modal height?

b What is the median height?

c Work out the mean height of the plants.

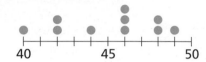

What do you think?

1 The table shows the numbers of cats owned by households in a cat-loving town.

Number of cats	Frequency
0–2	182
3–5	43
6–10	70
10+	29

a Explain why you cannot find the modal number of cats from the table.

b What is the modal class of the number of cats?

> "**Modal class**" means the class (0–2, or 3–5, and so on) that has the highest frequency.

2 Find the mode of this set of fractions. Explain your answer.

$\dfrac{24}{36}$ $\dfrac{3}{4}$ $\dfrac{5}{10}$ $\dfrac{8}{12}$ $\dfrac{10}{15}$ $\dfrac{9}{16}$ $\dfrac{60}{80}$ $\dfrac{2}{3}$ $\dfrac{8}{16}$ $\dfrac{18}{24}$ $\dfrac{16}{24}$ $\dfrac{1}{2}$

3 Mr Jones has five children, including a pair of twins who are $3\frac{1}{2}$ years old.

His other children are 8 years old, $6\frac{1}{4}$ years old and 16 months old.

Find the mean, median and mode of the ages of Mr Jones' children.

In the last exercise, you practised finding all the different types of **average**. When working with data in real situations, sometimes one of the averages is more suitable than the others.

> Remember that the purpose of an average is to represent the data set, so if an average isn't representative then it's not appropriate to use it.

Example 3

The total sales in a shop from Monday to Saturday one week were

£150   £130   £220   £210   £320   £960

a   Find the mean, median and mode of the daily sales.

b   Which average is the most representative of the daily sales? Give a reason for your answer.

a   Mean = $\dfrac{\text{total}}{\text{number of items}} = \dfrac{1990}{6}$

        = £331.67 to the nearest penny

For this question, you could also give the mean to the nearest pound.

£130   £150   (£210   £220)   £320   £960

Remember to put the data in order.

Median = $\dfrac{210 + 220}{2}$ = £215

There is no mode.

b   The median is the best average as it is close to most of the values over the week

The mean is only close to one of the values and has been affected by one value being much larger than the others. Values like this are called **outliers** and you will learn more about them in Chapter 17.3

You cannot use the mode here as there isn't one!

Practice 17.1B

1   Here are the marks scored by six students in a test.

    0   0   35   37   39   40

  a   Work out the mean, median and mode of the scores.

  b   Which average is the most representative of the scores? Explain your answer.

2   The maximum daily temperatures in a town one week were

    15°C   17°C   18°C   20°C   19°C   22°C   23°C

  a   Explain why the mode would not be a good average to use to represent the data.

  b   Find the mean and the median temperatures.

  c   Is the mean or the median a better average to represent this data? Explain your answer.

3   The table shows how often each of the vowels occur in an English book.

  a   Kate says, "The letter 'e' is the vowel that is the mode." Do you agree?

  b   Could you find the mean or median of the vowels? Why or why not?

Vowel	Frequency
a	8.2%
e	13%
i	7%
o	7.5%
u	2.8%

4 The ages of nine people (in years) at a birthday party were

2   2   3   3   3   3   3   36   38

a Find the mean, median and mode of the ages of the people at the birthday party.

b Which averages represent the data well and which do not?

5 The shoe sizes of six children are

5     $5\frac{1}{2}$ 6 $6\frac{1}{2}$ $6\frac{1}{2}$ 7

Why might the median size not be a useful average for this set of data?

6 The masses of the children in a family are 30 kg, 35 kg, 38 kg and 40 kg

a Find the mean, median and mode of the masses.

b The family have a new baby whose mass is 3 kg. How does the new baby affect each of the average masses you found in part **a**?

What do you think?

1 Look at the statements on these cards.

Easy to find	Not affected by extreme values

Sometimes not a member of the data set	May not be one

Uses all the values	Ignores many of the values	Can be difficult to calculate

a Which average do you think each statement describes? Do any describe more than one average?

b Which of the statements do you think are advantages and which are disadvantages?

2 Which averages are and which aren't useful for certain sets of data? Here are some sets to start your discussion, but you should think of some of your own to discuss.

Ages	Heights	Favourite sports	Marks in a test	Wages

Consolidate – do you need more?

1 Find the mean, median and mode of each set of numbers.

a 8   10   16   16   24   **b** 7   9   9   13

c 7   7   9   13   **d** 8   2   6   5   9   7   4

2 Here are the heights, in centimetres, of eight people.

157   174   176   165   176   182   173   176

a Find the mode.   **b** Find the median.   **c** Calculate the mean.

3 Here are the prices of a mobile phone in five different shops.

£675   £599   £710   £640   £659

a Which average of the prices cannot be found?

b Find the other two averages of the prices.

4 Here are Jackson's marks in a series of spelling tests.

10   9   10   9   10   10   7   10

a Find the mean, median and mode of Jackson's marks.

b Which average do you think is most representative of Jackson's marks?

5 The data shows the numbers of people who went to a town's cinema each day for one week.

65   87   45   91   248   250   89

a Find the mean, median and mode of the number of people who went to the cinema that week.

b Which average do you think is most representative of the cinema's attendance?

6 The vertical line chart shows the number of children in each house on a street.

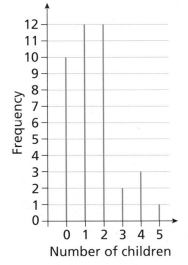

Number of children

Which average can you tell just from looking at the graph?

Stretch – can you deepen your learning?

1 Collect some discrete data from your class and find the mean, median and mode for each set of data.

Which average is most representative of the data? Which average is easiest/most difficult to find?

2 The mean of three integers is 8. A fourth number is added to the list and the mean is now 7

a What is the fourth number?

b Investigate whether it is possible that both the mode and median do not change, or that only one changes when the fourth number is added.

3 Write a set of six integers.

What happens to the mean, median and mode of a set of numbers

a if all the numbers are doubled

b if three of the numbers are doubled and three are halved?

c Investigate other ways of changing the numbers.

Reflect

Explain the difference between the mean, median and mode of a set of data.
When is it appropriate to use each average?

Small steps

- Find the mean from an ungrouped frequency table H
- Find the mean from a grouped frequency table H

Key words

Estimate – an approximate answer or to give an approximate answer

Frequency – the number of times something happens

Midpoint – the point halfway between two others

Subtotal – the total of part of a larger set of numbers

Are you ready?

1 Find the mean of each set of data.

    a 9   10   17   20   35

    b 5   8   8   8   8   12   7

2 Huda counts the number of sweets in 12 bags of sweets. There are 420 sweets altogether. Find the mean number of sweets in a bag.

3 There were 19 adults and 11 children at a party. Each adult had 2 pieces of cake and each child had 3 pieces of cake.

    a How many pieces of cake

        i   did the adults have altogether   **ii** did the children have altogether

        iii were eaten altogether?

    b What was the mean number of pieces of cake eaten?

4 Find the number halfway between

    a 0 and 10   **b** 20 and 29   **c** 60 and 65

Models and representations

Frequency table

Number of goals	Frequency
0	5
1	8
2	6
3	4
4	1
Total	24

Grouped frequency table

Height, h (cm)	Frequency
$150 < h \leqslant 160$	10
$160 < h \leqslant 170$	27
$170 < h \leqslant 180$	29
$180 < h \leqslant 190$	22
$190 < h \leqslant 200$	12
Total	100

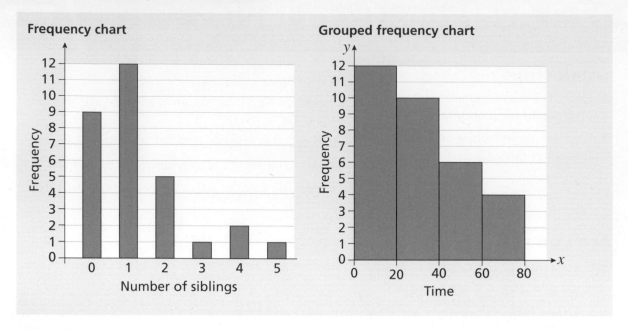

Frequency chart

Grouped frequency chart

When you have a large amount of data, it is better to organise it into a table rather than leave it in a list. You can find the mean of data in a table without having to deal with every individual piece of data.

Example 1

In a class of 30 students, 12 students have one pet, 7 have two pets, 5 have three pets and the rest have no pets.

a Show this information in a frequency table.

b Find the total number of pets owned by the class.

c Find the mean number of pets owned by the students.

a $30 - (12 + 7 + 5) = 6$ ○——— Work out how many students have no pets.

Number of pets	Frequency
0	6
1	12
2	7
3	5
Total	30

Put the information in a table.

Include a row for the total.

b

Number of pets	Frequency	Subtotals
0	6	$0 \times 6 = 0$
1	12	$1 \times 12 = 12$
2	7	$2 \times 7 = 14$
3	5	$3 \times 5 = 15$
Total	30	41

Total number of pets = 41

Add a third column to your table and fill it in with the total number of pets owned by the students with 3, 2, 1 and no pets.

Multiply each number of pets by the corresponding frequency.

You find the total number of pets by adding up the **subtotals** in your table.

c

Number of pets	Frequency	Subtotals
0	6	$0 \times 6 = 0$
1	12	$1 \times 12 = 12$
2	7	$2 \times 7 = 14$
3	5	$3 \times 5 = 15$
Total	30	41

This is a much better way of finding the total than by adding up each number individually:
$0 + 0 + 0 + 0 + 0 + 0 + 1 + 1 + \ldots$ etc. would take a long time and be prone to error.

$$\text{Mean} = \frac{\text{total}}{\text{number of items}} = \frac{41}{30} = 1.37$$

(to 3 significant figures)

The mean is the total number of pets divided by the total number of students in the class.

Practice 17.2A

1 The table shows the numbers of goals scored by a football team in 24 matches.

   a Copy and complete the table.

   b Find the mean number of goals scored.

Number of goals	Frequency	Subtotals
0	5	
1	8	
2	6	
3	4	
4	1	
Total	24	

2 The table shows the ages of the members of a youth club.

Work out the mean age of the members of the youth club.

Age	Frequency
12	8
13	12
14	15
15	7
16	3

3 In a times-tables test, 13 students scored 10 out of 10, nine students scored 9 out of 10, six students scored 8 out of 10, and four students scored 7 out of 10.

   a Show this information in a frequency table.

   b Find the mean score obtained in the test. Give your answer to 2 decimal places.

4 The table shows the shoe sizes of a group of people.

Shoe size	7	8	9	10	11
Number of people	4	10	15	8	3

   a Write down the modal shoe size.

   b Find the mean shoe size.

5 The bar chart shows the numbers of siblings of the members of a class.

a Show the information as a frequency table.

b Find the mean number of siblings per student.

c Find the mean number of siblings for the students in your class.

d How does the modal number of siblings in your class compare with the mode of the number of students in the class represented in the bar chart?

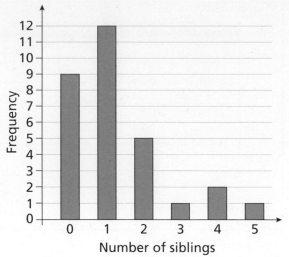

6 In a game, you can score –2, –1, 0, 1 or 2 points. Here are Sven's results from playing the game 80 times.

a Do you think Sven's mean score will be positive or negative? Why?

b Calculate Sven's mean score and see if you were correct.

Score	Number of times
–2	12
–1	18
0	30
1	13
2	7

What do you think? 💭

1 **a** What mistakes have been made in working out the mean from these frequency tables?

i

Number	Frequency	Subtotals
46	8	368
47	5	235
48	11	528
49	3	147
	27	1278

$$\text{Mean} = \frac{1278}{4} = 319.5$$

ii

Number	Frequency	Subtotals
0	10	10
1	8	8
2	7	14
3	6	18
4	8	32
	39	82

$$\text{Mean} = \frac{82}{39} = 2.1$$

b Copy the tables and correct the mistakes. Then work out the correct mean value for each table.

2

Number of goals	Frequency
0	5
1	8
2	6
3	4
4	1

The median number of goals scored is 2 because it's in the middle of 0, 1, 2, 3 and 4

a Explain why Marta is wrong.

b By writing out the full list of the numbers of goals scored, find the correct median.

c How can you find the median from a frequency table without writing out the data in a list?

If data has been put into groups, you do not know the values of all the data items and so you cannot find the exact mean of the data. You can use the **midpoint** of each group to give an **estimate** of each subtotal and then use these to work out an estimate of the mean.

The midpoint is the value in the middle of each class interval.

Example 2

The table shows the masses of some eggs.

Mass (g)	Frequency
$40 < m \leqslant 50$	12
$50 < m \leqslant 55$	18
$55 < m \leqslant 60$	20
$60 < m \leqslant 70$	13

Find an estimate for the mean mass of the eggs.

The midpoints are found by adding the end points of each class and then dividing by 2 Add columns for the midpoints and subtotals to your table.

Mass (g)	Frequency	Midpoint	Midpoint × frequency
$40 < m \leqslant 50$	12	45	540
$50 < m \leqslant 55$	18	52.5	945
$55 < m \leqslant 60$	20	57.5	1150
$60 < m \leqslant 70$	13	65	845
Total	63		3480

The estimates of the subtotals are found by multiplying each midpoint by the corresponding frequency e.g. 12 × 45 = 540

$$\text{Estimate of the mean} = \frac{\text{total}}{\text{number of items}}$$

$$= \frac{3480}{63} = 55.2 \, g$$

The estimate of the mean is the total of all the subtotals divided by the total number of eggs.

Practice 17.2B

1 The table shows the heights of 100 students.

Height (cm)	Frequency	Midpoint	Midpoint × frequency
$150 < h \geqslant 160$	10		
$160 < h \geqslant 170$	27		
$170 < h \geqslant 180$	29		
$180 < h \geqslant 190$	22		
$190 < h \geqslant 200$	12		
Total	100		

  a Copy and complete the table to find an estimate of the mean height of the students.

  b Why is there no space for the total of the Midpoint column?

2 The table shows information about the ages of visitors to an art gallery one day.

Age, a (years)	Frequency
$0 < a \leqslant 18$	30
$18 < a \leqslant 30$	25
$30 < a \leqslant 50$	61
$50 < a \leqslant 65$	72
$65 < a \leqslant 90$	58

  a Estimate the mean age of the visitors to the art gallery.

  b State the modal class.

3 As part of a job interview, 200 candidates took a Maths test. The times taken to complete the test are shown in the table.

Time taken, t (min)	$0 < t \leqslant 15$	$15 < t \leqslant 30$	$30 < t \leqslant 60$	$60 < t \leqslant 120$
Number of people	94	57	38	11

a State the modal class.

b Estimate the mean time taken to complete the test.

c For which classes do you think using the midpoint is likely to be reliable, and for which is it likely to be less reliable? Why?

4 A vet recorded the mass of each animal that came to the surgery one day. The results are shown in the table.

Mass, m (kg)	Frequency
$0 < m \leqslant 1$	7
$1 < m \leqslant 3$	14
$3 < m \leqslant 5$	23
$5 < m \leqslant 10$	24
$10 < m \leqslant 20$	8

Estimate the mean mass of the animals that came to the surgery that day.

5 The diagram shows the times spent by a class of students on their Maths homework.

Use the information in the diagram to find an estimate for the mean amount of time that these students spent on their homework.

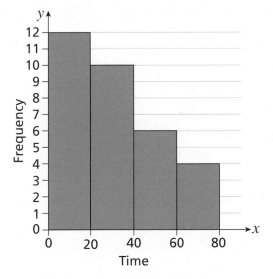

What do you think? 🗨

1 Why do you think the midpoint is used to provide estimates for the subtotals in grouped frequency tables?

2 The table shows the ages of people at a concert.

Why is it more difficult to find an estimate of the mean age for this data?

Compare strategies with a partner. How similar or different are your estimates of the mean?

Age, a (years)	Frequency
$16 < a \leqslant 25$	9
$25 < a \leqslant 35$	25
$35 < a \leqslant 45$	120
$45 < a \leqslant 60$	65
$60 < a$	32

Consolidate – do you need more?

1 The table shows the numbers of points scored by the players in a game.

Number of points	Number of players	Subtotals
0	4	
1	5	
2	12	
3	15	
4	0	
5	3	
6	1	
Total	40	

Find the mean score of these players.

2 This spinner is spun 120 times.

Score	1	4	5	8	10
Number of people	18	32	25	26	19

a Find the mean score obtained on the spinner.

🗨 **b** Do you think the spinner is fair?

3 The table shows how long some runners took to complete a marathon.

Time, t (hours)	Frequency
$2 < t \leqslant 3$	3
$3 < t \leqslant 4$	10
$4 < t \leqslant 5$	28
$5 < t \leqslant 7$	12

Work out an estimate of the mean time taken by these runners to complete the marathon.

4 The table shows the times that 50 patients waited at the dentist before their appointment.

a Find an estimate for the mean time that these patients waited.

b Explain why your answer is an estimate.

Time, t (min)	Frequency
$0 < t \leqslant 5$	3
$5 < t \leqslant 10$	8
$10 < t \leqslant 15$	21
$15 < t \leqslant 20$	16
$20 < t \leqslant 30$	2

Stretch – can you deepen your learning?

1 Collect some data from your class that is suitable to be put into groups.

Put the data into groups of different sizes and compare the mean from each method of grouping with the actual mean.

Calculate the percentage error from the actual mean for each of your estimates.

Do groups with smaller class sizes always produce more accurate estimates?

2 a How can you find which group in a table of data will contain the median?

b Suggest ways of estimating the median and range from a table. Compare your estimates with the actual medians and ranges using your own data from question **1**

Reflect

Explain why you can find the exact mean of a set of data in an ungrouped frequency table but only an estimate for the mean of a set of data in a grouped frequency table.

Small steps

- Identify outliers
- Compare distributions using averages and the range

Key word

Outlier – a value that differs significantly from the others in a data set

Are you ready?

1 **a** How do you find the range of a set of data?

b Is the range an average? Why or why not?

2 Find the mean, median, mode and range of each of these sets of data.

a 6 4 10 10 2

b 3 7 9 1 4 3

c 6.2 7.9 5.3 1.6

3 The tallest student in Class 8C is 175 cm tall. The range of the heights of the students in Class 8C is 31 cm. Work out the height of the shortest student in Class 8C.

Models and representations

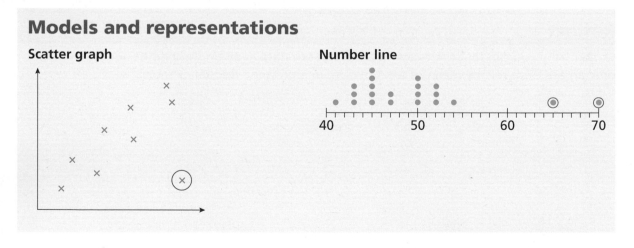

Scatter graph

Number line

Graphs and charts often have points plotted that do not seem to fit with the rest of the data. These are called **outliers** as they lie outside the expected range of the data.

Example 1

Here are the scores obtained by a group of students in a test.

56% 71% 63% 81% 2% 59% 70% 68% 61% 90% 82% 66%

a Find the mean and the range of the scores.

b Which data value might be regarded as an outlier?

c How do the mean and range of the scores change if the outlier is excluded?

a Mean $= \dfrac{\text{total}}{\text{number of items}} = \dfrac{769}{12}$

$= 64.1\%$ (1 decimal place)

Range $= 90\% - 2\% = 88\%$ ○——— To find the range, you subtract the smallest value from the largest value.

b The 2% score is the outlier. ○——| 2% is very different from the rest of the scores.

c New mean $= \dfrac{\text{total}}{\text{number of items}} = \dfrac{767}{11}$

$= 69.7\%$ (1 decimal place)

The mean has gone up by more than 5%

Range $= 90\% - 56\% = 34\%$

The range is less than half the size it was before. ○——| One outlier can have a big effect on the mean and the range.

Practice 17.3A

1 Here is a set of data.

2   6   35   7   6   4

  a Find the mean, median and mode of the data.

  b Which data item is an outlier?

  c Find the mean, median and mode of the data after the outlier is removed. What effect does removing the outlier have on these averages?

2 Filipo records the number of text messages he sends every day for a week.

  45   62   51   0   85   67   54

  a Find the mean, median and range of the number of texts he sent that week.

  b Which number is an outlier?

  c Find the mean, median and range of the number of texts he sent after the outlier is removed.

3 The lengths (in centimetres) of eight leaves are recorded as part of a science experiment.

  3.7   4.6   4.8   5.3   4.7   56   7.2   4.9

  a Which measurement is an outlier?

  b What value do you think Abdullah would replace the outlier with?

I think this outlier is due to an error in recording.

4 The table shows a company's sales over a 12-month period.

Month	Jan	Feb	Mar	Apr	May	Jun	Jul	Aug	Sep	Oct	Nov	Dec
Sales (£1000s)	23	11	9	14	12	16	14	11	10	18	26	94

a In which month are the sales very different from the others?
Why do you think this may be?

b Compare the range of monthly sales with and without the outlier.

c Which of the averages would change if the outlier were excluded from the calculation?

5 The times taken, in seconds, for ten students to solve a puzzle were recorded.

45   49   12   62   54   45   39   56   65   72

If the outlier is excluded

a by how much would the range change

b by how much would the mean change?

6 The maximum daily temperature (in °C) is recorded each day for a week.

8   7   −5   4   5   6   4

a Find the mean and range of the temperatures.

b

I think that one day was particularly cold.

I think that one of the temperatures was recorded incorrectly.

What do you think? Can you be sure?

c How would the mean and range change if

i the temperature of −5 °C was excluded

ii the temperature of −5 °C was replaced with 5 °C?

7 The chart shows the masses (in kilograms) of some leopards.

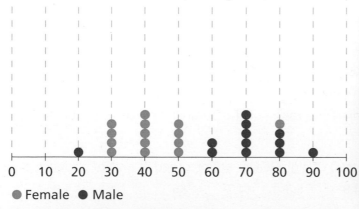

● Female ● Male

Do you think there are any outliers? Do you think they are genuine outliers
or could they be due to errors in recording the data?

8 A class of 30 students guess how many sweets there are in a jar.

The mean guess is 648

One student's guess was 9000

If this outlier is removed

a find the new mean of the guesses

b work out the percentage change in the mean.

What do you think? 💭

1 The mean mark for a class of 32 students in a Maths test is 68%. If the results of two students are excluded, the mean rises to 72%. Find the mean mark of the two students whose results were excluded.

2

> An outlier will always be the maximum or minimum value of a set of data.

Do you agree?

3 Look back at the scatter graphs you studied and drew in Block 5
Do you think that any of the values on any of the graphs are outliers?

4 How far away from the rest of the values in a set of data does a value have to be before it is considered to be an outlier?

In Chapter 16.4 you compared data sets by looking at charts and the range. You can also use averages to compare sets of data.

Example 2

The table shows some information about the masses of apples grown on two farms.

	Mean (g)	Range (g)
Farm A	170	20
Farm B	176	61

Compare the masses of the apples grown on the two farms.

On average, the apples from farm B are heavier than the apples from farm A, because the mean mass of the apples is greater.

The range of the masses of the apples from farm A is smaller, so they are more similar in size.

You can compare the masses using an average; this is usually the mean or the median.

The range tells you how spread out the data is. The smaller the range, the closer together the masses are.

Practice 17.3B

1 Class 8C sit tests in English, Maths and Science. The table shows some summary information about their results.

a In which subject did students do best on average?

b In which subject were the students' marks most similar?

Subject	Median mark	Range of marks
English	56	35
Maths	61	47
Science	59	22

2 Rhys and Ali both play cricket. The table shows the mean and the range of the numbers of runs they scored last season.

Compare their performances last season, explaining your reasoning.

	Mean	Range
Rhys	56	32
Ali	61	57

3 Here are the times taken, in seconds, by two sprinters in their practice runs for a race.

Sprinter A	12.6	12.8	13.5	12.7	13.9	12.6	13.1	14.2	12.4

Sprinter B	12.7	11.9	15.1	11.8	15.2	14. 4	11.8

a Which sprinter is quicker on average?

b Which sprinter is more consistent?

4 Here are the ages (in years) of two seven-a-side football teams.

Beckett's Brigade	35	37	28	39	51	25	28

Tommo's Team	42	43	41	49	47	43	41

a Use the median and the range to compare the ages of the two teams.

b Would your conclusion be different if you used the mean instead of the median?

5 Some Year 10 and Year 11 students took part in an experiment on reaction times. The table shows some information about the results.

	Quickest time (seconds)	Median (seconds)	Slowest time (seconds)
Year 10	0.3	1.2	5.6
Year 11	0.5	1.1	3.4

a Compare the reaction times of the Year 10 students with those of the Year 11 students.

b The slowest time for Year 10 is treated as an outlier and replaced with the next slowest time of 2.6 seconds. How does this affect your conclusions in part **a**?

6. The table shows information about the times taken, in seconds, to complete a shape puzzle by some people in different age groups.

   a   Samira thinks that the over 60s did best because they have the greatest mean. Explain why Samira is not correct.

   b   Put the age groups in order, from most consistent to least consistent.

Age (years)	Mean (seconds)	Range (seconds)
12–16	25	18
16–30	23	16
30–60	31	22
Over 60	44	11

What do you think? 💡

1   Is the range a reliable measure for comparing the spread of two data sets? What might you do to make it more reliable? Use the data sets in Practice 17.3A to help you decide.

Consolidate – do you need more?

1   Which value or values on the diagram do you think are outliers?

2   Here is a set of data.

   61   63   16   63   62

   a   Find the mean, median and range of the data.

   b   Which item of data is an outlier?

   c   Find the mean, median and range of the data when the outlier is removed.

   d   The outlier was recorded incorrectly and its digits should be reversed. Which measures will change when this is corrected?

3   Which average is most affected by outliers: the mean, the median or the mode?

4   The table shows some information about how boys and girls performed in a test.

	Median mark	Range of marks
Boys	56	75
Girls	61	23

Compare the boys' results with the girls' results.

5 A Year 4 teacher runs a two-week programme with her class on learning their times tables. Information about the class's performance in a times-tables test before and after the two-week programme is shown in the table.

	Lowest mark	Mean mark	Highest mark
Before	3	12.5	19
After	9	14.5	20

Do you think that the programme was effective? What other information might be useful to help you to decide?

Stretch – can you deepen your learning?

1 Investigate "frequency polygons" on spreadsheets or data programs. How do they represent the data? How can they be used to compare data?

2 Investigate some popular holiday resorts in the UK and abroad. Find data for several years and compare average temperatures and amounts of rainfall. Which destinations have higher average temperatures and higher amounts of rainfall on average? Which are more consistent? Which holiday destinations would you recommend and why?

Reflect

1 Explain how can you find out whether an outlier is an error or just a piece of data that is very different from the rest.

2 Describe how you can compare two sets of data. What measures would you use? What diagrams would you use?

I have become **fluent in...**	I have developed my **reasoning skills by...**	I have been **problem-solving through...**
■ calculating the mean, median and mode of a list of data ■ finding the range of a set of data ■ calculating the mean from a frequency table Ⓗ ■ estimating the mean of grouped data. Ⓗ	■ deciding which average is appropriate to use to summarise data ■ interpreting the range ■ identifying outliers ■ considering the causes of outliers ■ comparing sets of data.	■ interpreting data in different forms ■ using charts and graphs to find averages ■ investigating how changing a data set changes averages and the range ■ adjusting data sets to deal with outliers.

Check my understanding

1 For each data set, find the mode.

  a 7 5 6 6 4 5 5 7 4 5   **b** 2 9 4 6 7 5 3 8

  c 6 2 7 2 6 8 2 6 8

2 The numbers of words on the first six pages of a book are

  522   510   518   506   512   516

  a Find the mean and the median of the number of words on the pages.

  b Find the range of the number of words on the pages.

3 Ten students estimated the size of an angle. Here are the results.

  52°   31°   54°   56°   50°   50°   46°   58°   55°   52°

  a Which estimate is an outlier?

  b What is the effect on the median if the outlier is removed?

  c Find the mean and range of the data if the outlier is excluded.

4 A vet records the masses of the cats and dogs that come to the surgery during a month. The table summarises the data.

Compare the masses of the cats and the dogs.

Animal	Mean (kg)	Range (kg)
cat	4.5	3.2
dog	8.7	7.8

5 The table shows the times taken by teachers at a school to travel to work each day. Ⓗ

Time taken, t (min)	$0 < t \leqslant 10$	$10 < t \leqslant 20$	$20 < t \leqslant 30$	$30 < t \leqslant 60$
Number of teachers	8	24	61	5

Find an estimate for the mean time taken.

Glossary

Adjacent – next to each other

Adjacent sides/angles – sides or angles that are next to each other

Algebraic fraction – a fraction whose numerator and/or denominator are algebraic expressions

Alternate angles – a pair of angles between a pair of parallel lines on opposite sides of a transversal

Area – the amount of space inside a 2-D shape

Average – a number representing the typical value of a set of data

Axis – a reference line on a graph

Base – the number that gets multiplied when using a power/index

Binomial – expression with two terms

Bisect – to cut in half

Bisector – a line that divides something into two equal parts

Capacity – how much space a container holds

Centi – one hundredth

Check – find out if you are correct

Circumference – the distance around the edge of a circle

Class interval – the range of data in each group

Coefficient – a number in front of a variable, for example for $4x$ the coefficient of x is 4

Co-interior angles – a pair of angles between a pair of parallel lines on the same side of a transversal

Commutative – when an operation can be in any order

Compare – to evaluate two or more things to find the differences between them

Comparison bar models – two or more bar models to represent the parts in a problem

Compound shape – also known as a composite shape, this is a shape made up of two or more other shapes

Congruent – exactly the same size and shape, but possibly in a different orientation

Constant – not changing

Construct – draw accurately using a ruler and compasses

Continuous – data that is measured

Conversion graph – a graph used to change from one unit to another

Convert – change from one form to another, for example a percentage to a decimal

Coordinate – an ordered pair used to describe the position of a point

Correlation – a connection between two or more things

Corresponding angles – a pair of angles in matching positions compared with a transversal

Cube root – the cube root of a number is a value that, when multiplied by itself three times, gives that number

Curve – a line on a graph showing how one quantity varies with respect to another

Decimal places – the number of digits to the right of the decimal point in a number

Decrease – to make something smaller

Degree of accuracy – how precise a number is

Denominator – the bottom number in a fraction; it shows how many equal parts one whole has been divided into

Diagonal – a line that is neither horizontal nor vertical; in quadrilaterals, a line segment that joins two opposite vertices

Diameter – the distance from one point on a circle to another point on the circle through the centre

Direct proportion – two quantities are in direct proportion when as one increases or decreases, the other increases or decreases at the same rate

Discrete – data that can only take certain values

Distribution – the way in which data is spread out

Divide – to split into equal groups or parts

Divide in a ratio – share a quantity into two or more parts so that the shares are in a given ratio

Double number line – two lines used to represent ratio problems

Enlargement – making a shape bigger, or smaller

Equation – a statement with an equal sign, which states that two expressions are equal in value

Equidistant – at the same distance from

Equivalent – numbers or expressions that are written differently but are always equal in value

Error interval – the range of values a number could have taken before being rounded

Estimate – an approximate answer or to give an approximate answer

Exchange rate – the value of a currency compared to another

Expand – multiply to remove brackets from an expression

Express – write, often in a different form

Expression – a collection of terms involving mathematical operations

Exterior angle – an angle between the side of a shape and a line extended from the adjacent side

Factor – a positive integer that divides exactly into another positive integer

Factorise – find the factors you need to multiply to make an expression

Formula – a rule connecting variables written with mathematical symbols

Fraction – a number that compares equal parts of a whole

Frequency – the number of times something happens

General term – an expression that generates the terms of a sequence

Geometric sequence – a sequence is geometric if the value of each successive term is found by multiplying or dividing the previous term by the same number

Give a reason – state the mathematical rule(s) you have used, not just the calculations you have done

Gradient – the steepness of a line

Grouped data – data that has been ordered and sorted into groups called classes

Highest common factor (HCF) – the greatest number that is a factor of every one of a set of numbers

Hypothesis – an idea to investigate that might be true or false

Identity – a statement that is true no matter what the values of the variables are

Improper fraction – a fraction in which the numerator is greater than the denominator

Increase – to make something larger

Index (plural: **indices**) – an index number (or power) tells you how many times to multiply a number by itself

Inequality – a comparison between two quantities that are not equal to each other

Integer – a whole number

Interior angle – an angle on the inside of a shape

Isosceles – having two sides the same length

Key – used to identify the categories present in a graph. A key on a pictogram tells you how many items/people each picture stands for

Kilo – one thousand

Line graph – this has connected points and shows how a value changes over time

Line of symmetry – a line that cuts a shape exactly in half

Line segment – part of a line that connects two points

Linear – forming a straight line

Linear sequence – a sequence whose terms are increasing or decreasing by a constant difference

Loss – if you buy something and then sell it for a smaller amount,
loss = amount paid – amount received

Map – a diagram of a place, such as a town or a country

Mean – the result of sharing the total of a set of data equally between them

Median – the middle number in an ordered list

Midpoint – the point halfway between two others

Milli – one thousandth

Misleading graphs – a graph that suggests an incorrect conclusion or assumption

Mixed number – a number presented as an integer and a proper fraction

Modal class – the class in a set of grouped data that contains the highest frequency

Mode – the item which appears most often in a set of data

Multiple bar chart – a way to represent several, related sets of data

Multiplier – a number you multiply by

Negative – less than zero

Non-linear – not forming a straight line

Non-linear sequence – a sequence whose terms are not increasing or decreasing by a constant difference

Non-unit fraction – a fraction with a numerator that is not 1

Numerator – the top number in a fraction that shows the number of parts

Opposite sides/angles – sides or angles that are not next to each other

Order of operations – the rules that tell you the order in which to perform each part of a calculation

Origin – the point where the x-axis and y-axis meet

Original value – a value before a change takes place

Outcome – the possible result of an experiment

Outlier – a value that differs significantly from the others in a data set

Parallel – always the same distance apart and never meeting

Perpendicular height – the height of a shape measured at a right angle to the base

Pi – pronounced "pie" and written using the symbol π. It is the ratio of the circumference of a circle to its diameter

Polygon – a closed 2-D shape with straight sides

Power (or exponent) – this is written as a small number to the right and above the base number, indicating how many times to use the number in a multiplication. For example, the 5 in 2^5

Primary data – data you collect yourself

Probability – how likely an event is to occur

Profit – if you buy something and then sell it for a higher amount,
profit = amount received – amount paid

Proof – an argument that shows that a statement is true

Proportion – a part, share or number considered in relation to a whole

Quadrant – one of the four sections made by dividing an area with an x-axis and a y-axis

Qualitative – data that describes characteristics

Quantitative – numerical data

Questionnaire – a list of questions to gather information

Radius – the distance from the centre of a circle to a point on the circle

Random – happening without method or conscious decision; each outcome is equally likely to occur

Range – the difference between the greatest value and the smallest value in a set of data

Ratio – a ratio compares the sizes of two or more values

Reciprocal – the result of dividing 1 by a given number. The product of a number and its reciprocal is always 1

Reduce – to make something smaller

Reflection – a transformation resulting in a mirror image

Regular – a shape that has equal sides and equal angles

Regular polygon – a polygon whose sides are all equal in length and whose angles are all equal in size

Reverse percentage – a problem where you work out the original value

Sample – a selection taken from a larger group

Sample space – the set of all possible outcomes or results of an experiment

Satisfy – make an equation or inequality true

Scale – the ratio of the length in a drawing or a model to the actual object

Scale drawing – a diagram that represents a real object with accurate sizes reduced or enlarged by a ratio

Scale factor – how much a shape has been enlarged by

Secondary data – data already collected by someone else

Sector – a part of a circle formed by two radii and a fraction of the circumference

Sequence – a list of items in a given order, usually following a rule

Significant figure – the most important digits in a number that give you an idea of its size

Similar – two shapes are similar if their corresponding sides are in the same ratio

Simplify – rewrite in a simpler form, for example rewrite $8 \times h$ as $8h$

Solution – a value you can substitute in place of the unknown in an equation to make it true

Solution set – a range of values for which a statement is true

Solve – find a value that makes an equation true

Square root – a square root of a number is a value that, when multiplied by itself, gives the number

Standard form – a number written in the form $A \times 10^n$ where A is at least 1 and less than 10, and n is an integer

Substitute – to replace letters with numerical values

Subtotal – the total of part of a larger set of numbers

Symmetrical – when one half of a shape is the mirror image of the other

Term – in algebra, a single number or variable, or a number and variable combined by multiplication or division; in sequences, one of the members of a sequence

Term-to-term rule – a rule that describes how you get from one term of a sequence to the next

Transversal – a line that crosses at least two other lines

Two-way table – this displays two sets of data in rows and columns

Unit fraction – a fraction with a numerator of 1

Unknown – a variable (letter), whose value is not yet known

Variable – a numerical quantity that might change, often denoted by a letter, for example x or t

Venn diagram – a diagram used for sorting data

Vertex (plural: **vertices**) – a point where two line segments meet; a corner of a shape

Vertically opposite angles – angles opposite each other when two lines cross

Volume – the amount of space taken up by a 3-D shape

y-intercept – the point at which a graph crosses or intersects the y-axis

Block 1 Ratio and scale

Chapter 1.1

Are you ready?

1 a $\frac{1}{3}$ **b** $\frac{3}{4}$ **c** $\frac{3}{4}$ **d** $\frac{5}{6}$

2 a e.g.

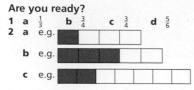

b e.g.

c e.g.

3 a $\frac{1}{3}$, e.g. as it is the unit fraction with the smallest denominator

b $\frac{4}{5}$, e.g. as they have the same denominator and $\frac{4}{5}$ has the greatest numerator

c $\frac{4}{5}$, compare methods as a class, e.g. convert to percentages

Practice 1.1A

1 a $3:2$ **b** $2:3$

2 a $4:1$ **b** $1:4$ **c** No, $4:5$
   d $1:5$

3 a $3:7$ **b** $7:5$ **c** $3:5$ **d** $7:15$

4 a
| Boys | Girls | Teachers |

b C and G are false, the rest are true

c C The ratio of girls to boys is $3:5$
   G The ratio of teachers to girls to boys is $2:3:5$

5 a $2:3$ **b** $2:5$ **c** $5:3$

What do you think?

1 Yes, when comparing two equal quantities

2 Both are correct

3 a $3:5$ **b** $5:8$ **c** $8:2$ or $4:1$
   d $9:10$

4 No, it could be any size where for every 3 boys there are 2 girls, e.g. a class of 25 students with 15 boys and 10 girls

5 Compare answers as a class

Practice 1.1B

1 Seb has not looked at the total number of parts. $\frac{2}{5}$ is white

2 a $\frac{4}{7}$ **b** $4:3$ **c** $3:7$

3 a $5:4$ **b** $\frac{5}{12}$ **c** $3:4$ **d** $\frac{1}{4}$

4 a $\frac{5}{7}$ **b** $\frac{2}{7}$

5 a $\frac{2}{5}$
   b She has compared to the total. The ratio should be $3:2$

What do you think?

1 a $1:4$ **b** Weaker
   c $2:3$, as $\frac{2}{5}$ is the greatest fraction

2 A is false, the rest of the statements are true

3 Compare answers as a class

4 Compare answers as a class

Consolidate

1 a $5:3$ **b** $3:5$
   c e.g.

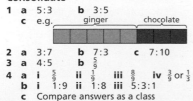
| ginger | chocolate |

2 a $3:7$ **b** $7:3$ **c** $7:10$

3 a $4:5$ **b** $\frac{5}{9}$

4 a i $\frac{5}{9}$ **ii** $\frac{1}{9}$ **iii** $\frac{8}{9}$ **iv** $\frac{3}{9}$ or $\frac{1}{3}$
   b i $1:9$ **ii** $1:8$ **iii** $5:3:1$
   c Compare answers as a class

---

5 e.g.

6 $1:3$

Stretch

1 £1 = 2 × 50p, so the ratio is $2:1$

2 a e.g. For the ratio $1:2$ the parts are $\frac{1}{3}$ and $\frac{2}{3}$ of the whole but for the ratio $2:3$ the parts are $\frac{2}{5}$ and $\frac{3}{5}$ of the whole
   b e.g. $4:6$, $40:60$, $10:15$, etc.

3 a i $3:1$ **ii** $1:3$ **iii** $4:2$ or $2:1$
   b $\frac{1}{3}$ **c** $\frac{3}{4}$
   d The answers to **a** would become
   i $1:3$, **ii** $3:1$, **iii** $4:2$ or $2:1$,
   b would be $\frac{3}{1}$ (or 3) and **c** would be $\frac{1}{4}$

4 Compare answers as a class

Chapter 1.2

Are you ready?

1 a $1:2$ **b** $2:1$

2 a $\frac{4}{7}$ **b** $\frac{3}{7}$

3 a £24 **b** £56
   c £96 **d** £4000

4 a £1.80 **b** £54
   c Compare answers as a class

Practice 1.2A

1 a 80 **b** 5

2 a $\frac{1}{8}$
   b i 35 kg **ii** 9 kg

3 a 300 ml **b** 40 ml **c** $\frac{5}{6}$

4 a $1:4$ **b** 120 **c** 10

5 a 64 cm **b** 5 cm **c** 120 cm
   d $\frac{1}{10}$

What do you think?

1 Compare answers as a class

2 a 12 cm **b** 18 cm **c** 30 cm
   d 3 cm **e** 1.5 cm **f** 1 cm

Practice 1.2B

1 a 8 **b** 18 **c** 45

2 a 25 **b** 12
   c To find one part, 10 is not divisible by 3 but it is divisible by 5

3 a $4:3$ **b** 280

4 a $3:2$ **b** £20 **c** £6

5 8 hours admin and 20 hours marketing

What do you think?

1 a

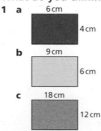

b

c

2 $1:4$

Consolidate

1 a 300 **b** 12 **c** $\frac{5}{6}$

2 a i 18 cups **ii** 4 cups
   b Chloe will need 3 cups of flour, Jakub will need 24 cups of milk

3 a 20 litres **b** 7.5 litres

4 78 cm

5 a $2:3:5$ **b i** 20 **ii** 12
   c i 75 **ii** 50
   d Compare answers as a class

Stretch

1 Abdullah, as 30, 60 and 90 sum to 180

2 a 40°, 50° and 90° **b** 40°, 32° and 108°

3 a 78 **b** 15
   c 4100 **d** 67.5

4 Yes – exactly the right amount of fuel, assuming consumption at the same rate

Chapter 1.3

Are you ready?

1 48

2 2000

3 a £100 **b** $\frac{3}{8}$

4 a 8 **b** 16 **c** 36
   d 36 **e** 150

Practice 1.3A

1 a £20, £40 **b** £15, £45
   c £12, £48 **d** £24, £36
   e £10, £50 **f** £10, £20, £30
   g £30, £30

2 a 50 cm, 100 cm
   b 37.5 cm, 112.5 cm
   c 60 cm, 90 cm **d** 45 cm, 105 cm
   e 70 cm, 80 cm

3 5 litres of red and 10 litres of white

4 5 cm

5 £16, £24 and £32

6 £64

What do you think?

1 a Rhys only had 5 parts, not all 8, so the first calculation should be £40 ÷ 5
   b £24

2 Compare answers as a class. Answers include $1:29$, $2:3$, $3:7$, etc.

3 The exact shares are £3.33$\frac{1}{3}$ and £6.66$\frac{2}{3}$

4 a $2:5$ **b** $3:5$

Practice 1.3B

1 a i B **ii** C **iii** A **iv** D
   b i £45, £75 **ii** £72, £192
   iii £200, £320
   iv £180, £300, £480

2 a 150 cm **b** 155 cm

3 a 48 cm, 64 cm, 80 cm
   b 9 cm, 12 cm, 15 cm
   c 5.4 cm, 7.2 cm, 9 cm
   d 27 cm, 36 cm, 45 cm

4 a i £90, £150 **ii** £60, £180
   iii £112.50, £127.50
   b Each share is £6$\frac{2}{3}$
   Write each share as £6.67.
   Compare strategies as a class
   c £130 and £110

5 a 28 and 12 **b** 70 and 30

6 a 6 cm **b** 16.5 cm
   c 12 cm **d** 72 cm
   e 4.5 cm

What do you think?

1 a 120 ml **b** 0.625 litres
   c 60 ml and 500 ml

2 a She needs another 200 ml of yellow and another 50 ml of white
   b 1350 ml, or 1.35 litres

Consolidate

1 a £30, £60 **b** £22.50, £67.50
   c £18, £72 **d** £36, £54
   e £15, £75 **f** £15, £30, £45
   g £45, £45

2 15, 20 and 25

3 a Rhys needs to get three times as much as Huda, so it should be 1:2:6
b Emily £400, Huda £800, Rhys £2400
4 255
5 a 15   **b** 32

Stretch
1 a The angles are 30°, 60° and 90° and 90° is a right angle
b 40°, 60° and 80°
c The middle angle is always 60°
2 a There are 8 parts in total, so 5 parts is $\frac{5}{8}$ of the total and 3 parts is $\frac{3}{8}$ of the total
b i $\frac{3}{7}, \frac{4}{7}$ **ii** $\frac{3}{4}, \frac{1}{4}$
iii $\frac{1}{6}, \frac{1}{3}, \frac{1}{2}$ **iv** $\frac{1}{4}, \frac{1}{3}, \frac{5}{12}$
v $\frac{a}{a+b}, \frac{b}{a+b}$
3 Both are correct
4 Compare answers as a class

Chapter 1.4
Are you ready?
1 a $\frac{1}{2}$ **b** $\frac{4}{5}$ **c** $\frac{3}{5}$ **d** $\frac{3}{4}$
2 $\frac{1}{15}$
3 a 1000   **b** 100
c 1000
4 a 1, 2, 3, 4, 6, 12   **b** 1, 2, 4, 5, 10, 20
c 4

Practice 1.4A
1 Compare answers as a class
2 5:3 with 25:15, 4:3 with 20:15
3:5 with 12:20, 4:5 with 20:25
3 a 4:5   **b** 5:6   **c** 5:1
d 5:6   **e** 1:3:6   **f** 2:3:9
g 6:9:10   **h** 7:10:16
4 4:1
5 4:3
6 a 15:2   **b** 20:3   **c** 1:5   **d** 5:2
7 e.g. 20 × 4:4 × 4 ≡ 20:4 ≡ 5:1
8 a All simplify to 3:2
b All are 1.5
c The ratios are equivalent, so the answers are the same

What do you think?
1 a All are £100 and £200
b The ratios are equivalent
2 a 2:3
b i They did not invest the same amount
ii £2640 and £3960
iii £3000 and £3600
3 a 2:5   **b** $\frac{2}{5}$
c Compare answers as a class

Practice 1.4B
1 a 1:0.6   **b** 1:2.5   **c** 1:1.33
d 1:0.24   **e** 1:3.33
2 a 1.67:1   **b** 0.4:1   **c** 0.75:1
d 4.25:1   **e** 0.3:1
3 a 1:0.4   **b** 2.5:1
4 350 g × 3 ≠ 1 kg. He needs to add 50 g more flour
5 a 10   **b** 15   **c** 18
6 a 1:1.5   **b i** 60   **ii** 24

What do you think?
1 a 0.625:1   **b** $\frac{4}{3}$:1   **c** 1:$\frac{a}{b}$
2 2.5:1
3 Compare answers as a class

Consolidate
1 a 5:1   **b** 1:10   **c** 1:4
d 4:5   **e** 4:5:6
2 a 3:5   **b** 1:3   **c** 6:1
d 5:6   **e** 3:5   **f** 1:5
3 a 25   **b** 30   **c** 5

4 a 5:2   **b** 1:0.4   **c** 2.5:1
5 a i 1:0.67   **ii** 1.5:1
b i 1:2.5   **ii** 0.4:1
c i 1:0.625   **ii** 1.6:1
d i 1:1.6   **ii** 0.625: 1

Stretch
1 a 40g of flour and 60 ml of milk
b 15g of flour, 22.5 ml of milk and $\frac{3}{8}$ of an egg
c It's hard to work with fractions of an egg
d Compare answers as a class
2 a 1:1.78   **b** 1:1.56   **c** 1:1
3 a i 3:1   **ii** 3:2   **iii** 4:1
b i 1:3   **ii** 1:6   **iii** 4:3

Chapter 1.5
Are you ready?
1 a 22 cm   **b** 24 cm   **c** 28 cm
2 a 31   **b** 31   **c** 31
d 31   **e** 32   **f** 32
3 a All are 2:5   **b** All are 0.4
c Compare answers as a class

Practice 1.5A
1 a 3.14   **b** 3.142   **c** 3.14
2 a 88 cm   **b** 355 mm   **c** 154 mm
3 a 70 mm   **b** 220 mm
4 a He has not included the diameter as part of the perimeter
b 18 cm
5 a 144 cm   **b** 100 cm   **c** 229 mm
6 Circumference is the the name given to the perimeter of a circle.

What do you think?
1 a It also doubles. The ratio of circumference to diameter is constant
b i 180 cm   **ii** 15 cm
2 a 1:2   **b** 1:2π
c Yes, because $d = 2r$

Practice 1.5B
1 a 1:3   **b** 1:1   **c** 2:3   **d** 1:4
e 3:4   **f** 4:3   **g** 4:1
2 a i 1:2   **ii** 2
b i 1:$\frac{1}{2}$ **ii** $\frac{1}{2}$
c i 1:5   **ii** 5
d i 1:1   **ii** 1
e i 1:$\frac{1}{3}$ **ii** $\frac{1}{3}$
f i 1:$\frac{1}{4}$ **ii** $\frac{1}{4}$
g i 1:2   **ii** 2
h i 1:$\frac{2}{3}$ **ii** $\frac{2}{3}$
i i 1:$\frac{5}{3}$ **ii** $\frac{5}{3}$
3 Compare answers with a partner
4 Compare answers with a partner

What do you think?
1 The first line has gradient 0, the second line has infinite gradient
2 Discuss answers as a class
3 −2

Consolidate
1 a 66 mm   **b** 710 mm   **c** 176 mm
2 a 40 mm   **b** 126 mm
3 257 cm
4 1:4
5 a 1   **b** 1.5   **c** $\frac{1}{3}$
d $\frac{5}{2}$ **e** $\frac{1}{6}$

Stretch
1 a Compare answers as a class
b i 264 cm **ii** 4.4 m
2 a i π is approximately 3 **ii** Less
b i 11 cm   **ii** 90 mm   **iii** 3 m
c i 5.5 cm   **ii** 45 mm   **iii** 1.5 m

3 6942, using 3.14 for π
4 Compare answers as a class

Check my understanding
1 a 80   **b** 90
2 a 120 cm   **b** 11:10
3 $\frac{2}{9}$
4 a 115
b Filipo would not have an integer number of cards
5 a £32 and £40   **b** £25
6 a Multiply the diameter by π
b Double the radius and multiply the result by π
7 e.g.

Block 2 Multiplicative change

Chapter 2.1
Are you ready?
1 a 1:3   **b** 5:8   **c** 4:5
2 a 80   **b** 21   **c** 11
3 a 5   **b** 20   **c** 30
d 7   **e** 7.5
4 a £37.50   **b** £375   **c** £750

Practice 2.1A
1 a £5.50   **b** £16.50
c £22   **d** £27.50
2 a 10.8lb   **b** 108lb
c 120ft   **d** 1.8lb
3 a 12 tablespoons butter, 3 onions, 1800 ml vegetable stock, 1500g tomatoes
b Compare answers as a class
c i 20
ii 20 tablespoons butter, 5 onions, 2500g tomatoes
4 a 800   **b** 15   **c** 25
5 a £11.60   **b** £69.60   **c** £29
c Milly's, e.g. 10 litres costs £11.40
6 a £3.60   **b** 200
c 1 costs 4.5p, so any odd number would cost an amount involving 0.5p

What do you think?
1 a Yes   **b** Yes   **c** No   **d** No
2 a No
b Compare answers as a class
3 a The first and third tables show direct proportion relationships
b First table: **i** 240   **ii** 41$\frac{2}{3}$
Third table: **i** 250   **ii** 40

Practice 2.1B
1 a £2.80, £4.20, £5.60
b Compare answers as a class
c

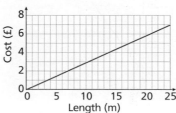

d Roughly £3.40
e Roughly 12.5m
f Yes, roughly 17m

2 a The ratio of number of copies to the cost is constant
b Compare answers as a class
c

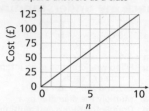

d i £37.50 **ii** £150
e 11
3 a 1.6 or $\frac{8}{5}$ for all pairs
b The ratio is constant
c

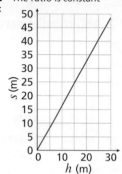

d i 12.5 m **ii** 19.2 m
4 a

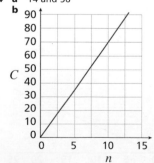

b i No; the graph does not go through (0, 0)
ii $\frac{l}{m}$ is not constant

What do you think?
1 You need (0, 0) and one other point
2 B and F

Consolidate
1 a 32 **b** 16 **c** 192 **d** 80
2 a £9.60 **b** £0.96 **c** £38.40 **d** 120
3 a 64 **b** 25
4 a 14 and 56
b

c £84

Stretch
1 A, B, E and G always and C sometimes
2 No
3 a Yes **b** Yes **c** No
4 Yes

Chapter 2.2

Are you ready?
1 A (3, 5), B (0, 7), C (3, 0), D (6, 1)
2

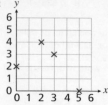

3 a 19 **b** 32 **c** 32 **d** 404
4 89p, 91p, £3.46 and £18.13

Practice 2.2A
1 a 45 litres **b** 6.7 gallons
c Discuss accuracy as a class
2 a Yes, 0 kg = 0 stones
b

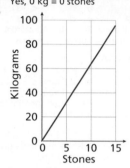

c i Roughly 51 kg
ii Roughly 12.6 stones
d Ed
3 a Yes **b** (80, 100)
c

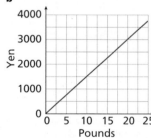

d i 75% **ii** 40% **iii** 65%
4 a i 21°C **ii** −12°C **iii** 29°C
b i 86°F **ii** 32°F **iii** 14°F
c They are not directly proportional

What do you think?
1 Jakub – easier to plot and draw a line through
2 Not all conversion graphs go through (0, 0)
3 a

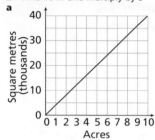

b i $28.30 **ii** $23.70
4 a e.g. Find £40 and multiply by 10
b e.g. Find £20 and multiply by 5

Practice 2.2B
1 a i $90 **ii** $360 **iii** $900
b The answer should be written as £5.40

c i £100 **ii** £400 **iii** £25.50
2 a i 1200 CZK **ii** 2700 CZK
iii 7500 CZK
b i £300 **ii** £0.50 **iii** £580
3 a 40 **b** 4000
c 180 **d** 10 000
4 a 63 000 **b** As 90 is less than 93
c 65 100
d 65 100 is not divisible by 200
e 325
5 a £290 **b** UK
6 £49 cheaper in Denmark

What do you think?
1 £90
2 a Both offices are £320 000
b Students' own check
3 London, e.g. £100 buys €110, but in Paris £100 buys €108.70

Consolidate
1 a i 19.5 ft **ii** 9.1 m
b Find 10 m and multiply by 5
2 a

b i 32 000 m² **ii** 2.5 acres
3 a 3000 yen
b

c i 2250 yen **ii** £18
4 a i 2400 birr **ii** 3840 birr
iii 96 000 birr
b i 200 **ii** £250 **iii** £5
5 £7292

Stretch
1 Compare answers as a class
2 Compare answers as a class

Chapter 2.3

Are you ready?
1 a 10 **b** 10 **c** 5 **d** 5
2 a 36 **b** 100
3 a 3:5 **b** 3:8 **c** 2:3
4 a $\frac{3}{5}$ **b** $\frac{5}{8}$ **c** $\frac{3}{5}$

Practice 2.3A
1 Rectangles 8 cm by 6 cm, 10 cm by 7.5 cm and 12 cm by 9 cm
2 a 1:2 **b** 2:5
3 Compare answers as a class
4 $a = 10$ cm, $b = 4$ cm, $c = 12$ cm, $d = 20$ cm, $e = 12$ cm
5 a Equilateral triangle with sides 5 cm
b 120 cm
c 6 cm
d Every angle in the triangles is 60°
6 a 8 mm **b** 100 mm

c 40 mm d 2:5

What do you think?
1 Otherwise images will not be in the same proportion
2 No, it could still be an enlargement
3 No, it is not an enlargement
4 Angles in corresponding positions in the shapes are equal

Practice 2.3B
1 a PQ b WX c DE
  d EF e XY f BC
2 Compare answers as a class
3 A, C, G and I increase the size
  B, D, E, F and H decrease the size
4 a 87 b 14 c 10
  d 6 e 7.5 f 19.2
5 All except C. Note, some photographs may have been rotated.

What do you think?
1 Faith's answer is exact but Jakub's is rounded
2 All of them except $9 \times 3 \div 7$
3 a 5 b $\frac{1}{4}$ or 0.25 c 1.25 or $\frac{5}{4}$
  d $\frac{5}{7}$ e $\frac{5}{2}$ or 2.5

Consolidate
1 Compare answers as a class
2 A and I, B and E, C and F, D and G, H and J
3 $x = 5$, $y = 21$, $z = 10$, $p = 13\frac{1}{3}$, $q = 8$, $k = 13\frac{1}{3}$
4 a Yes, enlargement by scale factor 5
  b No c No
  d Yes, enlargement by scale factor 1.25

Stretch
1 A, D and E are all true. Compare examples/non-examples as a class
2 Perimeter doubles, but area is multiplied by 4
3 Yes, 45°, 75° and 60°
4 a 37°, 103°, 40° b 10 cm
5 a PS b BC
  c DAB = 79°, PQR = 88°
  d AD = 15
  e 2 more sides (that are not a corresponding pair)

Chapter 2.4

Are you ready?
1 a 100 b 600 c 650 d 24
2 a 4 m b 0.4 m c 0.04 m d 1 m
3 a 600 b 29
4 a 160 ml b 62.5 ml

Practice 2.4A
1 a Rectangles of size i 8 cm by 4 cm, ii 4 cm by 2 cm, iii 2 cm by 1 cm
  b i 2 m by 1 m ii 8 mm by 4 mm
  c Flo's workings would give a scale of 1 cm to 2 m, not 2 cm to 1 m. The correct measurements are 16 cm by 8 cm
2 a Rectangle 9 cm by 5.5 cm
  b Smaller, it would be $\frac{1}{200}$ of the real size, which is less than $\frac{1}{100}$ of the real size
3 a 3 cm by 10 cm b 12 cm by 40 cm
4 a 1:50 b 1:400
  c 1:50 d 1:20
  e 1:250 f 1:250
5 a 9 cm by 6 cm b 20 m
6 Compare answers as a class
7 a Compare answers as a class
  b i 50 m ii 101 m iii 241 m

What do you think?
1 Compare answers as a class

2 a The same, as 200 cm = 2 m
  b i 1 cm to 4 m ii 2 cm to 3 m
    iii 1:250
3 There are 12 inches in a foot, so the scale is 1:12
4 Compare answers as a class

Practice 2.4B
1 a 0.8 cm b 40 km
  c More, the road will not be straight all the way
  d Compare answers as a class
  e 1 cm:50 km = 50000 m = 5000000 cm
2 a 10 km b 4 cm c 40 km
  d More accurate as the map is more detailed
  e Compare answers as a class
3 a 5 km b 8.3 cm c 41.5 km
  d Discuss as a class
  e Compare answers as a class
4 a 28.8 km b 25 cm
5 a 1:1000 b 1:200000
  c 1:4000000 d 1:500
  e 1:40000 f 1:250000

What do you think?
1 Jackson is correct because $\frac{1}{25000}$ is greater than $\frac{1}{50000}$
2 a 1 cm to 20 m b 1:2000
3 a 9 cm² b 1.2 km
  c 1.44 km² d 2304000

Consolidate
1 a Rectangles of size
    i 5 cm by 6 cm,
    ii 2.5 cm by 3 cm
  b i 12.5 cm by 15 cm
    ii 0.5 cm by 0.6 cm
  c Compare answers as a class
2 a 80 cm by 30 cm b 40 cm by 15 cm
  c 20 cm by 7.5 cm
3 a 10 000 cm, 100 m, 0.1 km
  b 60 000 cm, 600 m, 0.6 km
  c 1000 cm, 10 m, 0.01 km
  d 10 000 inches
4 1:200

Stretch
1 Yes, all are enlargements of the classroom by different scale factors
2 a 1 cm to 2 km b 1 cm to 5 km
  c 1 cm to 0.5 km d 1 cm to 0.25 km
  e 1 cm to 0.04 km
3 9.6 cm. Compare methods as a class
4 Compare answers as a class

Check my understanding
1 a £16 b 1000
2 a Roughly 4 inches
  b Roughly 25 cm
  c Roughly 125 cm
3 a £200 b 1820 dinar
  c 230 dinar
4 No, shortest and longest sides are enlarged by scale factor 1.5, but other side is enlarged by scale factor 1.56 (2 decimal places)
5 a 28 km b 15 cm

Block 3 Multiplying and dividing fractions

Chapter 3.1

Are you ready?
1 a 72 b 63 c 45
  d 88 e 130
2 a $\frac{1}{2}$ b $\frac{3}{8}$ c $\frac{3}{4}$
  d $\frac{7}{10}$ e $\frac{1}{9}$ f $\frac{5}{6}$

3 a $\frac{3}{4}$ b $\frac{3}{7}$ c $\frac{6}{11}$ d $\frac{3}{3}$ or 1
4 a $2\frac{2}{5}$ b $4\frac{1}{4}$ c $9\frac{1}{3}$ d $7\frac{10}{11}$
5 a $\frac{17}{5}$ b $\frac{25}{4}$ c $\frac{25}{3}$ d $\frac{83}{11}$

Practice 3.1A
1 There are 3 bar models, each split into sevenths with one shaded. In total there are 3 sevenths shaded.
2 a $4 \times \frac{1}{4}$ b $4 \times \frac{3}{4}$ c $7 \times \frac{1}{10}$ d $7 \times \frac{1}{13}$
3 a $4 \times \frac{1}{9} = \frac{4}{9}$ b $4 \times \frac{2}{9} = \frac{8}{9}$
  c $2 \times \frac{3}{10} = \frac{6}{10} \left(= \frac{3}{5}\right)$
  d Either $3 \times \frac{2}{10} = \frac{6}{10}$ or $6 \times \frac{1}{10} = \frac{6}{10} \left(= \frac{3}{5}\right)$
4 a 15 b 15 ones c 15 tens
  d 15 thousands e 15 cm
  f 15 trees g 15 eighteenths
  h $\frac{15}{18}$ i $15x$
5 a $\frac{2}{5}$ b $\frac{2}{7}$ c $\frac{3}{16}$ d $\frac{7}{19}$
  e $\frac{9}{100}$ f $\frac{4}{5}$ g $\frac{6}{7}$ h $\frac{15}{16}$
  i $\frac{14}{19}$ j $\frac{99}{100}$
6 There are six fifths in total. This is equivalent to one whole and one fifth.
7 a $\frac{3}{2} = 1\frac{1}{2}$ b $\frac{4}{3} = 1\frac{1}{3}$
  c $\frac{7}{4} = 1\frac{3}{4}$ d $\frac{8}{5} = 1\frac{3}{5}$
  e $\frac{15}{10} = 1\frac{1}{2}$ f $\frac{72}{10} = 7\frac{1}{5}$
  g $\frac{70}{11} = 6\frac{4}{11}$ h $\frac{35}{6} = 5\frac{5}{6}$
  i $\frac{105}{25} = 4\frac{1}{5}$ j $\frac{792}{100} = 7\frac{23}{25}$
8 a 5 b 7 c 3 d 9

What do you think?
1 a $\frac{12}{20}$ is equivalent to $\frac{3}{5}$. The answer cannot be $\frac{3}{5}$ because it has been multiplied by a number not equal to 1 so its value must change.
  b He has multiplied the denominator by 4 too. This is not helped by the fact that he has drawn his bar models touching so it looks like the diagram shows twentieths.
  c $\frac{12}{5} = 2\frac{2}{5}$
2 a $6 \times \frac{2}{3} = \frac{12}{3} = 4$
  b Compare answers as a class
3 a Beca and Abdullah are focusing on the shaded parts, with Beca seeing multiplication and Abdullah seeing repeated addition. Chloe and Seb are looking at the unshaded parts with Chloe looking at multiplication and Seb looking at repeated addition.
  b Beca and Abdullah both get an answer of $\frac{15}{8}$ or $1\frac{7}{8}$ and Chloe and Seb both get an answer of $\frac{25}{8}$ or $3\frac{1}{8}$. The pairs get the same answer because multiplication is repeated addition.
4 3

Consolidate – do you need more?
1 a $3 \times \frac{1}{4} = \frac{3}{4}$ b $5 \times \frac{1}{8} = \frac{5}{8}$
  c $4 \times \frac{2}{9} = \frac{8}{9}$ d $2 \times \frac{3}{7} = \frac{6}{7}$
2 Any correct diagram for each part
3 a $\frac{8}{9}$ b $\frac{5}{7}$ c $\frac{8}{9}$ d $\frac{4}{7}$
  e $\frac{16}{9} = 1\frac{7}{9}$ f $\frac{8}{7} = 1\frac{1}{7}$
  g $\frac{18}{9} = 2$ h $\frac{7}{7} = 1$

Stretch
1 $3\frac{3}{20}$ m
2 $22\frac{1}{2}$ hours
3 15
4 $\frac{5}{6}$ m
5 a $x = \frac{18}{5}$ or $3\frac{3}{5}$ b $x = \frac{63}{5}$ or $12\frac{3}{5}$
  c $x = \frac{108}{5}$ or $21\frac{3}{5}$
6 $\frac{ab}{c}$
7 a $\frac{9}{4}$ b $15\frac{3}{4}$

Chapter 3.2

Are you ready?

1 a $\frac{5}{7}$ b $\frac{4}{9}$ c $\frac{11}{15}$ d $\frac{99}{100}$

  e $\frac{10}{17}$ f $\frac{10}{17}$ g $\frac{45}{49}$ h $\frac{39}{100}$

2 Unit fractions: $\frac{1}{17}$, $\frac{1}{5}$, $\frac{1}{517}$

  Non-unit fractions: $\frac{5}{17}$, $\frac{17}{5}$, $5\frac{1}{17}$, $\frac{5}{1}$, $17\frac{1}{5}$

3 a $\frac{3}{4}$ b $\frac{4}{5}$ c $\frac{3}{5}$ d $\frac{2}{5}$

4 a 16 b 49 c 8

5 a −15 b 12 c −24

Practice 3.2A

1 a $\frac{1}{2} \times \frac{1}{4} = \frac{1}{8}$ b $\frac{1}{3} \times \frac{1}{3} = \frac{1}{9}$

  c $\frac{1}{5} \times \frac{1}{3} = \frac{1}{15}$ d $\frac{1}{7} \times \frac{1}{2} = \frac{1}{14}$

2 a $\frac{1}{2} \times \frac{1}{2} = \frac{1}{4}$ b $\frac{1}{4} \times \frac{1}{3} = \frac{1}{12}$

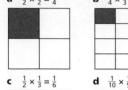

  c $\frac{1}{2} \times \frac{1}{3} = \frac{1}{6}$ d $\frac{1}{10} \times \frac{1}{5} = \frac{1}{50}$

3 a It would be time-consuming to draw b $\frac{1}{300}$

4 a $\frac{1}{45}$ b $\frac{1}{24}$ c $\frac{1}{80}$ d $\frac{1}{42}$

  e $\frac{1}{48}$ f $\frac{1}{540}$ g $\frac{1}{880}$ h $\frac{1}{950}$

5 a 10 b 7 c 1 d 1600

6 a $\frac{1}{25}$ b $\frac{1}{36}$ c $\frac{1}{24}$ d $\frac{1}{320}$

  e $\frac{1}{64}$ f $\frac{1}{400}$ g $\frac{1}{100}$ h $\frac{1}{10000}$

7 a $-\frac{1}{12}$ b $\frac{1}{45}$ c $\frac{1}{196}$

What do you think?

1 Yes, as the numerator will always be 1 × 1 = 1

2 $\frac{1}{10}$ of an hour

Practice 3.2B

1 a $\frac{1}{2} \times \frac{3}{4} = \frac{3}{8}$ b $\frac{2}{3} \times \frac{1}{3} = \frac{2}{9}$

  c $\frac{3}{5} \times \frac{2}{3} = \frac{6}{15}$ d $\frac{5}{7} \times \frac{1}{2} = \frac{5}{14}$

2 a $\frac{1}{2} \times \frac{3}{4} = \frac{3}{8}$ b $\frac{1}{4} \times \frac{2}{3} = \frac{2}{12}$ or $\frac{1}{6}$

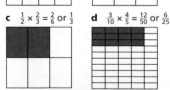

  c $\frac{1}{2} \times \frac{2}{3} = \frac{2}{6}$ or $\frac{1}{3}$ d $\frac{3}{10} \times \frac{4}{5} = \frac{12}{50}$ or $\frac{6}{25}$

3 a Discussion along the lines of multiply the numerators and multiply the denominators, then simplify where possible

  b $\frac{77}{300}$

  c It would take too long

4 a $\frac{8}{45}$ b $\frac{7}{24}$ c $\frac{13}{80}$ d $\frac{25}{42}$

  e $\frac{7}{48}$ f $\frac{247}{540}$ g $\frac{35}{880}$ h $\frac{143}{950}$

5 a $\frac{7}{8} \times \frac{1}{10} = \frac{7}{80}$ b $\frac{5}{63} = \frac{1}{7} \times \frac{5}{9}$

  c $\frac{11}{75} = \frac{11}{25} \times \frac{1}{3}$ d $\frac{7}{20} \times \frac{7}{8} = \frac{49}{160}$

6 a $\frac{2}{15}$ b $\frac{1}{4}$ c $\frac{21}{80}$ d $\frac{1}{14}$

  e $\frac{11}{16}$ f $\frac{17}{135}$ g $\frac{3}{10}$ h $\frac{3}{10}$

---

7 a $\frac{4}{25}$ b $\frac{25}{36}$ c $\frac{1}{4}$ d $\frac{5}{16}$

  e $\frac{27}{64}$ f $\frac{9}{100}$ g $\frac{9}{100}$ h $\frac{49}{10000}$

8 a $-\frac{14}{27}$ b $\frac{1}{2}$ c $\frac{1}{225}$

What do you think?

1 a He thinks the numerator does not change

  b 2 × 2 = 4 not 2. Students could draw a diagram to support this.

  c $\frac{4}{25}$

2 $\frac{3}{4}$ because $\frac{2}{2}$ is equivalent to 1

3 a $\frac{5}{21}$ b $\frac{2}{27}$ c $\frac{10}{63}$

  d $\frac{10}{189}$

4 $\frac{7}{18} = \frac{14}{36}$ so the calculation can be rewritten as $\frac{7}{12} \times \frac{7}{7} = \frac{14}{36}$; therefore the missing fraction is $\frac{2}{3}$

Consolidate

1 a $\frac{1}{8} \times \frac{1}{4} = \frac{1}{32}$ b $\frac{1}{6} \times \frac{1}{7} = \frac{1}{42}$

  c $\frac{1}{4} \times \frac{1}{5} = \frac{1}{20}$ d $\frac{1}{9} \times \frac{1}{3} = \frac{1}{27}$

  e $\frac{5}{8} \times \frac{3}{4} = \frac{15}{32}$ f $\frac{5}{6} \times \frac{4}{7} = \frac{20}{42}$ or $\frac{10}{21}$

  g $\frac{1}{4} \times \frac{4}{5} = \frac{4}{20}$ or $\frac{1}{5}$ h $\frac{7}{9} \times \frac{2}{3} = \frac{14}{27}$

2 a $\frac{1}{21}$ b $\frac{1}{10}$ c $\frac{1}{12}$ d $\frac{1}{12}$

  e $\frac{10}{21}$ f $\frac{3}{10}$ g $\frac{6}{12}$ (or $\frac{1}{2}$)

  h $\frac{5}{12}$

3 a $\frac{1}{150}$ b $\frac{1}{84}$ c $\frac{1}{120}$ d $\frac{1}{400}$

  e $\frac{7}{75}$ f $\frac{1}{6}$ g $\frac{13}{24}$ h $\frac{297}{400}$

Stretch

1 $\frac{ac}{bd}$

2 a $\frac{7}{72}$ m b $\frac{5}{9}$ m

3 a Both calculations are equal to $\frac{16}{625}$ because 0.16 is equivalent to $\frac{4}{25}$

  b Compare answers as a class

4 a $\frac{18}{25}$ km²

  b 7200 tubs, £20 088

5 a Students are used to finding a common denominator when calculating with fractions

  b The fractions will be equivalent so the answer will be equivalent

  c e.g. A pro is that they do not have to remember when they need a common denominator and when they do not. A con is that the answer will need to be further simplified.

Chapter 3.3

Are you ready?

1 a $\frac{2}{5}$ b $\frac{7}{50}$ c $\frac{11}{100}$

  d $\frac{4}{9}$ e $\frac{4}{5}$ f $\frac{21}{50}$

  g $\frac{99}{100}$ h $\frac{8}{9}$

2 a $\frac{1}{15}$ b $\frac{1}{18}$ c $\frac{1}{50}$ d $\frac{1}{28}$

3 a 40 ÷ 4 = 10, 40 ÷ 10 = 4

  b 40 ÷ 5 = 8, 40 ÷ 8 = 5

  c 54 ÷ 9 = 6, 54 ÷ 6 = 9

  d 8 ÷ 1 = 8, 8 ÷ 8 = 1

4 a $\frac{5}{5}$ b $\frac{7}{7}$ c $\frac{10}{10}$

  d $\frac{19}{19}$ e $\frac{135}{135}$ f $\frac{999}{999}$

5 a $\frac{14}{16}$ b $\frac{15}{27}$ c $\frac{1}{6}$

  d $\frac{9}{10}$ e $\frac{15}{20}$ f $\frac{7}{10}$

Practice 3.3A

1 a i 5 ii 10 iii 20 iv 50

  b i 5 ii 10 iii 20 iv 50

2 Any correct diagram

3 a 6 × 5 b 9 × 4 c 4 × 9

  d 10 × 8 e 24 × 7 f 15 × 6

---

  g 100 × 26 h 14 × 15

4 a 30 b 36 c 36 d 80

  e 168 f 90 g 2600 h 210

5 a 20 b 10

6 a There are 4 quarters in one whole so there are 9 lots of four quarters in 9 wholes

  b 12 c 12

7 Any correct diagram

8 a 15 b 12 c 9 d 16

  e 28 f 18 g 520 h 30

What do you think?

1 Both calculations are equal to 9. This happens because the same calculations are used to work them out.
  $5 \div \frac{5}{9} = 5 \times 9 \div 5$ and $5 \times \frac{9}{5} = 5 \times 9 \div 5$

2 Benji has found a common denominator and then divided the numerators and the denominators. This will always work but will not always be efficient.

3 25

Practice 3.3B

1 a 3 b 3

2 Any correct diagram

3 a i 3 ii 2 iii 6 iv 10

  b i 3 ii 2 iii 6 iv 10

4 a $\frac{6}{10}$ b 6

5 a 14 b 15 c 5 d 9

  e 15 f 35 g 35 h 95

6 16 m

What do you think?

1 a 6 b 18 c 48 d 36

2 $\frac{2}{3} \div \frac{1}{9}$ $\frac{3}{10} \div \frac{1}{50}$ $\frac{7}{15} \div \frac{1}{60}$ $\frac{15}{20} \div \frac{1}{40}$

Consolidate

1 a 8 b 16 c 24

  d 56 e 80 f 120

2 a 7 b 10 c 12

  d 63 e 30 f 30

3 a i 2 ii 6 iii 3

    iv 6 v 2 vi 8

  b i 2 ii 6 iii 3

    iv 6 v 2 vi 8

Stretch

1 a $a = 60$ b $t = 90$ c $p = 8$

  d $q = 18$ e $y = 60$

2 a 900 b 100 c 9

3 $a = 2160$ $b = 15$ $ab = 32\ 400$

4 a ab b $\frac{ab}{c}$ c $\frac{xb}{y}$

Chapter 3.4

Are you ready?

1 a 1 b 1 c 1 d 1

  e 1 f 1 g 1 h 1

  i 1 j 1

2 a 2 b 3 c 4 d 12

  e 20

3 a 6 b 15 c 8 d 120

  e 300

4 a 4 b 10 c 4 d 9

  e 25

5 a $2\frac{2}{5}$ b $1\frac{1}{2}$ c $3\frac{3}{14}$ d $2\frac{1}{4}$

Practice 3.4A

1 a i 8 ii 7 iii $\frac{1}{15}$

    iv $\frac{1}{3}$ v 5 vi $\frac{1}{5}$

  b i 8 ii 7 iii $\frac{1}{15}$

    iv $\frac{1}{3}$ v 5 vi $\frac{1}{5}$

2 a i $\frac{3}{2}$ ii $\frac{5}{4}$ iii $\frac{3}{4}$

    iv $\frac{8}{7}$ v $\frac{10}{9}$ vi $\frac{9}{10}$

  b i $\frac{3}{2}$ ii $\frac{5}{4}$ iii $\frac{3}{4}$

    iv $\frac{8}{7}$ v $\frac{10}{9}$ vi $\frac{9}{10}$

c i $\frac{3}{2}$ ii $\frac{5}{4}$ iii $\frac{3}{4}$
iv $\frac{8}{7}$ v $\frac{10}{9}$ vi $\frac{9}{10}$

3 a i 2, 2 ii 8, 8 iii 3, 3
iv 8, 8
They give the same answer
b To divide by a fraction, you multiply by its reciprocal

4 a $5 \times \frac{4}{3}$ b $7 \times \frac{3}{2}$ c 10×9
d $21 \times \frac{5}{2}$ e $\frac{4}{5} \times 3$ f $\frac{3}{4} \times 2$
g $\frac{5}{7} \times 10$ h $\frac{2}{5} \times 5$ i $\frac{4}{5} \times \frac{5}{2}$
j $\frac{3}{4} \times \frac{8}{5}$ k $\frac{5}{6} \times \frac{10}{3}$ l $\frac{5}{6} \times \frac{5}{2}$

5 a $\frac{20}{3}$ b $\frac{21}{2}$ c 90 d $\frac{105}{2}$
e $\frac{12}{5}$ f 6 g $\frac{50}{21}$ h $\frac{25}{6}$
i $\frac{6}{5}$ j $\frac{6}{5}$ k $\frac{50}{21}$ l $\frac{25}{24}$

6 a $2\frac{2}{5}$ b $1\frac{1}{2}$ c $3\frac{3}{14}$ d $2\frac{1}{4}$
e $1\frac{1}{6}$ f $3\frac{3}{5}$ g $1\frac{11}{24}$ h $6\frac{8}{15}$

7 a $x = \frac{15}{14}$ b $y = \frac{5}{9}$
c $p = \frac{40}{7}$ d $m = \frac{33}{8}$

What do you think?
1 a $\frac{2}{5}$ b $\frac{5}{2}$ c $\frac{15}{14}$
2 Yes. A number multiplied by its reciprocal is positive 1, so if a number is negative its reciprocal must also be negative, in order for the two numbers to have a positive product.
3 a $\frac{3}{10}$ b $\frac{10}{3}$ c $-\frac{4}{15}$
d $-\frac{15}{4}$ e $-\frac{8}{45}$
4 Method A uses a common denominator approach.
Method B multiplies by the reciprocal.
Method C turns the question into an equation that can be solved using inverse operations.

Consolidate
1 a i 5 ii 9 iii 11
iv 10 v 3 vi 2
b i 5 ii 9 iii 11
iv 10 v 3 vi 2
2 a $\frac{10}{31}$ b $\frac{27}{29}$ c $\frac{44}{51}$
d $\frac{1}{2}$ e 4 f $\frac{5}{3}$
3 a i $\frac{5}{4}$ ii $\frac{9}{4}$ iii $\frac{11}{7}$
iv $\frac{10}{3}$ v $\frac{3}{2}$ vi $\frac{8}{3}$
b i $\frac{5}{4}$ ii $\frac{9}{4}$ iii $\frac{11}{7}$
iv $\frac{10}{3}$ v $\frac{3}{2}$ vi $\frac{8}{3}$
4 a $\frac{15}{8}$ b $\frac{3}{2}$ c $\frac{55}{42}$
d $\frac{8}{3}$ e $\frac{21}{16}$ f $\frac{4}{3}$

Stretch
1 a $-w$ b $\frac{1}{p}$ c $\frac{1}{pq}$
d $\frac{k}{j}$ e $\frac{z}{xy}$ f $-\frac{xy}{z}$
2 a $\frac{2}{5}$ b $\frac{4}{5}$ c 8 d $\frac{100}{83}$
e $\frac{4}{23}$ f $\frac{5}{34}$
3 a ab b $\frac{ac}{b}$ c $\frac{ac}{b}$ d $\frac{ad}{bc}$
4 a $1:\frac{15}{2}$ b $1:\frac{20}{21}$ c $1:\frac{3}{4}$ d $1:\frac{5}{3}$
5 a Any three correct pairs of values such that $\frac{10q}{p}$ is an integer
b $10 \div \frac{q}{p}$ is an integer when $10q$ is a multiple of p
6 a Any correct values such that $ad \geqslant bc$
b Any correct values such that $ad < bc$
c $\frac{a}{b} \div \frac{c}{d}$ is improper when $ad \geqslant bc$

Chapter 3.5

Are you ready?
1 a $3\frac{2}{5}$ b $1\frac{2}{3}$ c $1\frac{2}{3}$ d $2\frac{1}{7}$
2 a $\frac{5}{2}$ b $\frac{21}{4}$ c $\frac{19}{3}$ d $\frac{9}{2}$

3 a $\frac{2}{15}$ b $\frac{32}{63}$ c $\frac{3}{40}$ d $\frac{21}{30}$ or $\frac{7}{10}$
e $\frac{10}{3}$ f $\frac{9}{14}$ g $\frac{2}{15}$ h $\frac{96}{105}$
4 a ab^2c b $7a$
c $6a^2$ d $9ab + 4a^2b$

Practice 3.5A
1 a $7\frac{1}{2}$ b $36\frac{3}{4}$ c $12\frac{2}{3}$ d $22\frac{1}{2}$
e 15 f 21 g $\frac{95}{3}$ or $31\frac{2}{3}$
h $\frac{63}{2}$ or $31\frac{1}{2}$
2 a Any correct explanation
b $\frac{4}{5}$
3 a $\frac{5}{6}$ b $1\frac{1}{20}$ c $3\frac{1}{6}$ d $\frac{9}{16}$
4

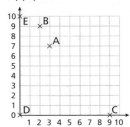

5 a $\frac{143}{12}$ or $11\frac{11}{12}$ b $\frac{17}{3}$ or $5\frac{2}{3}$
c $\frac{195}{36}$ or $5\frac{5}{12}$ d $\frac{249}{14}$ or $17\frac{11}{14}$
6 There are 16 lots of $\frac{1}{3}$ in $5\frac{1}{3}$
7 a 21 b 13 c 17 d 23
e 42 f 52 g 34 h 115
8 a $\frac{7}{2}$ b $\frac{14}{5}$
c i $\frac{35}{28}$ or $1\frac{1}{4}$ ii $\frac{28}{35}$ or $\frac{4}{5}$
9 a $\frac{52}{33}$ or $1\frac{19}{33}$ b $\frac{51}{25}$ or $2\frac{1}{25}$
c $\frac{117}{60}$ or $1\frac{19}{20}$ d $\frac{150}{581}$

What do you think?
1 In the first calculation you are working out $(5 + \frac{6}{7}) \times \frac{41}{5}$ whereas in the second you are working out $(5 \times \frac{6}{7}) \times \frac{41}{5} = \frac{30}{7} \times \frac{41}{5}$
2 a $\frac{69}{5}$ b $\frac{51}{10}$
c $\frac{47}{9}$ d $\frac{5}{2}$

Practice 3.5B
1 a $\frac{xy}{12}$ b $\frac{5xy}{12}$ c $\frac{5x^2y}{12}$ d $\frac{10x^2y}{3}$
2 a $\frac{4x}{3y}$ b $\frac{4x}{15y}$ c $\frac{4}{15y}$ d $\frac{32}{15y}$
3 a $\frac{12}{xy}$ b $\frac{12}{5xy}$ c $\frac{12}{5x^2y}$ d $\frac{3}{10x^2y}$
4 a $\frac{3y}{4x}$ b $\frac{15y}{4x}$ c $\frac{15y}{4}$ d $\frac{15y}{32}$
5 a $\frac{3b}{5}$ b $\frac{p^2}{2q}$ c $\frac{5y}{42}$ d $\frac{7j}{36k^2}$

What do you think?
1 No, $5 \times \frac{1}{5} = 1$ so $\frac{1}{5}$ is the reciprocal of 5
$\frac{w}{9} \div 5 = \frac{w}{9} \times \frac{1}{5} = \frac{w}{45}$
2 a $\frac{5}{24}$ b $\frac{x}{24}$ c $\frac{x}{4y}$ d $\frac{x}{zy}$
3 Compare methods as a class

Consolidate
1 a $\frac{16}{3}$ b $\frac{13}{5}$ c $\frac{13}{5}$ d $\frac{41}{10}$
2 a 32 b $\frac{39}{20}$ or $1\frac{19}{20}$
c $\frac{117}{35}$ or $3\frac{12}{35}$ d $\frac{328}{15}$ or $21\frac{13}{15}$
3 a 16 b 26
c $\frac{52}{21}$ or $2\frac{10}{21}$ d $\frac{287}{130}$ or $2\frac{27}{130}$
4 a $\frac{ab}{35}$ b $\frac{a}{35}$ c $\frac{5a^2}{3b}$ d $\frac{5a^2}{7b^2}$

Stretch
1 $3\frac{3}{5}$ m²
2 $\frac{36x^2}{25}$ cm²
3 a $x = \frac{15}{16}$ b $x = \frac{21}{50}$

c $x = \frac{319}{63}$ or $5\frac{4}{63}$
4 $-\frac{418}{255}$ or $-1\frac{163}{255}$
5 9 m

Check my understanding
1 $3 \times \frac{2}{5} = \frac{6}{5} = 1\frac{1}{5}$
2 a $\frac{2}{3}$ b $\frac{9}{10}$ c $\frac{15}{4} = 3\frac{3}{4}$
3 a $\frac{1}{35}$ b $\frac{8}{27}$
4 There are four whole ones and each is split into two halves. There are 8 halves altogether.
5 a 12 b 15 c $17\frac{1}{2}$
6 $\frac{9}{20}$
7 a $\frac{14}{5} = 2\frac{4}{5}$ b $\frac{105}{68} = 1\frac{37}{68}$
8 a $\frac{3}{5y}$ b $\frac{5x^2}{3y}$ c $\frac{7y}{8x^2}$

Block 4 Working in the Cartesian plane

Chapter 4.1

Are you ready?
1 A (3, 5), B (6, 8), C (10, 10), D (8, 0), E (0, 4)
2

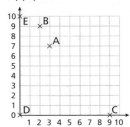

3 (2, 0)
4 (0, −5)

Practice 4.1A
1 A (3, 7), B (8, 1), C (8, −1), D (3, −4), E (−2, −7), F (−9, −4), G (−7, 7), H (−2, 0), I (0, −9), J (3, 0)
2

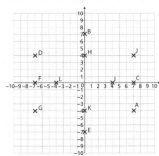

3 a Beca is correct
b Chloe has got the coordinates the wrong way around. Ed has ignored the fact that the 3 is negative; he has just counted squares across.
4 No, he has not noticed the scale used on the y-axis. The coordinates of point Q are (−2, −6)
5 a Quadrilateral b Rectangle

What do you think?
1 a First b Fourth
c Fourth d Third
e First f Second
g Third h Fourth
i First j Second
2 Either (−1, 11) and (6, 11) or (−1, −3) and (6, −3) or (2.5, 1.5) and (2.5, 7.5)

3 a i e.g. (–2, 7) **ii** e.g. (4, 9)
iii The coordinates of point B are
(b, 7) where b is negative
b i e.g. (5, –9) **ii** e.g. (4, –9)
iii The coordinates of point C are
(5, c) where c is negative

Practice 4.1B

1 a (4, –3), (4, 0.5), (4, 175)
b Any three coordinates of the form
(4, a)
c (9, –1), (7, –1), (5.2, –1)
d Any three coordinates of the form
(b, –1)

2 a i
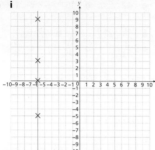

ii They lie in a straight line
iii At every point on the line
x = –6 the x value is equal to 6

b i
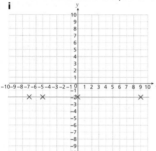

ii They lie in a straight line
iii At every point on the line
y = –2 the y value is equal to –2

3 a A Any three coordinates of the
form (a, –4), B Any three
coordinates of the form (b, –1),
C Any three coordinates of the
form (c, 3), D Any three
coordinates of the form (d, 7),
E Any three coordinates of the
form (e, –9)

b
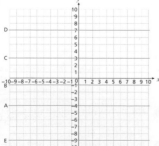

c They are all horizontal lines. They
are all parallel to the x-axis.
d Any line of the form y = a is a
horizontal line that goes through a
on the y-axis. Any line of the form
y = a is parallel to the x-axis.
At any point on the line y = a, the
y value is equal to a

4 a A Any three coordinates of the
form (–8, a), B Any three
coordinates of the form (–4, b),
C Any three coordinates of the
form (2, c), D Any three
coordinates of the form (5, d), E
Any three coordinates of the form
(–3, e)

b
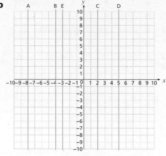

c They are all vertical lines. They are
all parallel to the y-axis.
d Any line of the form x = b is a
vertical line that goes through b
on the x-axis . Any line of the form
x = b is parallel to the y-axis. At
any point on the line x = b, the
x value is equal to b

5 a i y = 3 **ii** x = 3 **iii** y = –3
iv x = 6 **v** y = –6 **vi** x = –3
b y = 3, y = –3 and y = –6 are parallel
because they are of the form
y = __, and x = 3, x = 6 and x = –3
are parallel because they are of
the form x = __

What do you think?

1 a (–1, 7)
b The y value will be equal to 7 and
x will be –1
c (–8, 17)
2 a The equation of the line is y = __
b Each square on the y-axis could be
worth 6
c There is no scale given on the axis
3 a It is the x-axis **b** It is the y-axis

Consolidate

1 A (9, –9), B (3, 9), C (8, 0), D (–6, 8),
E (1, –5), F (–9, –9), G (8, 6), H (–7, 0),
I (0, 7), J (–9, 5)

2
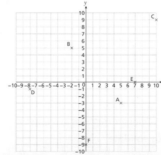

3 a Any five coordinates of the form
(5, a)
b Vertical line going through 5 on
the x-axis
c Any five coordinates of the form
(b, –2)
d Horizontal line going through –2
on the y-axis
4 a x = –8 **b** y = 5 **c** y = –9
d x = 1 **e** y = –3 **f** x = 10

Stretch

1 (–5, 0) and (–1, –4)

2 Any pair of coordinates of the form
(a, 9) or (b, –1)
3 A (1, 0), B (3, 0), C (8, 0), D (1, –5),
E (6, –7), F (8, –7)
4 a y = 4
b Compare answers as a class
c Compare answers as a class

Chapter 4.2

Are you ready?

1 a Any five points of the form (a, 5)
b Any five points of the form (–3, a)
c Any five points of the form (a, –3)
d Any five points of the form (5, a)
2 a x = 8 **b** y = –11
c x = –25 **d** y = 19
3 a **b** **c** **d**
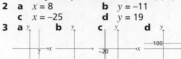

4 a i 12 **ii** –12
b i –2 **ii** 2
c i –6 **ii** 6
d i 40 **ii** –40
e i –200 **ii** 200

Practice 4.2A

1 a (0, 0), (4, 4) and (–9, –9)
b Any three coordinates of the form
(a, a)
c Any five coordinates of the form
(a, a)

2 a

x	–2	–1	0	1	2
y	–2	–1	0	1	2

b
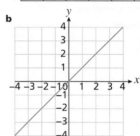

3 a (2, 4), (–3, –6), (0, 0), (19, 38), (–20,
–40)
b Any three coordinates of the form
(a, 2a)
c Any five coordinates of the form
(a, 2a)

4 a

x	–2	–1	0	1	2
y	–4	–2	0	2	4

b
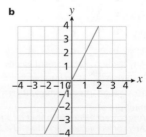

5 a i y = 3x

x	–2	–1	0	1	2
y	–6	–3	0	3	6

ii y = 5x

x	−2	−1	0	1	2
y	−10	−5	0	5	10

iii $y = 0.5x$

x	−2	−1	0	1	2
y	−1	−0.5	0	0.5	1

iv $y = 6x$

x	−2	−1	0	1	2
y	−12	−6	0	6	12

b

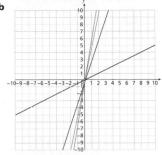

c All of the lines go through the origin

What do you think?

1 a It is not a straight line
b (0, 2)

2

3 a In each pair of coordinates, the y value is 10 times the x value
b $y = 10x$
4 Yes, because at the origin $x = 0$ so kx will always be equal to 0

Practice 4.2B

1 a

Number of bags	0	1	2	4	5	10	20
Cost (£)	0	2	4	8	10	20	40

b

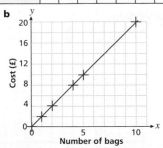

c It is a straight line through the origin
d $y = 2x$

2 a

Number of tennis balls	0	1	2	4	5	10	20
Cost (£)	0	0.5	1	2	2.5	5	10

b

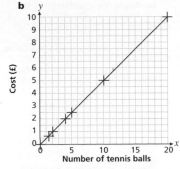

c It is a straight line through the origin
d $y = 0.5x$ (or $y = 50x$ if pence are used for cost)

3 a

Number of pens	0	1	2	3	4	5	10	20
Cost (pence)	0	24	48	72	96	120	240	480

b

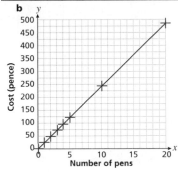

c It is a straight line through the origin
d $y = 24x$ (or $y = 0.24x$ if £ is used for cost)

4 a It is a straight line through the origin
b £160   **c** £16   **d** $y = 16x$

What do you think?

1 a 25 because 4 people make 100 so each person makes 100 ÷ 4 = 25
b $y = 25x$

Practice 4.2C

1 a

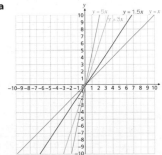

b Each graph is a straight line that goes through the third and first quadrants. Each line goes through the origin (0, 0). The lines have different gradients.
c $y = x$, $y = 1.5x$, $y = 3x$, $y = 5x$. Yes, by comparing gradients.

2 a i

x	−2	−1	0	1	2
y	−4	−2	0	2	4

ii The y-value increases by 2 each time. The gradient of the line is 2

b i

x	−2	−1	0	1	2
y	−20	−10	0	10	20

ii The y-value increases by 10 each time. The gradient of the line is 10
c The gradient of a line is how far up/down it goes for every 1 unit across. Having a gradient of 2 means that for every 1 unit across, the line goes up 2 units. Having a gradient of 10 means that for every 1 unit across, the line goes up 10 units. The y-values are increasing at a faster rate for $y = 10x$ so the line is steeper.

3 a i Any sketch that is less steep than the one shown
ii Any sketch that is steeper than the one shown
b Students should compare and discuss who is more likely to be accurate
4 On a graph of the form $y = kx$, the number k is the gradient of the line. The greater the value of k, the steeper the line. The lower the value of k, the less steep the line.

What do you think?

1 a None of the lines are at a 45-degree angle to the axes
b She does not know the scale used on the axes

Consolidate

1 a At each point on the line $y = 3x$, the y value is 3 times the x value
b

x	−2	−1	0	1	2
y	−6	−3	0	3	6

c

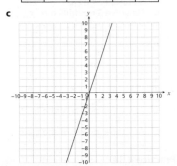

2 a At each point on the line $y = 8x$, the y value is eight times the x value
b

x	−2	−1	0	1	2
y	−16	−8	0	8	16

c

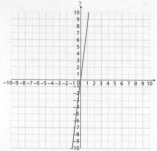

3 a

Number of bags	0	1	2	3	4	5	10
Cost (£)	0	5	10	15	20	25	50

b

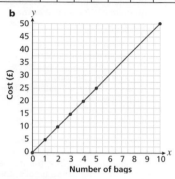

c It is a straight line through the origin

d $y = 5x$

Stretch

1 $y = 3.5x$

2 a $y = 0.5x$ **b** $y = 2x$ **c** $y = 2x$

3 $y = 36$

4 a $b = 7a$ **b** $a = \frac{1}{7}b$

5 8.5

Chapter 4.3

Are you ready?

1 a 5 multiplied by a

b Negative 5 multiplied by a

c a plus 5

d a minus 5

e 5 plus a

f 5 minus a

2 a 15 **b** −15 **c** 8 **d** −2

e 8 **f** 2

3 a −15 **b** 15 **c** 2 **d** −8

e 2 **f** 8

4 a Any three points of the form $(a, 5a)$

b Any three points of the form $(a, -5a)$

Practice 4.3A

1 a i The y value is 1 more than the x value

ii The y value is 1 less than the x value

iii The y value is 4 less than the x value

iv The y value is 4 more than the x value

b i Any three coordinates of the form $(a, a + 1)$

ii Any three coordinates of the form $(a, a - 1)$

iii Any three coordinates of the form $(a, a - 4)$

iv Any three coordinates of the form $(a, a + 4)$

2 a (4, 19), (−15, 0), (23, 38), (−17, −2). In each of the coordinates the y-value is 15 more than the x-value

b (19, 4), (0, −15), (15, 0), (100, 85), (12, −3). In each of the coordinates the y value is 15 less than the x value

3 a i $y = x + 1$

x	−2	−1	0	1	2
y	−1	0	1	2	3

ii $y = x + 2$

x	−2	−1	0	1	2
y	0	1	2	3	4

iii $y = x + 3$

x	−2	−1	0	1	2
y	1	2	3	4	5

iv $y = x + 4$

x	−2	−1	0	1	2
y	2	3	4	5	6

b

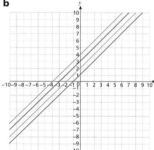

c They are all parallel but go through the y-axis at different points

d Correct line drawn parallel to the others going through (0, 9)

4 a i $y = x - 1$

x	−2	−1	0	1	2
y	−3	−2	−1	0	1

ii $y = x - 2$

x	−2	−1	0	1	2
y	−4	−3	−2	−1	0

iii $y = x - 3$

x	−2	−1	0	1	2
y	−5	−4	−3	−2	−1

iv $y = x - 4$

x	−2	−1	0	1	2
y	−6	−5	−4	−3	−2

b

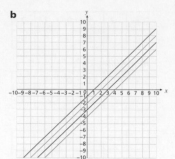

c They are all parallel but go through the y-axis at different points

d Correct line drawn parallel to the others going through (0, −7)

5 a i Any three coordinates from the line

ii At each point on the line, the y value is 1 less than the x value

iii $y = x - 1$

b i Any three coordinates from the line

ii At each point on the line, the y value is 3 more than the x value

iii $y = x + 3$

c i Any three coordinates from the line

ii At each point on the line, the y value is 4 less than the x value

iii $y = x - 4$

6 a $y = x - 8$ **b** $y = x + 7$ **c** $y = x$

What do you think?

1

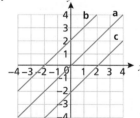

2 They are both correct because addition is commutative

Consolidate

1 a −1, 0, 1, 2, 3 **b** −3, −2, −1, 0, 1

c 2, 3, 4, 5, 6 **d** −6, −5, −4, −3, −2

2 a i $y = x + 1$

x	−2	−1	0	1	2
y	−1	0	1	2	3

ii $y = x - 1$

x	−2	−1	0	1	2
y	−3	−2	−1	0	1

iii $y = x + 4$

x	−2	−1	0	1	2
y	2	3	4	5	6

iv $y = x - 4$

x	−2	−1	0	1	2
y	−6	−5	−4	−3	−2

b

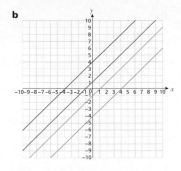

Stretch

1 Negative as the graph intersects the y-axis where y is negative

2

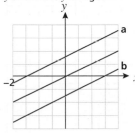

3 18 square units

4 $b = -18$, $c = 0$

Chapter 4.4

Are you ready?

1 a 16   **b** 40   **c** 10
   d 3   **e** −16   **f** −40
   g −6   **h** −13

2 a $j = 8$ **b** $k = 23$
   c $l = -10$ **d** $m = -2$

3

x	−2	−1	0	1	2
y	−30	−15	0	15	30

Practice 4.4A

1 a

x	−2	−1	0	1	2
y	−2	−1	0	1	2

b and **d**

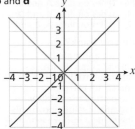

c

x	−2	−1	0	1	2
y	2	1	0	−1	−2

e They both go through the origin but have different gradients

f $y = -x$ as it slopes downwards

2 a i

x	−2	−1	0	1	2
y	8	4	0	−4	−8

ii

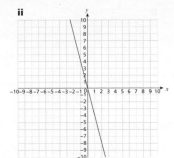

b i

x	−2	−1	0	1	2
y	7	6	5	4	3

ii

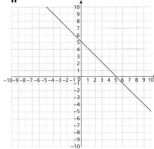

c i

x	−2	−1	0	1	2
y	8	7	6	5	4

ii

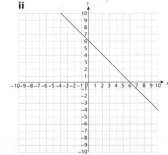

d Compare answers as a class

3 a i $y = 7 - x$

x	−2	−1	0	1	2
y	9	8	7	6	5

ii $y = -5x$

x	−2	−1	0	1	2
y	10	5	0	−5	−10

iii $x + y = 8$

x	−2	−1	0	1	2
y	10	9	8	7	6

iv $10 = x + y$

x	−2	−1	0	1	2
y	12	11	10	9	8

v $y = 1 - x$

x	−2	−1	0	1	2
y	3	2	1	0	−1

b

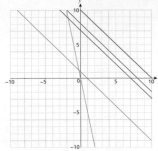

c You can tell from the graph because the lines slope down. You can tell from the equation because the coefficient of x is negative.

4 B, C, E and G have a negative gradient. They slope downwards.

5 $y = 2 - 3x$, $y + 3x = 2$, $2y = -3x$, $y + x = \frac{3}{2}$ all have a negative gradient

What do you think?

1 The line is sloping upwards

2 a The line is sloping downwards
   b The coefficient of x is negative
   c As x is increasing, y is decreasing

Consolidate

1 a 10, 5, 0, −5, −10 **b** 9, 8, 7, 6, 5
   c 14, 13, 12, 11, 10
   d 8, 4, 0, −4, −8

2 a $y = -5x$

x	−2	−1	0	1	2
y	10	5	0	−5	−10

b $y = 7 - x$

x	−2	−1	0	1	2
y	9	8	7	6	5

c $y = 12 - x$

x	−2	−1	0	1	2
y	14	13	12	11	10

d $y = -4x$

x	−2	−1	0	1	2
y	8	4	0	−4	−8

3

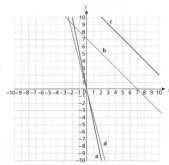

Stretch

1 a Negative because the graph slopes downward

Answers

b

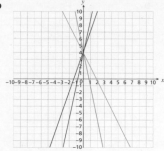

b

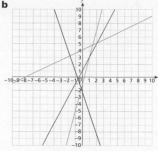

b

2 Sometimes, e.g. $y = -x$ does not but $x + y = 15$ does

3 No, the x values are decreasing too. As x increases, y increases so the gradient is positive.

4 **a** $y = 4 + x$, **b** $y = x$, **c** $y = 4$, **d** $x + y = 7$, **e** $y = 4 - x$, **f** $y = -x$, **g** $y = -x - 3$, **h** $x = -5$

Chapter 4.5

Are you ready?

1 **a** 11 **b** −1 **c** −2 **d** −9
  e 4 **f** −9
2 **a** 13 **b** 5 **c** 7 **d** −8
  e 5
3 **a** Non-linear – it is not decreasing by the same amount each time
  b Linear – it is increasing by 5 each time
  c Non-linear – it is not increasing by the same amount each time
  d Linear – it is decreasing by 3 each time
4 **a** Divide the previous term by 2
  b Add 5 to the previous term
  c Multiply the previous term by 2
  d Subtract 3 from the previous term

Practice 4.5A

1 **a** Yes. y is equal to 5 multiplied by x so at each point on the line $y = 5x$ the y value is 5 times the x value.
  b Yes. y is equal to x plus 3 so at each point on the line $y = x + 3$ the y value is 3 more than the x value.
  c The y value is 3 more than 5 times the x value
  d **i** The y value is 1 more than twice the x value
    ii The y value is 1 less than 4 times the x value
    iii The y value is 2 less than negative 3 times the x value
    iv The y-value is 4 more than half the x value

2 **a** **i** $y = 2x + 1$

x	−2	−1	0	1	2
y	−3	−1	1	3	5

ii $y = 4x - 1$

x	−2	−1	0	1	2
y	−9	−5	−1	3	7

iii $y = -3x - 2$

x	−2	−1	0	1	2
y	4	1	−2	−5	−8

iv $y = \frac{1}{2}x + 4$

x	−2	−1	0	1	2
y	3	3.5	4	4.5	5

3 **a** **i** $y = 3x + 1$

x	−2	−1	0	1	2
y	−5	−2	1	4	7

ii $y = 4 + 3x$

x	−2	−1	0	1	2
y	−2	1	4	7	10

iii $y = 3x - 1$

x	−2	−1	0	1	2
y	−7	−4	−1	2	5

iv $y = 3x - 7$

x	−2	−1	0	1	2
y	−13	−10	−7	−4	−1

b

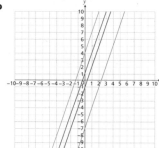

c The lines are all parallel

4 **a** **i** $y = 5x + 4$

x	−2	−1	0	1	2
y	−6	−1	4	9	14

ii $y = -2x + 4$

x	−2	−1	0	1	2
y	8	6	4	2	0

iii $y = 3x + 4$

x	−2	−1	0	1	2
y	−2	1	4	7	10

iv $y = 4 - 5x$

x	−2	−1	0	1	2
y	14	9	4	−1	−6

c The lines all intersect the y-axis at (0, 4)

5 **a** A $y = 3x - 7$

x	−2	−1	0	1	2
y	−13	−10	−7	−4	−1

B $y = -2x + 3$

x	−2	−1	0	1	2
y	7	5	3	1	−1

C $y = 1 - x$

x	−2	−1	0	1	2
y	3	2	1	0	−1

D $y = \frac{1}{2}x - 7$

x	−2	−1	0	1	2
y	−8	−7.5	−7	−6.5	−6

b

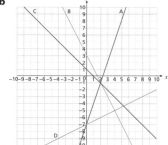

c (0, −7) lies on lines A and D.
  (4, −5) lies on lines B and D
d (2, −1) lies on lines A, B and C

What do you think?

1 **a** It is not a straight line
  b Correct graph drawn
  c Samira plotted (0, 4), (1, 6), (2, 8) and (3, 10) correctly
  d She plotted the points with negative x values incorrectly
  e She made a mistake with the negative numbers

2

x	−2	−1	0	1	2
y	7	5	3	1	−1

x	−2	−1	0	1	2
y	−4	−3	−2	−1	0

x	−2	−1	0	1	2
y	3	3.5	4	4.5	5

Practice 4.5B

1 a

b One square is added each time

c

Term	1	2	3	4	5	6
Position	2	3	4	5	6	7

d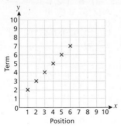

2 a

x	1	2	3	4	5	6
y	2	3	4	5	6	7

b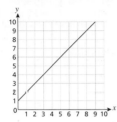

3 The tables of values are the same apart from the row titles. The line in question 2 goes through the same points that are plotted in question 1

4 a The sequence is linear
b There are two triangles added each time

5 a The sequence is linear
b There are two triangles added each time (2) and there are three triangles in the first term (3)

6 a The sequence is linear
b There are 6 triangles in the first term and 1 is removed each time

What do you think?

1 $y = 3x - 1$
2 a $y = 3x + 4$ is a straight line
b Increasing because $y = 3x + 4$ has a positive gradient
c A sequence with 7 as the first term that adds 3 each time
3 The positions in a sequence can only take integer values so the intermediate values on the graph have no meaning

Consolidate

1 a −9, −5, −1, 3, 7 **b** 7, 5, 3, 1, −1
c 3, 5, 7, 9, 11 **d** 2, 2.5, 3, 3.5, 4
2 a $y = 4x - 1$

x	−2	−1	0	1	2
y	−9	−5	−1	3	7

b $y = 2x + 3$

x	−2	−1	0	1	2
y	7	5	3	1	−1

c $y = 2x + 7$

x	−2	−1	0	1	2
y	3	5	7	9	11

d $y = \frac{1}{2}x + 3$

x	−2	−1	0	1	2
y	2	2.5	3	3.5	4

3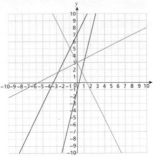

Stretch

1 a $y = 5x + 7$ **b** $y = 5x - 4$
c $y = \frac{1}{2}x$ **d** $y = -2x - 1$
e $y = 7 - 3x$ **f** $y = 5x + 2$
2 a Correct graph drawn
b The graph is only in the fourth quadrant between $x = 0$ and $x = \frac{3}{8}$
3 −1. When $x = 0$, $y = -1$
4 a 14.2 **b** $-\frac{3}{9}$
5 −8

Chapter 4.6

Are you ready?

1 a Linear – it is increasing by the same amount each time
b Non-linear – it is not increasing by the same amount each time
c Non-linear – it is not increasing by the same amount each time
d Linear – it is decreasing by the same amount each time
2 a 31, 35, 39 **b** 80, 160, 320
c 25, 36, 49 **d** 0, −2.5, −5
3 a 25 **b** 27 **c** 4
d −1 **e** 225
4 a 36 **b** 2 **c** 23
d 100 **e** 108

Practice 4.6A

1 a

x	−2	−1	0	1	2
y	−2	−1	0	1	2

b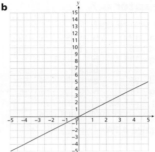

c The graph is a straight line

2 a

x	−2	−1	0	1	2
y	4	1	0	1	4

b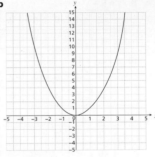

c The graph is not a straight line

3 a

x	−2	−1	0	1	2
y	−8	−1	0	1	8

b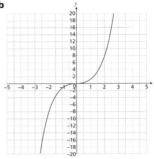

c No, it is not a straight line

4 a **i** $y = 2x^2$

x	−2	−1	0	1	2
y	8	2	0	2	8

ii $y = 5x^2$

x	−2	−1	0	1	2
y	20	5	0	5	20

iii $y = -x^2$

x	−2	−1	0	1	2
y	−4	−1	0	−1	−4

iv $y = -3x^2$

x	−2	−1	0	1	2
y	−12	−3	0	−3	−12

b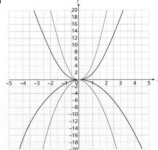

c They are all a curves with one turning point that goes through (0, 0)

5 a **i** $y = x^2 + 3$

x	−2	−1	0	1	2
y	7	4	3	4	7

ii $y = 5 + x^2$

x	−2	−1	0	1	2
y	9	6	5	6	9

iii $y = -x^2 + 4$

x	−2	−1	0	1	2
y	0	3	4	3	0

iv $y = -2 - x^2$

x	−2	−1	0	1	2
y	−6	−3	−2	−3	−6

b

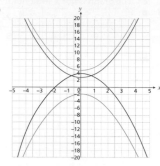

c They are all the same shape but the negatives are the other way up and they all intercept the y-axis at different points

6 a i $y = 2x^3 + 3$

x	−2	−1	0	1	2
y	−13	1	3	5	19

ii $y = x^3 + 4$

x	−2	−1	0	1	2
y	−4	3	4	5	12

iii $y = 0.5x^3$

x	−2	−1	0	1	2
y	−4	−0.5	0	0.5	4

iv $y = -6 + x^3$

x	−2	−1	0	1	2
y	−14	−7	−6	−5	2

b

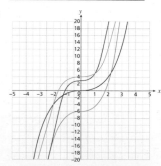

c They are all a similar shape. When the coefficient of x^3 changes the width of the graph changes. When something is added or subtracted to x^3 the graph moves up or down.

What do you think?

1 He has incorrectly calculated with negative numbers. A negative number squared gives a positive answer. He has also joined his points up with straight line segments rather than using a smooth curve.

2 $y = \dfrac{x^2}{2}$, $y = 3x^8 - 10$, $y = \dfrac{15}{4x}$
The power of x is not 1

3 A, B, D and E as they are not straight lines

Consolidate

1 a 5, 2, 1, 2, 5
b 16, 4, 0, 4, 16
c −14, −7, −6, −5, 2
d 16, 2, 0, −2, −16

2 a $y = x^2 + 1$

x	−2	−1	0	1	2
y	5	2	1	2	5

b $y = 4x^2$

x	−2	−1	0	1	2
y	16	4	0	4	16

c $y = x^3 - 6$

x	−2	−1	0	1	2
y	−14	−7	−6	−5	2

d $y = -2x^3$

x	−2	−1	0	1	2
y	16	2	0	−2	−16

3

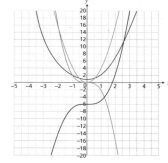

Stretch

1 A $y = 5 - x^2$, B $y = x^2$, C $y = 5x^2$, D $y = -x^2$
2 A $y = -x^3$, B $y = 5 + x^3$, C $y = x^3$, D $y = x^3 - 7$
3 3

Chapter 4.7

Are you ready?

1 a 17   **b** 17.5   **c** 18
d 28   **e** 18   **f** 30.5
g 117   **h** −8
2 a 4   **b** 8.5   **c** 15
d 38   **e** 43.5   **f** 57
g −1   **h** 0   **i** 15
j −11.5
3 Any three correct methods

Practice 4.7A

1 a i A (7, 9), B (7, 7)
    ii (7, 8)
b i C (−10, 9), D (−4, 9)
    ii (−7, 9)
c i E (−2, 6), F (2, 6)
    ii (0, 6)
d i G (2, 3), H (2, −3)
    ii (2, 0)

e i I (−5, −5), J (7, −5)
    ii (1, −5)
f i K (1, −9), L (10, −9)
    ii (5.5, −9)
g i M (−8, −3), N (−8, −10)
    ii (−8, −6.5)
h i O (9, 3), P (9, -6)
    ii (9, −1.5)

2 a (7, 6)   **b** (−3, 2)   **c** (6, 2)
d (−9, 3)   **e** (3, 0)   **f** (5, 0)
g (8.5, −1)   **h** (−2, 0.5)
3 a i A (7, 9), B (5, 5)
    ii (6, 7)
b i C (−10, 9), D (−4, 1)
    ii (−7, 5)
c i E (−5, −9), F (0, 1)
    ii (−2.5, −4)
d i G (1, −2), H (10, −5)
    ii (5.5, −3.5)
4 a (5, 8)   **b** (1, 3)   **c** (7, 5)
d (4, 3)   **e** (−0.5, −4) **f** (4.5, 1.5)
g (−4.5, 6.5) **h** (0, 1.5)
5 a (6, 4)   **b** (8.5, −4)
c (15, 42)   **d** (15, 17)
e (17.5, 22)   **f** (18, 28)
g (17, 31.5)   **h** (−1, −11.5)
i (57, 15)   **j** (15, 38)
k (117, −8)   **l** (43.5, 57)
6 AB and CD both have midpoint at (−1.5, −5)

What do you think?

1 a (6, 3)   **b** (−6, 21)   **c** (4, 7)
d (−34, 37) **e** (−194, −203)
2 Any three pairs of coordinates (a, b) and (c, d) such that $\dfrac{a+c}{2} = 17$ and $\dfrac{b+d}{2} = -21$

Consolidate

1 a (3, 4)   **b** (2, 7.5)   **c** (3, −3)
d (−9, −6) **e** (−7, −4.5) **f** (7.5, 8)
2 a (3, 5)   **b** (5, 7.5)   **c** (3, −4)
d (−4, −6) **e** (−5.5, −4.5)
f (7.5, 4)

Stretch

1 a (3.25, −2)   **b** (7.125, 0.7)
c (−3.1, 4.85)   **d** (2.75, 1.5)
e (−1.85, 0.375) **f** (8.25, −6.75)
2 a $(c + 10, d - 3)$ **b** $(3c, 6.5d)$
c $(3c + 10, 6.5d - 3)$
d $(2.5c + 11, 3.5d + 5.5)$
3 $\left(\dfrac{p+w}{2}, \dfrac{q+z}{2}\right)$
4 (31.5, 34.5)
5 a 36 square units   **b** 50%

Check my understanding

1

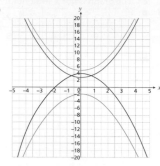

2

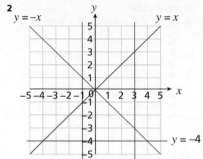

$y = -x$ $y = x$ $y = -4$ $x = -1$ $x = 3$

3 a i $y = 3x$

x	-2	-1	0	1	2
y	-6	-3	0	3	6

ii $y = -4x$

x	-2	-1	0	1	2
y	8	4	0	-4	-8

iii $y = x + 2$

x	-2	-1	0	1	2
y	0	1	2	3	4

b

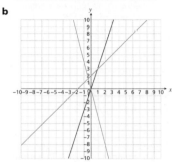

4 a i $y = 3x + 1$

x	-2	-1	0	1	2
y	-5	-2	1	4	7

ii $y = 2 - 4x$

x	-2	-1	0	1	2
y	10	6	2	-2	-6

iii $y = 2x - 3$

x	-2	-1	0	1	2
y	-7	-5	-3	-1	1

b

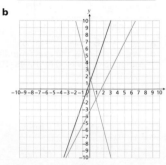

5 a Graph of $y = x^2$

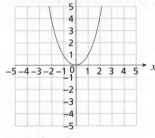

b Graph of $y = x^3$

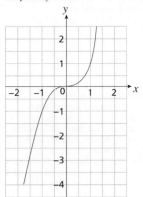

Block 5 Representing data

Chapter 5.1

Are you ready?

1 A (3, 11)   B (6, 4)
C (0, 7)   D (15, 15)   E (12, 13.5)

2

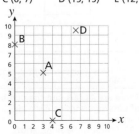

Practice 5.1A

1 a

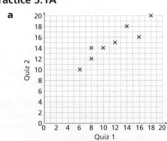

b The greater the score in quiz 1, the greater the score in quiz 2
c Compare answers as a class, e.g. they are both out of 20 marks and quiz 1 is harder than quiz 2
2 a £10000   **b** £6000   **c** 7 years
d The older the car, the lower the value

3 a i

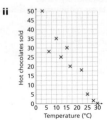

ii

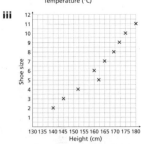

iii

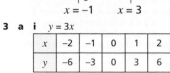

b i The higher the score in test 1, the higher the score in test 2
ii The higher the temperature, the fewer hot chocolates are sold
iii The taller the person, the bigger their shoe size
4 a The distance from the city centre has no impact on the house number
b Compare answers as a class, e.g. number of siblings and favourite number

What do you think?

1 Compare answers as a class
2 The point at (80, 40) does not follow the trend. There could be a number of reasons for this, e.g. the student feeling ill on the day of the calculator paper, or forgetting their calculator.

Practice 5.1B

1 a B   **b** C   **c** A
2 a Negative   **b** Positive
c Positive   **d** No correlation
3 Lydia, as her line shows the general trend of the data. Huda's line goes against the trend of the data. Filipo's line isn't a straight line, it is a collection of lines joining the points like a dot-to-dot picture.
4 a Positive
b Any correct line of best fit e.g.

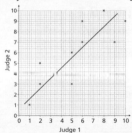

c Any correct reading from the line of best fit. Approximately 8 or 9 points.

5 a Negative
b Approximately 38 or 39 km
c No. At a distance of 100 km the extended graph would suggest it would cost nothing.

What do you think?

1 a Approximately 78 kg
b Yes; their total weight is 646 kg
c Mean = 146.7 cm;
Median = 145 cm; range = 50 cm

2 a Graphs A and C
b Compare answers as a class

Consolidate

1 a

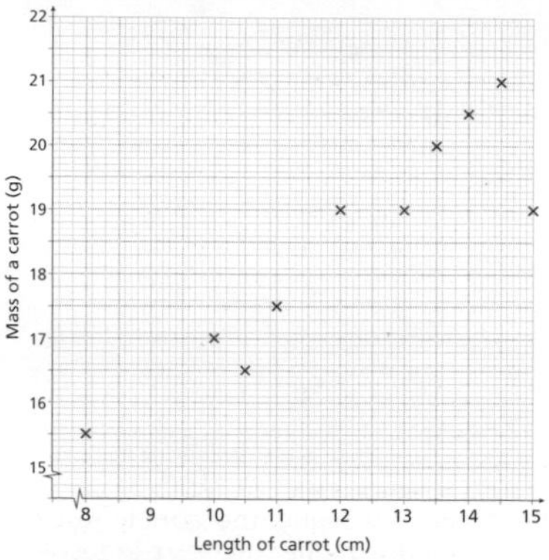

b The longer the carrot, the greater its mass
c Positive
d It might be thinner

2 a i B, D, E, H ii C, F
iii A, G
b F c D
d Compare answers as a class

Stretch

1 Approximately £7.50

2 a

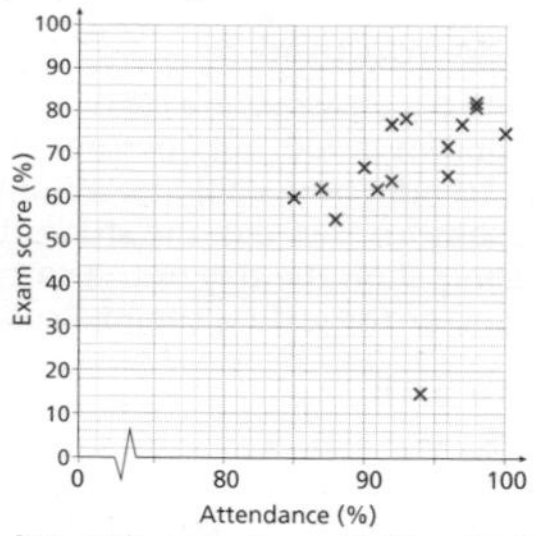

b (94, 15), as it doesn't fit with the trend of the data
c Approximately 70% (any correct reading from the line of best fit)
d There is no data in this range
e Yes. There is a positive correlation, although it isn't a strong positive so it isn't fully convincing.

Chapter 5.2

Are you ready?

1

Animal	Tally
Dog	卌 卌 ||||
Cat	卌 ||||
Horse	卌 卌 |
Other	卌 |
Total	

2

Age	Tally	Frequency
12	卌	5
13	||||	4
14	卌 ||	7
15	||||	4
Total		20

Practice 5.2A

1 a Discrete b Discrete
c Continuous d Continuous
e Continuous f Discrete
g Continuous h Continuous

2 Any two correct examples of discrete data, e.g. age in years, number of people in the class

3 Any two correct examples of continuous data, e.g. height of a tree, mass of a biscuit

4 a Quantitative b Qualitative
c Qualitative d Quantitative
e Qualitative f Quantitative

5 a Qualitative
b Discrete quantitative
c Continuous quantitative

6 a 90 b black
c For cars that don't fit into the other colours. It saves having too many rows.
d No, the numbers are there to record the numbers of cars of each colour. It is the colour of each car that he is interested in.
e 16
f e.g. pie chart or bar chart

7 a

Shoe size	Tally	Frequency
2	|	1
3	卌	5
4	卌	5
5	卌 卌	10
6	卌 |||	8
7		0
8	|	1
Total		30

8 a There are 44 cars, but the number of passengers inside each car varies.
b 102. Multiply the number of passengers by the frequency and find the total.

9 a 69 b 1

What do you think?

1 a e.g., age, number of text messages, height, number of days absent as one group and then surname and method of travel to school as the other group
b Answer varies depending on how the data is sorted. The example given is sorted into quantitative and qualitative.

2 The length of the longest word.
The number of words that Mario counts.
The most likely length of a word.
The total number of letters counted.
The number of words that had more than six letters.

Consolidate

1 a ... you can count it and the total will be an integer
b ... you can't count it
c ... it is a word not a number
d ... it is numerical

2 a 12 b strawberries
c 6 d 58

3 a 30
b $(0 \times 10) + (1 \times 13) + (2 \times 5) + (3 \times 2) = 0 + 13 + 10 + 6 = 29$
c Beth's class. It has 42 pets compared with 29 in Sven's.

Stretch

1 Zach might think age is discrete because it is counted in years. Ed might be considering age to include months, days, hours, minutes and even seconds.

2

Number of goals scored	Number of games
0	3
1	7
2	6
3	4

3 a 24
b Compare answers as a class, e.g. $\frac{5000}{25} = 200$; $200 - (32 + 45 + 58 + 41) = 24$, or, 32 lengths is 800 m, 45 lengths is 1125 m, 58 lengths is 1450 m, 41 lengths is 1025 m so there are 600 m left to swim which is 24 lengths, or, $(32 + 45 + 58 + 41) \times 25 = 4400$, $5000 - 4400 = 600$, $\frac{600}{25} = 24$

Chapter 5.3

Are you ready?

1 Any correct example of continuous data, e.g. length

2 Any correct example of discrete data, e.g. number of people in a room

3 17 cm

4 a t is greater than 4
b h is less than 163
c w is less than or equal to 15.1
d a is less than or equal to 0.2

5 $151 < h < 158$

Practice 5.3A

1 Tables B and C

2 a frequency table
b frequency table
c grouped frequency table
d grouped frequency table

3

Height	Frequency
700 to 799	7
800 to 899	7
900 to 999	6

4 a 36 b 145 c 65
d We don't know the exact number of laps each person in the 80–100 group completed.

What do you think?

1 a Answers will vary depending on intervals chosen. This is an example.

Age	Frequency
0–10	0
11–20	7
21–30	12
31–40	10
41–50	4
51–60	3
61–70	2
71–80	2

b Various answers depending on students' tables

c e.g. There was nobody under 10; most people were between 21 and 30; there was nobody older than 80. Answers may vary slightly depending on students' tables.

Practice 5.3B

1 a C   **b** A

2 a 145, 150 and 155 can be recorded in two different rows

b

Height, h (cm)	Frequency
$140 \leqslant h < 145$	
$145 \leqslant h < 150$	
$150 \leqslant h < 155$	
$155 \leqslant h < 160$	

c You don't know the heights of the people in each interval. For example, in the middle two rows, some or all of the 22 people could be 150 cm tall. You don't know where to record them from this information.

3 a $2 < t \leqslant 4$ **b** 40
   c 120   **d** $\frac{78}{120} = \frac{13}{20}$

4 a

Height, h, (cm)	Frequency
$0 < h \leqslant 10$	3
$10 < h \leqslant 20$	5
$20 < h \leqslant 30$	7
$30 < h \leqslant 40$	2
$40 < h \leqslant 50$	3

b $20 < h \leqslant 30$ **c** 15%

What do you think?

1 a You don't know exactly how long each of the 85 children in the $15 < t \leqslant 20$ minute interval took to run the race, so you don't know how many of them took longer than 18 seconds

b 120

2 Flo's table isn't suitable because the height of a plant is continuous data. For example, the height could be 10.6 cm and there is nowhere in her table to record this. She should instead use inequality symbols for her intervals to allow full coverage, e.g. $0 < h \leqslant 10$, $10 < h \leqslant 20$, $20 < h \leqslant 30$, $30 < h \leqslant 40$, where the values are all in centimetres.

Consolidate

1 a 65   **b** 265   **c** 0–30

2 a $10 \leqslant t < 20$
   b $10 < t \leqslant 20$
   c $10 < t \leqslant 20$

3 $5 < h \leqslant 15$

4

Mass, m (kg)	Frequency
$30 < m \leqslant 40$	6
$40 < m \leqslant 50$	4
$50 < m \leqslant 60$	5
$60 < m \leqslant 70$	5

Stretch

1

Number of emails received	Frequency
$0 < x \leqslant 10$	30
$10 < x \leqslant 20$	15
$20 < x \leqslant 30$	9
$30 < x \leqslant 40$	24
$40 < x \leqslant 50$	2

2 Mass is continuous data. You should use the second table using inequality symbols to ensure all possible values are accounted for. The first table does not include all possible values.

Chapter 5.4

Are you ready?

1 a 45   **b** 390   **c** 162
2 a 123   **b** 45   **c** 114
3 a 250   **b** 210   **c** 465

Practice 5.4A

1 a 11   **b** 20   **c** 7   **d** 10

2 a

	Girls	Boys	Total
Film	7	10	17
Bowling	9	4	13
Total	16	14	30

b 16   **c** 30

3 a

	Year 7	Year 8	Total
Left-handed	6	13	19
Right-handed	42	39	81
Total	48	52	100

b

	Football	Hockey	Netball	Total
Boys	7	11	6	24
Girls	5	3	8	16
Total	12	14	14	40

c

	Maths	English	Total
Year 7	81	19	100
Year 8	27	31	58
Year 9	67	75	142
Total	175	125	300

4 a

	Wear glasses	Do not wear glasses	Total
Male	15	8	23
Female	9	18	27
Total	24	26	50

b $\frac{15}{24}$ **c** 48%

5 a

	Scarf	No scarf	Total
Adult	13	32	45
Child	6	9	15
Total	19	41	60

b 32   **c** $\frac{6}{15} = \frac{2}{5}$

What do you think?

1 You can work out:
- The number of adults on the beach on Saturday morning (= 73)
- How many more children were on the beach on Saturday morning compared with Sunday morning (= 34)

You cannot work out:
- The total number of children on the beach on both days, because you do not have any information about the afternoons or evenings.
- The number of adults that were on the beach on both days, as you don't know specific details about which people were on the beach; you only know the total number of people.
- The number of adults on the beach at 9 a.m., as the data isn't broken down by time.
- The number of children on the beach on Saturday afternoon, as the information given is only about the morning.

2 It appears as though boys like the game more than girls do, as the numbers in the "Boys" column is greater in each row. However, 121 out of 644 boys gave the game 5 stars and 59 out of 100 girls gave the game 5 stars, so more than half of the girls gave the game 5 stars. This suggests that the girls like the game more than boys do.

3 Always

Consolidate

1 a

	Circle	Square	Total
Red	4	3	7
Blue	2	5	7
Total	6	8	14

b 14   **c** $\frac{6}{14} = \frac{3}{7}$ **d** $\frac{3}{8}$

2 a 91 people said football was their favourite sport and 41 of them were boys; the number of girls is $91 - 41$

b 19 girls and 11 boys said that netball was their favourite sport; the total number of students is $19 + 11$

c

	Rugby	Football	Netball	Badminton	Other	Total
Girls	31	50	19	41	49	190
Boys	48	41	11	35	25	160
Total	79	91	30	76	74	350

Stretch

1

	Walk	Don't walk	Total
Boys	52	28	80
Girls	26	14	40
Total	78	42	120

52 boys walk to school

2 a Monday (Monday: $\frac{415}{1292} = 32\%$,

Tuesday: $\frac{480}{1547} = 31\%$)

b

	Single scoop	Double scoop	Triple scoop	Total
Monday	313	312	118	743
Tuesday	391	386	128	905
Total	704	698	246	1648

3

	Under 60	Over 60	Total
Females	120	180	300
Males	105	270	375
Total	225	450	675

675 members in total

Check my understanding

1 a No correlation **b** Positive
   c Negative
2 a e.g. eye colour **b** e.g. age
   c e.g. number of siblings
   d e.g. height
3 19
4 So that all possible values can be included without having too many classes
5

	Year 7	Year 8	Year 9	Total
Boys	95	95	103	293
Girls	115	100	92	307
Total	210	195	195	600

Block 6 Tables and probability

Chapter 6.1

Are you ready?

1 a $\frac{1}{6}$ **b** $\frac{3}{6}$ **c** $\frac{3}{6}$
   d 0   **e** $\frac{4}{6}$
2 a $\frac{8}{20}$ **b** $\frac{11}{20}$ **c** $\frac{1}{20}$
3 0.2
4 Ratio

Practice 6.1A

1 a

	10p	5p	2p	1p
50p	50p, 10p	50p, 5p	50p, 2p	50p, 1p
20p	20p, 10p	20p, 5p	20p, 2p	20p, 1p
10p	10p, 10p	10p, 5p	10p, 2p	10p, 1p

b

	10p	5p	2p	1p
50p	60p	55p	52p	51p
20p	30p	25p	22p	21p
10p	20p	15p	12p	11p

c i $\frac{1}{12}$ **ii** $\frac{6}{12}$ **iii** $\frac{1}{12}$
   iv $\frac{4}{12}$ **v** 0

d Parts **i–iii** are easier to answer using the table from part **a** whereas parts **iv** and **v** are easier to answer using the table from part **b**

2 a The amount of space covered by "Win" sections is greater on spinner A than B. The probability of winning on spinner A is $\frac{1}{2}$ whereas on B it is only $\frac{2}{5}$

b

	Win	Win	Lose	Lose	Lose
Win	W, W	W, W	W, L	W, L	W, L
Lose	L, W	L, W	L, L	L, L	L, L

c $\frac{2}{10}$

3 a

+	1	2	3	4	5	6
1	2	3	4	5	6	7
2	3	4	5	6	7	8
3	4	5	6	7	8	9
4	5	6	7	8	9	10
5	6	7	8	9	10	11
6	7	8	9	10	11	12

b $\frac{5}{36}$ **c** $\frac{6}{36}$
d Scoring a 7 is more likely. The probability of scoring a 7 is $\frac{6}{36}$ and the probability of scoring less than 4 is $\frac{3}{36}$

4 a

	1	2	3	4	5
X	(X, 1)	(X, 2)	(X, 3)	(X, 4)	(X, 5)
Y	(Y, 1)	(Y, 2)	(Y, 3)	(Y, 4)	(Y, 5)
Z	(Z, 1)	(Z, 2)	(Z, 3)	(Z, 4)	(Z, 5)

b i $\frac{2}{15}$ **ii** $\frac{2}{15}$ **iii** $\frac{7}{15}$

5 a They have included duplicates. It is not possible for them both to get the same number as Rhys keeps hold of his card so it will not be an option for Amina.
   b 20   **c** $\frac{2}{20}$

What do you think?

1 a H, T
   b

	H	T
H	(H, H)	(H, T)
T	(T, H)	(T, T)

c

	1	2	3	4	5	6
H	H, 1	H, 2	H, 3	H, 4	H, 5	H, 6
T	T, 1	T, 2	T, 3	T, 4	T, 5	T, 6

2 a Results of a game of rock paper scissors
   b The probability of each person winning; the probability of the game being a tie, etc.

3 a No, if you did not choose at random, then the game would not be fair as you would just make sure that you picked a 0 card each time
   b No, there are 25 possible outcomes and 9 of them contain a 0
   c No, if you do not replace the card, then there are only 20 possible outcomes and this affects the probabilities

Consolidate

1 a $\frac{1}{4}$ **b** $\frac{2}{4}$ **c** $\frac{3}{4}$
2 a

	5	6	7	8
5	10	11	12	13
6	11	12	13	14
7	12	13	14	15
8	13	14	15	16

b $\frac{10}{16}$

Stretch

1 a $\frac{46}{180}$ **b** $\frac{59}{180}$ **c** $\frac{64}{180}$ **d** $\frac{29}{180}$
2 Compare answers as a class, e.g. the first spinner might have 1, 3 and 6 and the second spinner might have 1, 3, 5 and 5
3 $\frac{1}{8}$

Chapter 6.2

Are you ready?

1 a

	Red	Blue	Total
Square	7	13	20
Triangle	8	4	12
Total	15	17	32

b 32   **c** 20   **d** 15   **e** 4
2 a D   **b** E and F
   c A, B, C, E and F   **d** G and H
3 FP, FC, FL, RR, FT, SP, SC, SL, SR, ST, HP, HC, HL, HR, HT, MP, MC, ML, MR, MT

Practice 6.2A

1 a $\frac{18}{30}$ **b** $\frac{15}{25}$ **c** $\frac{10}{28}$
   d i $\frac{28}{55}$ **ii** $\frac{25}{55}$ **iii** $\frac{12}{55}$
2 a $\frac{96}{191}$ **b** $\frac{76}{191}$ **c** $\frac{75}{191}$
   d i $\frac{75}{200}$ **ii** $\frac{105}{200}$
3 a

	Yellow	Green	Red	Total
Pencil case A	17	29	29	75
Pencil case B	30	26	19	75
Total	47	55	48	150

b i $\frac{75}{150} = \frac{1}{2} = 0.5$
   ii $\frac{48}{150}$ **iii** $\frac{103}{150}$ **iv** $\frac{103}{150}$

c $\frac{17}{47}$ He only had 47 pencils to choose from and 17 of them were in A.

d $\frac{94}{150}$ Common mistakes are adding the total for red and the total for pencil case A, which includes the red pencils in A twice.

4 a George

b Jamil does not own either

c i $\frac{1}{10}$ ii $\frac{1}{10}$ iii $\frac{4}{10}$ iv $\frac{3}{10}$

5 a $\frac{53}{200}$ b $\frac{14}{200}$ c $\frac{125}{200}$ d $\frac{75}{200}$

6 There is nobody in the overlap

7 $\frac{35}{100}$

What do you think?

1 Answers vary depending on preference

2 a Compare answers as a class

b i $\frac{19}{30}$

ii Change – it would be out of 18 not out of 30

c i $\frac{37}{30} > 1$

ii He has added the total of box A and the total of orange

iii He should have subtracted 15 from 37 so that orange from A isn't duplicated

Practice 6.2B

1 15

2 For each of the 5 possible boys, there are 7 possible girls. $5 \times 7 = 35$

3 228

4 $3 \times 4 \times 5 = 60$; there are 50 days in 10 school weeks; $60 > 50$

5 12

What do you think?

1 1 scarf and 48 pairs of gloves, 2 scarves and 24 pairs of gloves, 3 scarves and 16 pairs of gloves, 4 scarves and 12 pairs of gloves, 6 scarves and 8 pairs of gloves, 8 scarves and 6 pairs of gloves, 12 scarves and 4 pairs of gloves, 16 scarves and 3 pairs of gloves, 24 scarves and 2 pairs of gloves, 48 scarves and 1 pair of gloves

2 a 25 b 20

c In **a** there are still 5 options for the second pick but in **b** there are only 4

3 No, because $8 \times 7 \times 6 = 336$

Consolidate

1 a $\frac{12}{40}$ b $\frac{23}{40}$ c $\frac{16}{40}$ d $\frac{35}{40}$

2 a 40 b 23 c 5

d i $\frac{8}{40}$ ii $\frac{27}{40}$ iii $\frac{5}{40}$

Stretch

1 $\frac{15}{33}$

2 a 7 b $\frac{75}{100}$

c You cannot make exact predictions.

d $\frac{7}{100}$

3 a 720 b 45

Check my understanding

1

	Tomato	Mayo	Lettuce
Cheese	C T	C M	C L
Egg	E T	E M	E L
Salmon	S T	S M	S L

2 a

	A	B	C	D	E
1	1A	1B	1C	1D	1E
2	2A	2B	2C	2D	2E
3	3A	3B	3C	3D	3E
4	4A	4B	4C	4D	4E
5	5A	5B	5C	5D	5E
6	6A	6B	6C	6D	6E

b $\frac{3}{30}$ or $\frac{1}{10}$

3 $\frac{19}{100}$

4 $\frac{26}{108}$ or $\frac{13}{54}$

5 60

Block 7 Brackets, equations and inequalities

Chapter 7.1

Are you ready?

1 a $3m$ b $3m$ c $\frac{m}{3}$

d $\frac{3}{m}$ e m^2

2 a 4 b 400 c 10 d 23

e 40 f 1600 g 0

3 a -2 b 8 c 2 d -8

4 a -16 b -16 c 16 d -4

e 4

Practice 7.1A

1 a $n+5$ b $n-2$ c $2n-3$

d $2(n+1)$ or $2n+2$

e $\frac{n}{3}$ f $\frac{n}{3}+2$ g $\frac{n}{2}-1$

2 a i 14 ii $12+x$

b i $y+2$ ii $y+x$ iii $2y$

c 7 and $y-5$

d $y+18$

3 a i $5x$ ii $3y$ iii $5x+3y$

b $5x-3y$

4 a $a-15$ b $12a$ c $\frac{a}{2}$

d $6a$

5 a $5t$ cm²

b i ab ii ba

c d^2

d i $6g$ ii $2hk$ iii $20mn$

iv $25d^2$ v πa^2 vi $\pi 4b^2$

6 a $p+q$ b pq c $\frac{p}{q}$

d $\frac{q}{p}+5$ e $5pq$

7 a $8a+11b$ b $a+7b$ c $7x^2-x$

d $3x$ e $2pq+2p+2q$

What do you think?

1 a $\frac{a-b}{2}$ b $xy+5$ c $12-2mn$

d $5f^2g$

2 a $\frac{mp+np}{4}$

3 a $15t+5$ b $\frac{15p}{2}+5$ c $15x+5$

4 a Should be 5

b Should be p

c Should be 1

Practice 7.1B

1 a i 26 ii 14 iii -14

iv -26

b i -14 ii -26 iii 26

iv 14

2 a -4 b -20 c -96

d 32 e -3 f -48

g 14 h 10 i -5

3 a The expression means x multiplied by (y squared), not (x multiplied by y) and then squared

b 48

c i Yes ii No iii Yes

4 a $g=15$ b $g=25$ c $g=-25$

d $g=-15$ e $g=4$ f $g=-4$

g $g=100$ h $g=-100$

5 a $2a$ b $-2a$ c a^2

d $-a^2$ e $2ab$ f $-2ab$

g $2ab$

6 a $x+2y$ b $-3x-4y$ c $5x-9y$

d $-3x-3y$ e $4x$ f $-x^2-2x$

g $-x^2-4x$ h x^2-4x

What do you think?

1 Compare answers as a class

2 a A and D

b i Even ii Odd iii Odd

3 Compare answers as a class

Consolidate

1 a $n-7$ b $n+3$ c $2n+2$

d $2(n-3)$ or $2n-6$

e $\frac{n}{4}$ f $\frac{n}{2}+2$ g $\frac{n}{3}-1$

2 a

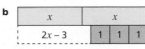

b

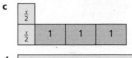

c

d

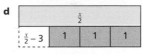

3 a $7p$

b i $6m$ ii $9m$ iii mn

4 a i 18 ii 2 iii -18 iv -2

b i 2 ii 18 iii -2 iv -18

5 a -14 b 14 c 0

d -50 e -4 f -16

g 10 h 11 i 2.5

Stretch

1 a i $15p+12q$ ii $25p+21q$

b i $15p+12q-5$

ii $25p+21q-9$

2 a i $\frac{5}{6}$ ii $\frac{2p+3q}{6}$

iii $\frac{p+q}{pq}$ iv $\frac{q-p}{pq}$

b Compare answers as a class

3 Compare answers as a class, e.g.

a $x^2, 3x^4$ b $-x^2, -5x^4$

Chapter 7.2

Are you ready?

1 a 11 b 14 c $2x+3$

d $2(x+3)$

2 a $5a+20$ b $5a-20$

3 a $3p$ b $6p$ c $-3p$ d $-3p$

e $-6p$ f p^2 g $-p^2$ h $2p^2$

i $-2p^2$ j $6p^2$

4 a 3 b 3 c 6

d 4 e 6

Practice 7.2A

1 a $3x+6$ b $5x-5$

c $8x+4$ d $6x-9$

2 a Compare answers as a class

b i $8x+72$ ii $8y-8$

c Part **b i** and $x=90$ links to Jakub's method. Part **b ii** and $y=100$ links to Chloe's method.

3 a £59.88

b i $12a-12$ ii $12a+12$

iii $12a + 36$ **iv** $12a + 12b - 24$
v $12a - 12b + 24$
vi $12a - 12b - 24$
vii $24 - 12a + 12b$
viii $24 - 12a - 12b$

4 a $3x + 12$ **b** $3x - 12$
c $6x + 12$ **d** $6x - 12$
e $6x - 15y + 12$

5 a $a^2 + 4a$ **b** $ab + 4a$
c $ab - 4a$ **d** $4a - ab$
e $4a - a^2$ **f** $3a^2 - 4a$
g $4a - 3a^2$ **h** $3a^2 - 12a$
i $6a^2 - 12a$
j $12a - 6a^2 + 15ab$

6 a $5x + 12$ **b** $5x$
c x **d** $x + 12$
e $p^2 + 8p + 15$ **f** $2p^2$
g $9p - 5p^2$

What do you think?

1 a Jackson is correct
b i $3x - 3$ **ii** $3x + 3$
iii $1 - 5x$ **iv** $5 - 5x$
v $1 + 5x$ **vi** $1 - 5x$

2 a $7x + 7a$
b Possible answers include
i 100 and 3
ii 100 and -1
iii 10 and -0.2
iv 20 and -0.1
v 10 and 8.9

Practice 7.2B

1 a $3x$ **b** $2x$ **c** x
d $2x$ **e** $3x$

2 a 5 cm and 3 cm, 15 cm and 1 cm
b $6y$ cm and 1 cm, $3y$ cm and 2 cm, y cm and 6 cm, etc.
c $x + 3$ cm

3 a 4 **b** 4 **c** $4b$
d $4a$ **e** $4a$

4 a $2(x + 3)$ **b** $4(x + 2)$ **c** $5(4 + x)$
d $6(x - 1)$ **e** $8(x + 2)$ **f** $x(x + 3)$
g $x(x - 3)$

5 a $3(2a - 6b + 3c)$ **b** $6(a + 3b - 2c)$
c $a(10 + b)$ **d** $a(10 + b + a)$
e $a(5a + 6b + 7c)$ **f** $5q(p - 2r + 3q)$
g $5(pq - 2qr + 6rs)$

6 a $x(8 - y)$ **b** $2x(4 - y)$
c $4x(2 - y)$ **d** $x(8 - x)$
e $x(x - 8)$ **f** $3x(x - 4)$
g $x(3x - 12 + y)$ **h** $3x(x - 4 + 2y)$
i $3x(3x + 1 - 4y)$

What do you think?

1 a 5700 **b** 830 **c** 6200 **d** 2400
2 a $(a + b)(x + 4)$ **b** $(x + 3)(x - 5)$
c $11(p + 5)$
3 The terms have no common factor
4 a $5x - 1$ **b** $6x + 3$

Consolidate

1 a 618 **b** 693
2 a $5a + 15$ **b** $5a + 10b$
c $10a - 15b$ **d** $14a - 21b + 28$
e $x^2 + 2x$ **f** $y^2 - 3y$
g $p^2 + pq + pr$ **h** $2q^2 - 3rq$
3 a $5x + 12$ **b** $5x + 10$ **c** $7x - 7$
4 a $2(4x - 3)$ **b** $3(2 + 3x)$
c $2(2 - x)$ **d** $x(x + 6)$
e $x(5 - x)$
5 a $5b(3a + 4c)$ **b** $2x(4y + 5x)$
c $4q(3p - 4r)$ **d** $3d(2e + 4d - 1)$

Stretch

1 a i C **ii** A **iii** B
b Compare answers as a class
2 a $4x^2 + 11x$ **b** $x^2 - 6x$
c $5x^2 - 6x$ **d** $x^2 + 10x$
e $x^2 + y^2 + 2xy$ **f** $8x^2 - y^2$
g $2y^2 - 3x^2$

---

3 a 47.5
b Compare answers as a class

Chapter 7.3

Are you ready?

1 a $3x + 6$ **b** $x^2 + 2x$ **c** $3x - 6$
d $x^2 - 2x$
2 a $7a$ **b** $-a$ **c** a **d** $-7a$
3 a $-3a$ **b** $3a$ **c** $6x$
d $-20x$ **e** $20x$
4 a $x^2 + 7x$ **b** $x^2 + 7x + 5$
c $x^2 + x + 5$ **d** $x^2 - 7x - 5$

Practice 7.3A

1 480, 42, 480 + 42 = 522, 3480 + 522 = 4002, 4002
2 a $ab + 3a + 2b + 6$
c $ab + 2a + 5b + 10$, $ab + 5a + 2b + 10$. The coefficients of a and b are swapped.
3 a $pq + 6p + 4q + 24$
b $pq + 6p - 4q - 24$
c $pq - 6p - 4q + 24$
4 a i $x^2 + 8x + 12$ **ii** $x^2 + 8x + 12$
iii $x^2 - 4x - 12$ **iv** $x^2 + 4x - 12$
5 a $-3 \times -4 = +12$, so expansion is $x^2 - 7x + 12$
b i $x^2 - 6x + 5$
ii $x^2 - 8x + 12$
iii $x^2 - 18x + 80$
c The coefficient of the x term changes from positive to negative

What do you think?

1 a The coefficient of the first x does not affect all four terms; expansion is $2x^2 + 14x + 20$
b i $2x^2 + 7x + 3$
ii $2x^2 + 5x + 3$
iii $4x^2 + 8x + 3$
iv $2x^2 - 5x - 3$
v $2x^2 - x - 3$
vi $4x^2 + 4x - 3$
vii $4x^2 - 4x - 3$
c i $4x^2 - 8x + 3$
ii Because multiplication is commutative
2 a The correct answer is $x^2 + 6x + 9$
b i $x^2 + 8x + 16$
ii $x^2 - 8x + 16$
iii $x^2 + 2ax + a^2$
iv $4x^2 + 20x + 25$
c $8x$

Consolidate

1 a 4674
b The answer is 6586. Compare methods as a class.
2 a $mn + 6m + 3n + 18$
b i $mn + 6m + 4n + 24$
ii $mn + 7m + 2n + 14$
c The coefficients of m and n will be different but the other terms will be the same
3 a $a^2 + 12a + 32$ **b** $x^2 + 12x + 32$
c $y^2 + 12y + 32$ **d** $p^2 + 10p + 21$
e $q^2 + 12q + 20$
4 a $a^2 - 4a - 32$ **b** $x^2 + 4x - 32$
c $y^2 - 12y + 32$ **d** $p^2 + 4p - 21$
e $q^2 - 8q - 20$

Stretch

1 a i $x^2 - 9$ **ii** $x^2 - 25$
iii $x^2 - 64$
In general, $(x + a)(x - a) = x^2 - a^2$
b i $x^2 - 100$ **ii** $x^2 - 49$
c 400
2 $a = 2$, $b = -3$, $c = -1$, $d = -5$. The values of f and g depend on the choice of e

---

3

$x^2 + 2x + 1$	$x^2 + 3x + 2$	$x^2 + 4x + 3$	$x^2 + 5x + 4$
$x^2 + 3x + 2$	$x^2 + 4x + 4$	$x^2 + 5x + 6$	$x^2 + 6x + 8$
$x^2 + 4x + 3$	$x^2 + 5x + 6$	$x^2 + 6x + 9$	$x^2 + 7x + 12$

Compare answers as a class
4 Compare methods as a class. The answers are $x^3 + 6x^2 + 11x + 6$ and $x^3 + 9x^2 + 27x + 27$

Chapter 7.4

Are you ready?

1 a $x = 20$ **b** $x = 80$ **c** $x = 38$
d $x = -38$ **e** $x = 42$ **f** $x = 38$
2 a $y = 17$ **b** $y = 16.75$ **c** $y = 19.75$
d $y = 19.75$
3 a $z = 45$ **b** $z = 57$ **c** $z = 33$
d $z = 63$ **e** $z = 57$
4 a $4x + 12$ **b** $5y - 15$ **c** $21 - 7p$
d $10a - 22$

Practice 7.4A

1 Compare answers as a class
2 a $a = 3$ **b** $b = 13$ **c** $c = 1.5$
3 The coefficient of x will change. The solutions are $x = 1$ and $x = \frac{1}{2}$
4 a $x = 1.4$ **b** $x = 5$
c $x = 3$ **d** $x = \frac{5}{3}$
5 a $y = 24$ **b** $y = 8$
c $y = -\frac{4}{3}$ **d** $y = \frac{80}{3}$
6 a Compare answers as a class
b i $x = 1$ **ii** $x = 2$
iii $x = 0$ **iv** $x = 8$

What do you think?

1 a $5(n + 4) = 70$
b $n = 10$
2 a Different; they have performed the operations in a different order
b Ed's number is 8 and Faith's is 4.5
c -1.5
d Compare answers as a class

Practice 7.4B

1 13
2 14
3 a i There will be no fractions in the expressions
ii 15 blue, 30 red, 90 green
b 20 yellow, 22 white, 66 pink
4 a / b Both methods give 12 £5 notes and 15 £10 notes
5 a $y = 7$ **b** $x = 6$
6 165 cm²

What do you think?

1 38
2 Compare methods as a class
a $x = 14$ **b** $x = 40$

Consolidate

1 a $x = 7$ **b** $x = 4.5$ **c** $x = 8$
d $x = 10.5$
2 a $y = 5$ **b** $p = 2.5$ **c** $q = -2$
3 a $a = 5$ **b** $b = 25$ **c** $c = 3.75$
4 Flo scores 26, Emily scores 21 and Kath scores 52

Stretch

1 a $a = 9$ **b** $b = 3$ **c** $c = -5$
d $d = 2.125$
2 a Compare methods as a class
Answers are $n = 0.5$ and $n = 1.5$
b Compare methods as a class
Answers are $n = -1$ and $n = 3$
3 a -5 **b** -10

Chapter 7.5

Are you ready?

1 **a** $-5, 3, 7$ **b** $-6, -4, 2$
   c $-8, -3, 3, 8$
2 **a** Any number below 5
   b Any number above 5
   c Any number below -3
   d Any number below -4
3 **a** $p = 25$ **b** $q = 8$
   c $r = 50$ **d** $t = \frac{1}{2}$
4 **a** $m = 13$ **b** $n = 8$
   c $x = -6$ **d** $y = 23$

Practice 7.5A

1 **a** **i** a is greater than 7
      ii b is less than 4
      iii c is less than or equal to 0
      iv d is greater than or equal to -2
   b **i** $x < 6$
      ii $y \geqslant -3$
      iii $10 > z$
2 **a** **i** Any integer more than 7
      ii Any integer below 3
      iii Any integer below 6
      iv Any integer more than 5
   b Compare answers as a class
3 **a** **i** $x > 5$
      ii $y < 7$
      iii $p + 5 > 13$
      iv $2q + 1 > 30$
   b Compare answers with a partner
   c It can show greater than, or it can show equal to, but it would be hard to show both possibilities
4 $5 < x$
5 **a** $x \leqslant 3$ **b** $y \leqslant -3$ **c** $p > 60$
   d $q \leqslant 15$ **e** $w \geqslant 4.5$ **f** $m < 3$
   g $n > -17$ **h** $m < 4\frac{1}{5}$
6 **a** **i** 7 **ii** 6 **iii** -1
   b **i** 6 **ii** 5 **iii** -2

What do you think?

1 **a** **i** 4 **ii** 2 **iii** 7 **iv** 34
   b **i** 3 **ii** -1 **iii** 0 **iv** 2
2 **b** $3 > x$
   c **i** $x > -6$ **ii** $x \leqslant -1$ **iii** $x \leqslant 6$

Practice 7.5B

1 **a** $a \leqslant 8$ **b** $b > 0$ **c** $c \geqslant 6.4$
2 45 or more
3 **a** Each side is greater than 12.5 cm
   b The area is greater than 156.25 cm²
4 **a** $4x + 8 < 30$, so $x < 5.5$
   b $x \leqslant 5.5$
5 **a** The 24th term, 101
   b The 248th term, 997
6 $2x + 3 > 50$, $x > 23.5$
7 18 months

What do you think?

1 e.g. $x = 4.1$ is a solution to the first inequality but not the second
2 **a** $-1, 0, 1, 2, 3, 4$
   b $-3, -2, -1, 0, 1$
   c **i** 16 **ii** 17, 19, 23
      iii 15, 18, 21, 24
3 **a** The shorter sides will not meet
   b $a + b > c, a + c > b, b + c > a$

Consolidate

1 Compare values as a class
2 $x > 4, 5 \leqslant x, 4 < x, x \leqslant 5$
3 **a** $a < 11$ **b** $b < 4$ **c** $c \geqslant 9$
   d $d < 48$ **e** $e \geqslant 180$ **f** $f > 6$
   g $g \leqslant 7.5$ **h** $h \leqslant 11$
4 **a** $p > 1$ **b** $q \leqslant 0.8$
   c $x < 2.6$ **d** $y \geqslant -1\frac{5}{6}$
5 8

Stretch

1 e.g. true for $a = 3$ and $b = 4$, but false for $a = -4$ and $b = 3$ or $a = -4$ and $b = -3$
2 **a** $3.5 < a < 6$ **b** 4, 5
3 $x \geqslant 24$

Chapter 7.6

Are you ready?

1 **a** $x = 310$ **b** $y = 1020$
   c $p = -5$ **d** $q = 18$
2 **a** $7a - 21$ **b** $40 + 8b$
   c $21 - 3c$ **d** $8d - 20$
3 **a** $p = 6.5$ **b** $x = 32$
   c $y = 2$ **d** $w = \frac{1}{3}$
4 $2(2x + 3) = 34$, $x = 7$

Practice 7.6A

1 **a** $6x + 2 = 4x + 9$
   b $x = 3.5$
2 **a** **i** $5x = 3x + 18$
      ii $5y + 12 = 7y + 2$
   b **i** $x = 9$
      ii $y = 5$
3 **a** **i** $x = 2$ **ii** $x = 1$ **iii** $x = 3$
   b Compare answers as a class
4 **a** $a = 4$ **b** $b = \frac{1}{2}$ **c** $c = 2\frac{1}{2}$
5 **a** $w = -3$ **b** $x = 3$ **c** $y = -3$
6 **a** $x = 3$ and $x = 1$
   b **i** $p = 8$ **ii** $q = -14$ **iii** $m = 5$

What do you think?

1 **a** $5(n + 4) = 7n$ **b** $n = 10$
2 $4n + 10 = 2n + 18$, $n = 4$
3 Compare answers as a class. $x = 8$, $x < 8$
4 **a** -1 **b** 5 **c** 1

Practice 7.6B

1 **a** $x = 4$ **b** $y = 1$ **c** $z = -2$
   d $a > 1.5$ **e** $b < \frac{1}{2}$ **f** $c \leqslant 2$
2 **a** $x = 7$
   b 90 units², 46 units
3 **a** 11 **b** 140
4 30 units²
5 **a** 16 **b** 41

What do you think?

1 It should be $5x + 5 = 0$, so $x = -1$
2 Compare methods as a class; answer is $x = 7$
3 Compare answers as a class

Consolidate

1 **a** $3y + 18 = 4y + 7$
   b $y = 11$
2 **a** **i** $6x + 8 = 4x + 18$
      ii $5y + 9 = 7y$
   b **i** $x = 5$ **ii** $y = 4\frac{1}{2}$
3 **a** $x = 7$ **b** $x = 12$ **c** $x < -7$
   d $y = 10$ **e** $y > 7.5$ **f** $y = 12.5$
   g $n = 4$ **h** $n = 10$ **i** $n = -10$
   j $p > 10$ **k** $p = 6$ **l** $p = -2$
4 11
5 20 units² and 40 units²

Stretch

1 **a** There is not enough information. As it is, the equation has an infinite number of solutions.
   b $a = 5$, $b = 15$
2 $x = 2.5$, $y = 9$
3 **a** 13 and 7 **b** 28.5 and 11.5

Chapter 7.7

Are you ready?

1 **a** 43 **b** 144 **c** 7
2 **a** $7a - 21$ **b** $21 - 7a$ **c** $20t - 40b$
3 **a** $P = 2a + 2b$ or $P = 2(a + b)$
   b $A = ab$
4 $A = 140$

Practice 7.7A

1 A, B, E and F are formulae; C and D are equations
2 **a** $7a$ **b** $8c$ **c** $4x + 3y$
   d $6p$ **e** $4q$
3 **a** 52 **b** 4.5
4 **a** 90 **b** 8
5 **a** $P = 3a + 3b$ or $P = 3(a + b)$
   b 30 **c** 5
6 **a** It is the sum of sides of length p, p and p
   b $6p + 6q$ or $6(p + q)$
   c $P = 6p + 6q$ or $P = 6(p + q)$
   d The expression does not include "$P = $"
7 $a = 18$ and $b = 7$

What do you think?

1 Discuss as a class
2 All are correct, it is the same expression in different forms
3 **a** 2.84 s
   b Period when $l = 4$ m is 4.01 s; this is not 2×2.84 s

Consolidate

1 B, E and G are expressions; C, D, H and I are formulae; A and F are equations
2 **a** $7a$ **b** $6b - 10$ **c** $2x + y$
   d $45p^2$ **e** $5m$
3 **a** $P = 4l$ **b** 92 units **c** $5\frac{3}{4}$ units
4 **a** $W = 800 + 150n$
   b £2600 **c** 7
5 Compare answers as a class. The formulae could be e.g. $P = a + b + c$, $P = 2a + b$, $P = 3a$

Stretch

1 **a** 18 cm **b** $w = \frac{A}{l}$ **c** $l = \frac{A}{w}$
2 **a** $d = 2r$ **b** $r = \frac{d}{2}$
   c $d = \dfrac{\text{circumference}}{\pi}$
   d $r = \dfrac{\text{circumference}}{2\pi}$
3 **a** 75 **b** 9 **c** 6
4 **a** 5050 **b** 20 100 **c** 15 050
   d It gives the nth triangular number

Check my understanding

1 **a** **i** $n + 3$ **ii** $3n - 5$ **iii** $\frac{n^2}{2}$
   b **i** -1 **ii** -17 **iii** 8
2 **a** $3(2a - 4b + 5) \equiv 6a - 12b + 15$
   b 8217
   c $18b - 21c$
   d $24 - 5x - x^2$
3 **a** **i** $2(5x + 6y)$ **ii** $x(10 + y)$
      iii $2x(5x + 6y)$
   b Both terms have more than a single factor
4 **a** 18.5 **b** 17
5 **a** **i** $x > 68$ **ii** $y \geqslant 171$ **iii** $z \geqslant -2.5$
   b **i** 69 **ii** 171 **iii** -2
6 $x = 2$
7 A is a formula, B is an expression and C is an equation

Block 8 Sequences

Chapter 8.1

Are you ready?

1 **a** 18
   b Each term is 3 more than the previous one
   c 27
2

Answers

3 A and B linear, C and D non-linear

4 Same: both go up in 2s. Different: the even numbers are all multiples of 2, but none of the odd numbers are.

Practice 8.1A

1 **a** 130 **b** 70
  c 100 000 **d** $\frac{1}{10}$

2 **a**

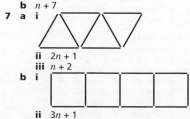

  b Add 1 to the previous term
  c Add 2 to the previous term
  d Add 3 to the previous term

3 **a** Subtract 3 from the previous term
  b 6

4 **a** You don't know any of the terms
  b **i** Start at 50, add 50 to the previous term
     ii Start at 50, double the previous term
     iii Start at 50, subtract 20 from the previous term

5 **a** Multiply the previous term by 10
  b 20 000

6 **a** 23, 47, 95
  b 6, 13, 27, 55, 111
  c **i** 5, 12, 26, 54, 110
     ii 6, 14, 30, 62, 126

7 **a** 2, 4, 10, 28, 82
  b 0, –2, –8, –26, –80
  c 1, 1, 1, 1, 1

8 **a** (7, 8,) 15, 23, 38, 61, 99
  b **i** 3, 6, 9
     ii All are multiples of 3 and this will continue

What do you think?

1 Compare answers as a class
2 Compare answers as a class
3 **a** **i** The difference between terms starts at 1 and then increases by 1 every time
     ii Multiply the previous term by 2, then by 3, then by 4, etc.
     iii Divide the previous term by 4
  b **i** Neither
     ii Neither
     iii Geometric

Practice 8.1B

1 **a** **i** 4, 7, 10, 13, 16
     ii 3, 8, 13, 18, 23
     iii 4, 8, 12, 16, 20
     iv 11, 17, 23, 29, 35
  b **i** Start at 4, add 3 every time
     ii Start at 3, add 5 every time
     iii Start at 4, add 4 every time
     iv Start at 11, add 6 every time
  c Compare answers as a class

2 **i** 301 **ii** 498 **iii** 400 **iv** 605

3 **a** 7, 4, 1, –2, –5
  b Linear
  c **i** –140 **ii** –590

4 **a** $3n - 1 = 128$, 43rd
  b 200 is the 67th term; the 133rd term is 398 and the 134th is 401 so 400 isn't in the sequence

5 **a** 253
  b **i** –67, –64, –61
     ii The 23rd term is –1

6 **a** 2 or 7
  b **i** 3 or 8
     ii a or $a + 5$
  c e.g. $10n + 1$

What do you think?

1 Compare answers as a class. Possible answers include
  a $2n, 4n, 12n + 8$

---

b $2n + 1, 8n + 13$
c $10n + 6, 30n + 6$

2 **a** **i** 5, 9, 13, 17, 21 and 41
     ii 3, 6, 9, 12, 15 and 30
     iii 7, 13, 19, 25, 31 and 61
     iv 8, 16, 24, 32, 40 and 80
  b $3n$ and $8n$ because these give times tables, but the others do not
  c Compare answers as a class. Possible answers include
     i n **ii** $n - 1$ **iii** $n + 1$

Consolidate

1 **a** 17, 14, 11
  b 40, 80, 160
  c 10, 5, 2.5
  d 50, 80, 110

2 **a** 1, 12, 34, 78, 166
  b 0, 10, 30, 70, 150
  c 10, 30, 70, 150, 310

3 **a** **i** 14, 20
     ii 32, 128
     iii 10, 18
  b **i** –4, –10
     ii $\frac{1}{2}, \frac{1}{8}$
     iii 10, 12

4 **a** **i** 5, 7, 9, 11, 13
     ii 3, 6, 9, 12, 15
     iii –3, 1, 5, 9, 13
     iv 5, 2, –1, –4, –7
  b **i** Start at 5, go up in 2s
     ii Start at 3, go up in 3s
     iii Start at –3, go up in 4s
     iv Start at 5, go down in 3s

5 **a** 35th
  b The 125th term, 1004
  c The 24th term is 196 and the 25th is 204

Stretch

1 The next difference is 2 not 1
2 **a** 9 **b** 8, 10
  c 7.5, 9, 10.5
3 **a** **i** $5 + x$
     ii $2x + 5 = 17, x = 6$
     iii 11 and 28
  b **i** (3), 15, 18, (33)
     ii (2), 8, 10, 18, (28)
     iii –2, (7), 5, 12, (17)
4 **a** M (first letter of the months of the year)
  b T (first letter of the days of the week)
  c 11 (prime numbers)
  d 25 (square numbers)
  e S (first letter of one, two, three, four…)

Chapter 8.2

Are you ready?

1 **a** 25 **b** 36 **c** $\frac{1}{5}$ **d** 5
2 **a** 95 **b** 25 **c** 25
3 **a** 19 **b** 49 **c** –46
4 **a** **i** 3 **ii** 2 **iii** 6
  b 6, 12, 18, 24
  c They are the same

Practice 8.2A

1 **a** 1, 4, 9, 16, 25
  b 4, 10, 18, 28, 40
  c –2, –2, 0, 4, 10
  d 28, 35, 42, 49, 56
  e 4, 10, 18, 28, 40
  f 20, 30, 42, 56, 72
  g –12, –10, –6, 0, 8
  h 1, 8, 27, 64, 125
2 **a** 7600 **b** 7600 **c** 7600
3 **a**   **b**

---

c 4950
d For example, the nth triangular number has factor $(n + 1)$

4 **a** 3, 12, 27, 48, 75 and 9, 36, 81, 144, 225
  b 9, 16, 25, 36, 49 and 3, 6, 11, 18, 27

5 A, C, E, H, as they can all be written in the form $an + b$

6 **a** 2, 4, 8, 16, 32
  b 1, 3, 7, 15, 31
  c 1, 2, 4, 8, 16

7 **a** $\frac{1}{n + 1}$ is the same sequence without the first term
  b No

What do you think?

1 **a** **i** $\frac{1}{2}, \frac{2}{3}, \frac{3}{4}, \frac{4}{5}, \frac{5}{6}$
     ii $\frac{1}{3}, \frac{2}{5}, \frac{3}{7}, \frac{4}{9}, \frac{5}{11}$
  b **i** The terms get closer and closer to 1
     ii The terms get closer and closer to $\frac{1}{2}$
  c **i** The terms get closer and closer to $\frac{1}{3}$
     ii The terms get closer and closer to $\frac{1}{a}$
  d Compare answers as a class
2 Compare answers as a class

Practice 8.2B

1 **a** **i** 4, 7, 10, 13, 16
     ii 1, 4, 7, 10, 13
     iii 0, 3, 6, 9, 12
  b **i** Start at 4, go up in 3s
     ii Start at 1, go up in 3s
     iii Start at 0, go up in 3s
  c All the rules have the term $3n$, they all go up in 3s

2 **a** **i** 7, 9, 11, 13, 15
     ii 4, 10, 16, 22, 28
     iii 2, 9, 16, 23, 30
  b **i** Add 2 to the previous term
     ii Add 6 to the previous term
     iii Add 7 to the previous term
  c The constant difference is the same as the coefficient of n in the rule for the sequence

3 **a** Different starting numbers, but the same term-to-term rule
  b **i** Start at 6, go up in 4s
     ii Start at 7, go up in 4s
     iii Start at 1, go up in 4s
     iv Start at 14, go up in 4s
     v Start at $4 + a$, go up in 4s
     vi Start at $4 - a$, go up in 4s

4 **a** 17
  b Add 3 to the previous term
  c 3 **d** $3n + \_\_$ **e** $3n + 2$

5 **a** **i** $2n$ **ii** $2n + 3$ **iii** $6n$
     iv $6n - 2$ **v** $5n + 2$ **vi** $3n + 5$
     vii $10n - 3$ **viii** $7n - 11$
  b It is not a linear sequence

6 **a** e.g. the first term given by Rob's rule is 2, not 8
  b $n + 7$

7 **a** **i**

     ii $2n + 1$
     iii $n + 2$
  b **i**

     ii $3n + 1$
     iii $2n + 2$

What do you think?

1 **a** $4n + 5$

b The sequence is decreasing, not increasing

c i $29 - 4n$
 ii $11 - n$
 iii $14 - 8n$
 iv $115 - 15n$

2 a $5 + 4n - 1$ simplifies to $4n + 4$, which is of the form $an + b$
 b Yes
 c Both start at 0 but $4(n - 1)$ is increasing whilst $4(1 - n)$ is decreasing

3 a First differences are 5, 7, 9…, second differences are all 2
 b First differences are 3, 5, 7…, second differences are all 2
 c First differences are 6, 10, 14…, second differences are all 4
 d First differences are 12, 20, 28…, second differences are all 8

Consolidate

1 a 5, 8, 13, 20, 29
 b 0, 2, 6, 12, 20
 c −15, −10 −5, 0, 5
 d 5, 11, 19, 29, 41
 e −2, −2, 0, 4, 10
 f 12, 20, 30, 42, 56
 g 2, 0, 0, 2, 6
 h 5, 20, 45, 80, 125

2 a 1100   **b** 145   **c** 2220

3 a

c 9900

4 a 5
 b i $5n + 1$ **ii** $3n + 10$ **iii** $2n - 2$
 iv $4n + 5$ **v** $n + 6$ **vi** $8n + 4$
 vii $n + 18$ **viii** $8n - 5$

5 Compare answers as a class

Stretch

1 a $n^2 + 1$
 b i $2n^2$ **ii** $n^2 - 1$
 iii $n^2 + 10$ **iv** $100 - n^2$
 v $(n + 1)^2$

2 a $2n + 1$ **b** $3n + 1$
 c $\dfrac{2n + 1}{3n + 1}$
 d i $\dfrac{n + 3}{4n + 1}$
 ii $\dfrac{2n + 1}{2n + 6}$
 iii $\dfrac{2n - 1}{n^2 + 1}$

Check my understanding

1 a i Starts at 40 and goes up in 40s
 ii Starts at 40 and doubles
 iii Starts at 40 and halves
 iv Starts with 40 and 20, then each term is the sum of the previous two
 b i 240   **ii** 1280   **iii** 1.25
 iv 220

2 a 1626, 8131
 b 64.8, 12.76

3 a 50   **b** 300   **c** 590
 d 600   **e** 390

4 A and C

5 The 638th term is 2001

6 a $8n - 1$
 b $10n$
 c $21 - n$

Block 9 Indices

Chapter 9.1

Are you ready?

1 a −1   **b** −5   **c** 1

2 A ($4x$ and $3x$), B ($7y$ and y) and C ($2a$ and $5a$)

3 a $4x$ **b** $2x + 2y$ **c** $6x$
 d $5y - 3x$ **e** $15a - 5b$ **f** $-b + 10$

Practice 9.1A

1 a i 6   **ii** 4   **iii** 10
 b i $6x$ **ii** $4x$ **iii** $10x$
 c i $6x^2$ **ii** $4x^2$ **iii** $10x^2$
 d The number is the same in each; the first set have no letters, the second set just have x and the third set all have x^2

2 $5x^2$ and $3x^2$ can be simplified to $8x^2$ but $8x^2$ and $5x$ are not like terms so they cannot be simplified

3 a $2x^2 + 3x$ **b** $3x^2 + 2x$ **c** $4x^2 + 5x$

4 There are five lots of x^2 which is the same as $5x^2$

5 a $3x^2$ **b** $4x^2$ **c** $2x^2$
 d $10x^2$ **e** $2y^2$ **f** $7p^2$
 g t^2 **h** $35m^2$

6 a $4x^3$ **b** $6x^3$ **c** $10x^3$

7 a $14x^3$ **b** $3y^3$ **c** $5p^3$
 d x^3 **e** $12y^3$ **f** $24t^3$
 g $36k^3$ **h** $0.36k^3$

8 a They are not the same type of object so they cannot be collected together
 b They are not like terms

9 a $8x^3 + 2x^2$ **b** $6x^2 + 10x^3$
 c $18x^3 - 15x^2$ **d** $9c^2$
 e $k^3 + k^2$ **f** $h^3 + 21h^2$

10 a $12a^5 + 21a^2 - 4a$
 b $8p^2 - 15p + 15p^5$
 c $x^3 + 8x^4 - 6x^5$
 d $y - 13y^5 + 2y^3$
 e $21j^4$
 f $-4q^4 - 10q^3 + q^2$

11 a $p^3 + 2p^2 + 2m^3 + m^2$
 b $14j^4 - 10k^3 - 15j^3$
 c $9a^3 + 5a^2 + 27b^3$
 d $12x^3 - 7y^2 + 24x$

What do you think?

1 $5x, -2x, -7.5x, 3x$
 $x^2, -3x^2, -7x^2, \dfrac{x^2}{3}, 2x^2$
 $x^3, 5x^3, -4x^3$

2 a They are both squared
 b They don't have the same variable

3 Any correct expression, e.g.
 $10p^3 - 2p + p^3 - 5p^3$

Consolidate

1 a $3x^2$ **b** $2a^2$ **c** $4y^2$
2 a $5x^2$ **b** $12y^2$ **c** $2m^2$
3 a $3c^2 + 3d^2$ **b** $11p^2 - 7q$ **c** 1
4 a $7y^3 + 2x^2$ **b** $4x^3 - 4x^2$
 c $8x^3 - 5x^2$ **d** $7c^2 - 2c^3$
 e $4m^3 + m^2$ **f** $19h^3 + 21h^2$

Stretch

1 a i $2x^3$ **ii** $9x^3$ **iii** $7x^3$ **iv** x^2
 b Yes, e.g. $7x^3 - (7x^3 - 7x^2) \equiv 7x^2$

2 a Yes. Multiplying by one half is the same as dividing by 2, and multiplying by one third is the same as dividing by 3
 b $\dfrac{5x^3}{6}$

3 a $12f^2g^3$ **b** $13x^2y + 8xy^2$
 c $-6m^5 + 10m^3 - 4m^2 + m - 1$

Chapter 9.2

Are you ready?

1 a 4^3 **b** g^6 **c** 7^5 **d** M^8
2 a $3g$ **b** ab **c** $4gh$ **d** $2y^2$

Practice 9.2A

1 a $p^2 = p \times p$, $p^5 = p \times p \times p \times p \times p$
 so $p^2 \times p^5 = p \times p \times p \times p \times p \times p \times p$
 $= p^7$
 b i $m^3 \times m^4 = m \times m \times m \times m \times m \times m \times m = m^7$
 ii $q^3 \times q \times q^2 = q \times q \times q \times q \times q \times q = q^6$
 c The power in the answer is the sum of the powers in the question

2 a i c^6 **ii** c^7 **iii** c^8 **iv** c^{10}
 b i y^4 **ii** y^5 **iii** y^6 **iv** y^{11}

3 a x^7 **b** y^7 **c** a^{11}
 d 4^8 **e** 5^4 **f** b^9
 g e^{13} **h** b^7 **i** n^{12}
 j x^{22} **k** p^{77}

4 a Seb added the 2 and the 4 because he knows to add the powers
 b $2q^4 \times 4q^5 = 2 \times q^4 \times 4 \times q^5 = 2 \times 4 \times q^4 \times q^5 = 8 \times q^9 = 8q^9$

5 a i x^6 **ii** $-x^6$ **iii** x^6
 b i $2m^5$ **ii** $2m^5$ **iii** $6m^5$
 iv $20m^5$

6 a 2   **b** 4   **c** 12   **d** 1
 e 5   **f** 14   **g** 3

7 a 2^{4x} **b** $5^{10x - 1}$ **c** $r^{3y + 1}$

8 The base isn't the same in each part of the expression

What do you think?

1 a Jackson has multiplied the powers
 b Ed thinks that w means w to the power 0 when it really means w to the power 1

2 a i Any pair of cards whose powers sum to 7, e.g. a^2 and a^5
 ii Any pair of cards whose powers sum to 10, e.g. a^7 and a^3
 b Any three cards whose powers sum to 10, e.g. a^2, a^3 and a^5

3 a 6^5 **b** 5^5 **c** 3^4 **d** 4^6

Consolidate

1 a q^{10} **b** r^7 **c** n^4 **d** b^9
 e $2g^8$ **f** $6x^9$ **g** $30t^9$
2 a 2   **b** 6   **c** 7

Stretch

1 a $4 = 2^2$, $8 = 2^3$ so $4 \times 8 = 2^2 \times 2^3 = 2^5$
 b i 3^5 **ii** 2^{10} **iii** 5^6
 iv 10^{11} **v** 2^{21}

2 a f^7h^7 **b** $3m^{10}n$ **c** a^5b^5
 d x^8y^6 **e** m^3n^7 **f** $3c^{16}d^{10}$
 g $12gh^7$ **h** $7a^3b^8c^8$ **i** x^4y^3
 j $u^{12}v^{13}$ **k** $(x + 3)^{11}$ **l** $(b + c)^7$

3 a $x = 5$ **b** $x = 4$ **c** $x = 6$
 d $x = 9$ **e** $x = 3$

4 a $6x^{12} + 33x^5$
 b $12a^3b^{10} - 48a^{11}$
 c $30x^3y^6 + 18x^9y - 6x^2y^2$
 d $x^8 - 7x^3 + 4x^5 - 28$
 e $6y^9 - 8y^4 + 3y^5 - 4$
 f $10x^3y^4 - 2x^4 + 20y^5 - 4xy$

Chapter 9.3

Are you ready?

1 a p^4 **b** g^5 **c** n^9
 d m^4 **e** t^{12}
2 a 2   **b** 3.5   **c** 2.5
3 a $\frac{1}{4}$ **b** $\frac{2}{3}$ **c** $\frac{3}{4}$
 d $\frac{4}{5}$ **e** $\frac{2}{3}$ **f** $\frac{1}{10}$

Practice 9.3A

1 a Flo has written the numerator and denominator in expanded form and then cancelled common factors from the numerator and denominator

b i $\dfrac{m \times m \times m \times m \times m \times m \times m \times m}{m \times m \times m}$
$\equiv m \times m \times m \times m \times m \equiv m^5$

ii $\dfrac{y \times y \times y}{y \times y} \equiv y$

iii $\dfrac{6 \times 6 \times 6 \times 6 \times 6 \times 6}{6 \times 6 \times 6} = 6 \times 6 \times 6 = 6^3$

2 a Beca has divided the powers rather than subtract them

b $\dfrac{u^8}{u^2} \equiv u^6$

3 They represent the same calculation

4 a i d^3 **ii** d^2 **iii** d
b i y^2 **ii** y^3 **iii** y^4

5 a n^4 **b** x^4 **c** 7^2
d q **e** 5^3 **f** y^{48}
g e^3 **h** a^9 **i** x^4
j n^6 **k** 7^4 **l** r^{20}

6 a i x^6 **ii** $-x^6$ **iii** x^6
b i $10m^4$ **ii** $5m^6$ **iii** $5m^4$

7 a i g^6 **ii** g^{-6}
b i $4k^2$ **ii** $4k^{-2}$
c i $2p^5$ **ii** $\frac{1}{2}p^{-5}$
d i $\frac{4}{3}x$ **ii** $\frac{3}{4}x^{-1}$

8 a 3 **b** 8 **c** 0 **d** 6
9 a x^5 **b** y^3 **c** 4^6 **d** a^6
e p^{-2} **f** t^4 **g** 3^4 **h** y^4
i 2^{-3}

What do you think?

1 The base isn't the same
2 a $12y^3$ and 12 or $10y^4$ and $10y$
b $18y^6$ and $6y^8$
3 Emily has divided the base, but the base should stay the same
4 a The numerator is equal to the denominator so the fraction is equal to 1
b e.g. $\frac{h^5}{h^5} = h^{5-5} = h^0 = 1$
5 a 512 **b** 2048

Consolidate

1 a d^8 **b** y^3 **c** w
d u^5 **e** h^2 **f** l^3
g 1 **h** $8y^4$ **i** $6p^{18}$
2 a d^8 **b** y^3 **c** w
d u^5 **e** h^2 **f** l^3
3 Any correct expressions, e.g. $h^4 \times h^6$ or $\dfrac{h^{12}}{h^2}$

Stretch

1 a Any pair of values such that $x - y = 3$, e.g. $x = 10$ and $y = 7$
b i 1 **ii** -1
2 a f^6g **b** f^6g^{-1} **c** $f^{-6}g$ **d** $f^{-6}g^{-1}$
3 a $5m^5n$ **b** $\dfrac{3a^3b^2}{2}$ **c** $\dfrac{2h^4}{5}$ **d** $10xy^{-2}$
4 a 7^{3a} **b** y^{3x+3} **c** y^{m-2}
5 a x^{a+b} **b** x^{a-b}

Chapter 9.4

Are you ready?

1 a 6^5 **b** k^8 **c** y^3 **d** h^{10}
2 a 7^8 **b** y^{10} **c** p^6
d $5m^3$ **e** $24R^6$
3 a $s^3 + s^6$ **b** $x^8 - x^{12}$
c $6u^5 + 2u^7$
4 a 12^4 **b** y^4 **c** g

Practice 9.4A

1 a $(4^2)^3 = 4^2 \times 4^2 \times 4^2 = 4^6$
b $(5^3)^4 = 5^3 \times 5^3 \times 5^3 \times 5^3 = 5^{12}$
c $(d^4)^3 \equiv d^4 \times d^4 \times d^4 = d^{12}$
d $(v^5)^6 \equiv v^5 \times v^5 \times v^5 \times v^5 \times v^5 \times v^5 = v^{30}$

2 a i 5^6 **ii** 5^9 **iii** 5^{12}
iv 5^{15} **v** 5^{18}
b i e^{10} **ii** e^{15} **iii** e^{20}
iv e^{25} **v** e^{30}
c You might write each out in full or start to spot patterns
d The power in the answer is the product of the powers in the question
3 a x^{28} **b** p^8 **c** v^{24}
d d^{18} **e** t^{40} **f** x^{10}
g p^{30} **h** u^{12} **i** 4^8
j 5^{27} **k** 22^{15} **l** 176.4^{1000}
4 Multiplication is commutative so $273 \times 95 = 95 \times 273$
5 a 4 **b** 7 **c** 10
6 a Faith has multiplied 2 by 3 rather than calculating 2^3 (2 cubed)
b $2x^4 \times 2x^4 \times 2x^4 = 2^3 \times x^{12} = 8x^{12}$
7 a $9x^{10}$ **b** $16x^{12}$ **c** $125p^9$
d $\frac{1}{8}g^{12}$
8 a 5^{26} **b** m^{14} **c** x^{14}
d n^{16} **e** m^{23} **f** g^4

What do you think?

1 Possible answers are
$x = 1, y = 24; x = 2, y = 12; x = 3, y = 8;$
$x = 4, y = 6; x = 6, y = 4; x = 8, y = 3;$
$x = 12, y = 2; x = 24, y = 1$
2 $(10y^6)^5$
$10^5 = 100\,000$, so the power of the bracket must be 5, leaving the other power to be 6
3 $g^3(g^2 + g^5)$ matches $g^5 + g^8$ and $(g^2)^3 + (g^5)^3$ matches $g^6 + g^{15}$
$g^3(g^2 \times g^5)$ and $(g^2)^3 \times (g^5)^3$ both match to g^{21}

Consolidate

1 a $(a^4)^5 = a^4 \times a^4 \times a^4 \times a^4 \times a^4$
b $(h^3)^4 = h^3 \times h^3 \times h^3 \times h^3$
c $(s^9)^5 = s^9 \times s^9 \times s^9 \times s^9 \times s^9$
d $(p^{100})^2 = p^{100} \times p^{100}$
e $(6^9)^3 = 6^9 \times 6^9 \times 6^9$
f $(21^5)^8 = 21^5 \times 21^5 \times 21^5 \times 21^5 \times 21^5 \times 21^5 \times 21^5 \times 21^5$
g $(17^3)^4 = 17^3 \times 17^3 \times 17^3 \times 17^3$
h $(101^{99})^5 = 101^{99} \times 101^{99} \times 101^{99} \times 101^{99} \times 101^{99}$

2 a a^{20} **b** h^{12} **c** s^{45} **d** p^{200}
e 6^{27} **f** 21^{40} **g** 17^{12} **h** 101^{495}
3 a $(2a^4)^5 \equiv 2a^4 \times 2a^4 \times 2a^4 \times 2a^4 \times 2a^4$
b $(3h^3)^4 \equiv 3h^3 \times 3h^3 \times 3h^3 \times 3h^3$
c $(4s^9)^5 \equiv 4s^9 \times 4s^9 \times 4s^9 \times 4s^9 \times 4s^9$
d $(12p^{100})^2 \equiv 12p^{100} \times 12p^{100}$
4 a $32a^{20}$
b $81h^{12}$
c $1024s^{45}$
d $144p^{200}$
5 Any correct expressions, e.g. $(p^5)^6 \equiv p^{30}$

Stretch

1 a a^{bc} **b** $A^c a^{bc}$
2 a $(x^3y^2)^3 \equiv x^3y^2 \times x^3y^2 \times x^3y^2 = x^9y^6$
b i a^8b^6 **ii** $f^{12}g^{24}$
iii t^3m^{18} **iv** $16p^{20}q^8$
v $27m^{30}n^{18}$ **vi** $a^{15}b^{25}c^{35}$
3 a Abdullah has used the fact that $8 = 2^3$ to rewrite the calculation and then used laws of indices to simplify
b i 8 **ii** 10 **iii** 9 **iv** 12
c $x = 15$
4 a i $p^{\frac{3}{8}}$ **ii** $h^{\frac{37}{40}}$ **iii** $x^{\frac{3}{80}}y^{\frac{5}{48}}$
b $x = \frac{55}{27}$
5 a p^{2x^2-x} **b** $h^{15y-27y^2}$
c $m^{15w^5+10w^2}$ **d** y^{t^2-4x-5}
e $t^{12a^5+a^3-a}$

Check my understanding

1 a $7x^2 + 5x$ **b** $10y - 2y^3$
c $p^3 + 3p^2$ **d** $-2a^3 - 12a^4$
e $8b^4 - 6b^3$ **f** $-12h^3 + 3h^2$
2 a j^9 **b** s^{10} **c** 5^8 **d** h^{13}
e 2^{20} **f** y^{200} **g** 8^{16} **h** d^{116}
3 a h^3 **b** p^5 **c** 17^5 **d** w^{90}
e p^{12} **f** 16^8 **g** 1 **h** 125^{450}
4 a h^8 **b** x^{10} **c** 2^7 **d** y^{80}
e a^{18}
5 a $6a^{12}$ **b** $28x^{14}$ **c** $7y^3$ **d** $3x$
e $25p^{20}$
6 a x^{20} **b** y^{18} **c** t^8 **d** h^{98}
e $8k^{15}$

Block 10 Fractions and percentages

Chapter 10.1

Are you ready?

1 a 12 **b** 20 **c** 2 **d** 12.5
2 a 0.4 **b** 0.44 **c** 0.04 **d** 0.024
3 a 0.3 **b** $\frac{3}{10}$
4 a 0.17 **b** $\frac{17}{100}$
5 True: $10\% = \frac{1}{10}$ and $25\% = \frac{1}{4}$
The rest are false

Practice 10.1A

1 Compare answers as a class
2 a To change a percentage to a decimal, divide by 100; to change a decimal to a percentage multiply by 100
b i 0.65 **ii** 0.80 **iii** 0.08 **iv** 0.12
v 0.01 **vi** 0.015 **vii** 0.175
c i 72% **ii** 6% **iii** 30%
iv 67% **v** 67.5%
vi 0.2% **vii** 34.5%
3 a 45% < 50% **b** $\frac{9}{20}$
c A half is $\frac{10}{20}$ and $\frac{9}{20}$ is less than $\frac{10}{20}$
4 65% and $\frac{13}{20}$, 4% and $\frac{1}{25}$, 74% and $\frac{37}{50}$, 72% and $\frac{18}{25}$, 35% and $\frac{7}{20}$
5 a 21 **b** 7.5 **c** 63 **d** 50
e 12
6 a i 80 **ii** 40
b i 160 **ii** 120 **iii** 280
c Take 5% from the whole amount
7 Benji, Marta, Abdullah
8 a 0.225, 23%, $\frac{1}{4}$, 0.3, $\frac{1}{3}$
b $\frac{7}{10}$, 72%, $\frac{3}{4}$, 0.77, 0.8

What do you think?

1 Compare answers as a class
2 a 12.5%, 37.5%, 50%, 62.5%, 75%, 87.5%
b i $\frac{1}{40}$ **ii** $\frac{3}{40}$
3 a 8 **b** 80
c Compare answers as a class
4 a i 3 **ii** 4.5 **iii** 0.27
b e.g. find 25% and 1% and add them together

Practice 10.1B

1 a 55.5cm **b** £238 **c** 12.6g
d £38.88 **e** 1.608km **f** 3.05m
g 1.56kg **h** £121.93
2 a £280 **b** £280
c £160 + £80 + £40 = £280
3 a 0.27 = 27%, 2.7% = 0.027
b 8.1kg **c** £0.72 or 72p
4 a £14.60 **b** £8.32
c £0.75 or 75p **d** £545.92
5 a £139.65 **b** £33.25 **c** £6.30
6 a 480m **b** 18cm **c** 207g

What do you think?

1 42% of 68, as 28.56 > 28.35

2 Protein bar: 4.4g of fat, 16g of carbohydrate, 7.1g of protein and 6.5g of other
Chocolate bar: 7.2g of fat, 26.4g of carbohydrate, 2g of protein and 4.4g of other
Discuss comparisons as a class

Consolidate

1 a 25% b 10% c 30%
  d 60% e 22% f 55%
2 a 0.45 b 0.38 c 0.07
  d 0.12 e 0.125
3 a $\frac{33}{50}$ b $\frac{4}{5}$ c $\frac{3}{20}$ d $\frac{19}{25}$ e $\frac{17}{20}$
4 a 12 b 36 c 18 d 9
  e 117 f 40.5 g 600 h 60
  i 180 j 128
5 a £12.35 b £37.80 c £18.78
  d £19.09 e £28.91
6 Compare answers as a class

Stretch

1 Yes, they will be numbers greater than 1
2 a i 28 ii 28
  b They are the same
  c Yes e.g. $\frac{a}{100} \times b = \frac{b}{100} \times a = \frac{ab}{100}$
  d i 9 ii 8.5 iii 8 iv 60
3 Compare answers as a class

Chapter 10.2

Are you ready?

1 a 0.8 b 0.08 c 0.35 d 0.23
  e 0.17
2 40%
3 82%
4 90%
5 a 1 b 1.6 c 1.06 d 1.26

Practice 10.2A

1 a 80% b 0.8 c £72
2 a 82% b 0.82 c £14760
3 a 98% b 0.98 c £262640
4 a 70% b £1960
5 a The calculation gives 40%, not a 40% reduction
  b £54
6 a 90% b 85% c 65%
  d 48% e 40% f 94%
  g 25% h 10% i 83%
7 a 0.9 b 0.85 c 0.65
  d 0.48 e 0.4 f 0.94
  g 0.25 h 0.1 i 0.83

What do you think?

1 a i, iii, iv, vi
  b i 0.88 iii 0.7 iv 0.89 vi 0.96
2 a 0.95 b 0.975 c 0.985 d 0.995
  e 0.999 f 0.989 g 0.899

Practice 10.2B

1 a 118% b 1.18
  c i £767 ii £1121
2 a 107% b 1.07
  c i £25680 ii £40660
3 £984
4 a 110% b 115% c 135%
  d 152% e 160% f 106%
  g 175% h 190% i 117%
5 a 1.1 b 1.15 c 1.35
  d 1.52 e 1.6 f 1.06
  g 1.75 h 1.9 i 1.17
6 a Increase by 23% b Increase by 8%
  c Decrease by 15% d Increase by 1%
  e Decrease by 38% f Increase by 92%
  g Decrease by 82%
7 a £1380 b £1020

What do you think?

1 a ii and v b ii 1.17 or 72p v 1.09

---

2 a The second 20% is of a larger number than the first
  b 1.44
3 a 50%
  b Compare answers as a class
4 £3000 × 0.27

Consolidate

1 a 120% b 80% c 102% d 98%
  e 117% f 83% g 165% h 35%
2 a 1.2 b 0.8 c 1.02 d 0.98
  e 1.17 f 0.83 g 1.65 h 0.35
3 a £45.60 b £165.60 c £74.40
4 660
5 3780
6 25 increased by 40% is 5 greater than 40 decreased by 25%

Stretch

1 a Less than the original as × 1.1 then × 0.9 is the same as ×0.99 which is a 1% reduction
  b The same, as multiplication is commutative
  c 20%
2 a It depends on the context, e.g. are negatives allowed?
  b i × 1.5 would be a 50% increase
    ii 2.5
  c 320%
3 £26500

Chapter 10.3

Are you ready?

1 a 50% b 10% c 70% d 33$\frac{1}{3}$%
  e 75% f 38% g 24%
2 a 70% b 71% c 17% d 7%
3 a 12% b 35% c 68% d 90%
  e 8%
4 $\frac{4}{5}$ because $\frac{4}{5}$ = 80% > 75%

Practice 10.3A

1 a i $\frac{7}{20}$ ii 35%
  b i $\frac{33}{50}$ ii 66%
  c i $\frac{13}{25}$ ii 52%
  d i $\frac{3}{20}$ ii 15%
  e i $\frac{27}{50}$ ii 54%
2 a $\frac{3}{5}$ b 60%
3 a $\frac{7}{10}$, 0.7, 70% b $\frac{3}{10}$, 0.3, 30%
4 a $\frac{3}{5}$ b 60%
5 a $\frac{60}{400}$ or any equivalent fraction
  b 15% c 30%
6 a $\frac{3}{4}$ b 75%
7 66$\frac{2}{3}$%
8 30%

What do you think?

1 Compare answers as a class
2 a 40% b 60% c 30%
  d 50% e 25% f 66$\frac{2}{3}$%
3 a 50% b 40% c 5%
  d 100% e 200% f 600%
  g 150% h 133$\frac{1}{3}$%

Practice 10.3B

1 a $\frac{55}{80}$ or $\frac{11}{16}$ b 0.6875 c 69%
2 23%
3 a 48% b 52%
4 31.25%
5 11%
6 44%
7 26%
8 a Maths, History, French, Science, English
  b 73%
9 a i 91% ii 81%
  b Compare answers as a class

What do you think?

1 Flo (68% success)
2 82%
3 a 2% b 10% c 71% d 171%
  e 58% f 1% g 233$\frac{1}{3}$%
  h 19%

Consolidate

1 a i $\frac{7}{10}$ ii 70%
  b i $\frac{23}{25}$ ii 92%
  c i $\frac{17}{50}$ ii 34%
  d i $\frac{11}{20}$ ii 55%
  e i $\frac{28}{50}$ or $\frac{14}{25}$ ii 56%
2 a $\frac{3}{5}$ b 60%
3 a 65% b 60% c 80% d 62.5%
4 22.5%
5 a 125g b 68%
6 a 63% b 82% c 75%

Stretch

1 a 8%
  b Compare answers as a class

Chapter 10.4

Are you ready?

1 a 26% b 84% c 45% d 70%
2 a 46% b 37° c 17% d 24%
  e 80% f 6%
3 a 9 b 9 c 28 d 27
4 a 19% b 18% c 42% d 41%
  e 2%

Practice 10.4A

1 a £20 b 25%
2 a £180 b 60%
3 a 80% b 20% c 40%
4 The fraction should be £3/£12 not £3/£15; the denominator should be the original value
5 25%
6 40%
7 a 39% profit b 31% loss
  c 88% profit d 15% loss
  e 43% profit f 189% profit
8 a 106% b 37%

What do you think?

1 Compare answers as a class
2 a No, the population cannot go below zero
  b Compare answers as a class

Practice 10.4B

1 a 180 b 20%
2 a £9750 b 22%
3 a 89.25kg b 11%
4

	Price paid	Selling price	Percentage profit or loss
a	£30	£43.50	45% profit
b	£80	£68	15% loss
c	£16	£20	25% profit
d	£20	£16	20% loss
e	£5.99	£7.49	25% profit
f	£80	£16	80% loss

5 a Because multiplying by 1.08 and 0.82 does not give an overall result of multiplying by 1
  b £645.84 c 99.36%
6 a

	Left-handed	Right-handed	Total
Boys	122	460	582
Girls	84	519	603
Total	206	979	1185

b 21% **c** 59% **d** 53% **e** 51%
f 17% **g** 89% **h** 27
7 a Green Phones **b** 90%
c 111%
8 About 169 412

What do you think?

1 a 16.9% **b** 23.4%
2 a Three days out of a working week
of 5 days is $\frac{3}{5}$ which is equal to 0.6
b i 12.8% **ii** 13.1%

Consolidate

1 a 24 **b** 16%
2 a 36 **b** 60%
3 a 25% profit **b** 20% loss
c 30% loss **d** 18% profit
e 14% profit **f** 12% loss
4 41%
5 £14.45
6 a £780 **b** £455
7 2180 cm^3

Stretch

1 a £500 **b** £750
2 Decreasing the size by 10%

Chapter 10.5

Are you ready?

1 a 0.45 **b** 1.45 **c** 0.09 **d** 1.09
2 a 1.35 **b** 0.65 **c** 0.35 **d** 1.65
e 0.35 **f** 0.65
3 A decrease of 65% is the same as
finding 35% and a decrease of 35% is
the same as finding 65%

Practice 10.5A

1 a 24 **b** 12 **c** 96 **d** 144
e 240
2 a 750 **b** 600
3 a 60 **b** 40 **c** 800
d 200 **e** 20 **f** 16
4 £70
5 £3200
6 He has found 88% of the new wage,
but £168 is 112% of the original wage.
The correct answer is £150
7 2.5 litres
8 1.35 m
9 £37 500

What do you think?

1 £16 000
2 1 m 50 cm
3 £64

Practice 10.5B

1 £84
2 £342.30
3 £240
4

	Price paid	Selling price	Percentage profit or loss
a	£30	£24	20% loss
b	£25	£30	20% profit
c	£65	£75	15% profit
d	£80	£100	25% profit
e	£200	£100	50% loss
f	£85	£69.70	18% loss
g	£700	£665	5% loss
h	£550	£649	18% profit

5 420
6 £182 000
7 10%

What do you think?

1 1 060 000. The percentages are not
exact.
2 $33\frac{1}{3}$%

Consolidate

1 a 21 **b** 10.5 **c** 168 **d** 63
e 210
2 a £600 **b** £2000 **c** £850
3 a 1 hour
b 2 hours (1 hour 57 minutes)
4

	Price paid	Selling price	Percentage profit or loss
a	£40	£50	25% profit
b	£50	£40	20% loss
c	£60	£75	25% profit
d	£64	£80	25% profit
e	£100	£80	20% loss
f	£200	£10	95% loss
g	£75	£97.50	30% profit
h	£500	£350	30% loss

Stretch

1 Compare answers as a class
2 a Decrease by 20%
b Compare answers as a class, e.g.
decrease one by $33\frac{1}{3}$% and
increase the other by 50%
c i Compare answers as a class,
e.g. increase one by 25% and
increase the other by 20%
ii Compare answers as a class,
e.g. decrease one by 40% and
increase the other by 50%
d Compare answers as a class
e Compare answers as a class –
should be the same

Check my understanding

1 e.g. $\frac{1}{5} = \frac{2}{10} = \frac{20}{100} = 20\%$
2 a e.g. 0.45×120
b e.g. 50% = $\frac{1}{2}$ of 120 = 60 and 5% =
$\frac{1}{10}$ of 50% = 60 ÷ 10 = 6.
Subtract 5% from 50%
3 D, A, B, C
4 £917
5 £126 000
6 a 15% **b** 736
7 £280

Block 11 Standard index form

Chapter 11.1

Are you ready?

1 a 10^3 **b** 10^6 **c** 10^2
d 10^2 **e** 10^3 **f** 10^4
2 a i 60 **ii** 600 **iii** 6000
b i 52 **ii** 520 **iii** 5200
c i 16.5 **ii** 165 **iii** 1650
3 a 6000 **b** 140 000 **c** 8100
d 10 000 **e** 400 000 **f** 28 000 000

Practice 11.1A

1

10^2	10 × 10	100
10^3	10 × 10 × 10	1000
10^4	10 × 10 × 10 × 10	10 000
10^5	10 × 10 × 10 × 10 × 10	100 000
10^6	10 × 10 × 10 × 10 × 10 × 10	1 000 000

2 a 10^7 **b** 10^8 **c** 10^6
d 10^{10} **e** 10^{14} **f** 10^9
3 Yes
4 a 10^{100} **b** 10^{1000}
5 a i 7×10^2 **ii** 7×10^3
iii 7×10^4 **iv** 7×10^5
b i 9×10^4 **ii** 2×10^4
iii 5×10^4 **iv** 6×10^4
c i 3×10^6 **ii** 5×10^3
iii 6×10^7 **iv** 4×10^8
6 The greatest place value of each
number is ten thousand so each
number will be $A \times 10^4$ but the value
of A will change. In the first number it
will be an integer, but in the others it
won't be.
7 a 2.6×10^4 **b** 6.7×10^2
c 2.19×10^5 **d** 6.3×10^7
e 7.54×10^6 **f** 3.5×10^1
g 7.18×10^8 **h** 9.03×10^4
i 6.612×10^{10} **j** 1.2×10^6
k 7.155×10^3 **l** $7.300\,000\,25 \times 10^{11}$
8 a Set A: 140 000, 632 000, 718 200,
302 610
Set B: 1320, 13 200, 132 000,
1 320 000
b Compare answers as a class
9 a 20 000 **b** 1500 **c** 382 000
d 17 900 **e** 9 910 000 **f** 650
g 227 400 **h** 658 230 000
i 800 000

What do you think?

1 a He added four zeros on the end
b 23 000
c Multiply 2.3 by 10 four times
2 a i 1.26×10^{10} **ii** 1.26×10^{10}
b The second one because it was
easier to count the digits
3 No, none of them are. In the first two
the value of A isn't between 1 and 10
and in the third one the power of 10
isn't an integer.
4 Yes; $1.3 \times 10^6 = 1.3 \times 1\,000\,000 =$
$1.3 \times 100 \times 10\,000 = 1.3 \times 10^4 \times 100$

Consolidate

1 a 7×10^3 **b** 9×10^4 **c** 6×10^8
d 8×10^2 **e** 1×10^{10}
2 a 2.5×10^4 **b** 1.1×10^5
c 9.2×10^2 **d** 4.51×10^5
e 7.65×10^7 **f** 6.3×10^6
3 a 200 000 **b** 32 000 **c** 6200
d 79 100 **e** 7 910 000 **f** 8 032 000

Stretch

1 a 1.2×10^{28} **b** 6.75×10^{22}
c 5.3×10^{21} **d** 1.7635×10^{26}
2 No; it will change all the time
3 a $3.709\,180\,64 \times 10^8$
b You still have to write the same
number of digits, so it actually
makes the number harder to write.
c Round it to two or three
significant figures
4 3.6×10^{10}
5 a B (1.6×10^5) **b** C (1.7×10^4)
c 1.6×10^7

Chapter 11.2

Are you ready?

1 a 0.1 b 0.01 c 0.001 d 0.0001
2 a i 6 ii 60 iii 600 iv 6000
  b i 0.6 ii 0.06 iii 0.006
3 a 64 b 81
  c 125 d 1 000 000

Practice 11.2A

1 a

Power of 10	Ordinary number
10^5	100 000
10^4	10 000
10^3	1000
10^2	100
10^1	10
10^0	1
10^{-1}	0.1 or $\frac{1}{10}$
10^{-2}	0.01 or $\frac{1}{100}$
10^{-3}	0.001 or $\frac{1}{1000}$
10^{-4}	0.0001 or $\frac{1}{10000}$
10^{-5}	0.00001 or $\frac{1}{100000}$

  b As the power reduces by 1 the answer is divided by 10
  c All of them
  d None of them
2 a 10^{-3} b 10^{-4} c 10^{-6}
  d 10^{-7} e 10^{-9}
3 The power is the negative of one more than the number of zeros before the 1
4 a i 3×10^{-4} ii 6×10^4
    iii 7×10^{-4}
  b i 2×10^{-1} ii 2×10^{-2}
    iii 2×10^{-3}
5 a 5×10^{-6} b 7×10^{-3}
  c 3×10^{-7} d 2×10^{-6}
6 a i 1.2×10^{-1} ii 1.2×10^{-2}
    iii 1.2×10^{-3} iv 1.2×10^{-4}
  b They are all of the form 1.2×10^n but the value of n is different
7 a i 3×10^{-5} ii 3.2×10^{-5}
    iii 3.17×10^{-5}
  b They are all of the form $A \times 10^{-5}$ but the value of A is different
8 0.72 isn't between 1 and 10
9 a 1.4×10^{-3} b 7.8×10^{-4}
  c 5.27×10^{-2} d 3.2×10^{-6}
  e 5.032×10^{-3} f 5.68×10^{-5}
  g 9.6×10^{-3} h 1.056×10^{-2}
  i 3.5×10^{-1} j 7.1×10^{-8}
10 Set A: 0.14, 0.014, 0.0014, 0.00014
   Set B: 0.003, 0.0036, 0.00365, 0.003651
11 a 0.0024 b 0.000156
   c 0.081 d 0.000007685
   e 0.000000135 f 0.76
   g 0.0005564 h 0.000002367
   i 0.006

What do you think?

1 They will all have the same number of zeros before the first digit
2 a She counted the number of zeros before the non-zero digit
  b She has used a negative power when it should be positive
  c When the number is between 0 and 1

3

Numbers in standard form	Numbers not in standard form
3.61×10^{-4}	0.42×10^{-2}
7×10^4	0.7×10^4
3.276×10^3	12.6×10^5
	0.003
	$2.3 \times 10^{-3.5}$

4 a A (4.2×10^{-4}) b B (5×10^{-5})

Consolidate

1 a 7×10^{-3} b 9×10^{-2} c 6×10^{-6}
  d 8×10^{-5} e 5×10^{-9}
2 a 4.18×10^{-3} b 7.3×10^{-4}
  c 4.132×10^{-2} d 5.381×10^{-3}
  e 2.3×10^{-2} f 1×10^{-3}
3 a 0.002 b 0.0042
  c 0.0756 d 0.000063 5
  e 0.000791 f 0.000004308

Stretch

1 a 4.02×10^{-16} b 3.18×10^{-12}
  c 7.3184×10^{-18} d 9.00041×10^{-12}
  e 1×10^{-6} f 1×10^{-9}
2 a 3.72×10^{-4} b 7.48×10^{-5}
  c 5×10^{-5}
3

Number	8.2×10^{-4}	1.39×10^{-1}	6.04×10^{-3}
10 times larger	8.2×10^{-3}	1.39	6.04×10^{-2}
100 times larger	8.2×10^{-2}	1.39×10^1	6.04×10^{-1}
1000 times larger	8.2×10^{-1}	1.39×10^2	6.04
10 times smaller	8.2×10^{-5}	1.39×10^{-2}	6.04×10^{-4}
100 times smaller	8.2×10^{-6}	1.39×10^{-3}	6.04×10^{-5}
1000 times smaller	8.2×10^{-7}	1.39×10^{-4}	6.04×10^{-6}

Chapter 11.3

Are you ready?

1 a 2×10^4 b 9.6×10^3
  c 3.56×10^6 d 5×10^3
  e 3.06×10^{-2} f 8.215×10^{-6}
2 a 6000 b 32 000 c 415 000
  d 3050 e 0.0006 f 0.071
  g 0.00001842
3 a < b > c <
  d < e < f =
4 a i 7000 ii 2500
    iii 2050 iv 2005
  b i 600 ii 6000
    iii 60 000 iv 600 000
  c i 0.7 ii 0.6
    iii 0.97 iv 0.96

Practice 11.3A

1 a 70 000, 5000 b 70 000
  c She can look at the power of 10
2 A (2.7×10^5), D (6.61×10^6)
3 B (6.5×10^2), E (3.85×10^{-1})

4

Numbers greater than 1 million	Numbers less than 1 million
6×10^6	6×10^4
2.356×10^7	6×10^5
1.6×10^{11}	1.8×10^5
	1.9×10^5
	1.2×10^5
	4.8×10^{-7}

5 a > b > c <
  d < e > f <
6 A 2.8×10^4 B 3.92×10^4
  C 1.46×10^5 D 1.605×10^6
7 Jakub has confused negative numbers. The greatest number is 1.3×10^{-1}
8 3.97×10^5, 4.2 million, $73 \times 100 000$
9 a Venus; the power of 10 is greater
  b Earth; the power of 10 is the same and 5.972 is greater than 4.867
10 Discuss answers with a partner

What do you think?

1 a 7 is greater than 2
  b He has ignored the power of 10
  c 10^6 is greater than 10^4
2 Zach is correct, Faith is not
3 $x \leqslant 4$ or $x < 5$

Practice 11.3B

1 a i 7×10^9 ii 7×10^{10}
    iii 7×10^{11} iv 7×10^{17}
  b i 5×10^4 ii 5×10^5
    iii 5×10^6 iv 5×10^7
  c i 7.9×10^{-5} ii 7.9×10^{-4}
    iii 7.9×10^{-3} iv 7.9×10^{-2}
  d i 8×10^5 ii 2×10^7
    iii 7.2×10^9 iv 3.26×10^{10}

2

Numbers written in standard form	Numbers not written in standard form
2×10^7	0.2×10^5
2.54×10^2	20×10^{-3}
5.3×10^3	64×10^{11}
	256×10^{-9}
	16.4×10^{305}
	$20 000 \times 10^{19}$
	0.756×10^{-1}
	208×10^{10}
	10×10^3

3 a i 3×10^4 ii 3×10^5
    iii 3×10^6 iv 3×10^{28}
  b i 8×10^8 ii 8×10^9
    iii 8×10^{10} iv 8×10^{12}
  c i 2×10^4 ii 1.2×10^4
    iii 2.3×10^4 iv 1.05×10^5
  d i 8×10^{15} ii 9.7×10^{15}
    iii 2.3×10^{16} iv 1.05×10^{15}
  e i 1.25×10^0 ii 1.25×10^{-1}
    iii 1.25×10^{-2} iv 1.25×10^{-4}
4 a i 4×10^3 ii 6×10^3
    iii 8×10^3 iv 1×10^4
  b i 3×10^3 ii 2×10^3
    iii 1.5×10^3 iv 1.2×10^3

What do you think?

1 a D and E, because the answer won't need to be adjusted to be in standard form

b A and C, because the calculations are hard to perform mentally

c Not giving the answer in correct standard form

2 a Any pair of numbers that have a product of 60

b Any pair of numbers such that the first divided by the second is equal to 1.5

Consolidate

1 a > **b** > **c** < **d** <
e < **f** < **g** >

2 A (5×10^6), C (2×10^5), D (3.1×10^4)

3 A (2×10^{-3}), C (1.5×10^{-2}), E (9.8×10^{-3})

4 a 6×10^3 **b** 1.7×10^8
c 6.4×10^3 **d** 4.5×10^{-3}

Stretch

1 a A ($x > y$) **b** B ($M > N$)

2 a B is greater as the power of 10 is greater

b No, because the number in the circle could be anything

3 a i 7, 8 or 9 **ii** Any digit

b i 4, 5, 6, 7, 8 or 9
ii 6, 7, 8 or 9

4 a $x < 4$ **b** $x \leqslant 4$ or $x < 5$
c $x < 4$

Chapter 11.4

Are you ready?

1 a 4×10^4 **b** 3.6×10^4
c 1.75×10^5 **d** 2×10^6
e 2.871×10^8 **f** 6×10^{-4}
g 6.2×10^{-2} **h** 1.934×10^{-4}

2 a 1600 **b** 7520
c 285 000 **d** 19 000 000
e 0.0007 **f** 0.0081
g 0.036 56 **h** 0.18

3 a 10^7 **b** 10^7 **c** 10^2 **d** 10^6
e 10^4 **f** 10^{22} **g** 10^7 **h** 10^{99}

Practice 11.4A

1 a i 1.2×10^5 **ii** 1.2×10^6
iii 1.2×10^7

b i 2.6×10^4 **ii** 2.6×10^5
iii 2.6×10^6

2 a i 3×10^1 **ii** 3×10^2 **iii** 3×10^3
b i 7×10^0 **ii** 5.2×10^1
iii 5.6×10^{-1}

3 Discuss answers as a class

4 a 8×10^3 **b** 2.6×10^4
c 2.06×10^5

5 a 3.5×10^4 **b** 1.67×10^4
c 3.15×10^6 **d** 9.68×10^5
e 7.486×10^4 **f** 2.068×10^4
g 3.313×10^5 **h** 7.983×10^3
i 4.4×10^{-2} **j** 2.3068×10^{-1}
k 2.2164×10^5

6 a 7.7×10^3 **b** 6.93×10^5
c 2.6×10^5 **d** 1.92×10^3
e 3.8419×10^6 **f** 7.146×10^4
g 3.7×10^{-2} **h** 1.53×10^{-1}
i 9.7835×10^{-3}

7 1.293×10^5 cm

8 a 7.97×10^8 miles
b 2.707×10^9 miles

9 8.75×10^4 mm

What do you think?

1 He has added five zeros on to 2.5 rather than multiplying it by 10 five times.

2 a 7.1×10^4 **b** 3.16×10^5
c 1.803×10^6

3 a It would take a long time and could introduce mistakes
b Rewrite 2.3×10^{23} as 0.23×10^{24}

4 $4.3 \times 10^7 > 7.2 \times 10^4$

Practice 11.4B

1 a 6×10^5 **b** 6×10^6 **c** 6×10^7
d 6×10^9 **e** 6×10^{-3}

2 1.2×10^7

3 a 2.4×10^8 **b** 2.1×10^9
c 2×10^7 **d** 2.8×10^8
e 8.1×10^8 **f** 1.36×10^{10}
g 1.725×10^{10} **h** 3.04×10^{12}
i 8.4×10^3 **j** 1.95×10^4
k 5.16×10^4 **l** 1.17×10^4
m 5.58×10^{12} **n** 1.98×10^5

4 a 4×10^5 **b** 4×10^4 **c** 4×10^3
d 4×10^2 **e** 4×10^{11}

5 a 6 **b** 8 **c** 5 **d** -1

6 a i $4 \div 8 = 0.5$, $10^8 \div 10^5 = 10^3$
ii $0.5 \times 10^3 = 5 \times 10^{-1} \times 10^3 = 5 \times 10^2$
b $0.5 \times 10^{-3} = 5 \times 10^{-4}$
c i 5×10^8 **ii** 5×10^{-3} **iii** 5×10^{12}

7 a 2×10^3 **b** 4×10^2
c 1.2×10^2 **d** 1.5×10^3
e 1.1×10^1 **f** 6×10^1
g 3×10^3 **h** 9.2×10^2
i 1.5×10^2 **j** 7.5×10^3
k 1.6×10^{-11} **l** 1.4×10^{-2}
m 2×10^5 **n** 2.3×10^8

What do you think?

1 $0.8 \times 10^3 = 8 \times 10^{-1} \times 10^3 = 8 \times 10^2$

2 a Any pair of integers that sum to 6
b Any pair of integers such that x is 6 more than y
c Any pair of integers that have a product of 6
d Any pair of integers that have a product of 12

Consolidate

1 a 8×10^4 **b** 6.4×10^5
c 6.04×10^5 **d** 1.73×10^4
e 7.8085×10^8 **f** 4.59×10^{-2}
g 4.32×10^{-3} **h** 6.998×10^8
i 6.7952×10^5

2 a 6×10^{11} **b** 1.8×10^8
c 8.6×10^{19} **d** 7.2×10^4
e 2.4×10^{11} **f** 1.58×10^7
g 3×10^3 **h** 4×10^{-3}
i 1.2×10^{12} **j** 5×10^{-4}
k 1.1×10^{-8} **l** 2×10^{10}

Stretch

1 1.8 m

2 a $5 \times 10^4 + 4.85 \times 10^5 = 5.35 \times 10^5$
b $1.6 \times 10^3 - 3.9 \times 10^2 = 1.21 \times 10^3$
c $4.85 \times 10^5 - 3.9 \times 10^2 = 4.8461 \times 10^5$

3 1.989×10^{12} cm^2

Chapter 11.5

Are you ready?

1 a 8 **b** 16 **c** 32 **d** 64
e 9 **f** 27 **g** 25 **h** 125

2 a 3125 **b** 216 **c** 14 641
d 0.216

3 a 2^5 **b** 2^3 **c** 2^7 **d** 2^8

4 5, 625, 25, 125, 1

5 a 10 **b** 7 **c** 2 **d** 0.6
e 3 **f** 5 **g** 10 **h** 2

Practice 11.5A

1 2^{-1} and 0.5
3^{-1} and $\frac{1}{3}$
4^{-1} and $\frac{1}{4}$
5^{-1} and $\frac{1}{5}$
8^{-1} and $\frac{1}{8}$
10^{-1} and 0.1

2

Power	Fraction
2^{-1}	$\frac{1}{2}$
2^{-2}	$\frac{1}{2^2} = \frac{1}{4}$
2^{-3}	$\frac{1}{2^3} = \frac{1}{8}$
2^{-4}	$\frac{1}{2^4} = \frac{1}{16}$
2^{-5}	$\frac{1}{2^5} = \frac{1}{32}$

3 It shouldn't be negative

4 a $\frac{1}{9}$ **b** $\frac{1}{100}$
c $\frac{1}{5^2} = \frac{1}{25}$ **d** $\frac{1}{6^2} = \frac{1}{36}$

5 a $\frac{1}{27}$ **b** $\frac{1}{64}$ **c** $\frac{1}{125}$ **d** $\frac{1}{36}$
e $\frac{1}{1000}$ **f** $\frac{1}{50}$ **g** $\frac{1}{1331}$ **h** 8
i $-\frac{1}{2}$ **j** $\frac{1}{4}$ **k** $-\frac{1}{8}$

6 No; $(-5)^{-2} = \frac{1}{25}$ but $-5^{-2} = -\frac{1}{25}$

7 a 0.125 **b** 0.25 **c** 0.01
d 0.04 **e** 0.04 **f** 0.01

8 a $8 = 2^3$ and $\frac{1}{8} = 2^{-3}$
b $16 = 4^2$ and $\frac{1}{16} = 4^{-2}$
c $16 = 2^4$ and $\frac{1}{16} = 2^{-4}$
d $25 = 5^2$ and $\frac{1}{25} = 5^{-2}$
e $125 = 5^3$ and $\frac{1}{125} = 5^{-3}$
f $625 = 5^4$ and $\frac{1}{625} = 5^{-4}$
g $1\,000\,000 = 10^6$ and $\frac{1}{1\,000\,000} = 10^{-6}$

What do you think?

1 a C (0.125) and D ($\frac{1}{8}$)
b -8 has made it negative, $\frac{1}{6}$ has multiplied 2 by 3

2 a i If I know that $2^{10} = 1024$, then 2^{-10} is equal to $\frac{1}{1024}$
ii If I know that $3^{-5} = \frac{1}{243}$, then 3^5 is equal to 243

3 A **i** 2 **ii** 3 **iii** 4 **iv** 5
B **i** $\frac{3}{2}$ **ii** $\frac{4}{3}$ **iii** $\frac{2}{5}$ **iv** $\frac{5}{2}$
C **i** 2 **ii** 4 **iii** 8 **iv** 16

Practice 11.5B

1

Expression with powers	Working
$16^{\frac{1}{2}}$	$\sqrt{16} = 4$
$25^{\frac{1}{2}}$	$\sqrt{25} = 5$
$36^{\frac{1}{2}}$	$\sqrt{36} = 6$
$64^{\frac{1}{2}}$	$\sqrt{64} = 8$
$100^{\frac{1}{2}}$	$\sqrt{100} = 10$

2 He needs to find the cube root not divide by 3

3 a 2 **b** 2 **c** 2 **d** 10
e 10 **f** 7 **g** 5 **h** 5
i 0.5 **j** -5 **k** -2

4 $64^{\frac{1}{2}}$ and 8, $64^{\frac{1}{3}}$ and 4, $64^{\frac{1}{6}}$ and 2, 64^0 and 1

5 a $\frac{1}{2}$ **b** $\frac{1}{3}$ **c** $\frac{1}{3}$ **d** $\frac{1}{2}$
e $\frac{1}{4}$ **f** $\frac{1}{7}$ **g** $\frac{1}{5}$ **h** $\frac{1}{3}$

6 a 2.645 75 **b** 1.912 93
c 4.641 59 **d** 2.352 16
e 0.871 78 **f** $-1.259 92$

What do you think?

1 a $x^{\frac{1}{2}} \times x^{\frac{1}{2}} = x^{\frac{1}{2} + \frac{1}{2}} = x^1 = x$
b $(x^{\frac{1}{2}})^2 = x$ so $x^{\frac{1}{2}} = \sqrt{x}$

2 a Because $25^{\frac{1}{2}} = 5$

b No number when squared gives a negative answer so −25 doesn't have a square root

c This is the negative of $25^{\frac{1}{2}}$ rather than negative 25 to the power of half

d It changes the answer

3 a

Can calculate the answer	Can't calculate the answer
$(-20)^{\frac{1}{3}}$	$(-20)^{\frac{1}{2}}$
$(-20)^{\frac{1}{5}}$	$(-20)^{\frac{1}{4}}$
$(-10)^{\frac{1}{3}}$	$(-7)^{\frac{1}{2}}$
$(-20)^{\frac{1}{5}}$	$(-17)^{\frac{1}{4}}$

b When the denominator in the power is odd, you can calculate the answer

c A power of a half can be calculated for positive numbers

Consolidate

1 a $\frac{1}{8}$ **b** $\frac{1}{9}$ **c** $\frac{1}{6}$ **d** $\frac{1}{49}$
  e $\frac{1}{125}$ **f** $\frac{1}{9}$ **g** $\frac{1}{32}$ **h** $-\frac{1}{343}$
2 a 3 **b** 5 **c** 2 **d** 5
  e 3 **f** 2 **g** −10 **h** 1.2

Stretch

1 a 4.5 **b** $\frac{5}{8}$ **c** $\frac{5}{6}$
  d $\frac{1}{6}$ **e** 2.5 **f** $\frac{3}{4}$
2 Yes; 0.5 is equivalent to one half
3 a 9 **b** 12 **c** 2 **d** 2
  e 5 **f** $\frac{1}{2}$ **g** −3 **h** −3
4 a E $\left(\frac{1}{4}\right)$
  b $16^{-\frac{1}{2}} = \frac{1}{16^{\frac{1}{2}}} = \frac{1}{4}$
5 a $\frac{1}{5}$ **b** $\frac{1}{5}$ **c** $\frac{1}{4}$ **d** $\frac{1}{3}$
  e $\frac{1}{2}$ **f** $\frac{1}{7}$ **g** 25 **h** $\frac{25}{4}$
  i $\frac{4}{9}$ **j** $\frac{2}{3}$

Check my understanding

1 a 2×10^4 **b** 3×10^5 **c** 5×10^3
  d 7×10^7 **e** 4.1×10^5 **f** 1.9×10^4
  g 6.5×10^2 **h** 8.703×10^9
2 a 30 000 **b** 700 000 000
  c 410 **c** 605 000 000 000
3 a 2×10^{-3} **b** 9×10^{-7}
  c 3.35×10^{-4} **d** 7.8×10^{-1}
  e 8.01×10^{-6} **f** $8.200\,59 \times 10^{-3}$
4 a 0.0005 **b** 0.000 000 000 67
  c 0.000 000 199 **d** 0.0204
5 3.7×10^6
6 a 4.5×10^4 **b** 1.84×10^4
  c 3.75×10^6
7 a 1.2×10^6 **b** 3.5×10^7 **c** 1.8×10^{10}
8 a $\frac{1}{8}$ **b** $\frac{1}{16}$ **c** $\frac{1}{125}$
9 a 3 **b** 2 **c** 10

Block 12 Number sense

Chapter 12.1

Are you ready?

1 a 872 **b** 870 **c** 900
2 a 9 **b** 7 **c** 3 **d** 6
3 a 900 **b** 0.7 **c** 3
  d 0.0007
4 $C = \pi d$

Practice 12.1A

1 a 80 **b** 380 **c** 3080

  d 3880 **e** 50 **f** 50
  g 50 **h** 840
2 a 156 000 **b** 2000 **c** 19 000
  d 20 000
3 a 900 **b** 90 **c** 0.9
  d 10 000 **e** 0.1 **f** 0.02
  g 6 000 000
4 a 4200 **b** 70 **c** 8100
  d 100 **e** 0.008
  Various answers, e.g. part **c** is an overestimate as both numbers in the calculation are overestimates
5 a 36 cm **b** 72 cm **c** 90 m
  d 100 cm
6 a 10 cm **b** 5 cm
7 a i 2400
    ii £50
  b i Overestimate
    ii Underestimate
8 1100 cm²
Discuss accuracy as a class

What do you think?

1 a 90 is almost halfway between 9² and 10²
  b 150 is much closer to 12² than 13²
2 a i 19 **ii** 37 000
    iii 230 **iv** 47 000 000
  b i 3 **ii** 3.1
    iii 3.14 **iv** 3.142
3 Discuss answers as a class

Practice 12.1B

1 a 50 **b** 20 **c** 5 **d** 14
  e 52 **f** 28 **g** 7 **h** 13
  i 72 **j** 258 **k** 7 **l** 12
2 a The pairs are $\frac{20}{0.5}$ and $\frac{200}{5}$;
    $\frac{20}{0.05}$ and $\frac{2000}{5}$, $\frac{2}{0.4}$ and $\frac{20}{4}$,
    $\frac{200}{0.4}$ and $\frac{2000}{4}$
  b i 30 **ii** 200
    iii 200 **iv** 150
    v 30 **vi** 15 000
3 a 750 **b** 2000 **c** 40
  d 10 **e** 10 000 **f** 20 000
4 a 5000
  b i 500
    ii Underestimate
5 a 1090 **b** 92.5 **c** 7 **d** 120
  e 7 **f** 20 **g** 475
Discuss accuracy used as a class

What do you think?

1 Compare answers as a class
2 80%
3 a 7 **b** 400 **c** £1800

Consolidate

1 a 800 **b** 2800 **c** 16 700
  d 12 000 **e** 19 900 **f** 20 000
  g 100
2 a 100 **b** 2 **c** 20
  d 0.6 **e** 0.001 **f** 600 000
3 a 400 **b** 20 **c** 200
  d 10 000 **e** 80 000
4 a 2 **b** 18 **c** −10 **d** −28
  e 7 **f** 1.6 **g** 35
5 a 5 **b** 50 **c** 50 **d** 50
6 a 2.7 **b** 30 **c** 720

Stretch

Compare answers as a class

Chapter 12.2

Are you ready?

1 a 3 tens **b** 3 tenths **c** 3 ones
  d 3 hundredths
2 a 7 **b** 19 **c** 3
  d 8 **e** 5
3 a 4.22 **b** 4.75 **c** 5.08 **d** 5.88

Practice 12.2A

1 a 3.4 **b** 3.8 **c** 3.9
2 a iii 5.8
  b i 7.6 and 7.7 **ii** 7.6
  c i 9.1 **ii** 23.7
    iii 9.6 **iv** 0.7
3 Discuss the mistakes as a class. Correct answers are
  a 6.2 **b** 4.3 **c** 9.0
4 a 5.71 **b** 5.76 **c** 5.79
5 a iii 4.24
  b i 18.36 and 18.37
    ii 18.37
  c i 8.58 **ii** 0.37
    iii 9.47 **iv** 62.35
6 a 33.72 **b** 1.83 **c** 3.49 **d** 1.89
7 a 24.50 cm **b** 31.8 cm
8 a £30.70 **b** £87.72
  c £2193 **d** £877 000
Compare the "sensible" degrees of accuracy as a class

What do you think?

1 Exactly the same
2 a i 940 **ii** 942.49
  b i 0.14 **ii** 0.14
  c i 610 **ii** 613.68
  d i 0.019 **ii** 0.02
3 a 120 cm² **b** 122.01 cm²
  c 2.01 cm² **d** 1.6%

Practice 12.2B

1 a 7.4, 7.403, 6.91, 6.51, 7.06
  b $6.5 \leqslant x < 7.5$
2 a Compare answers as a class
  b 38°
3 a 6.35
  b i $29.5 \leqslant p < 30.5$
    ii $15.35 \leqslant q < 15.45$
    iii $2.715 \leqslant r < 2.725$
  c $7.5 \leqslant x < 8.5$, $7.95 \leqslant x < 8.05$
4 a $39.5 \leqslant x < 40.5$
  b $35 \leqslant x < 45$
5 a 49.995 m
  b $49.995 \leqslant l < 50.005$
6 a $7999.5 \leqslant a < 8000.5$
  b $7995 \leqslant b < 8005$
  c $7950 \leqslant c < 8050$
  d $7500 \leqslant d < 8500$
7 The number of people must be an integer. This could be written as $5500 \leqslant n \leqslant 6499$ where n is an integer. The area of the field in m² could take any value, and be written as $5500 \leqslant A < 6500$
8 a $350 \leqslant n < 450$
  b $395 \leqslant n < 405$
  c $399.5 \leqslant n < 400.5$

What do you think?

1 a 77 cm **b** 75 cm
  c 9 cm **d** 7 cm
2 a $17.5 \leqslant x < 22.5$
  b $19 \leqslant y < 21$
  c $20 - \frac{a}{2} \leqslant z < 20 + \frac{a}{2}$

Consolidate

1 a 9.3 **b** 41.4 **c** 8.4
  d 2.1 **e** 9.4
2 a 45.67 **b** 9.39 **c** 12.82 **d** 48.48
3 a 33.9 cm² **b** 39.6 cm² **c** 28.3
4 a 150.53 **b** 0.17 **c** 5.74
  d 83.36 **e** 16.49 **f** 3.78
  g 0.02
5 a £54.64 **b** £1093 **c** £27 322
Compare the "sensible" degrees of accuracy as a class

Stretch

1 a From 8 − 0.5 cm to 8 + 0.5 cm
  b i $7.5 \leqslant x \leqslant 8.5$

ii $30 \leqslant P \leqslant 34$
iii $56.25 \leqslant A \leqslant 72.25$
2 a $7.97 \leqslant s \leqslant 8.03$
b $7.96 \leqslant s \leqslant 8.04$
c $7.93 \leqslant x \leqslant 8.07$
3 Compare answers as a class

Chapter 12.3

Are you ready?
1 100
2 a 10 **b** 1000
3 1000
4 a 93.6 **b** 936 **c** 4700
d 0.893 **e** 12.67 **f** 0.038

Practice 12.3A
1 a 400 cm **b** 450 cm **c** 8 cm
d 72.5 cm **f** 103 cm **f** 0.6 cm
g 0.65 cm **h** 600 cm **i** 300 cm
j 300 000 cm
2 a $600 \div 12$ and $6 \div 0.12$
b 50
3 a 4 kg **b** 0.3 kg **c** 80 g
d 6000 g **e** 0.7 kg
4 a 20 **b** 3 g
5 a 5 litres **b** 0.5 litres
c 0.75 litres **d** 0.33 litres
e 0.02 litres **f** 80 litres
6 a 16 **b** 11 **c** 8
7 330 ml > 20 cl (which is 200 ml)
8 60

What do you think?
1 a 6 **b** 5
c Yes, the ratio is constant
2 a 0.1175 g **b** 3
c No, you would need to drink 80 glasses of milk each day

Practice 12.3B
1 a Because 85p = £0.85
b £1.30
c Yes, the total cost is £4.35
2 a £93 **b** £4 less
3 300 small and 50 large is £31 cheaper
4 a 6
b Three: a 5p, a 2p and a 1p
5 £26.94
6 £46.10
7 £260.71

What do you think?
1 3
2 15
3 8 hours

Practice 12.3C
1 a 168 **b** 288 **c** 744
2 a 5 **b** Sunday
c December 10th **d** Saturday
3 a 45 **b** 40 **c** 270
d 1440 **e** 5040
4 9 p.m.
5 42 minutes
6 a 8.7%
b 3 minutes 30 seconds is 3.5 minutes, not 3.3 minutes. Correct answer is 34 songs.
7 23:15
8 3.2×10^7

What do you think?
1 No; there are 52 weeks and 1 day (or 2 days in a leap year)
2 18628
3 Compare answers as a class

Consolidate
1 a 80 **b** 236 **c** 1000 **d** 2500
2 54 cm²
3 3.5 kg

4 a 7 **b** £6.23
5 a £43.80 **b** £13.72
6 a 91 **b** 2160
7 Compare answers as a class

Stretch
1 200 g
2 Compare answers as a class
3 a 5 **b** 12
c 210 days is more than 6 months
4 a If the number of the year divides exactly by 4 but not by 100
b 21 or 22

Chapter 12.4

Are you ready?
1 a 184 cm² **b** 1120 cm²
c 18 000 cm²
2 a 10 **b** 1000
c 1 000 000
3 40
4 9600 cm³

Practice 12.4A
1 a 800 mm² **b** 9000 mm²
c 20 cm² **d** 0.4 cm²
2 a i 28 cm² **ii** 2800 mm²
b i 15 cm² **ii** 1500 mm²
c i 6.75 cm² **ii** 675 mm²
3 a 16 m² **b** 160 000 cm²
c 1 m² = 10 000 cm²
4 a 60 000 cm² **b** 92 000 cm²
c 2 500 000 cm²
5 a 0.8 m² **b** 7.5 m² **c** 100 m²
6 a 2000 **b** 25
7 1 000 000

What do you think?
1 a 0.5 km **b** 0.25 km²
c 225 000 000 cm²
2 D (1 000 000 m²)
3 a i 10 000 m²
ii 100 000 000 cm²
b 100

Practice 12.4B
1 a 9000 mm³ **b** 200 mm³
c 2.6 cm³ **d** 0.05 cm³
2 a 48 cm³ **b** 48 000 mm³
c 48 000 ÷ 48 = 1000
3 a 8 m³ **b** 8 000 000 cm³
c 1 m³ = 1 000 000 cm³
4 a 6 000 000 cm³ **b** 250 000 cm³
c 25 000 cm³
5 a 0.2 m³ **b** 50 m³ **c** 300 m³
6 a 750
b i 150 000 cm³ **ii** 0.15 m³
7 1×10^9 mm³

What do you think?
1 646. The lengths of the container may not be divisible exactly by the lengths of the box.
2 a 1 000 000
b A 2 cm cube is 8 times the volume of a 1 cm cube
c 125 000 **d** 1000 **e** 8000
3 1000 of each make a litre, but 1 cm³ measures volume whereas 1 ml measures capacity

Consolidate
1 a 30 000 cm² **b** 5000 cm²
c 160 000 cm² **d** 2500 cm²
e 5 cm² **f** 80 cm²
g 0.25 cm²
2 a 8 m² **b** 0.5 m²
c 0.03 m² **d** 64.2 m²
3 a 2 500 000 cm³ **b** 750 000 cm³
c 5 cm³ **d** 0.1 cm³
e 62 000 000 cm³

4 a 2 m³ **b** 32 m³
c 0.05 m³ **d** 0.85 m³
5 a 45 litres **b** 0.8 litres
c 1000 litres **d** 500 litres
e 0.6 litres

Stretch
1 a 45.7 cm **b** 116 cm²
c 11 600 mm² **d** 0.0116 m²
2 B (5 litres)
3 35
4 a 904.8 litres
b Compare answers as a class

Check my understanding

1 $8 \times 8 - 2 \times 2 = 64 - 4 = 60$
2 a 850 000 000 **b** 0.16
3 189.3 cm²
4 400
5 £597
6 a 85
b 1 hour 40 minutes
7 $10.145 \leqslant x < 10.155$
8 2500

Block 13 Angles in parallel lines and polygons

Chapter 13.1

Are you ready?
1 a Obtuse **b** Reflex **c** Acute
d Obtuse **e** Acute **f** Reflex
2 ∠CBD or ∠DBC, ∠SQR or ∠RQS, ∠LON or ∠NOL
3 Check accuracy with a partner
4 89°, 30°, 61°

Practice 13.1A
1 $a = 25°$, $b = 40°$, $c = 60°$, $d = 278°$, $e = 80°$, $f = 223°$, $g = 38°$, $h = 76°$, $i = 104°$, $j = 48°$
Discuss possible reasons as a class
2 $a = 25°$, $b = 35°$, $c = 55°$, $d = 100°$, $e = 30°$, $f = 149°$, $g = 252°$, $h = 48°$, $i = 10°$
Discuss possible reasons as a class
3 ∠PQS + ∠SQR ≠ 180°

What do you think?
1 a 135°, 45° **b** 120°, 60°
c 67.5°, 112.5°
2 a Kite, delta
b Isosceles trapezium, parallelogram, rhombus
3 No, it could be a trapezium or a kite

Practice 13.1B
1 $a = 118°$, $b = 62°$, $c = 56°$, $d = 59°$, $e = 59°$, $f = 121°$, $g = 60°$, $h = 60°$, $i = 60°$, $j = 120°$, $k = 21°$, $l = 21°$, $m = 21°$, $n = 69°$, $p = 28°$, $q = 28°$, $r = 124°$
Discuss possible reasons as a class
2 $a = 30°$, $b = 10°$, $c = 130°$, $d = 65°$, $e = 19°$, $f = 22.5°$, $g = 110°$, $h = 70°$, $x = 30°$, $y = 15°$, $z = 40°$
Discuss possible reasons as a class
3 53°

What do you think?
1 All the angles are equal and sum to 180°, so each is 180° ÷ 3 = 60°
2 a Yes; 40° and 100°, or 70° and 70°
b i 4° and 172°, or 88° and 88°
ii 35° and 35°
iii 45° and 45°
iv 10° and 10°

Consolidate

1 168°

2 $a = 99°$, $b = 180°$, $c = 27°$, $d = 197°$, $e = 30°$, $f = 141°$, $g = 112°$, $h = 23°$, $i = 45°$

3 56° and 56°, or 68° and 44°

4 $a = 52°$, $b = 118°$, $c = 31°$, $d = 31°$, $w = 66°$, $x = 40°$, $y = 11°$, $z = 20.5°$

Stretch

1 Because any quadrilateral can be split into two triangles

2 a 135°   **b** 1080°

3 The sides are not all equal in length

4 Compare answers as a class. The angles should be 36°, 72°, 108° and 144°, and possible answers include a trapezium.

Chapter 13.2

Are you ready?

1 Diagram C

2 $a = 66°$, $b = 135°$, $c = 45°$, $d = 28°$, $e = 64°$

3 $a = 55°$, $b = 125°$, $c = 69°$, $d = 111°$, $e = 68°$, $f = 44°$

Practice 13.2A

1 a f **b** e **c** e
   d h **e** f **f** h

2 a h **b** e **c** f
   d g **e** e **f** e

3 a a and r, m and p, c and t, q and w
   b c and p, r and q

4 a $a = 112°$, $b = 112°$, $c = 112°$, $d = 68°$, $e = 68°$, $f = 68°$, $g = 68°$
   b Angles around a point add up to 360°, vertically opposite angles are equal

5 $a = 105°$, $b = 105°$, $c = 74°$, $d = 106°$, $e = 74°$, $f = 68°$, $g = 112°$, $h = 68°$, $i = 125°$, $j = 125°$, $k = 74°$, $l = 74°$

6 a 74°   **b** 74°   **c** 106°

7 a The lines are not parallel
   b b, d and e

What do you think?

1 a 36°   **b** 72°

2 a 122°   **b** 58°   **c** 80°

3 $x = 11$, $y = 8$

Practice 13.2B

1 a   **b**   **c**

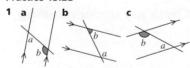

2 a i c **ii** a **iii** d **iv** z
   b i s **ii** q **iii** p **iv** z
   c i t **ii** w **iii** a **iv** n

3 a h and r, i and s
   b h and s, i and r
   c f and r, h and q, g and s, i and p

4 $a = 122°$, $b = 58°$, $c = 122°$, $d = 58°$, $e = 122°$, $f = 122°$, $g = 58°$

5 $a = 44°$, $b = 136°$, $c = 109°$, $d = 71°$, $e = 109°$, $f = 84°$, $g = 96°$, $h = 84°$, $i = 57°$, $j = 123°$, $k = 57°$, $l = 123°$, $m = 39°$, $n = 71°$, $o = 109°$, $p = 109°$

What do you think?

1 $a + b = 180°$ (angles on a straight line add up to 180°)
$c = a$ (corresponding angles are equal)
So $c + b = 180°$ (substituting for a in the first equation)

2 $p + s = 180°$, $q + r = 180°$

3 a Yes, the co-interior angles add to 180°
   b $x = 12°$ so $9x = 108°$; yes, the

---

co-interior angles add to 180°
   c $\angle CRQ = 75°$, so $\angle AQP = \angle CRQ$ and are corresponding, so the lines are parallel
   d Not parallel as e.g. $\angle BQR \neq \angle CRQ$

Consolidate

1 a $\angle DRS$ **b** $\angle QRD$ **c** $\angle QRD$
   d $\angle QRC$ **e** $\angle AQR$

2 $a = 149°$, $b = 149°$, $c = 149°$, $d = 31°$, $e = 149°$, $f = 31°$, $g = 31°$

3 $a = 115°$, $b = 65°$, $c = 133°$, $d = 47°$, $e = 133°$, $f = 81°$, $g = 81°$, $h = 77°$, $i = 77°$, $j = 103°$, $k = 115°$, $l = 65°$, $m = 57°$, $n = 123°$, $o = 57°$, $p = 123°$

Stretch

1 Compare answers as a class

2 a 85°   **b** 110°   **c** $(x + y)°$

3 $\angle ABC = 72°$, $\angle ACB = 57°$, $\angle ECF = 57°$, $\angle CFE = 51°$

Chapter 13.3

Are you ready?

1 a Two sides are equal and two angles are equal
   b All sides and angles are different
   c All sides and angles are equal

2 Check accuracy with a partner

3 Check accuracy with a partner

4 Discuss as a class. Your answers should include kite, rhombus, square, rectangle, trapezium and parallelogram.

Practice 13.3A

1–4 Check accuracy with a partner

5 a Check accuracy with a partner
   b i 108° and 108°
     ii 72° and 72°
   The opposite angles are equal

6 a Check accuracy with a partner
   b i 49° and 49°
     ii 131° and 131°
   The opposite angles are equal

7 a Check accuracy with a partner
   b Check accuracy with a partner
   c Yes

What do you think?

1 a Alternate angles are equal
   b Alternate angles are equal
   c They are both equal to the sum of the same two angles
   d They are equal
   e They are co-interior angles

2 a It has two pairs of adjacent equal angles, and two pairs of adjacent angles that add to 180°
   b It has two pairs of adjacent angles that add to 180°

3 One pair of opposite angles are equal

Practice 13.3B

1 $x = 126°$, $y = 54°$, $z = 126°$

2 a 58°   **b** 61°   **c** 61°

3 $m = 40°$, $n = 140°$, $o = 140°$, $p = 40°$, $e = 153°$, $f = 153°$, $g = 27°$, $h = 27°$

4 a i 20°   **ii** 80°
   b They are equal; triangle BCD is isosceles

5 $a = 128°$, $b = 52°$, $c = 72°$, $d = 72°$, $e = 108°$, $f = 108°$, $g = 50°$, $h = 100°$

What do you think?

1 Two non-equal sides and one angle

2 Yes

3 $a = 65°$, $b = 65°$

Practice 13.3C

1 b They are equal
   c They are equal

---

   d e.g. ABX, BCX, CDX, ADX
   e 45°, 45° and 90°

2 They are equal in length, and bisect each other, but do not bisect the angles at the vertices

3 Statements **b**, **e** and **f** are false, and **g** is false unless the trapezium is isosceles. The rest are true.

4 $a = 41°$, $b = 98°$

5 $a = 90°$, $b = 42°$, $c = 48°$, $p = 90°$, $q = 21°$

6 Check accuracy with a partner

What do you think?

1 Faith is correct

2 Seb is wrong

3 A: square and rectangle
B: square, rectangle, parallelogram, rhombus
C: square, rhombus, kite
D: square, rhombus

Consolidate

1 and 2 Check accuracy with a partner

3 $a = 112°$, $b = 132°$, $x = 125°$, $y = 64°$, $p = 112°$, $q = 68°$, $r = 112°$, $s = 68°$, $l = 75°$, $m = 75°$, $n = 105°$, $g = 115°$, $h = 65°$, $f = 57°$, $e = 90°$

Stretch

1 a Students' own drawings
   b $\dfrac{n(n-3)}{2}$

2 Compare answers as a class

Chapter 13.4

Are you ready?

1 a 180°   **b** 360°   **c** 180°

2 $a = 33°$, $b = 85°$, $c = 42°$, $d = 138°$

3 a Pentagon   **b** Octagon
   c Decagon

4 All the shapes are hexagons

Practice 13.4A

1 $a = 150°$, $b = 91°$, $c = 61°$, $d = 119°$, $e = 65°$, $f = 96°$, $g = 82°$, $h = 79°$, $i = 101°$

2 a 60°   **b** 120°   **c** 360°

3 a 90°   **b** 90°   **c** 360°

4 a i 94°, 132°, 134°
    ii 90°, 55°, 97°, 50°, 68°
    iii 90°, 90°, 55°, 125°
    iv 90°, 20°, 80°, 55°, 75°, 40°
   b 360° for all the shapes

5 Compare answers as a class

6 $a = 140°$, $b = 53°$, $c = 127°$, $d = 72°$, $e = 108°$

What do you think?

2 The pencil completes a full turn, which is 360°

3 a The total is 360° and as they are all equal, each will be $360° \div 6 = 60°$
   b i 72°   **ii** 36°   **iii** 12°
   c They decrease in size
   d 24

Practice 13.4B

1 A and C

2 $a + b + c = 180°$, $p + q + r = 180°$, $x + y + z = 180°$
The angles in the pentagon are b, $a + p + x$, y, $r + z$ and $c + q$
Their total will be $a + b + c + p + q + r + x + y + z = 3 \times 180° = 540°$

3 a Students' own drawings
   b 4   **c** 720°

4 b Heptagon 5, octagon 6
   c Heptagon 900°, octagon 1080°

5

Shape	No. of sides	No. of triangles formed	Sum of interior angles
Hexagon	6	4	$4 \times 180° = 720°$
Heptagon	7	5	$5 \times 180° = 900°$
Octagon	8	6	$6 \times 180° = 1080°$

6 a 7   **b** 1260°

What do you think?

1 There are two fewer triangles than sides

2 a $n - 2$ **b** $(n - 2) \times 180°$

Practice 13.4C

1 a 1080°   **b** 1440°

2 $a = 100°$, $b = 270°$, $c = 70°$, $d = 110°$, $e = 140°$, $f = 130°$, $g = 75°$, $h = 260°$

3 120°

4 a 108°   **b** 135°   **c** 162°

5 30°

6 90°, 120°, 110°, 90° and 130°

7 7 sides

What do you think?

1 a 18   **b** 36

2 a No; 150 isn't a factor of 360

   b Yes; that would give an exterior angle of 30° and 30 is a factor of 360

3 130°

4 Equilateral triangle

Consolidate

1 Diagram C

2 360°

3 $(n - 2) \times 180°$

4 $a = 140°$, $b = 130°$, $c = 120°$, $d = 128°$, $e = 108°$, $f = 135°$

Stretch

1 a $e = \dfrac{360°}{n}$

  b $i = 180° - \dfrac{360°}{n}$ and $i = \dfrac{(n-2)180°}{n}$

  c $i + e = 180°$

2 a Equilateral triangle, square and regular hexagon; the size of the exterior angles is a factor of 360

  b Yes; for example, all quadrilaterals will tessellate

Chapter 13.5

Are you ready?

1 Alternate angles are equal, corresponding angles are equal, and co-interior angles add up to 180°

2 a d **b** g **c** h **d** f

3 $a = 85°$, $b = 37°$, $c = 143°$, $d = 72°$, $e = 72°$, $f = 80°$, $g = 28°$

Practice 13.5A

1 $a = 42°$, $b = 138°$, $x = 32°$

2 $p = 20°$, $q = 80°$, $w = 64°$, $x = 64°$, $y = 48°$, $z = 30°$

3 $a = 50°$, $b = 35°$, $c = 95°$

4 $x = 153°$

5 $w = 85°$, $x = 95°$, $y = 95°$, $z = 23°$

6 40°

What do you think?

1 135°

2 a 77°   **b** 103°

Practice 13.5B

1 $\angle BFE = 35°$ (alternate angles are equal)

  $\angle EBF = 90°$ (angles in a triangle add up to 180°), which is a right angle, so triangle BEF is right-angled

2 a $\angle BEF = 110°$ (co-interior angles

add up to 180°)

  $\angle AEF = 80°$ (corresponding angles are equal)

  $\angle BEC = 110° - 80° = 30°$

  b $\angle BEF = 180° - x$ (co-interior angles add up to 180)

  $\angle AEF = y$ (corresponding angles are equal)

  $\angle BEC = (180° - x) - y = 180° - (x + y)$

3 a They add up to 180°

  b x and p, y and r

  c $x + q + y = p + q + r = 180°$

4 a Any parallelogram ABCD drawn such that angles ABC and ADC are 65°, and angles DAB and DCB are 115°

  b $\angle DAB + \angle ABC = 180°$ (co-interior angles add up to 180°) so $\angle DAB = 180° - \angle ABC$

  $\angle DCB + \angle ABC = 180°$ (co-interior angles add up to 180°) so $\angle DCB = 180° - \angle ABC$

  $\angle DAB = \angle DCB$ as they are both equal to $180° - \angle ABC$

  The same proof can be used for the other pair of opposite angles

5 a $\angle PRQ = 180° - a$ (angles on a straight line add up to 180)

  $\angle QPR = 180° - a$ (base angles in an isosceles triangle are equal)

  $\angle PQR = 180° - (180° - a + 180° - a) = (2a - 180°)$ (angles in a triangle add up to 180°)

  b $\angle PRQ = \dfrac{180 - x}{2}$ (base angles in an isosceles triangle are equal)

  $\angle QRS = 180° - \dfrac{180 - x}{2} = 90 + \dfrac{x}{2}$ (angles on a straight line add up to 180°)

What do you think?

1 $\angle PQS = a$ (base angles in an isosceles triangle are equal)

  $\angle PSQ = 180° - 2a$ (angles in a triangle add up to 180°)

  $\angle QSR = 180° - (180° - 2a) = 2a$ (angles in a triangle add up to 180°)

  $\angle QRS = 2a$ (base angles in an isosceles triangle are equal)

2 $\angle ACB = 180° - (90° + 90° - 2x) = 2x$ (angles in a triangle add up to 180°)

  $\angle ACD = 180° - 2x$ (angles on a straight line add up to 180°)

  $180° - 2x + x + \angle CAD = 180°$ (angles in a triangle add up to 180°)

  $\angle CAD = x$

  $\angle CAD = \angle CDA$, so triangle ACD is isosceles

Consolidate

1

2 $a = 65°$, $b = 65°$, $c = 75°$, $d = 40°$, $e = 140°$

3 a 70°   **b** 42°   **c** 133°   **d** 5°

Stretch

1 a 85°   **b** 15 cm

2 $x = 36$, $y = 20$

3 $5a + 7a = 180°$ (co-interior angles add up to 180°)

  $a = 15°$

  $3(a + 10)° = 75°$ and $5a = 75°$

  The corresponding angles are equal, so WY is parallel to XZ

Chapter 13.6

Are you ready?

Check the accuracy of diagrams in this section with a partner

Practice 13.6A

Check the accuracy of diagrams in this section with a partner

1 The lengths of the radii from each arm must be equal

4 The answer is the angle bisector of ABC

5 Ed is wrong

What do you think?

Check the accuracy of diagrams in this section with a partner

2 You have already found the centre after two bisectors are drawn

Practice 13.6B

Check the accuracy of diagrams in this section with a partner

4 AXBY is a rhombus. The diagonals of a rhombus bisect each other at right angles.

What do you think?

Check the accuracy of diagrams in this section with a partner

1 You have already found the centre after two bisectors are drawn

Consolidate

Check the accuracy of diagrams in this section with a partner

Stretch

1 Beca is correct

2 a Rhombus, square and one pair in kite

  b i No

    ii Some diagonals bisect the angles and other diagonals bisect, but some do not

Check my understanding

1 $a = 35°$, $b = 100°$, $c = 45°$

2 AB and CD are parallel so angles $\angle ABC$ and $\angle BCD$ are co-interior angles. Co-interior angles add up to 180°, so $\angle ABC = 180° - 85° = 95°$

3 45°

4 30

5 Check accuracy with a partner

6 The lower left angle in the top triangle equals a (corresponding angles). The top angle in that triangle equals $180 - (a + b)$ (angles in a triangle add up to 180°).

  $x = 180 - [180 - (a + b)]$ (angles on a straight line add up to 180°), so $x = a + b$

Block 14 Area of trapezia and circles

Chapter 14.1

Are you ready?

1 a Triangle   **b** Trapezium

  c Parallelogram

2 A 7 cm²   B 10 cm²   C 24 cm²
3 a 36   **b** 9   **c** 9   **d** 20

Practice 14.1A

1 a 10 cm²   **b** 12 cm²   **c** 48 mm²
   d 10 mm²   **e** 10 cm²   **f** 36 cm²
   g 210 m²   **h** 90 mm²   **i** 15 cm²
2 No, the area of the triangle is half of
   the area of the rectangle
3 a i 120 cm²   **ii** 30 cm²   **iii** 72 m²
   b Multiplying all of the lengths given
     rather than identifying the
     relevant ones
4 a 1 cm²   **b** $\frac{3}{4}$ cm²   **c** 5x cm²
5 a i 9 cm²   **ii** 0.0625 cm²
   iii 10x cm²
   b Compare methods with a partner
6 a 4 cm   **b** 2 cm   **c** 5 cm   **d** 8 cm
7 a Samira has substituted known
     values into the formula for the
     area of a triangle. This has then
     resulted in an equation that she
     can solve by using inverse
     operations
   b 0.5 × 12 × 8 = 48
8 a 14 cm   **b** 2 cm   **c** 12.4 cm

What do you think?

1 Any triangle whose base and height
   have a product of 28
2 16 cm by 12 cm

Practice 14.1B

1 a x and z **b** c and e
   c s and q **d** g and e
2 a 25 cm²   **b** 21 cm²
   c 42 mm²   **d** 68 m²
3 28 cm²
4 a 15 cm²   **b** 2.25 m²
5 a 12 cm   **b** 6 cm

What do you think?

1 Any trapezium such that
   $(a + b)h = 160$ where a and b are the
   lengths of the parallel sides and h is
   the perpendicular height
2 a $\frac{1}{2}$ × (5 + 8) × 4 = 26
   b There are two triangles and Benji
     can't tell what the base of each
     triangle is

Consolidate

1 a 15 cm²   **b** 1 cm²   **c** 80 mm²
   d 21 mm²   **e** 40 m²   **f** 3 mm²
   g 60 cm²   **h** 21 cm²   **i** 80 m²
2

Shape	Base (cm)	Perpendicular height (cm)	Area (cm²)
Rectangle	5	7	35
Triangle	5	7	17.5
Triangle	9	4	18
Parallelogram	5	7	35
Parallelogram	9	4	36
Rectangle	20	5	100
Parallelogram	3	12	36
Triangle	12	11	66

Stretch

1 a 13.76 cm   **b** 19.2 mm
2 a 10.5 cm   **b** 12 cm   **c** 4 cm
3 a Any pair of values that sum to 35
     where $a < b$
   b $a + b = 35$, $b > a > 0$
4 a $\frac{1}{2}(2x + 4)(x + 2) = 49 \Rightarrow$
     $(x + 2)(x + 2) = 49$
     $\Rightarrow x^2 + 4x + 4 = 49 \Rightarrow$
     $x^2 + 4x - 45 = 0$

   b $x = 5$

Chapter 14.2

Are you ready?

1
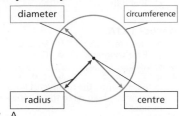

2 A
3 B, C
4 a The radius is half of the diameter
   b The diameter is double the radius
5

Radius	Diameter
4 cm	8 cm
50 mm	100 mm
12.5 cm	25 cm
3.4 m	6.8 m

6 a i 47.3   **ii** 47.28
   b i 0.9   **ii** 0.95
   c i 108.0   **ii** 108.00

Practice 14.2A

1 a 16π cm²   **b** π m²
   c 169π mm²   **d** 1.69π m²
2 a Ed has not noticed that 7 m is the
     diameter not the radius
   b $\frac{49}{9}\pi$ m², 12.25π m²
3 a 100π mm²   **b** 0.25π m²
   c 121π cm²   **d** 1.21π m²
4 a 36π cm²   **b** 18π cm²
   c 9π cm²
5 a 2π m²   **b** 8π m²   **c** 0.04π m²
6 a They are all square numbers
   b i 3 cm   **ii** 4 cm
     iii 10 cm   **iv** 6 cm

What do you think?

1 a $\frac{9}{4}\pi$ cm²   **b** $(21 - \frac{9}{4})\pi$ cm²
2 49$y^2\pi$ cm²
3 Diameter of circle = $\frac{12\pi}{\pi}$ = 12 cm so
   radius = 6 cm and area = π × 6 =
   36π cm²

Practice 14.2B

1 a 1661.9 cm²   **b** 0.8 m²
   c 929.4 mm²   **d** 93.3 m²
2 a 88.4 m²   **b** 25.1 cm²
   c 1157.9 mm²
3 28.3 inches²
4 a 4.7 cm   **b** 5.6 cm   **c** 0.4 m
5 Circle with a radius of 3 cm
6 3.6 m
7

Radius (cm)	Diameter (cm)	Area in terms of π (cm²)	Area to 1 dp (cm²)
8.5	17	72.25π	227.0
7	14	49π	153.9
8.7	17.4	75.69π	237.8

What do you think?

1 Deal 1 is better value as there is twice
   as much pizza than in deal 2 but for
   less than double the price
2 a Area of circle = 81π
     Area of square = 324
     $\frac{81\pi}{324}$ = 0.785 which is greater

than 75%
   b Whatever the value of a, the
     sticker will cover more than 75%
     because $\frac{\pi x^2}{4x^2} = \frac{\pi}{4} = 0.785$
3 153.9 units²
4 a 795.8 cm²
   b 35.45 cm
   c Yes, if the radius is 2

Consolidate

1 a i 9π cm²   **ii** 28.3 cm²
   b i 64π cm²   **ii** 201.1 cm²
   c i 1030.41π mm²
     ii 3237.1 mm²
   d i 1072.5625π cm²
     ii 3369.6 cm²
   e i 25π cm²   **ii** 78.5 cm²
   f i 49π cm²   **ii** 153.9 cm²
   g i 156.25π cm²   **ii** 490.9 cm²
   h i 2500π mm²   **ii** 7854.0 mm²
   i i 4π mm²   **ii** 12.6 mm²
   j i 289π m²   **ii** 907.9 m²
2 a 39.3 m²   **b** 14.1 cm²
   c 79.2 m²   **d** 19.6 m²
   e 50.3 cm²   **f** 1963.5 mm²

Stretch

1 £297.85
2 8.96 cm
3 a 100 cm²   **b** 50 cm

Chapter 14.3

Are you ready?

1 a 28 cm²   **b** 28 cm²
   c 14 cm²   **d** 14 cm²
   e Shapes **a** and **b** have the same area
     and so do **c** and **d**
   f You can only find the perimeter of
     the rectangle, because you don't
     know all of the side lengths in the
     other shapes
2 a 201.1 cm²   **b** 9.8 cm²
   c 44.2 cm²
3 a 24   **b** 8   **c** 28

Practice 14.3A

1 a
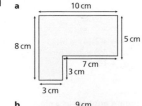

   b
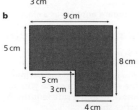

   c
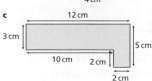

   d
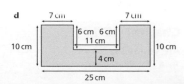

e 36 cm, 34 cm, 34 cm, 82 cm

2 a 59 cm²   **b** 57 cm²
   c 40 cm²   **d** 184 cm²

3 a Chloe is thinking about finding the area of the whole rectangle and subtracting the missing bit. This only requires calculating the areas of two rectangles rather than three.

   b 41 cm²

   c Some parts of the perimeters of the rectangles are within the compound shape and not part of its perimeter

4 a 124 cm²   **b** 596 mm²

   c Compare answers as a class

What do you think?

1 163.2 cm²

2 No, because we don't know how much they overlap

Practice 14.3B

1 a 45.5 cm²   **b** 260 mm²
   c 68.4 cm²

2 a Jakub's method: (12×6) $+ (0.5 \times 5 \times 4) = 72 + 10 = 82 \text{ cm}^2$
Marta's method: $(0.5 \times (10 + 6) \times 5)$ $+ (7 \times 6) = 40 + 42 = 82 \text{ cm}^2$

   b Compare answers as a class

3 a 112 m²   **b** 24 cm²   **c** 41 cm²
   d 357.0 mm²   **e** 401.7 cm²

4 a 5.9 m²   **b** £72   **c** £540

What do you think?

1 Various answers. Students could split the hexagon into 2 trapeziums, 6 equilateral triangles, a rectangle and two triangles, etc. The area is 93.6 cm²

2 50%

3 Compare answers as a class

4 Compare answers as a class

Consolidate

1 a 30 cm²   **b** 22 cm²
   c 59 cm²   **d** 16 cm²
   e 26 cm, 24 cm, 36 cm, 20 cm

2 a 103.5 cm²   **b** 49 cm²

3 a 12.6 m²   **b** 15.7 m²   **c** 31.4 m

Stretch

1 70.6%

2 $3y(y + 4) + (y + 2)(4y + 5)$
$= 3y^2 + 12y + 4y^2 + 5y + 8y + 10$
$= 7y^2 + 25y + 10$

3 a $\frac{\pi x^2}{2} + 2x^2$

   b $\pi x + 6x$

   c Compare answers as a class

Check my understanding

1 a 65 cm²   **b** 21 cm²
   c 500 mm²   **d** 12.5 m²
   e You can only find the perimeter of the rectangle, because you don't know all of the side lengths in the other shapes

2 a i 49π cm²   **ii** 153.9 cm²
   b i 20.25π cm²   **ii** 63.6 cm²

3 a 96 cm²   **b** 44 cm

Block 15 Line symmetry and reflection

Chapter 15.1

Are you ready?

1 a i Pentagon

---

   ii Isosceles trapezium
   iii Isosceles triangle
   iv Equilateral triangle
   v Hexagon   **vi** Hexagon

   b i, iv and v

   c ii and iii

2 C

Practice 15.1A

1 a Isosceles triangle   **b** Square
   c Parallelogram   **d** Circle

2 B and C

3 a 3   **b** 5   **c** 6   **d** 7
In these shapes, the number of lines of symmetry is equal to the number of sides

4 This shape is not regular. It has one line of symmetry.

5 a 1   **b** 8   **c** 1

6 a 1

   b There is one pair of equal angles, so it is an isosceles triangle

What do you think?

1 Any irregular pentagon will suffice. Students should recognise that regular pentagons have five lines of symmetry, but irregular pentagons do not.

2 a No, it has two lines of symmetry

   b Compare answers as a class

3 No, the pattern on the logos affects the number of lines of symmetry Logo A has no line of symmetry. Logos B and C each have one line of symmetry.

Consolidate

1 A, B, E, F, G, H

2 B, C, G, H

3

	No lines of symmetry	One line of symmetry	More than one line of symmetry
Triangle	E	C	D
Quadrilateral	F	H	A B G

Stretch

1 a **b**

   c

2 Scalene triangles have 0, isosceles triangles have 1 and equilateral triangles have 3

3 a 5   **b** 6   **c** 10   **d** n

4 None, as none of the angles are equal in size

Chapter 15.2

Are you ready?

1 B and C

2 A(2, 3), B(0, 4), C(−2, 1), D(3, −2)

3 a A and D   **b** B
   c A   **d** $x = −4$

Practice 15.2A

1 a 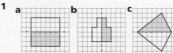 **b** **c**

---

2 a 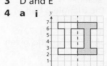 **b** **c**

3 D and E

4 a i **ii**

   iii

   b i (7, 7), (7, 6), (6, 6), (6, 3), (7, 3), (7, 2), (4, 2), (4, 3), (5, 3), (5, 6), (4, 6), (4, 7)
   ii (4, 2), (4, 7), (2, 7), (2, 4)
   iii (2, 3), (4, 1), (6, 2)

5 a (4, 2), (0, 2), (2, 4)
   b (1, −2), (2, −2), (2, −4), (3, −4)
   c (0, 0), (0, 3), (−2, 4), (−2, −1)

6 Both Benji and Beca are correct

7 a $x = 0$ (the y-axis)
   b $x = −2$ **c** $y = 1$

8 Amina has counted 2 from the y-axis instead of 2 from the mirror line

What do you think?

1 (−4, −1), (−4, −4), (0, −1), (0, −4)

2

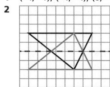

The image overlaps the object

Consolidate

1 a i is correct

   b ii The reflected triangle is the wrong size

     iii The shape has been drawn on the mirror line instead of being reflected in the mirror line

     iv The shape has been drawn on the other side of the mirror line rather than being reflected

2 a 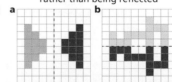 **b**

3 a ii and iii are correct

   b i The shape has been drawn on the other side of the mirror line rather than being reflected

     iv The shape has been rotated not reflected

Stretch

1 a The object sits exactly on top of the image

   b The vertices B and C have swapped position

   c For example: $x = −4$, $x = −2$ or $y = −1$

2 a (−5, 3), (−2, 7), (1, 7), (2, 3), (−2, −1), (1, −1)

   b 40 square units

3 $y = 6$, $y = 12$, $x = 1$ and $x = 8$

Chapter 15.3

Are you ready?

1 A and C

2

x	1	2	3	4	5
y	1	2	3	4	5

3 (3, 3), (–6, –6), (1.5, 1.5)

Practice 15.3A

1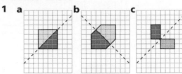

2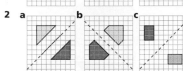

3 a Emily has one of the vertices in the correct place but has drawn the rest of the rectangle incorrectly. She should check every vertex.

b

4 c and **d** are correct. There are two options for diagrams **a** and **b**.

a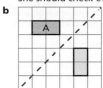

b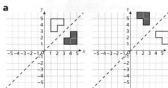

5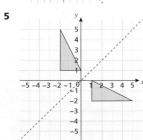

What do you think?

1 Jackson is wrong. The reflection would overlap the original.

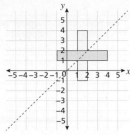

2 (1, 5), (1, 3), (–3, 3) and (–3, 5)

Consolidate

1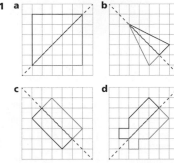

2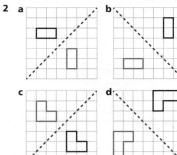

Stretch

1 a $y = 0$ (the x-axis)　**b** $x = –3$
　c $y = x$　　　　　　**d** $y = x + 2$
　e $y = –x + 1$　　　　**f** $y = x + 2$

2 $x = 4$
Discuss findings as a class

Check my understanding

1 C

2 a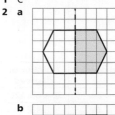

b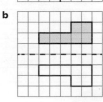

c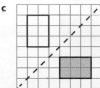

---

d

3 a $y = 1$　**b** The y-axis or $x = 0$

Block 16 The data handling cycle

Chapter 16.1

Are you ready?

1 a 5, 13, 8, 4
　b Dog　　　　**c** 30
2 a Continuous quantitative
　b Qualitative
　c Discrete quantitative
　d Continuous quantitative
　e Qualitative
　f Discrete quantitative

Practice 16.1A

1 a Any correct enquiry
　b Students' own answers
　c Students' own answers
　d Compare answers as a class
　e Compare answers as a class
2 a Students' own answers
　b Students' own answers
3 a Collect data
　b Scatter graph
4 a Any correct hypothesis
　b The more people he asks, the more accurate

What do you think?

1 He only asks his friends, so it is biased. There is no option other than those things.

Practice 16.1B

1 a No, it doesn't allow people to dislike the menu and it's not very specific
　b e.g. "How would you rate the food available on our menu?
Excellent　Good　Satisfactory
Poor　Inadequate"
2 a £10 and £20 are each in two categories. There are no options for less than £5 or more than £30. There's no time frame given.
　b e.g. "How much do you spend on music per month?
Less than £5　£5–£9.99
£10–£14.99　£15+"
3 a It's a leading question
　b Her friends are likely to have a similar opinion
　c Discuss answers with a partner
4 a e.g. "How many times per month do you go to the cinema?
0　1 to 3　4 to 6　7 to 9
10 or more"
　b e.g. "How much time do you spend doing homework per week?
Less than 1 hour
1–2 hours
More than 2 hours"
　c e.g. "How much money do you spend on food shopping per week?
Less than £10
£10–£19.99
£20–£29.99
£30+"
5 Discuss answers with a partner

What do you think?

1 a It's not very big. It's biased.
  b The whole school
  c Any suitable questionnaire

Consolidate

1 Any correct enquiry
2 a The options don't cover everything
  b e.g. "Excellent, good, satisfactory, poor, inadequate"
3 a No time frame given
  b No option for never

Stretch

1 a The first question is asking two things. The second one is a leading question.
  b e.g. "How do you feel about this statement? 'The internet helps with homework.'
    Strongly disagree
    Disagree
    Neither
    Agree
    Strongly agree"
  c e.g. "How long do you spend online per week?
    Less than 1 hour
    1–2 hours
    More than 2 hours"
2 a It's a leading question
  b There's no option to disagree
  c It's biased
3 a In Method 1 you are asking only the people available at a shopping centre on a Monday morning. Method 2 will not include people who do not have a landline. Method 3 only asks one part of town.
  b e.g. send a postal survey to the whole town

Chapter 16.2

Are you ready?

1 a Bar chart   **b** Pie chart
  c Pictogram
2 360
3 Check accuracy with a partner

Practice 16.2A

1 a 24   **b** 18   **c** 60
2 The circles aren't evenly spaced and there is no key
3 a Any correct pictogram with a key given. Answers vary depending on the key used.
  b Compare answers with a partner
4 a 9
  b No, they won 5 at the Winter Olympics and 10 at the Summer Olympics
  c Summer 73, Winter 28
5 a

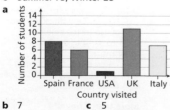

  b 7   **c** 5

6 a

Sport	Number of children	Number of degrees
tennis	7	7 × 10 = 70°
netball	8	8 × 10 = 80°
football	12	12 × 10 = 120°
hockey	5	5 × 10 = 50°
rugby	4	4 × 10 = 40°
Total	36	36 × 10 = 360°

b

7 a

Fuel type	Frequency	Number of degrees
diesel	11	11 × 9 = 99°
petrol	20	20 × 9 = 180°
electric battery	8	8 × 9 = 72°
hydrogen fuel cell	1	1 × 9 = 9°

b

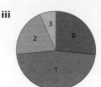

8 a 20%
  b i 6   **ii** 18

What do you think?

1 Chart C. There are fewer yellow circles than pink, so the choice is between charts A and C. There are three times as many pink circles as yellow circles, so chart C represents the data.

2

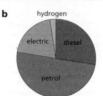

3 a

Meal choice	Frequency	Number of degrees
meat	20	20 × 6 = 120°
fish	16	96°
vegetarian	13	78°
vegan	7	7 × 6 = 42°
gluten free	4	4 × 6 = 24°
Total	60	360°

b

Consolidate

1 Class A and chart 4, Class B and chart 3, Class C and chart 1, Class D and chart 2
2 a Apple   **b** 4
  c 3   **d** 26
3 a i Any correct pictogram. Answers vary depending on key used.
   ii

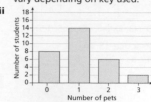

   iii

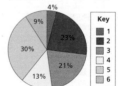

  b Answers will vary depending on student

Stretch

1 a 12   **b** 5   **c** 20%   **d** $\frac{3}{25}$
2 a The frequencies are high and aren't all multiples of 5 or 10, so it would be inefficient
  b The data sums to 500 which isn't a factor or multiple of 360. The angles would need rounding.
3 a Town B   **b** 495
  c

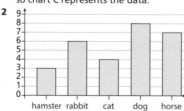

  d The pie charts are equal in size but represent different amounts.

Chapter 16.3

Are you ready?

1 a 8   **b** 1   **c** 2   **d** 19
2

3 a t is greater than 4
  b h is greater than or equal to 163
  c w is greater than or equal to 15 and less than or equal to 20

Practice 16.3A

1 a 5 teas or coffees
  b 20   **c** 45
  d Yes, more coffees: the café bar is only lower on one day and only by a little
  e Tea 127, coffee 201
2 a White Rose Athletic
  b 14

c Trinity United 32 White Rose Athletic 30

d Trinity United 30 points, White Rose United 52 points

3 a 2020 **b** 2016

c i 60% **ii** $\frac{2}{5}$

d 77

What do you think?

1 a

	Year 7	Year 8	Year 9	Year 10	Year 11	Total
School dinner	101	86	93	78	55	413
Packed lunch	19	34	27	52	55	187
Total	120	120	120	130	110	600

b Any correct chart with a key, e.g.

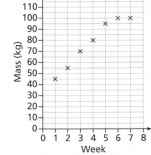

■ School dinner ■ Packed lunch

c Compare answers with a partner

Practice 16.3B

1 a 6 cm **b** Day 6

c About 4.5 cm **d** Day 7

2

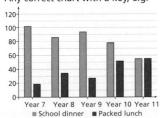

3 No, it did not rain on day 3

4 a 7000 **b** 1960

c 6000. It's an estimate because the points don't lie exactly on the line.

What do you think?

1 The line graph is more appropriate because the data is continuous

2 Various answers, e.g. distance from home versus time

Practice 16.3C

1

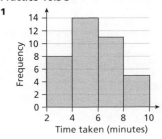

2 a

Time in hours (h)	Tally	Frequency
$5 < h \leqslant 10$	ΗΗ ΙΙ	7
$10 < h \leqslant 15$	ΗΗ ΗΗ	10
$15 < h \leqslant 20$	ΗΗ ΙΙ	7
$20 < h \leqslant 25$	ΗΗ Ι	6

b

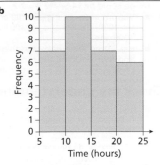

What do you think?

1 Continuous data

2 There are no gaps between the intervals in a frequency diagram. The bars in a bar chart relate to discrete data, which has gaps between data values.

3 He has used the $\leqslant$ symbol in both places in his inequalities which means that 10, 15 and 20 can be placed in two different rows

Consolidate

1

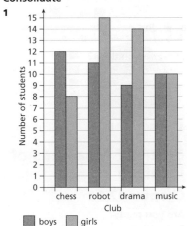

■ boys □ girls

2

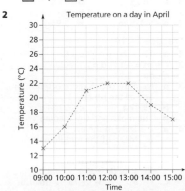

3 a 8 m **b** 2 **c** 9 m

d It may have been pruned/had the top cut off

Stretch

1 a

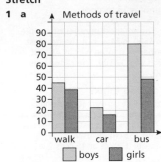

□ boys ■ girls

b i 103 **ii** 147

2

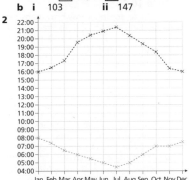

-•- sunrise -✱- sunset

3 Answers may vary, for example:

a More energy was used in winter months. The least amount of energy was used in August. There was some energy used every month.

b How much energy was used in total. How much more energy was used in March than in July. The average amount of energy used per month.

4 a e.g.

Distance (d) metres	Frequency
$0 < d \leqslant 200$	60
$200 < d \leqslant 400$	100
$400 < d \leqslant 600$	130
$600 < d \leqslant 800$	80
$800 < d \leqslant 1000$	10

b Differences could be in interpretation of the height of the bars that aren't on a line and also in where the $\leqslant$ symbol is placed

Chapter 16.4

Are you ready?

1 a Greatest 112, smallest 9

b Greatest 1050, smallest 150

c Greatest 13, smallest 12.4

2 a 57 **b** 22 **c** 33.5

Practice 16.4A

1 352 g

2 a Beth has described the range instead of calculating it and she used 102 instead of 120 as the highest value.

b 103 miles

3 26°

4 a Jackson has used the first and last numbers rather than the smallest and largest.

b Jackson hasn't converted the units

5 43
6 3 minutes 16 seconds
7 **a** Y5 **b** Y2 **c** 42
8 No. The smallest number of people was 6500 in May, but the greatest number of people was in August, which was 12 500. The range was 6000.

What do you think?

1 Benji hasn't converted the units. The range is 3800 ml or 3.8 litres.
2 **a** Their height is between 141 cm and 155 cm inclusive
   b Either 139 cm or 157 cm

Practice 16.4B

1 **a** Marta doesn't know how many cars are in each car park so there could be 1000 in A and only 10 in B.
   b The total number of cars in each car park, and the angles in the pie charts
   c There is a greater proportion of red cars in car park B than in A. There are greater proportions of black and white cars in A than in B. There is a similar proportion of blue and other colours
2 **a** The bar for class B is a lot taller
   b The vertical axis covers only a very small range, so there actually isn't that much difference in the scores
   c Class B performed best and class C performed worst, but all the scores were quite similar across all five classes, with the range being only 1.1%
3 **a** One chart being drawn larger implies that it represents more
   b Seb finished 1st more times than Emily. Emily and Seb finished in either 4th or 5th a similar number of times.

What do you think?

1 **a** The lower the time, the faster the run, so Ed's bar being tallest means he was the slowest
   b Emily was fastest, then Benji, then Zach, then Seb, then Faith and then Ed
   c It's misleading
2 Discuss as a class

Consolidate

1 **a** 7 **b** 70 **c** £60 **d** 65 g
2 42
3 60

Stretch

1 **a** The shapes are narrower at the top, which makes them look as though they represent a smaller frequency than they actually do. Also, it's drawn "3-D", so it's hard to see where the dividing line between the two colours lines up with the percentage axis.
   b Each region recycles more waste than it doesn't recycle. A greater proportion of waste in the South West is recycled than in any other region. The region that recycles the smallest proportion is the South East.
2 $x = 3$
3 Marta 1 is correct if all the data is the same, e.g. 5 5 5 5
   Benji is incorrect. Let a be the smallest and b be the greatest number in a set of data, where $b > a$

The range is given by $b - a$
If b is positive and a is positive then $b - a$ is positive because $b > a$
If b is positive and a is negative then $b - a$ is positive because you will add a to b
If b is negative then a must be more negative and $b - a > 0$ because a has a greater negative value
Faith is correct. For example, the range could be 14.5 − 12.3 = 2.2
4 **a** Various answers, e.g. $x = 100$ and $y = 1000$, $x = 103$ and $y = 1001$ or 1009, etc.
   b e.g. listing the numbers as they are and then deciding the range and finding values of x and y to suit.

Check my understanding

1 **a** She hasn't given a time frame; there are not enough options and they are too vague
   b

How far do you run per week?	
0 miles ☐	Up to 5 miles ☐
5 to 10 miles ☐	More than 10 miles ☐

2 **a**

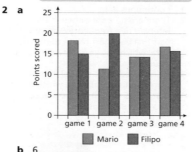

   b 6
3

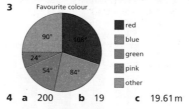

Favourite colour
□ red □ blue □ green □ pink □ other

4 **a** 200 **b** 19 **c** 19.61 m

Block 17 Measures of location

Chapter 17.1

Are you ready?

1 **a** 18.25 **b** 11.4
2 **a** 19 **b** 7
3 **a** 5.925 kg **b** 5.9 kg

Practice 17.1A

1 **a** 12 **b** 3 **c** 6 **d** 9
2 Mean = 4.1 (1 d.p.), median = 4, mode = 3
3 **a** No mass occurs more often than any of the others
   b 93.4 kg **c** 93.8 kg
   d The mean and the median
4 **a** **i** Third in the list
      ii Mean of the middle pair
      iii Middle pair are equal
   b **i** 2 modes **ii** No mode
      iii 1 mode
   c **i** 8 **ii** 7.75 **iii** 7.5
5 **a** Ed is correct, provided there is a mode
   b If the middle pair of values are not the same, then the median will not

be a member of the data set
6 The mode
7 **a** The "modal value" is another way of saying "the value that is the mode". She is correct, the modal value is 1
   b The mean
8 **a** 46 cm **b** 46 cm
   c 45.1 cm (1 d.p.)

What do you think?

1 **a** The data is grouped
   b 0–2 cats
2 $\frac{2}{3}$
3 Mean = 4.5 years (1 d.p.), median = $3\frac{1}{2}$ years, mode = $3\frac{1}{2}$ years

Practice 17.1B

1 **a** Mean = 25.2 (1 d.p.), median = 36, mode = 0
   b The median
2 **a** There is no mode
   b Mean = 19.1 °C, median = 19 °C
   c Either, as they are almost identical
3 **a** Yes
   b No, the data is not numerical
4 **a** Mean = 10.3 years (1 d.p.) Median = 3 years, mode = 3 years
   b The mean does not
5 $6\frac{1}{4}$ is not a shoe size you can buy
6 **a** Mean = 35.75 kg, median = 36.5 kg, there is no mode
   b The mean goes down to 29.2 kg and the median goes down to 35 kg There is still no mode.

What do you think?

1 Compare answers as a class
2 Compare answers as a class

Consolidate

1 **a** Mean = 14.8, median = 16, mode = 16
   b Mean = 9.5, median = 9, mode = 9
   c Mean = 9, median = 8, mode = 7
   d Mean = 5.9 (1 d.p.), median = 6, mode = none
2 **a** 176 cm **b** 175 cm
   c 172.4 cm (1 d.p.)
3 **a** The mode
   b Mean = £656.60, median = £659
4 **a** Mean = 9.4 (1 d.p.), median = 10, mode = 10
   b Discuss as a class
5 **a** Mean = 125, median = 89, mode = none
   b Compare answers as a class
6 The modes are 1 and 2

Stretch

1 Compare answers as a class
2 **a** 4
   b Discuss as a class
3 **a** All the measures are also doubled
   b It depends on which values are changed
   c Compare answers as a class

Chapter 17.2

Are you ready?

1 **a** 18.2 **b** 8
2 35
3 **a** **i** 38 **ii** 33 **iii** 71
   b 2.37 (2 d.p.)
4 **a** 5 **b** 24.5 **c** 62.5

Practice 17.2A

1 a

Number of goals	Frequency	Subtotals
0	5	0
1	8	8
2	6	12
3	4	12
4	1	4
Total	24	36

b 1.5
2 13.7 years (1 d.p.)
3 a

Score	Frequency
7	4
8	6
9	9
10	13
Total	32

b 8.97 (2 d.p.)
4 a 9   **b** 8.9
5 a

Score	Frequency
0	9
1	12
2	5
3	1
4	2
5	1
Total	30

b 1.27 (2 d.p.)
c Discuss as a class
d Discuss as a class
6 b −0.1875

What do you think?

1 a i In the first calculation of the mean, the mistake is using the number of different row entries (4) instead of the total frequency (27)
ii In the second table, the first subtotal should be 0, not 10
b i Mean = 47.3 (1 d.p.)
ii Mean = 1.8 (1 d.p.)
2 a She has not taken the frequencies into account
b 1 goal
c Work out the position of the median and count up in the frequency column

Practice 17.2B

1 a Mean = 174.9 cm
b The total of the midpoints is not needed to work out the mean
2 a 48.6 years (1 d.p.)
b $50 < a \leqslant 65$ years
3 a $0 < t \leqslant 15$ minutes
b 23.4 minutes
c May be less reliable for a wider interval
4 5.57 kg (2 d.p.)
5 $31\frac{1}{4}$ minutes

What do you think?

1 Discuss as a class
2 You don't know the endpoint of the last interval so you can't work out the midpoint. Compare answers as a class.

Consolidate

1 2.4 (1 d.p.)
2 a 5.6 (1 d.p.)
b Discuss as a class

3 4.5 hours (1 d.p.)
4 a 13.2 minutes
b You don't know the actual values

Stretch

1 Compare answers as a class
2 a Work out the position in the table of the $\frac{n+1}{2}$th term
b Compare answers as a class

Chapter 17.3

Are you ready?

1 a Subtract the smallest value from the largest value
b No, it is a measure of spread
2 a Mean = 6.4, median = 6, mode = 10, range = 8
b Mean = 4.5, median = 3.5, mode = 3, range = 8
c Mean = 5.25, median = 5.75, mode = none, range = 6.3
3 144 cm

Practice 17.3A

1 a Mean = 10, median = 6, mode = 6
b 35
c Mean = 5, median = 6, mode = 6 The mean has halved and the others have stayed the same
2 a Mean = 52, median = 54, range = 85
b 0
c Mean = 60.7 (1 d.p.), median = 58, range = 40
3 a 56 cm   **b** 5.6 cm
4 a December sales are greater, most likely due to Christmas
b £85 000 and £17 000
c Mean
5 a 27 s   **b** 4.2 s
6 a 4.1 °C (1 d.p.) and 13°
b Discuss as a class
c i 5.7 °C (1 d.p.) and 4°
ii 5.6 °C (1 d.p.) and 4°
7 Discuss as a class
8 a 360
b A 44% decrease when the outlier is removed

What do you think?

1 8%
2 There might be two or more outliers at the end(s) of a distribution
3 Compare answers as a class
4 Compare answers as a class

Practice 17.3B

1 a Maths   **b** Science
2 Ali scored more runs on average. Rhys was more consistent.
3 a A   **b** A
4 a Tommo's Team is older on average and their ages are less spread out
b No
5 a Year 11 are quicker and their times are closer together
b Year 10 results are now closer together
6 a The greater the mean time taken, the slower the reaction time
b Over 60s, 16–30 years, 12–16 years, 30–60 years

What do you think?

Discuss as a class – the range is not reliable if there are outliers

Consolidate

1 62 and 88
2 a Mean = 53, median = 62, range = 47

b 16
c Mean = 62.25, median = 62.5, range = 2
d Mean and range
3 Mean
4 The girls scored better on average and their marks were more consistent
5 Discuss as a class – the mean mark has risen and the range has got smaller, so probably yes, the programme was effective

Stretch

1 Discuss as a class
2 Compare answers as a class

Check my understanding

1 a 5   **b** No mode
c 2 and 6
2 a Both are 514   **b** 16
3 a 31°
b The median does not change
c Mean = 52.6° (1 d.p.), range = 12°
4 On average, the cats are lighter and have more similar masses
5 22 minutes (nearest minute)